ECOTOXICITY OF CHEMICALS TO *PHOTOBACTERIUM PHOSPHOREUM*

HANDBOOKS OF ECOTOXICOLOGICAL DATA

A series edited by

J. Devillers
Centre de Traitement de l'Information Scientifique, Lyon, France

---

Volume 1
Ecotoxicity of Chemicals to Amphibians
Edited by J. Devillers and J. M. Exbrayat

Volume 2
Ecotoxicity of Chemicals to *Photobacterium phosphoreum*
K. L. E. Kaiser and J. Devillers

This book is part of a series. The publisher will accept continuation orders which may be cancelled at any time and which provide for automatic billing and shipping of each title in the series upon publication. Please write for details.

ECOTOXICITY OF CHEMICALS TO *PHOTOBACTERIUM PHOSPHOREUM*

by

K. L. E. Kaiser
National Water Research Institute
Burlington, Ontario, Canada

and

J. Devillers
CTIS, Lyon, France

Routledge
Taylor & Francis Group
LONDON AND NEW YORK

First published 1994 by OPA B.V.

Published 2022 by Routledge
4 Park Square, Milton Park, Abingdon, Oxon OX14 4RN
605 Third Avenue, New York, NY 10017

Routledge is an imprint of the Taylor & Francis Group, an informa business

Publisher's Note

The publisher has gone to great lengths to ensure the quality of this reprint but points out that some imperfections in the original copies may be apparent.

Library of Congress Cataloging-in-Publication Data

is available for this book.

ISBN 978-2-881-24974-7 (hbk)

CONTENTS

PREFACE

Ecotoxicity of Chemicals to Photobacterium phosphoreum reviews data obtained with the Microtox® test system. It contains toxicity results on more than 1,250 chemicals obtained from literature or private sources. This book is essential reading for researchers and administrators in industry, university and government, in fields ranging from health and welfare, to environmental effects and controls of chemicals.

We acknowledge with pleasure the courtesy of N. A. Casseri (Occidental Chemical Corporation, New York, USA), M. T. D. Cronin (Liverpool John Moores University, Liverpool, UK), C. Reteuna (French Ministry of the Environment, Paris, France), R. Speece (Drexel University, Philadelphia, USA), and P. Vasseur (University of Metz, Metz, France) for providing information on their results. We are also indebted to A. A. Bulich (Microbics Corporation, Carlsbad, USA) for communicating taxonomic information related to the strain NRRL-B-11177.

The preparation of this handbook would not have been possible without the help of M. B. McKinnon, V. S. Palabrica (National Water Research Institute, Burlington, Canada), S. Bintein, and D. Domine (CTIS, Lyon, France).

Last, we would like to thank Ms S. Foubert for proofreading.

Klaus L. E. Kaiser
James Devillers

INTRODUCTION

Industrial countries today are facing serious ecotoxicological and toxicological problems resulting from the release of toxic substances into the marine and freshwater ecosystems. In response to these increased stresses on the aquatic environment, and in the belief that there is no single criterion by which to judge the potential hazard of a given chemical, a wide range of bioassays are developed, proposed, and used for biological monitoring and toxicity assessment. The use of microorganisms, bacteria in particular, as the assay agent allows to develop simple, inexpensive, rapid, and generally sensitive tests to determine and monitor the ecotoxicity of the effluents and chemicals discharged into the aquatic ecosystems.

One of the most commonly used microbial toxicity bioassays is the Microtox® test, which was developed in the late 1970s by Beckman Instruments, Inc. (USA). Using the knowledge that the flow of electrons in the respiratory chain is an indication of the metabolic state of the cell, the Microtox® test is based on rehydrating freeze-dried luminescent marine bacteria (strain NRRL-B-11177) and measuring the toxicant-induced reductions in bioluminescence in a photometer under defined experimental conditions of temperature, exposure time, and so on.

There is confusion in the naming of the bacteria used for the Microtox® test. Indeed, in the first papers related to this test, the strain NRRL-B-11177 was designated to be *Photobacterium fischeri*. Soon thereafter, the strain was reclassified as *Photobacterium phosphoreum* and this designation has widely been carried throughout the literature since the early 1980s. Even if in the Microtox® literature the strain NRRL-B-11177 is rarely related to *Vibrio fischeri* (the current name of *Photobacterium fischeri*), it has been confirmed that this is the correct taxonomic association. As for the majority of (eco)toxicologists the name *Photobacterium phosphoreum* is unquestionably linked to the Microtox® test and, to avoid confusion, we have also decided to use this name in this handbook.

Other microbial ecotoxicity tests are also based on bioluminescent reactions. Despite their interest, the results of these bioassays are not compiled in this book due to their scarcity. Only toxicity results on chemical substances following the principles of the Microtox® bioassay procedure are considered in this book. Indeed, even though the Microtox® test is widely used for determining the toxicity of effluents, it was impossible to incorporate the results of these tests nor those relating to the toxicity of sediments without considerably increasing the length of the book.

In preparing this second volume in the series of *Handbooks of Ecotoxicological Data*, we have followed the manner and style used in volume 1, *Ecotoxicity of Chemicals to Amphibians*, by providing both the original references and commentaries on the test conditions and data, where applicable.

It is our sincere hope that this handbook will benefit a wide variety of readers from various scientific disciplines, and become a reference book for estimating the ecotoxicity of chemicals to aquatic bacteria.

ECOTOXICOLOGICAL DATA

Chemicals are arranged alphabetically by their common names. Prefixes commonly used in organic chemistry which are not normally considered part of the name, such as *o*-, *m*-, *p*-, α-, β-, γ-, *n*-, *sec*-, *tert*-, *cis*-, *trans*-, *N*- have not been considered for alphabetical order. Other prefixes which normally are considered part of the name, such as bis-, iso-, di-, tri-, and tetra-, are used for alphabetical positioning.

Under each chemical heading are listed the Chemical Abstract Service Registry Number (CAS RN) and the synonym(s) for easy cross-reference.

The ecotoxicological results for each chemical are presented as follows:

Sample purity: and/or formulation
Temperature: This parameter is given only when it was clearly indicated in the original publication, or confirmed by the authors who performed the assays. However, we assume that when in the "Materials and Methods" of a paper, the authors indicate that they followed the procedure detailed in the Microtox® system operating manual, we can consider that their tests were carried out at 15°C.
Test parameter:
Effect:
Concentration:
Exposure time:
Comment: This section includes all the available information allowing the interpretation of the ecotoxicity results and the estimation of their validity. Due to the confusion related to the naming of the bacteria used for the Microtox® test, we also give in this section the strain designation and/or the origin of the bacteria used in the microbial toxicity test when standard material was not used.
Bibliographical reference:

In some cases, for practical purposes, data are listed in tables.

ACENAPHTHENE

CAS RN: 83-32-9

Temperature: 15°C
Test parameter: EC50
Effect: Reduction in light output
Concentration: 0.74 mg/l (0.58-0.93)
Exposure time: 15 min
Comment: Chemical was added to 5 ml DMSO and placed in a bath sonicator until all visible material had dissolved.

Bibliographical reference: Jacobs, M.W., Coates, J.A., Delfino, J.J., Bitton, G., Davis, W.M., and Garcia, K.L. (1993). *Arch. Environ. Contam. Toxicol.* **24**, 461-468.

ACENAPHTHYLENE

CAS RN: 208-96-8

Sample purity: 98%
Temperature: 15°C
Test parameter: EC50
Effect: Reduction in light output

Concentration: 0.23 mg/l
Exposure time: 5 min

Concentration: 0.24 mg/l
Exposure time: 15 min

Concentration: 0.28 mg/l
Exposure time: 30 min

Comment: Mean of four assays. Methanol (<10%) was used to prepare the stock solutions. EC50 values were calculated from nominal concentrations.

Bibliographical reference: Kaiser, K.L.E., and Palabrica, V.S. (1991). *Water Poll. Res. J. Canada* **26**, 361-431.

Temperature: 15°C
Test parameter: EC50

Effect: Reduction in light output
Concentration: 0.31 mg/l (0.25-0.39)
Exposure time: 15 min
Comment: Chemical was added to 5 ml DMSO and placed in a bath sonicator until all visible material had dissolved.

Bibliographical reference: Jacobs, M.W., Coates, J.A., Delfino, J.J., Bitton, G., Davis, W.M., and Garcia, K.L. (1993). *Arch. Environ. Contam. Toxicol.* **24**, 461-468.

ACETALDEHYDE

CAS RN: 75-07-0
Synonym: Ethanal

Test parameter: EC50
Effect: Reduction in light output
Concentration: 342 mg/l
Exposure time: 5 min
Comment: Concentrations in the test were measured.

Bibliographical reference: Curtis, C., Lima, A., Lozano, S.J., and Veith, G.D. (1982). In: *Aquatic Toxicology and Hazard Assessment: Fifth Conference, ASTM STP 766*, J.G. Pearson, R.B. Foster, and W.E. Bishop (eds.), American Society for Testing and Materials, Philadelphia, p. 170-178.

Sample purity: 99%
Temperature: 15°C
Test parameter: EC50
Effect: Reduction in light output

Concentration: 389 ± 14.5 mg/l
Exposure time: 5 min

Concentration: 320 ± 19.8 mg/l
Exposure time: 15 min

Concentration: 303 ± 22.4 mg/l
Exposure time: 25 min

Comment: Mean of five assays. The values were converted to mg/l from the original data expressed in µM and a rounded molecular weight of 44 given by the authors. Phenol solution was used for quality control/quality assurance. The 5-min EC50 value was 18.2 mg/l. The EC50 value at 15 min was 20.7 mg/l with a relative error of <5%.

Bibliographical reference: Chou, C.C., and Que Hee, S.S. (1992). *Ecotoxicol. Environ. Safety* **23**, 355-363.

4-ACETAMIDOACETOPHENONE

CAS RN: 2719-21-3

Sample purity: 98%
Temperature: 15°C
Test parameter: EC50
Effect: Reduction in light output

Concentration: 134 mg/l
Exposure time: 5 min

Concentration: 112 mg/l
Exposure time: 15 min

Concentration: 95 mg/l
Exposure time: 30 min

Comment: Mean of four assays. Methanol (<10%) was used to prepare the stock solutions. EC50 values were calculated from nominal concentrations.

Bibliographical reference: Kaiser, K.L.E., and Palabrica, V.S. (1992). National Water Research Institute, Burlington, Ontario, Canada, unpublished results.

4-ACETAMIDOBENZOIC ACID

CAS RN: 556-08-1

Sample purity: 98%
Temperature: 15°C
Test parameter: EC50
Effect: Reduction in light output

Concentration: 103 mg/l
Exposure time: 5 min

Concentration: 98.5 mg/l
Exposure time: 15 min

Concentration: 96.2 mg/l
Exposure time: 30 min

Comment: Mean of three assays. Methanol (<10%) was used to prepare the stock solutions. EC50 values were calculated from nominal concentrations.

Bibliographical reference: Kaiser, K.L.E., and Palabrica, V.S. (1991). *Water Poll. Res. J. Canada* **26**, 361-431.

2-ACETAMIDO-5-NITROTHIAZOLE

CAS RN: 140-40-9

Temperature: 15°C
Test parameter: EC50
Effect: Reduction in light output

Concentration: 6.34 mg/l
Exposure time: 5 min

Concentration: 1.49 mg/l
Exposure time: 15 min

Concentration: 0.86 mg/l
Exposure time: 30 min

Comment: Mean of three assays. Methanol (<10%) was used to prepare the stock solutions. EC50 values were calculated from nominal concentrations.

Bibliographical reference: Kaiser, K.L.E., and Palabrica, V.S. (1992). National Water Research Institute, Burlington, Ontario, Canada, unpublished results.

4-ACETAMIDOPHENOL

CAS RN: 103-90-2
Synonym: 4'-Hydroxyacetanilide

Sample purity: 98%
Temperature: 15°C
Test parameter: EC50
Effect: Reduction in light output

Concentration: 1120 mg/l
Exposure time: 5 min

Concentration: 1050 mg/l
Exposure time: 15 min

Concentration: 1000 mg/l*
Exposure time: 30 min

Comment: Mean of three assays. Methanol (<10%) was used to prepare the stock solutions. EC50 values were calculated from nominal concentrations.

Bibliographical references: Kaiser, K.L.E., and Palabrica, V.S. (1991). *Water Poll. Res. J. Canada* **26**, 361-431.
* Kaiser, K.L.E. (1987). In: *QSAR in Environmental Toxicology - II*, K.L.E. Kaiser (ed.), D. Reidel Publishing Company, Dordrecht, p. 169-188.

ACETIC ACID

CAS RN: 64-19-7

Temperature: 15°C
Test parameter: EC50
Effect: Reduction in light output
Concentration: 11 mg/l
Exposure time: 15 min

Bibliographical reference: Bulich, A.A., Tung, K.K., and Scheibner, G. (1990). *J. Biolumin. Chemilumin.* **5**, 71-77.

Sample purity: 99.99%

Temperature: 15°C
Test parameter: EC50
Effect: Reduction in light output

Concentration: 9.24 ± 0.38 mg/l
Exposure time: 5 min

Concentration: 9.60 ± 0.45 mg/l
Exposure time: 15 min

Concentration: 9.60 ± 0.58 mg/l
Exposure time: 25 min

Comment: Mean of six assays. The values were converted to mg/l from the original data expressed in µM and a rounded molecular weight of 60 given by the authors. Phenol solution was used for quality control/quality assurance. The 5-min EC50 value was 18.2 mg/l. The EC50 value at 15 min was 20.7 mg/l with a relative error of <5%.

Bibliographical reference: Chou, C.C., and Que Hee, S.S. (1992). *Ecotoxicol. Environ. Safety* **23**, 355-363.

ACETONE

CAS RN: 67-64-1
Synonyms: Dimethylketone; 2-Propanone

Test parameter: EC50
Effect: Reduction in light output
Concentrations: 21000 mg/l
22000 mg/l
Exposure time: 5 min
Comment: Toxicity values were calculated from nominal concentrations.

Bibliographical reference: Curtis, C., Lima, A., Lozano, S.J., and Veith, G.D. (1982). In: *Aquatic Toxicology and Hazard Assessment: Fifth Conference, ASTM STP 766*, J.G. Pearson, R.B. Foster, and W.E. Bishop (eds.), American Society for Testing and Materials, Philadelphia, p. 170-178.

Sample purity: >98%
Temperature: 15 ± 0.1°C

Test parameter: EC50
Effect: Reduction in light output

Concentration: 22270 mg/l
Exposure time: 5 min

Concentration: 28940 mg/l
Exposure time: 15 min

Comment: Test was performed in duplicate. Toxicity values were based on nominal concentrations.

Bibliographical reference: de Zwart, D., and Slooff, W. (1983). *Aquat. Toxicol.* **4**, 129-138.

Sample purity: >98%
Temperature: 15 ± 0.1°C
Test parameter: EC10
Effect: Reduction in light output

Concentration: 8700 mg/l
Exposure time: 5 min

Concentration: 7900 mg/l
Exposure time: 15 min

Comment: Test was performed in duplicate. Toxicity values were based on nominal concentrations.

Bibliographical reference: de Zwart, D., and Slooff, W. (1983). *Aquat. Toxicol.* **4**, 129-138.

Temperature: 15 ± 0.1°C
Test parameter: EC50
Effect: Reduction in light output
Concentration: 18250 mg/l
Exposure time: 5 min
Comment: Test was performed on *Photobacterium phosphoreum* NZ11D obtained from the Scripps Institute of Oceanography (La Jolla, CA).

Bibliographical reference: McFeters, G.A., Bond, P.J., Olson, S.B., and Tchan, Y.T. (1983). *Water Res.* **17**, 1757-1762.

Sample purity: Pesticide grade
Test parameter: EC50
Effect: Reduction in light output

Concentration: 12364 mg/l
Exposure time: 5 min

Concentration: 12601 mg/l
Exposure time: 15 min

Concentration: 13213 mg/l
Exposure time: 30 min

Comment: Two replicates of each concentration were tested.

Bibliographical reference: Greene, J.C., Miller, W.E., Debacon, M.K., Long, M.A., and Bartels, C.L. (1985). *Arch. Environ. Contam. Toxicol.* **14**, 659-667.

Temperature: 15°C
Test parameter: EC50
Effect: Reduction in light output
Concentration: 21100 mg/l
Exposure time: 15 min

Bibliographical reference: Hermens, J., Busser, F., Leeuwangh, P., and Musch, A. (1985). *Ecotoxicol. Environ. Safety* **9**, 17-25.

Temperature: 15°C
Test parameter: EC50
Effect: Reduction in light output
Concentration: 18.35 μl/ml
Exposure time: 15 min
Comment: Four concentrations of chemical were tested in duplicate.

Bibliographical reference: Schiewe, M.H., Hawk, E.G., Actor, D.I., and Krahn, M.M. (1985). *Can. J. Fish. Aquat. Sci.* **42**, 1244-1248.

Temperature: 15 ± 0.1°C
Test parameter: EC50
Effect: Reduction in light output

Concentration: 16000 mg/l

Exposure time: 5 min

Concentration: 16800 mg/l
Exposure time: 15 min

Concentration: 17000 mg/l
Exposure time: 30 min

Comment: The pH was not adjusted.

Bibliographical reference: Tarkpea, M., Hansson, M., and Samuelsson, B. (1986). *Ecotoxicol. Environ. Safety* **11**, 127-143.

Sample purity: 99%
Temperature: 15°C
Test parameter: EC50
Effect: Reduction in light output

Concentration: 20900 mg/l
Exposure time: 5 min

Concentration: 20050 mg/l*
Exposure time: 15 min

Bibliographical references: Cronin, M.T.D., Dearden, J.C., and Dobbs, A.J. (1991). *Sci. Total Environ.* **109/110**, 431-439.
* Cronin, M.T.D. (1993). Liverpool John Moores University, UK, private communication.

Temperature: 15°C
Test parameter: EC50
Effect: Reduction in light output
Concentration: 8600 mg/l
Exposure time: 5 min

Bibliographical reference: Kahru, A. (1993). *ATLA* **21**, 210-215.

ACETONITRILE

CAS RN: 75-05-8

Temperature: 15°C

Test parameter: EC50
Effect: Reduction in light output
Concentration: 22.3 µl/ml (21.1-23.5)
Exposure time: 15 min
Comment: Mean of 3 bioassays.

Bibliographical reference: Jacobs, M.W., Delfino, J.J., and Bitton, G. (1992). *Environ. Toxicol. Chem.* **11**, 1137-1143.

Temperature: 15°C
Test parameter: EC50
Effect: Reduction in light output
Concentration: 19500 mg/l
Exposure time: 5 min

Bibliographical reference: Kahru, A. (1993). *ATLA* **21**, 210-215.

ACETOPHENONE

CAS RN: 98-86-2

Temperature: 15°C
Test parameter: EC50
Effect: Reduction in light output
Concentration: 15.5 mg/l
Exposure time: 30 min
Comment: Mean of three assays. Methanol (<10%) was used to prepare the stock solutions. EC50 values were calculated from nominal concentrations.

Bibliographical reference: Kaiser, K.L.E., Palabrica, V.S., and Ribo, J.M. (1987). In: *QSAR in Environmental Toxicology - II*, K.L.E. Kaiser (ed.), D. Reidel Publishing Company, Dordrecht, p. 153-168.

4-ACETOXYBENZOIC ACID

CAS RN: 2345-34-8

Sample purity: 98%
Temperature: 15°C
Test parameter: EC50

Effect: Reduction in light output

Concentration: 53.2 mg/l
Exposure time: 5 min

Concentration: 47.4 mg/l
Exposure time: 15 min

Concentration: 41.3 mg/l
Exposure time: 30 min

Comment: Mean of three assays. Methanol (<10%) was used to prepare the stock solutions. EC50 values were calculated from nominal concentrations.

Bibliographical reference: Kaiser, K.L.E., and Palabrica, V.S. (1991). *Water Poll. Res. J. Canada* **26**, 361-431.

4-ACETOXYBIPHENYL

CAS RN: 117-34-0

Sample purity: 98%
Temperature: 15°C
Test parameter: EC50
Effect: Reduction in light output

Concentration: 2.55 mg/l
Exposure time: 5 min

Concentration: 2.73 mg/l
Exposure time: 15 min

Concentration: 3.14 mg/l
Exposure time: 30 min

Comment: Mean of three assays. Methanol (<10%) was used to prepare the stock solutions. EC50 values were calculated from nominal concentrations.

Bibliographical reference: Kaiser, K.L.E., and Palabrica, V.S. (1992). National Water Research Institute, Burlington, Ontario, Canada, unpublished results.

4-ACETOXYSTYRENE

CAS RN: 2628-16-2

Temperature: 15°C
Test parameter: EC50
Effect: Reduction in light output

Concentration: 3.60 mg/l
Exposure time: 5 min

Concentration: 3.50 mg/l
Exposure time: 15 min

Concentration: 3.40 mg/l
Exposure time: 30 min

Comment: Mean of four assays. Methanol (<10%) was used to prepare the stock solutions. EC50 values were calculated from nominal concentrations.

Bibliographical reference: Kaiser, K.L.E., and Palabrica, V.S. (1993). National Water Research Institute, Burlington, Ontario, Canada, unpublished results.

4-ACETYLBENZOIC ACID

CAS RN: 586-89-0

Sample purity: 98%
Temperature: 15°C
Test parameter: EC50
Effect: Reduction in light output

Concentration: 94.5 mg/l
Exposure time: 5 min

Concentration: 98.9 mg/l
Exposure time: 15 min

Concentration: 101 mg/l
Exposure time: 30 min

Comment: Mean of three assays. Methanol (<10%) was used to

prepare the stock solutions. EC50 values were calculated from nominal concentrations.

Bibliographical reference: Kaiser, K.L.E., and Palabrica, V.S. (1991). *Water Poll. Res. J. Canada* **26**, 361-431.

4-ACETYLBENZONITRILE

CAS RN: 1443-80-7

Sample purity: 99%
Temperature: 15°C
Test parameter: EC50
Effect: Reduction in light output

Concentration: 69.5 mg/l
Exposure time: 5 min

Concentration: 60.5 mg/l
Exposure time: 15 min

Concentration: 51.5 mg/l*
Exposure time: 30 min

Comment: Mean of three assays. Methanol (<10%) was used to prepare the stock solutions. EC50 values were calculated from nominal concentrations.

Bibliographical references: Kaiser, K.L.E., and Palabrica, V.S. (1991). *Water Poll. Res. J. Canada* **26**, 361-431.
* Kaiser, K.L.E., and Gough, K.M. (1989). In: *Aquatic Toxicology and Environmental Fate: Eleventh Volume, ASTM STP 1007*, G.W. Suter and M.A. Lewis (eds.), American Society for Testing and Materials, Philadelphia, p. 424-441.

1-ACETYL-2-PHENYLHYDRAZINE

CAS RN: 114-83-0

Temperature: 15°C
Test parameter: EC50
Effect: Reduction in light output

Concentration: 97.0 mg/l
Exposure time: 5 min

Concentration: 90.5 mg/l
Exposure time: 15 min

Concentration: 88.4 mg/l
Exposure time: 30 min

Comment: Mean of four assays. Methanol (<10%) was used to prepare the stock solutions. EC50 values were calculated from nominal concentrations.

Bibliographical reference: Kaiser, K.L.E., and Palabrica, V.S. (1993). National Water Research Institute, Burlington, Ontario, Canada, unpublished results.

4-ACETYLPHENYL ISOCYANATE

CAS RN: 49647-20-3

Sample purity: 97%
Temperature: 15°C
Test parameter: EC50
Effect: Reduction in light output

Concentration: 2.08 mg/l
Exposure times: 5 and 15 min

Concentration: 2.12 mg/l
Exposure time: 30 min

Comment: Mean of four assays. Methanol (<10%) was used to prepare the stock solutions. EC50 values were calculated from nominal concentrations.

Bibliographical reference: Kaiser, K.L.E., and Palabrica, V.S. (1992). National Water Research Institute, Burlington, Ontario, Canada, unpublished results.

2-ACETYLPYRIDINE

CAS RN: 1122-62-9

Sample purity: >99%
Temperature: 15°C
Test parameter: EC50
Effect: Reduction in light output

Concentration: 392 mg/l
Exposure time: 5 min

Concentration: 420 mg/l
Exposure times: 15 and 30 min

Comment: Mean of three assays. Methanol (<10%) was used to prepare the stock solutions. EC50 values were calculated from nominal concentrations.

Bibliographical reference: Kaiser, K.L.E., and Palabrica, V.S. (1991). *Water Poll. Res. J. Canada* **26**, 361-431.

3-ACETYLPYRIDINE

CAS RN: 350-03-8

Sample purity: 98%
Temperature: 15°C
Test parameter: EC50
Effect: Reduction in light output

Concentration: 127 mg/l
Exposure time: 5 min

Concentration: 106 mg/l
Exposure time: 15 min

Concentration: 118 mg/l
Exposure time: 30 min

Comment: Mean of four assays. Methanol (<10%) was used to prepare the stock solutions. EC50 values were calculated from nominal concentrations.

Bibliographical reference: Kaiser, K.L.E., and Palabrica, V.S. (1991). *Water Poll. Res. J. Canada* **26**, 361-431.

4-ACETYLPYRIDINE

CAS RN: 1122-54-9

Sample purity: 97%
Temperature: 15°C
Test parameter: EC50
Effect: Reduction in light output

Concentration: 236 mg/l
Exposure time: 5 min

Concentration: 231 mg/l
Exposure time: 15 min

Concentration: 247 mg/l
Exposure time: 30 min

Comment: Mean of three assays. Methanol (<10%) was used to prepare the stock solutions. EC50 values were calculated from nominal concentrations.

Bibliographical reference: Kaiser, K.L.E., and Palabrica, V.S. (1991). *Water Poll. Res. J. Canada* **26**, 361-431.

ACRIDINE

CAS RN: 260-94-6

Temperature: 15°C
Test parameter: EC50
Effect: Reduction in light output

Concentration: 6.66 mg/l
Exposure time: 5 min

Concentration: 6.97 mg/l
Exposure time: 15 min

Concentration: 7.47 mg/l
Exposure time: 30 min

Comment: Mean of three assays. Methanol (<10%) was used to prepare the stock solutions. EC50 values were calculated from nominal concentrations.

Bibliographical reference: Kaiser, K.L.E., and Palabrica, V.S. (1991). *Water Poll. Res. J. Canada* **26**, 361-431.

ACROLEIN

CAS RN: 107-02-8

Temperature: 15°C
Test parameter: EC50
Effect: Reduction in light output
Concentration: 0.67 mg/l
Exposure time: 5 min

Bibliographical reference: Microtox® Application Notes (1982). Beckman Instruments, Inc., Carlsbad, California.

Sample purity: 97%
Temperature: 15°C
Test parameter: EC50
Effect: Reduction in light output

Concentration: 0.32 mg/l
Exposure time: 5 min

Concentration: 0.12 mg/l*
Exposure time: 15 min

Bibliographical references: Cronin, M.T.D., Dearden, J.C., and Dobbs, A.J. (1991). *Sci. Total Environ.* **109/110**, 431-439.
* Cronin, M.T.D. (1990). Thesis, Liverpool Polytechnic, Liverpool, UK.

ACRYLONITRILE

CAS RN: 107-13-1

Temperature: 15°C
Test parameter: EC50
Effect: Reduction in light output
Concentration: 3910 mg/l
Exposure time: 5 min

Bibliographical reference: Microtox® Application Notes (1982). Beckman Instruments, Inc., Carlsbad, California.

Sample purity: >99%
Temperature: 15°C
Test parameter: EC50
Effect: Reduction in light output

Concentration: 495 mg/l
Exposure time: 5 min

Concentration: 367 mg/l
Exposure time: 15 min

Concentration: 254 mg/l
Exposure time: 30 min

Comment: Mean of four assays. Methanol (<10%) was used to prepare the stock solutions. EC50 values were calculated from nominal concentrations.

Bibliographical reference: Kaiser, K.L.E., and Palabrica, V.S. (1991). *Water Poll. Res. J. Canada* **26**, 361-431.

AFLATOXIN B_1

CAS RN: 1162-65-8

Temperature: 15°C
Test parameter: EC50
Effect: Reduction in light output

Concentration: 21.97 mg/l

Exposure time: 5 min

Concentration: 21.19 mg/l
Exposure time: 10 min

Concentration: 19.44 mg/l
Exposure time: 15 min

Concentration: 20.35 mg/l
Exposure time: 20 min

Comment: Freshly reconstituted bacterial suspensions. Chemical was dissolved in DMSO.

Bibliographical reference: Yates, I.E., and Porter, J.K. (1982). *Appl. Environ. Microbiol.* **44**, 1072-1075.

Temperature: 15°C
Test parameter: EC20
Effect: Reduction in light output

Concentration: 4.54 mg/l
Exposure time: 5 min

Concentration: 3.61 mg/l
Exposure time: 20 min

Comment: Freshly reconstituted bacterial suspensions. Chemical was dissolved in DMSO.

Bibliographical reference: Yates, I.E., and Porter, J.K. (1982). *Appl. Environ. Microbiol.* **44**, 1072-1075.

Temperature: 15°C
Test parameter: EC50
Effect: Reduction in light output

Concentration: 24.79 mg/l
Exposure time: 5 min

Concentration: 24.43 mg/l
Exposure time: 10 min

Concentration: 25.66 mg/l

Exposure time: 15 min

Concentration: 25.87 mg/l
Exposure time: 20 min

Comment: Performed on bacterial suspensions maintained at 3°C for 5 h after reconstitution. Chemical was dissolved in DMSO.

Bibliographical reference: Yates, I.E., and Porter, J.K. (1982). *Appl. Environ. Microbiol.* **44**, 1072-1075.

Temperature: 20°C
Test parameter: EC50
Effect: Reduction in light output

Concentration: 22.01 mg/l
Exposure time: 5 min

Concentration: 22.84 mg/l
Exposure time: 10 min

Concentration: 23.32 mg/l
Exposure time: 15 min

Concentration: 23.27 mg/l
Exposure time: 20 min

Comment: Chemical was dissolved in DMSO. Test was performed at pH = 7.0 units.

Bibliographical reference: Yates, I.E., and Porter, J.K. (1984). In: *Toxicity Screening Procedures Using Bacterial Systems*, D. Liu and B.J. Dutka (eds.), Marcel Dekker, New York, p. 77-88.

Test parameter: EC50
Effect: Reduction in light output

Concentration: 22.0 ± 1.5 mg/l
Exposure time: 5 min

Concentration: 21.2 ± 0.5 mg/l
Exposure time: 10 min

Concentration: 19.4 ± 0.4 mg/l

Exposure time: 15 min

Concentration: 20.4 ± 1.2 mg/l
Exposure time: 20 min

Comment: Chemical was solubilized in DMSO.

Bibliographical reference: Yates, I.E. (1985). *J. Microbiol. Meth.* **3**, 181-186.

AFLATOXIN B_2

CAS RN: 7220-81-7

Test parameter: EC50
Effect: Reduction in light output

Concentration: 55.0 ± 1.1 mg/l
Exposure time: 5 min

Concentration: 61.0 ± 1.3 mg/l
Exposure time: 10 min

Concentration: 59.3 ± 1.2 mg/l
Exposure time: 15 min

Concentration: 63.1 ± 2.2 mg/l
Exposure time: 20 min

Comment: Chemical was solubilized in DMSO.

Bibliographical reference: Yates, I.E. (1985). *J. Microbiol. Meth.* **3**, 181-186.

AFLATOXIN G_1

CAS RN: 1165-39-5

Test parameter: EC50
Effect: Reduction in light output

Concentration: 41.5 ± 0.8 mg/l

Exposure time: 5 min

Concentration: 39.1 ± 0.8 mg/l
Exposure time: 10 min

Concentration: 39.0 ± 0.6 mg/l
Exposure time: 15 min

Concentration: 35.6 ± 0.9 mg/l
Exposure time: 20 min

Comment: Chemical was solubilized in DMSO.

Bibliographical reference: Yates, I.E. (1985). *J. Microbiol. Meth.* **3**, 181-186.

AFLATOXIN G_2

CAS RN: 7241-98-7

Test parameter: EC50
Effect: Reduction in light output

Concentration: 68.7 ± 0.1 mg/l
Exposure time: 5 min

Concentration: 77.1 ± 0.6 mg/l
Exposure time: 10 min

Concentration: 74.0 ± 1.5 mg/l
Exposure time: 15 min

Concentration: 69.5 ± 1.3 mg/l
Exposure time: 20 min

Comment: Chemical was solubilized in DMSO.

Bibliographical reference: Yates, I.E. (1985). *J. Microbiol. Meth.* **3**, 181-186.

ALLYL ALCOHOL

CAS RN: 107-18-6
Synonym: 2-Propen-1-ol

Temperature: 15°C
Test parameter: EC50
Effect: Reduction in light output
Concentration: 1100 mg/l
Exposure time: 15 min

Bibliographical reference: Bulich, A.A., Tung, K.K., and Scheibner, G. (1990). *J. Biolumin. Chemilumin.* **5**, 71-77.

Sample purity: 99+%
Temperature: 15°C
Test parameter: EC50
Effect: Reduction in light output

Concentration: 608 mg/l
Exposure time: 5 min

Concentration: 342 mg/l
Exposure time: 15 min

Concentration: 216 mg/l
Exposure time: 30 min

Comment: Mean of three assays. Methanol (<10%) was used to prepare the stock solutions. EC50 values were calculated from nominal concentrations.

Bibliographical reference: Kaiser, K.L.E., and Palabrica, V.S. (1992). National Water Research Institute, Burlington, Ontario, Canada, unpublished results.

ALLYLAMINE

CAS RN: 107-11-9
Synonym: 2-Propenylamine

Sample purity: >98%
Temperature: 15 ± 0.1°C

Test parameter: EC50
Effect: Reduction in light output

Concentration: 19.9 mg/l
Exposure time: 5 min

Concentration: 16.3 mg/l
Exposure time: 15 min

Comment: Test was performed in duplicate. Toxicity values were based on nominal test concentrations.

Bibliographical reference: de Zwart, D., and Slooff, W. (1983). *Aquat. Toxicol.* **4**, 129-138.

Sample purity: >98%
Temperature: 15 ± 0.1°C
Test parameter: EC10
Effect: Reduction in light output

Concentration: 12.5 mg/l
Exposure time: 5 min

Concentration: 12.8 mg/l
Exposure time: 15 min

Comment: Test was performed in duplicate. Toxicity values were based on nominal test concentrations.

Bibliographical reference: de Zwart, D., and Slooff, W. (1983). *Aquat. Toxicol.* **4**, 129-138.

4-ALLYLANISOLE

CAS RN: 140-67-0
Synonym: Estragole

Sample purity: 98%
Temperature: 15°C
Test parameter: EC50
Effect: Reduction in light output

Concentration: 0.36 mg/l

Exposure time: 5 min

Concentration: 0.39 mg/l
Exposure time: 15 min

Concentration: 0.46 mg/l
Exposure time: 30 min

Comment: Mean of three assays. Methanol (<10%) was used to prepare the stock solutions. EC50 values were calculated from nominal concentrations.

Bibliographical reference: Kaiser, K.L.E., and Palabrica, V.S. (1992). National Water Research Institute, Burlington, Ontario, Canada, unpublished results.

ALLYL CYANIDE

CAS RN: 109-75-1
Synonyms: 3-Butenenitrile; Vinylacetonitrile

Temperature: 15°C
Test parameter: EC50
Effect: Reduction in light output

Concentration: 2700 mg/l
Exposure time: 5 min

Concentration: 2300 mg/l*
Exposure time: 15 min

Bibliographical references: Cronin, M.T.D., Dearden, J.C., and Dobbs, A.J. (1991). *Sci. Total Environ.* **109/110**, 431-439.
* Cronin, M.T.D. (1993). Liverpool John Moores University, UK, private communication.

1-ALLYLIMIDAZOLE

CAS RN: 31410-01-2

Sample purity: ~95%
Temperature: 15°C

Test parameter: EC50
Effect: Reduction in light output

Concentration: 160 mg/l
Exposure time: 5 min

Concentration: 139 mg/l
Exposure time: 15 min

Concentration: 130 mg/l
Exposure time: 30 min

Comment: Mean of four assays. Methanol (<10%) was used to prepare the stock solutions. EC50 values were calculated from nominal concentrations.

Bibliographical reference: Kaiser, K.L.E., and Palabrica, V.S. (1993). National Water Research Institute, Burlington, Ontario, Canada, unpublished results.

ALLYL ISOTHIOCYANATE

CAS RN: 57-06-7

Sample purity: 97%
Temperature: 15°C
Test parameter: EC50
Effect: Reduction in light output

Concentration: 0.11 mg/l
Exposure time: 5 min

Concentration: 0.099 mg/l
Exposure time: 15 min

Concentration: 0.106 mg/l
Exposure time: 30 min

Comment: Mean of four assays. Methanol (<10%) was used to prepare the stock solutions. EC50 values were calculated from nominal concentrations.

Bibliographical reference: Kaiser, K.L.E., and Palabrica, V.S. (1992). National Water Research Institute, Burlington, Ontario, Canada, unpublished results.

2-ALLYLPHENOL

CAS RN: 1745-81-9

Test parameter: EC50
Effect: Reduction in light output
Concentration: 10 mg/l
Exposure time: 5 min
Comment: Concentrations in the test were measured.

Bibliographical reference: Curtis, C., Lima, A., Lozano, S.J., and Veith, G.D. (1982). In: *Aquatic Toxicology and Hazard Assessment: Fifth Conference, ASTM STP 766*, J.G. Pearson, R.B. Foster, and W.E. Bishop (eds.), American Society for Testing and Materials, Philadelphia, p. 170-178.

ALUMINUM SULFATE OCTADECAHYDRATE

CAS RN: 7784-31-8

Sample purity: Analytical grade
Temperature: 15°C
Test parameter: EC50
Effect: Reduction in light output

Concentration: 1.62 mg/l Al^{+++}
Exposure time: 5 min

Concentration: 1.28 mg/l Al^{+++}
Exposure time: 10 min

Concentration: 1.10 mg/l Al^{+++}
Exposure time: 15 min

Concentration: 1.08 mg/l Al^{+++}
Exposure time: 20 min

Concentration: 1.04 mg/l Al^{+++}
Exposure time: 30 min

Comment: Results derived from the average of two replicates.

Bibliographical reference: Qureshi, A.A., Coleman, R.N., and Paran, J.H. (1984). In: *Toxicity Screening Procedures Using Bacterial Systems*, D. Liu and B.J. Dutka (eds.), Marcel Dekker, New York, p. 1-22.

4-AMINOACETANILIDE

CAS RN: 122-80-5

Sample purity: 99%
Temperature: 15°C
Test parameter: EC50
Effect: Reduction in light output

Concentration: 207 mg/l
Exposure time: 5 min

Concentration: 157 mg/l
Exposure time: 15 min

Concentration: 172 mg/l
Exposure time: 30 min

Comment: Mean of three assays. Methanol (<10%) was used to prepare the stock solutions. EC50 values were calculated from nominal concentrations.

Bibliographical reference: Kaiser, K.L.E., and Palabrica, V.S. (1991). *Water Poll. Res. J. Canada* **26**, 361-431.

4'-AMINOACETOPHENONE

CAS RN: 99-92-3

Temperature: 15°C
Test parameter: EC50
Effect: Reduction in light output

Concentration: 4.00 mg/l
Exposure time: 5 min

Concentration: 4.40 mg/l
Exposure time: 15 min

Concentration: 5.00 mg/l*
Exposure time: 30 min

Comment: Mean of four assays. Methanol (<10%) was used to prepare the stock solutions. EC50 values were calculated from nominal concentrations.

Bibliographical references: Kaiser, K.L.E., and Palabrica, V.S. (1992). National Water Research Institute, Burlington, Ontario, Canada, unpublished results.
* Kaiser, K.L.E. (1987). In: *QSAR in Environmental Toxicology - II*, K.L.E. Kaiser (ed.), D. Reidel Publishing Company, Dordrecht, p. 169-188.

9-AMINOACRIDINE

CAS RN: 90-45-9

Temperature: 15°C
Test parameter: EC50
Effect: Reduction in light output

Concentration: 40.1 ± 1.7 mg/l
Exposure time: 5 min

Concentration: 39.7 ± 1.6 mg/l
Exposure time: 10 min

Concentration: 37.7 ± 1.9 mg/l
Exposure time: 15 min

Concentration: 35.9 ± 1.4 mg/l
Exposure time: 20 min

Comment: Test was performed at least in triplicate. Chemical was solubilized in DMSO.

Bibliographical reference: Yates, I.E. (1985). *J. Microbiol. Meth.* **3**, 171-180.

4-AMINOAZOBENZENE

CAS RN: 60-09-3

Sample purity: 98%
Temperature: 15°C
Test parameter: EC50
Effect: Reduction in light output

Concentration: 2.48 mg/l
Exposure time: 5 min

Concentration: 2.54 mg/l
Exposure time: 15 min

Concentration: 2.66 mg/l*
Exposure time: 30 min

Comment: Mean of three assays. Methanol (<10%) was used to prepare the stock solutions. EC50 values were calculated from nominal concentrations.

Bibliographical references: Kaiser, K.L.E., and Palabrica, V.S. (1991). *Water Poll. Res. J. Canada* **26**, 361-431.
* Kaiser, K.L.E., Palabrica, V.S., and Ribo, J.M. (1987). In: *QSAR in Environmental Toxicology - II*, K.L.E. Kaiser (ed.), D. Reidel Publishing Company, Dordrecht, p. 153-168.

2-AMINOBENZIMIDAZOLE

CAS RN: 934-32-7

Sample purity: 97%
Temperature: 15°C
Test parameter: EC50
Effect: Reduction in light output

Concentration: 43.1 mg/l
Exposure time: 5 min

Concentration: 26.0 mg/l
Exposure time: 15 min

Concentration: 24.8 mg/l

Exposure time: 30 min

Comment: Mean of four assays. Methanol (<10%) was used to prepare the stock solutions. EC50 values were calculated from nominal concentrations.

Bibliographical reference: Kaiser, K.L.E., and Palabrica, V.S. (1991). *Water Poll. Res. J. Canada* **26**, 361-431.

4-AMINOBENZOIC ACID

CAS RN: 150-13-0

Temperature: 15°C
Test parameter: EC50
Effect: Reduction in light output
Concentration: 27.4 mg/l
Exposure time: 30 min
Comment: Mean of three assays. Methanol (<10%) was used to prepare the stock solutions. EC50 values were calculated from nominal concentrations.

Bibliographical reference: Kaiser, K.L.E. (1987). In: *QSAR in Environmental Toxicology - II*, K.L.E. Kaiser (ed.), D. Reidel Publishing Company, Dordrecht, p. 169-188.

4-AMINOBENZONITRILE

CAS RN: 873-74-5
Synonym: 4-Cyanoaniline

Temperature: 15°C
Test parameter: EC50
Effect: Reduction in light output
Concentration: 0.19 mg/l
Exposure time: 30 min
Comment: Mean of three assays. Methanol (<10%) was used to prepare the stock solutions. EC50 values were calculated from nominal concentrations.

Bibliographical reference: Kaiser, K.L.E. (1987). In: *QSAR in Environmental Toxicology - II*, K.L.E. Kaiser (ed.), D. Reidel

Publishing Company, Dordrecht, p. 169-188.

4-AMINOBENZOPHENONE

CAS RN: 1137-41-3

Sample purity: 98%
Temperature: 15°C
Test parameter: EC50
Effect: Reduction in light output

Concentration: 17.2 mg/l
Exposure time: 5 min

Concentration: 18.4 mg/l
Exposure time: 15 min

Concentration: 20.2 mg/l*
Exposure time: 30 min

Comment: Mean of three assays. Methanol (<10%) was used to prepare the stock solutions. EC50 values were calculated from nominal concentrations.

Bibliographical references: Kaiser, K.L.E., and Palabrica, V.S. (1991). *Water Poll. Res. J. Canada* **26**, 361-431.
* Kaiser, K.L.E. (1987). In: *QSAR in Environmental Toxicology - II*, K.L.E. Kaiser (ed.), D. Reidel Publishing Company, Dordrecht, p. 169-188.

4-AMINOBENZOTRIFLUORIDE

CAS RN: 455-14-1
Synonym: α,α,α-Trifluoro-*p*-toluidine

Temperature: 15°C
Test parameter: EC50
Effect: Reduction in light output
Concentration: 0.95 mg/l
Exposure time: 30 min
Comment: Mean of three assays. Methanol (<10%) was used to prepare the stock solutions. EC50 values were calculated from nominal

concentrations.

Bibliographical reference: Kaiser, K.L.E. (1987). In: *QSAR in Environmental Toxicology - II*, K.L.E. Kaiser (ed.), D. Reidel Publishing Company, Dordrecht, p. 169-188.

4-AMINOBENZYL CYANIDE

CAS RN: 3544-25-0

Sample purity: 99%
Temperature: 15°C
Test parameter: EC50
Effect: Reduction in light output

Concentration: 0.40 mg/l
Exposure time: 5 min

Concentration: 0.39 mg/l
Exposure time: 15 min

Concentration: 0.36 mg/l*
Exposure time: 30 min

Comment: Mean of four assays. Methanol (<10%) was used to prepare the stock solutions. EC50 values were calculated from nominal concentrations.

Bibliographical references: Kaiser, K.L.E., and Palabrica, V.S. (1991). *Water Poll. Res. J. Canada* **26**, 361-431.
* Kaiser, K.L.E. (1987). In: *QSAR in Environmental Toxicology - II*, K.L.E. Kaiser (ed.), D. Reidel Publishing Company, Dordrecht, p. 169-188.

2-AMINOBIPHENYL

CAS RN: 90-41-5
Synonym: 2-Biphenylamine

Temperature: 15°C
Test parameter: EC50
Effect: Reduction in light output

Concentration: 5.65 ± 0.30 mg/l
Exposure time: 5 min

Concentration: 6.21 ± 0.35 mg/l
Exposure time: 10 min

Concentration: 6.75 ± 0.40 mg/l
Exposure time: 15 min

Concentration: 6.67 ± 0.38 mg/l
Exposure time: 20 min

Comment: Test was performed at least in triplicate. Chemical was solubilized in DMSO.

Bibliographical reference: Yates, I.E. (1985). *J. Microbiol. Meth.* **3**, 171-180.

4-AMINOBIPHENYL

CAS RN: 92-67-1
Synonym: 4-Biphenylamine

Sample purity: 98%
Temperature: 15°C
Test parameter: EC50
Effect: Reduction in light output

Concentration: 5.11 mg/l
Exposure time: 5 min

Concentration: 5.48 mg/l
Exposure time: 15 min

Concentration: 6.74 mg/l*
Exposure time: 30 min

Comment: Mean of three assays. Methanol (<10%) was used to prepare the stock solutions. EC50 values were calculated from nominal concentrations.

Bibliographical references: Kaiser, K.L.E., and Palabrica, V.S. (1991). *Water Poll. Res. J. Canada* **26**, 361-431.
* Kaiser, K.L.E., Palabrica, V.S., and Ribo, J.M. (1987). In: *QSAR in*

Environmental Toxicology - II, K.L.E. Kaiser (ed.), D. Reidel Publishing Company, Dordrecht, p. 153-168.

2-AMINO-4-(4-CHLOROPHENYL)THIAZOLE

CAS RN: 2103-99-3

Sample purity: 98%
Temperature: 15°C
Test parameter: EC50
Effect: Reduction in light output

Concentration: 5.80 mg/l
Exposure time: 5 min

Concentration: 5.94 mg/l
Exposure time: 15 min

Concentration: 6.36 mg/l
Exposure time: 30 min

Comment: Mean of three assays. Methanol (<10%) was used to prepare the stock solutions. EC50 values were calculated from nominal concentrations.

Bibliographical reference: Kaiser, K.L.E., and Palabrica, V.S. (1991). *Water Poll. Res. J. Canada* **26**, 361-431.

4-AMINO-3,5-DICHLORO-2,6-DIFLUOROPYRIDINE

CAS RN: 2840-00-8

Sample purity: 99%
Temperature: 15°C
Test parameter: EC50
Effect: Reduction in light output

Concentration: 43.5 mg/l
Exposure time: 5 min

Concentration: 37.9 mg/l
Exposure time: 15 min

Concentration: 39.7 mg/l
Exposure time: 30 min

Comment: Mean of three assays. Methanol (<10%) was used to prepare the stock solutions. EC50 values were calculated from nominal concentrations.

Bibliographical reference: Kaiser, K.L.E., and Palabrica, V.S. (1991). *Water Poll. Res. J. Canada* **26**, 361-431.

2-AMINO-4,5-DIMETHOXYBENZONITRILE

CAS RN: 26961-27-3

Sample purity: 96%
Temperature: 15°C
Test parameter: EC50
Effect: Reduction in light output

Concentration: 11.0 mg/l
Exposure time: 5 min

Concentration: 12.0 mg/l
Exposure time: 15 min

Concentration: 14.5 mg/l
Exposure time: 30 min

Comment: Mean of three assays. Methanol (<10%) was used to prepare the stock solutions. EC50 values were calculated from nominal concentrations.

Bibliographical reference: Kaiser, K.L.E., and Palabrica, V.S. (1991). *Water Poll. Res. J. Canada* **26**, 361-431.

4-AMINO-*N*,*N*-DIMETHYLANILINE

CAS RN: 99-98-9

Sample purity: >98%
Temperature: 15°C

Test parameter: EC50
Effect: Reduction in light output

Concentration: 2.16 mg/l
Exposure time: 5 min

Concentration: 0.99 mg/l
Exposure time: 15 min

Concentration: 0.84 mg/l
Exposure time: 30 min

Comment: Mean of three assays. Methanol (<10%) was used to prepare the stock solutions. EC50 values were calculated from nominal concentrations.

Bibliographical reference: Kaiser, K.L.E., and Palabrica, V.S. (1991). *Water Poll. Res. J. Canada* **26**, 361-431.

2-AMINO-4,6-DIMETHYLPYRIMIDINE

CAS RN: 767-15-7

Sample purity: 95%
Temperature: 15°C
Test parameter: EC50
Effect: Reduction in light output

Concentration: 513 mg/l
Exposure time: 5 min

Concentration: 458 mg/l
Exposure time: 15 min

Concentration: 408 mg/l
Exposure time: 30 min

Comment: Mean of four assays. Methanol (<10%) was used to prepare the stock solutions. EC50 values were calculated from nominal concentrations.

Bibliographical reference: Kaiser, K.L.E., and Palabrica, V.S. (1991). *Water Poll. Res. J. Canada* **26**, 361-431.

4-AMINODIPHENYLAMINE

CAS RN: 101-54-2

Sample purity: 98%
Temperature: 15°C
Test parameter: EC50
Effect: Reduction in light output

Concentration: 0.27 mg/l
Exposure time: 5 min

Concentration: 0.26 mg/l
Exposure time: 15 min

Concentration: 0.33 mg/l*
Exposure time: 30 min

Comment: Mean of three assays. Methanol (<10%) was used to prepare the stock solutions. EC50 values were calculated from nominal concentrations.

Bibliographical references: Kaiser, K.L.E., and Palabrica, V.S. (1991). *Water Poll. Res. J. Canada* **26**, 361-431.
* Kaiser, K.L.E., Palabrica, V.S., and Ribo, J.M. (1987). In: *QSAR in Environmental Toxicology - II*, K.L.E. Kaiser (ed.), D. Reidel Publishing Company, Dordrecht, p. 153-168.

AMINODIPHENYLMETHANE

CAS RN: 91-00-9

Sample purity: 96%
Temperature: 15°C
Test parameter: EC50
Effect: Reduction in light output

Concentration: 51.6 mg/l
Exposure time: 5 min

Concentration: 47.1 mg/l
Exposure times: 15 and 30 min

Comment: Mean of four assays. Methanol (<10%) was used to prepare the stock solutions. EC50 values were calculated from nominal concentrations.

Bibliographical reference: Kaiser, K.L.E., and Palabrica, V.S. (1991). *Water Poll. Res. J. Canada* **26**, 361-431.

4-(2-AMINOETHYL)BENZENESULFONAMIDE

CAS RN: 35303-76-5

Sample purity: 99%
Temperature: 15°C
Test parameter: EC50
Effect: Reduction in light output

Concentration: 56.4 mg/l
Exposure time: 5 min

Concentration: 49.2 mg/l
Exposure times: 15 and 30 min

Comment: Mean of four assays. Methanol (<10%) was used to prepare the stock solutions. EC50 values were calculated from nominal concentrations.

Bibliographical reference: Kaiser, K.L.E., and Palabrica, V.S. (1991). *Water Poll. Res. J. Canada* **26**, 361-431.

2-AMINO-6-NITROBENZOTHIAZOLE

CAS RN: 6285-57-0

Sample purity: >97%
Temperature: 15°C
Test parameter: EC50
Effect: Reduction in light output

Concentration: 3.47 mg/l
Exposure time: 5 min

Concentration: 4.27 mg/l
Exposure time: 15 min

Concentration: 5.38 mg/l
Exposure time: 30 min

Comment: Mean of three assays. Methanol (<10%) was used to prepare the stock solutions. EC50 values were calculated from nominal concentrations.

Bibliographical reference: Kaiser, K.L.E., and Palabrica, V.S. (1991). *Water Poll. Res. J. Canada* **26**, 361-431.

5-AMINO-2-NITROBENZOTRIFLUORIDE

CAS RN: 393-11-3

Sample purity: 98%
Temperature: 15°C
Test parameter: EC50
Effect: Reduction in light output

Concentration: 0.49 mg/l
Exposure time: 5 min

Concentration: 0.53 mg/l
Exposure time: 15 min

Concentration: 0.59 mg/l
Exposure time: 30 min

Comment: Mean of four assays. Methanol (<10%) was used to prepare the stock solutions. EC50 values were calculated from nominal concentrations.

Bibliographical reference: Kaiser, K.L.E., and Palabrica, V.S. (1991). *Water Poll. Res. J. Canada* **26**, 361-431.

4-AMINO-2-NITROPHENOL

CAS RN: 119-34-6

Test parameter: EC50
Effect: Reduction in light output

Concentration: 35.9 mg/l
Exposure time: 5 min
Comment: Concentrations in the test were measured.

Bibliographical reference: Curtis, C., Lima, A., Lozano, S.J., and Veith, G.D. (1982). In: *Aquatic Toxicology and Hazard Assessment: Fifth Conference, ASTM STP 766*, J.G. Pearson, R.B. Foster, and W.E. Bishop (eds.), American Society for Testing and Materials, Philadelphia, p. 170-178.

Temperature: 15°C
Test parameter: EC50
Effect: Reduction in light output

Concentration: 43.4 mg/l
Exposure time: 5 min

Concentration: 42.5 mg/l
Exposure time: 15 min

Concentration: 38.7 mg/l
Exposure time: 30 min

Comment: Mean of three assays. Methanol (<10%) was used to prepare the stock solutions. EC50 values were calculated from nominal concentrations.

Bibliographical reference: Kaiser, K.L.E., and Palabrica, V.S. (1991). *Water Poll. Res. J. Canada* **26**, 361-431.

2-AMINO-5-NITROPYRIDINE

CAS RN: 4214-76-0

Sample purity: 99%
Temperature: 15°C
Test parameter: EC50
Effect: Reduction in light output

Concentration: 81.9 mg/l
Exposure time: 5 min

Concentration: 76.4 mg/l

Exposure time: 15 min

Concentration: 89.8 mg/l
Exposure time: 30 min

Comment: Mean of three assays. Methanol (<10%) was used to prepare the stock solutions. EC50 values were calculated from nominal concentrations.

Bibliographical reference: Kaiser, K.L.E., and Palabrica, V.S. (1991). *Water Poll. Res. J. Canada* **26**, 361-431.

5-AMINO-6-NITROQUINOLINE

CAS RN: 35975-00-9

Sample purity: 98%
Temperature: 15°C
Test parameter: EC50
Effect: Reduction in light output

Concentration: 3.29 mg/l
Exposure time: 5 min

Concentration: 1.57 mg/l
Exposure time: 15 min

Concentration: 1.04 mg/l
Exposure time: 30 min

Comment: Mean of three assays. Methanol (<10%) was used to prepare the stock solutions. EC50 values were calculated from nominal concentrations.

Bibliographical reference: Kaiser, K.L.E., and Palabrica, V.S. (1991). *Water Poll. Res. J. Canada* **26**, 361-431.

(±)-*exo*-2-AMINONORBORNANE

CAS RN: 7242-92-4
Synonym: *exo*-2-Norbornanamine

Sample purity: 99%
Temperature: 15°C
Test parameter: EC50
Effect: Reduction in light output

Concentration: 41.3 mg/l
Exposure time: 5 min

Concentration: 36.8 mg/l
Exposure time: 15 min

Concentration: 40.4 mg/l
Exposure time: 30 min

Comment: Mean of three assays. Methanol (<10%) was used to prepare the stock solutions. EC50 values were calculated from nominal concentrations.

Bibliographical reference: Kaiser, K.L.E., and Palabrica, V.S. (1991). *Water Poll. Res. J. Canada* **26**, 361-431.

4-AMINOPHENETHYL ALCOHOL

CAS RN: 104-10-9

Sample purity: 98%
Temperature: 15°C
Test parameter: EC50
Effect: Reduction in light output

Concentration: 4.24 mg/l
Exposure times: 5 and 15 min

Concentration: 3.87 mg/l*
Exposure time: 30 min

Comment: Mean of three assays. Methanol (<10%) was used to prepare the stock solutions. EC50 values were calculated from nominal concentrations.

Bibliographical references: Kaiser, K.L.E., and Palabrica, V.S. (1991). *Water Poll. Res. J. Canada* **26**, 361-431.
* Kaiser, K.L.E. (1987). In: *QSAR in Environmental Toxicology - II*, K.L.E. Kaiser (ed.), D. Reidel Publishing Company, Dordrecht, p. 169-188.

2-AMINOPHENOL

CAS RN: 95-55-6
Synonym: 2-Hydroxyaniline

Temperature: 15°C
Test parameter: EC50
Effect: Reduction in light output

Concentration: 207 mg/l
Exposure time: 5 min

Concentration: 135.5 mg/l
Exposure time: 30 min

Bibliographical reference: Speece, R. (1987). Drexel University, Philadelphia, USA, private communication.

4-AMINOPHENOL

CAS RN: 123-30-8
Synonym: 4-Hydroxyaniline

Temperature: 15°C
Test parameter: EC50
Effect: Reduction in light output

Concentration: 3.97 mg/l
Exposure time: 5 min

Concentration: 3.29 mg/l*
Exposure time: 30 min

* EC50 value from linear regression fit to data.

Bibliographical reference: Speece, R. (1987). Drexel University, Philadelphia, USA, private communication.

Sample purity: 98+%
Temperature: 15°C
Test parameter: EC50
Effect: Reduction in light output

Concentration: 0.91 mg/l
Exposure time: 5 min

Concentration: 0.81 mg/l
Exposure time: 15 min

Concentration: 0.77 mg/l
Exposure time: 30 min

Comment: Mean of three assays. Methanol (<10%) was used to prepare the stock solutions. EC50 values were calculated from nominal concentrations.

Bibliographical reference: Kaiser, K.L.E., and Palabrica, V.S. (1991). *Water Poll. Res. J. Canada* **26**, 361-431.

4-AMINOPHENYLACETIC ACID

CAS RN: 1197-55-3

Sample purity: 96%
Temperature: 15°C
Test parameter: EC50
Effect: Reduction in light output

Concentration: 170 mg/l
Exposure time: 5 min

Concentration: 126 mg/l
Exposure time: 15 min

Concentration: 182 mg/l*
Exposure time: 30 min

Comment: Mean of three assays. Methanol (<10%) was used to prepare the stock solutions. EC50 values were calculated from nominal concentrations.

Bibliographical references: Kaiser, K.L.E., and Palabrica, V.S. (1991). *Water Poll. Res. J. Canada* **26**, 361-431.
* Kaiser, K.L.E. (1987). In: *QSAR in Environmental Toxicology - II*, K.L.E. Kaiser (ed.), D. Reidel Publishing Company, Dordrecht, p. 169-188.

DL-1-AMINO-2-PROPANOL

CAS RN: 78-96-6
Synonym: Isopropanolamine

Test parameter: EC50
Effect: Reduction in light output
Concentration: 27.2 mg/l
Exposure time: 5 min
Comment: The EC50 value was calculated from nominal concentrations.

Bibliographical reference: Curtis, C., Lima, A., Lozano, S.J., and Veith, G.D. (1982). In: *Aquatic Toxicology and Hazard Assessment: Fifth Conference, ASTM STP 766*, J.G. Pearson, R.B. Foster, and W.E. Bishop (eds.), American Society for Testing and Materials, Philadelphia, p. 170-178.

2-AMINOPYRIDINE

CAS RN: 504-29-0

Sample purity: >99%
Temperature: 15°C
Test parameter: EC50
Effect: Reduction in light output

Concentration: 248 mg/l
Exposure time: 5 min

Concentration: 253 mg/l
Exposure time: 15 min

Concentration: 284 mg/l
Exposure time: 30 min

Comment: Mean of three assays. Methanol (<10%) was used to prepare the stock solutions. EC50 values were calculated from nominal concentrations.

Bibliographical reference: Kaiser, K.L.E., and Palabrica, V.S. (1991). *Water Poll. Res. J. Canada* **26**, 361-431.

3-AMINOPYRIDINE

CAS RN: 462-08-8

Sample purity: 99%
Temperature: 15°C
Test parameter: EC50
Effect: Reduction in light output

Concentration: 636 mg/l
Exposure time: 5 min

Concentration: 622 mg/l
Exposure time: 15 min

Concentration: 682 mg/l
Exposure time: 30 min

Comment: Mean of three assays. Methanol (<10%) was used to prepare the stock solutions. EC50 values were calculated from nominal concentrations.

Bibliographical reference: Kaiser, K.L.E., and Palabrica, V.S. (1991). *Water Poll. Res. J. Canada* **26**, 361-431.

4-AMINOPYRIDINE

CAS RN: 504-24-5

Sample purity: 99+%
Temperature: 15°C
Test parameter: EC50
Effect: Reduction in light output

Concentration: 26.5 mg/l
Exposure time: 5 min

Concentration: 23.6 mg/l
Exposure time: 15 min

Concentration: 25.9 mg/l
Exposure time: 30 min

Comment: Mean of three assays. Methanol (<10%) was used to

prepare the stock solutions. EC50 values were calculated from nominal concentrations.

Bibliographical reference: Kaiser, K.L.E., and Palabrica, V.S. (1991). *Water Poll. Res. J. Canada* **26**, 361-431.

5-AMINOQUINOLINE

CAS RN: 611-34-7
Synonym: 5-Quinolinamine

Sample purity: 97%
Temperature: 15°C
Test parameter: EC50
Effect: Reduction in light output

Concentration: 12.0 mg/l
Exposure time: 5 min

Concentration: 13.5 mg/l
Exposure time: 15 min

Concentration: 15.8 mg/l
Exposure time: 30 min

Comment: Mean of three assays. Methanol (<10%) was used to prepare the stock solutions. EC50 values were calculated from nominal concentrations.

Bibliographical reference: Kaiser, K.L.E., and Palabrica, V.S. (1991). *Water Poll. Res. J. Canada* **26**, 361-431.

4-AMINO-2,3,5,6-TETRAFLUOROPYRIDINE

CAS RN: 1682-20-8

Sample purity: 98%
Temperature: 15°C
Test parameter: EC50
Effect: Reduction in light output

Concentration: 159 mg/l

Exposure time: 5 min

Concentration: 155 mg/l
Exposure time: 15 min

Concentration: 151 mg/l
Exposure time: 30 min

Comment: Mean of three assays. Methanol (<10%) was used to prepare the stock solutions. EC50 values were calculated from nominal concentrations.

Bibliographical reference: Kaiser, K.L.E., and Palabrica, V.S. (1991). *Water Poll. Res. J. Canada* **26**, 361-431.

5-AMINO-1,3,4-THIADIAZOLE-2-THIOL

CAS RN: 2349-67-9

Sample purity: 98%
Temperature: 15°C
Test parameter: EC50
Effect: Reduction in light output

Concentration: 1.60 mg/l
Exposure time: 5 min

Concentration: 2.02 mg/l
Exposure time: 15 min

Concentration: 3.05 mg/l
Exposure time: 30 min

Comment: Mean of three assays. Methanol (<10%) was used to prepare the stock solutions. EC50 values were calculated from nominal concentrations.

Bibliographical reference: Kaiser, K.L.E., and Palabrica, V.S. (1991). *Water Poll. Res. J. Canada* **26**, 361-431.

2-AMINOTHIAZOLE

CAS RN: 96-50-4

Sample purity: 97%
Temperature: 15°C
Test parameter: EC50
Effect: Reduction in light output

Concentration: 1410 mg/l
Exposure time: 5 min

Concentration: 1120 mg/l
Exposure time: 15 min

Concentration: 956 mg/l
Exposure time: 30 min

Comment: Mean of three assays. Methanol (<10%) was used to prepare the stock solutions. EC50 values were calculated from nominal concentrations.

Bibliographical reference: Kaiser, K.L.E., and Palabrica, V.S. (1991). *Water Poll. Res. J. Canada* **26**, 361-431.

3-AMINO-1,2,4-TRIAZOLE

CAS RN: 61-82-5
Synonyms: 1*H*-1,2,4-Triazol-3-amine; Amitrole

Sample purity: Agrichemical grade
Temperature: 15 ± 0.1°C
Test parameter: EC50
Effect: Reduction in light output
Concentration: 181 mg/l
Exposure time: 5 min
Comment: Test was performed on *Photobacterium phosphoreum* NZ11D obtained from the Scripps Institute of Oceanography (La Jolla, CA).

Bibliographical reference: McFeters, G.A., Bond, P.J., Olson, S.B., and Tchan, Y.T. (1983). *Water Res.* **17**, 1757-1762.

Sample purity: 95%
Temperature: 15°C
Test parameter: EC50
Effect: Reduction in light output

Concentration: 335 mg/l
Exposure time: 5 min

Concentration: 431 mg/l
Exposure time: 15 min

Concentration: 582 mg/l
Exposure time: 30 min

Comment: Mean of four assays. Methanol (<10%) was used to prepare the stock solutions. EC50 values were calculated from nominal concentrations.

Bibliographical reference: Kaiser, K.L.E., and Palabrica, V.S. (1991). *Water Poll. Res. J. Canada* **26**, 361-431.

4-AMINO-1,2,4-TRIAZOLE

CAS RN: 584-13-4

Sample purity: 99%
Temperature: 15°C
Test parameter: EC50
Effect: Reduction in light output

Concentration: 922 mg/l
Exposure time: 5 min

Concentration: 901 mg/l
Exposure time: 15 min

Concentration: 841 mg/l
Exposure time: 30 min

Comment: Mean of four assays. Methanol (<10%) was used to prepare the stock solutions. EC50 values were calculated from nominal concentrations.

Bibliographical reference: Kaiser, K.L.E., and Palabrica, V.S.

(1991). *Water Poll. Res. J. Canada* **26**, 361-431.

2-AMINO-3,5,6-TRIFLUOROTEREPHTHALONITRILE

CAS RN: 133622-66-9

Sample purity: 98%
Temperature: 15°C
Test parameter: EC50
Effect: Reduction in light output

Concentration: 80.3 mg/l
Exposure time: 5 min

Concentration: 74.9 mg/l
Exposure time: 15 min

Concentration: 80.3 mg/l
Exposure time: 30 min

Comment: Mean of four assays. Methanol (<10%) was used to prepare the stock solutions. EC50 values were calculated from nominal concentrations.

Bibliographical reference: Kaiser, K.L.E., and Palabrica, V.S. (1991). *Water Poll. Res. J. Canada* **26**, 361-431.

AMMONIA

CAS RN: 7664-41-7

Temperature: 15°C
Test parameter: EC50
Effect: Reduction in light output
Concentration: 2.0 mg/l
Exposure time: 5 min

Bibliographical reference: Bulich, A.A., Greene, M.W., and Isenberg, D.L. (1981). In: *Aquatic Toxicology and Hazard Assessment: Fourth Conference, ASTM STP 737*, D.R. Branson and K.L. Dickson, (eds.), American Society for Testing and Materials, Philadelphia, p. 338-347.

AMYTAL

CAS RN: 57-43-2

Temperature: 15 ± 0.1°C
Test parameter: EC50
Effect: Reduction in light output
Concentration: 1000 mg/l
Exposure time: 5 min

Bibliographical reference: Chang, J.C., Taylor, P.B., and Leach, F.R. (1981). *Bull. Environ. Contam. Toxicol.* **26**, 150-156.

ANILINE

CAS RN: 62-53-3
Synonyms: Aminobenzene; Phenylamine

Sample purity: >98%
Temperature: 15 ± 0.1°C
Test parameter: EC50
Effect: Reduction in light output

Concentration: 425 mg/l
Exposure time: 5 min

Concentration: 488 mg/l
Exposure time: 15 min

Comment: Test was performed in duplicate. Toxicity values were based on nominal test concentrations.

Bibliographical reference: de Zwart, D., and Slooff, W. (1983). *Aquat. Toxicol.* **4**, 129-138.

Sample purity: >98%
Temperature: 15 ± 0.1°C
Test parameter: EC10
Effect: Reduction in light output

Concentration: 120 mg/l
Exposure time: 5 min

Concentration: 126 mg/l
Exposure time: 15 min

Comment: Test was performed in duplicate. Toxicity values were based on nominal test concentrations.

Bibliographical reference: de Zwart, D., and Slooff, W. (1983). *Aquat. Toxicol.* **4**, 129-138.

Temperature: 15°C
Test parameter: EC50
Effect: Reduction in light output

Concentration: 64.4 mg/l
Exposure time: 5 min

Concentration: 69.0 mg/l
Exposure time: 15 min

Concentration: 70.6 mg/l
Exposure time: 30 min

Comment: Mean of three assays. Methanol (<10%) was used to prepare the stock solutions. EC50 values were calculated from nominal concentrations.

Bibliographical reference: Ribo, J.M., and Kaiser, K.L.E. (1984). In: *QSAR in Environmental Toxicology*, K.L.E. Kaiser (ed.), D. Reidel Publishing Company, Dordrecht, p. 319-336.

Temperature: 15°C
Test parameter: EC50
Effect: Reduction in light output
Concentration: 139 mg/l
Exposure time: 5 min

Bibliographical reference: Kahru, A. (1993). *ATLA* **21**, 210-215.

4-ANISIDINE

CAS RN: 104-94-9
Synonym: 4-Methoxyaniline

Temperature: 15°C
Test parameter: EC50
Effect: Reduction in light output
Concentration: 14.5 mg/l
Exposure time: 30 min
Comment: Mean of three assays. Methanol (<10%) was used to prepare the stock solutions. EC50 values were calculated from nominal concentrations.

Bibliographical reference: Kaiser, K.L.E. (1987). In: *QSAR in Environmental Toxicology - II*, K.L.E. Kaiser (ed.), D. Reidel Publishing Company, Dordrecht, p. 169-188.

ANISOLE

CAS RN: 100-66-3

Temperature: 15°C
Test parameter: EC50
Effect: Reduction in light output
Concentration: 18.8 mg/l
Exposure time: 30 min
Comment: Mean of three assays. Methanol (<10%) was used to prepare the stock solutions. EC50 values were calculated from nominal concentrations.

Bibliographical reference: Kaiser, K.L.E., Palabrica, V.S., and Ribo, J.M. (1987). In: *QSAR in Environmental Toxicology - II*, K.L.E. Kaiser (ed.), D. Reidel Publishing Company, Dordrecht, p. 153-168.

4-ANISOYL CHLORIDE

CAS RN: 100-07-2
Synonym: 4-Methoxybenzoyl chloride

Sample purity: 99%
Temperature: 15°C
Test parameter: EC50
Effect: Reduction in light output
Concentration: 1.91 mg/l
Exposure times: 5, 15, and 30 min
Comment: Mean of three assays. Methanol (<10%) was used to

prepare the stock solutions. EC50 values were calculated from nominal concentrations.

Bibliographical reference: Kaiser, K.L.E., and Palabrica, V.S. (1991). *Water Poll. Res. J. Canada* **26**, 361-431.

AROCLOR 1016

CAS RN: 12674-11-2

Sample purity: Analytical reference standard grade
Temperature: 15 ± 0.1°C
Test parameter: EC50
Effect: Reduction in light output
Concentration: 2.05 mg/l
Exposure time: 5 min
Comment: Test was performed on *Photobacterium phosphoreum* NZ11D obtained from the Scripps Institute of Oceanography (La Jolla, CA).

Bibliographical reference: McFeters, G.A., Bond, P.J., Olson, S.B., and Tchan, Y.T. (1983). *Water Res.* **17**, 1757-1762.

AROCLOR 1242

CAS RN: 53469-21-9

Temperature: 15°C
Test parameter: EC50
Effect: Reduction in light output
Concentration: 0.7 mg/l
Exposure time: 5 min
Comment: Chemical was mixed with diluent for two days in a separatory funnel, and the aqueous phase was removed for testing.

Bibliographical reference: Bulich, A.A., Greene, M.W., and Isenberg, D.L. (1981). In: *Aquatic Toxicology and Hazard Assessment: Fourth Conference, ASTM STP 737*, D.R. Branson and K.L. Dickson, (eds.), American Society for Testing and Materials, Philadelphia, p. 338-347.

ARSENIC(III) OXIDE

CAS RN: 1327-53-3

Test parameter: EC50
Effect: Reduction in light output

Concentration: 73.73 mg/l
Exposure time: 5 min

Concentration: 43.56 mg/l
Exposure time: 15 min

Concentration: 33.39 mg/l
Exposure time: 30 min

Concentration: 31.43 mg/l
Exposure time: 60 min

Comment: Two replicates of each concentration were tested.

Bibliographical reference: Greene, J.C., Miller, W.E., Debacon, M.K., Long, M.A., and Bartels, C.L. (1985). *Arch. Environ. Contam. Toxicol.* **14**, 659-667.

ATRAZINE

CAS RN: 1912-24-9
Synonym: 2-Chloro-4-ethylamino-6-isopropylamino-1,3,5-triazine

Temperature: 15°C
Test parameter: EC50
Effect: Reduction in light output
Concentration: >86 mg/l
Exposure time: 5 min
Comment: Chemical was dissolved in DMSO (0.2%). EC50 value was based on nominal concentrations.

Bibliographical reference: Reteuna, C. (1988). Thesis, University of Metz, Metz, France.

AZOBENZENE

CAS RN: 103-33-3

Sample purity: 99%
Temperature: 15°C
Test parameter: EC50
Effect: Reduction in light output

Concentration: 0.93 mg/l
Exposure time: 5 min

Concentration: 1.05 mg/l
Exposure time: 15 min

Concentration: 1.29 mg/l*
Exposure time: 30 min

Comment: Mean of three assays. Methanol (<10%) was used to prepare the stock solutions. EC50 values were calculated from nominal concentrations.

Bibliographical references: Kaiser, K.L.E., and Palabrica, V.S. (1991). *Water Poll. Res. J. Canada* **26**, 361-431.
* Kaiser, K.L.E., Palabrica, V.S., and Ribo, J.M. (1987). In: *QSAR in Environmental Toxicology - II*, K.L.E. Kaiser (ed.), D. Reidel Publishing Company, Dordrecht, p. 153-168.

BENZALDEHYDE

CAS RN: 100-52-7

Sample purity: 98%
Temperature: 15°C
Test parameter: EC50
Effect: Reduction in light output

Concentration: 6.11 mg/l
Exposure time: 5 min

Concentration: 5.08 mg/l
Exposure time: 15 min

Concentration: 4.85 mg/l*

Exposure time: 30 min

Comment: Mean of three assays. Methanol (<10%) was used to prepare the stock solutions. EC50 values were calculated from nominal concentrations.

Bibliographical references: Kaiser, K.L.E., and Palabrica, V.S. (1991). *Water Poll. Res. J. Canada* **26**, 361-431.
* Kaiser, K.L.E., Palabrica, V.S., and Ribo, J.M. (1987). In: *QSAR in Environmental Toxicology - II*, K.L.E. Kaiser (ed.), D. Reidel Publishing Company, Dordrecht, p. 153-168.

BENZALKONIUM CHLORIDE

CAS RN: 8001-54-5
Synonym: Alkylbenzyldimethylammonium chloride

Temperature: 15°C
Test parameter: EC50
Effect: Reduction in light output
Concentration: 0.6 mg/l
Exposure time: 15 min

Bibliographical reference: Bulich, A.A., Tung, K.K., and Scheibner, G. (1990). *J. Biolumin. Chemilumin.* **5**, 71-77.

BENZAMIDE

CAS RN: 55-21-0

Sample purity: 99%
Temperature: 15°C
Test parameter: EC50
Effect: Reduction in light output

Concentration: 63.6 mg/l
Exposure time: 5 min

Concentration: 60.7 mg/l
Exposure time: 15 min

Concentration: 59.3 mg/l*

Exposure time: 30 min

Comment: Mean of three assays. Methanol (<10%) was used to prepare the stock solutions. EC50 values were calculated from nominal concentrations.

Bibliographical references: Kaiser, K.L.E., and Palabrica, V.S. (1991). *Water Poll. Res. J. Canada* **26**, 361-431.
* Kaiser, K.L.E., Palabrica, V.S., and Ribo, J.M. (1987). In: *QSAR in Environmental Toxicology - II*, K.L.E. Kaiser (ed.), D. Reidel Publishing Company, Dordrecht, p. 153-168.

BENZ[*a*]ANTHRACENE

CAS RN: 56-55-3
Synonym: 1,2-Benzanthracene

Temperature: 15°C
Test parameter: EC50
Effect: Reduction in light output
Concentration: 0.26 mg/l (0.26-0.29)
Exposure time: 15 min
Comment: Chemical was added to 5 ml DMSO and placed in a bath sonicator until all visible material had dissolved.

Bibliographical reference: Jacobs, M.W., Coates, J.A., Delfino, J.J., Bitton, G., Davis, W.M., and Garcia, K.L. (1993). *Arch. Environ. Contam. Toxicol.* **24**, 461-468.

BENZENE

CAS RN: 71-43-2

Temperature: 15°C
Test parameter: EC50
Effect: Reduction in light output
Concentration: 2.0 mg/l
Exposure time: 5 min

Bibliographical reference: Bulich, A.A., Greene, M.W., and Isenberg, D.L. (1981). In: *Aquatic Toxicology and Hazard Assessment: Fourth Conference, ASTM STP 737*, D.R. Branson and K.L. Dickson, (eds.), American Society for Testing and Materials, Philadelphia, p. 338-347.

Temperature: 15 ± 0.1°C
Test parameter: EC50
Effect: Reduction in light output
Concentration: 200 mg/l
Exposure time: 5 min

Bibliographical reference: Chang, J.C., Taylor, P.B., and Leach, F.R. (1981). *Bull. Environ. Contam. Toxicol.* **26**, 150-156.

Sample purity: >98%
Temperature: 15 ± 0.1°C
Test parameter: EC50
Effect: Reduction in light output

Concentration: 214 mg/l
Exposure time: 5 min

Concentration: 238 mg/l
Exposure time: 15 min

Comment: Test was performed in duplicate. Toxicity values were based on nominal test concentrations.

Bibliographical reference: de Zwart, D., and Slooff, W. (1983). *Aquat. Toxicol.* **4**, 129-138.

Sample purity: >98%
Temperature: 15 ± 0.1°C
Test parameter: EC10
Effect: Reduction in light output

Concentration: 36 mg/l
Exposure time: 5 min

Concentration: 64 mg/l
Exposure time: 15 min

Comment: Test was performed in duplicate. Toxicity values were based on nominal test concentrations.

Bibliographical reference: de Zwart, D., and Slooff, W. (1983). *Aquat. Toxicol.* **4**, 129-138.

Temperature: 15°C
Test parameter: EC50
Effect: Reduction in light output
Concentration: 73 ± 21 mg/l
Exposure time: 5 min
Comment: Mean of four assays. Toxicity values were calculated from nominal concentrations.

Bibliographical reference: Ferard, J.F., Vasseur, P., Danoux, L., and Larbaigt, G. (1983). *Rev. Fr. Sci. Eau* **2**, 221-237.

Temperature: 15 ± 0.1°C
Test parameter: EC50
Effect: Reduction in light output
Concentration: 4.11 mg/l
Exposure time: 5 min
Comment: Test was performed on *Photobacterium phosphoreum* NZ11D obtained from the Scripps Institute of Oceanography (La Jolla, CA).

Bibliographical reference: McFeters, G.A., Bond, P.J., Olson, S.B., and Tchan, Y.T. (1983). *Water Res.* **17**, 1757-1762.

Temperature: 15°C
Test parameter: EC50
Effect: Reduction in light output

Concentration: 81.8 mg/l
Exposure time: 5 min

Concentration: 78.8 mg/l
Exposure time: 15 min

Concentrations: 74.6 mg/l
103 mg/l*
Exposure time: 30 min

Comment: Mean of three assays. Methanol (<10%) was used to prepare the stock solutions. EC50 values were calculated from nominal concentrations.

Bibliographical references: Ribo, J.M., and Kaiser, K.L.E. (1983). *Chemosphere* **12**, 1421-1442.
* Kaiser, K.L.E., Palabrica, V.S., and Ribo, J.M. (1987). In: *QSAR in Environmental Toxicology - II*, K.L.E. Kaiser (ed.), D. Reidel Publishing Company, Dordrecht, p. 153-168.

Temperature: 15°C
Test parameter: EC50
Effect: Reduction in light output
Concentration: 160 mg/l
Exposure time: 15 min

Bibliographical reference: Hermens, J., Busser, F., Leeuwangh, P., and Musch, A. (1985). *Ecotoxicol. Environ. Safety* **9**, 17-25.

Temperature: 15°C
Test parameter: EC50
Effect: Reduction in light output
Concentration: 83.6 ± 31 mg/l
Exposure time: 10 min
Comment: Toxicity values were calculated from nominal concentrations.

Bibliographical reference: Vasseur, P., Bois, F., Ferard, J.F., Rast, C., and Larbaigt, G. (1986). *Tox. Assess.* **1**, 283-300.

Temperature: 15°C
Test parameter: EC50
Effect: Reduction in light output
Concentration: 84 mg/l
Exposure time: 5 min
Comment: EC50 value was calculated from nominal concentrations.

Bibliographical reference: Reteuna, C. (1988). Thesis, University of Metz, Metz, France.

Temperature: 15°C
Test parameter: EC50
Effect: Reduction in light output
Concentration: 531 mg/l
Exposure time: 5 min

Bibliographical reference: Kahru, A. (1993). *ATLA* **21**, 210-215.

1,4-BENZENEDIMETHANOL

CAS RN: 589-29-7

Sample purity: 99%
Temperature: 15°C
Test parameter: EC50
Effect: Reduction in light output

Concentration: 235 mg/l
Exposure time: 5 min

Concentration: 174 mg/l
Exposure time: 15 min

Concentration: 159 mg/l
Exposure time: 30 min

Comment: Mean of three assays. Methanol (<10%) was used to prepare the stock solutions. EC50 values were calculated from nominal concentrations.

Bibliographical reference: Kaiser, K.L.E., and Palabrica, V.S. (1991). *Water Poll. Res. J. Canada* **26**, 361-431.

BENZENESELENINIC ACID

CAS RN: 6996-92-5

Sample purity: 99%
Temperature: 15°C
Test parameter: EC50
Effect: Reduction in light output

Concentration: 30.0 mg/l
Exposure time: 5 min

Concentration: 21.7 mg/l
Exposure time: 15 min

Concentration: 17.2 mg/l*
Exposure time: 30 min

Comment: Mean of three assays. Methanol (<10%) was used to prepare the stock solutions. EC50 values were calculated from nominal concentrations.

Bibliographical references: Kaiser, K.L.E., and Palabrica, V.S. (1991). *Water Poll. Res. J. Canada* **26**, 361-431.
* Kaiser, K.L.E., Palabrica, V.S., and Ribo, J.M. (1987). In: *QSAR in Environmental Toxicology - II*, K.L.E. Kaiser (ed.), D. Reidel Publishing Company, Dordrecht, p. 153-168.

BENZENESULFONAMIDE

CAS RN: 98-10-2

Sample purity: 98%
Temperature: 15°C
Test parameter: EC50
Effect: Reduction in light output

Concentration: 217 mg/l
Exposure time: 5 min

Concentration: 249 mg/l
Exposure time: 15 min

Concentration: 243 mg/l*
Exposure time: 30 min

Comment: Mean of four assays. Methanol (<10%) was used to prepare the stock solutions. EC50 values were calculated from nominal concentrations.

Bibliographical references: Kaiser, K.L.E., and Palabrica, V.S. (1991). *Water Poll. Res. J. Canada* **26**, 361-431.
* Kaiser, K.L.E., Palabrica, V.S., and Ribo, J.M. (1987). In: *QSAR in Environmental Toxicology - II*, K.L.E. Kaiser (ed.), D. Reidel Publishing Company, Dordrecht, p. 153-168.

BENZENESULFONYL ISOCYANATE

CAS RN: 2845-62-7

Sample purity: 95%
Temperature: 15°C
Test parameter: EC50
Effect: Reduction in light output

Concentration: 13.3 mg/l
Exposure time: 5 min

Concentration: 9.61 mg/l
Exposure time: 15 min

Concentration: 6.65 mg/l
Exposure time: 30 min

Comment: Mean of three assays. Methanol (<10%) was used to prepare the stock solutions. EC50 values were calculated from nominal concentrations.

Bibliographical reference: Kaiser, K.L.E., and Palabrica, V.S. (1991). *Water Poll. Res. J. Canada* **26**, 361-431.

BENZHYDROL

CAS RN: 91-01-0
Synonym: Diphenylmethanol

Sample purity: 99%
Temperature: 15°C
Test parameter: EC50
Effect: Reduction in light output
Concentration: 55.6 mg/l
Exposure time: 30 min
Comment: Mean of two assays. Methanol (<10%) was used to prepare the stock solutions. EC50 values were calculated from nominal concentrations.

Bibliographical reference: Kaiser, K.L.E., Palabrica, V.S., and Ribo, J.M. (1987). In: *QSAR in Environmental Toxicology - II*, K.L.E. Kaiser (ed.), D. Reidel Publishing Company, Dordrecht, p. 153-168.

BENZIDINE

CAS RN: 92-87-5
Synonym: 4,4'-Diaminobiphenyl

Temperature: 15°C
Test parameter: EC50
Effect: Reduction in light output

Concentration: 44.2 mg/l
Exposure time: 5 min

Concentration: 50.7 mg/l
Exposure time: 15 min

Concentration: 63.9 mg/l
Exposure time: 30 min

Comment: Mean of four assays. Methanol (<10%) was used to prepare the stock solutions. EC50 values were calculated from nominal concentrations.

Bibliographical reference: Kaiser, K.L.E., and Palabrica, V.S. (1992). National Water Research Institute, Burlington, Ontario, Canada, unpublished results.

BENZIL

CAS RN: 134-81-6

Sample purity: 98%
Temperature: 15°C
Test parameter: EC50
Effect: Reduction in light output

Concentration: 0.53 mg/l
Exposure time: 5 min

Concentration: 0.58 mg/l
Exposure time: 15 min

Concentration: 0.63 mg/l*
Exposure time: 30 min

Comment: Mean of three assays. Methanol (<10%) was used to prepare the stock solutions. EC50 values were calculated from nominal concentrations.

Bibliographical references: Kaiser, K.L.E., and Palabrica, V.S. (1991). *Water Poll. Res. J. Canada* **26**, 361-431.
* Kaiser, K.L.E., Palabrica, V.S., and Ribo, J.M. (1987). In: *QSAR in Environmental Toxicology - II*, K.L.E. Kaiser (ed.), D. Reidel Publishing Company, Dordrecht, p. 153-168.

2,3-BENZOFURAN

CAS RN: 271-89-6

Sample purity: 99.5%
Temperature: 15°C
Test parameter: EC50
Effect: Reduction in light output

Concentration: 3.11 mg/l
Exposure time: 5 min

Concentration: 3.49 mg/l
Exposure time: 15 min

Concentration: 4.19 mg/l
Exposure time: 30 min

Comment: Mean of four assays. Methanol (<10%) was used to prepare the stock solutions. EC50 values were calculated from nominal concentrations.

Bibliographical reference: Kaiser, K.L.E., and Palabrica, V.S. (1991). *Water Poll. Res. J. Canada* **26**, 361-431.

BENZOIC ACID

CAS RN: 65-85-0

Temperature: 15°C
Test parameter: EC50
Effect: Reduction in light output

Concentration: 16.9 mg/l
Exposure time: 15 min
Comment: Mean of three assays. Methanol (<10%) was used to prepare the stock solutions. EC50 values were calculated from nominal concentrations.

Bibliographical reference: Kaiser, K.L.E., Palabrica, V.S., and Ribo, J.M. (1987). In: *QSAR in Environmental Toxicology - II*, K.L.E. Kaiser (ed.), D. Reidel Publishing Company, Dordrecht, p. 153-168.

BENZOIC ACID HYDRAZIDE

CAS RN: 613-94-5
Synonyms: Benzoylhydrazine; Benzhydrazide

Sample purity: 98%
Temperature: 15°C
Test parameter: EC50
Effect: Reduction in light output

Concentration: 82.0 mg/l
Exposure time: 5 min

Concentration: 65.2 mg/l
Exposure time: 15 min

Concentration: 76.6 mg/l*
Exposure time: 30 min

Comment: Mean of two assays. Methanol (<10%) was used to prepare the stock solutions. EC50 values were calculated from nominal concentrations.

Bibliographical references: Kaiser, K.L.E., and Palabrica, V.S. (1991). *Water Poll. Res. J. Canada* **26**, 361-431.
* Kaiser, K.L.E., Palabrica, V.S., and Ribo, J.M. (1987). In: *QSAR in Environmental Toxicology - II*, K.L.E. Kaiser (ed.), D. Reidel Publishing Company, Dordrecht, p. 153-168.

BENZONITRILE

CAS RN: 100-47-0

Temperature: 15°C
Test parameter: EC50
Effect: Reduction in light output
Concentration: 19 mg/l
Exposure time: 5 min

Bibliographical reference: Lebsack, M.E., Anderson, A.D., DeGraeve, G.M., and Bergman, H.L. (1981). In: *Aquatic Toxicology and Hazard Assessment: Fourth Conference, ASTM STP 737*, D.R. Branson and K.L. Dickson (eds.), American Society for Testing and Materials, Philadelphia, p. 348-356.

Sample purity: 99.9%
Temperature: 15°C
Test parameter: EC50
Effect: Reduction in light output

Concentration: 10.1 mg/l
Exposure time: 5 min

Concentration: 10.6 mg/l
Exposure time: 15 min

Concentration: 11.6 mg/l*
Exposure time: 30 min

Comment: Mean of four assays. Methanol (<10%) was used to prepare the stock solutions. EC50 values were calculated from nominal concentrations.

Bibliographical references: Kaiser, K.L.E., and Palabrica, V.S. (1991). *Water Poll. Res. J. Canada* **26**, 361-431.
* Kaiser, K.L.E., Palabrica, V.S., and Ribo, J.M. (1987). In: *QSAR in Environmental Toxicology - II*, K.L.E. Kaiser (ed.), D. Reidel Publishing Company, Dordrecht, p. 153-168.

BENZOPHENONE

CAS RN: 119-61-9

Temperature: 15°C
Test parameter: EC50
Effect: Reduction in light output

Concentration: 8.92 mg/l
Exposure time: 30 min
Comment: Mean of three assays. Methanol (<10%) was used to prepare the stock solutions. EC50 values were calculated from nominal concentrations.

Bibliographical reference: Kaiser, K.L.E., Palabrica, V.S., and Ribo, J.M. (1987). In: *QSAR in Environmental Toxicology - II*, K.L.E. Kaiser (ed.), D. Reidel Publishing Company, Dordrecht, p. 153-168.

1,4-BENZOQUINONE

CAS RN: 106-51-4

Temperature: 15°C
Test parameter: EC50
Effect: Reduction in light output
Concentrations: 0.08 mg/l
1.4 mg/l
Exposure time: 5 min

Bibliographical reference: King, E.F., and Painter, H.A. (1981). In: *Les Tests de Toxicité Aiguë en Milieu Aquatique. Acute Aquatic Ecotoxicological Tests*, H. Leclerc and D. Dive (eds.), INSERM 106, Paris, p. 143-153.

Temperature: 15°C
Test parameter: EC50
Effect: Reduction in light output
Concentration: 0.0085 mg/l
Exposure time: 5 min

Bibliographical reference: Lebsack, M.E., Anderson, A.D., DeGraeve, G.M., and Bergman, H.L. (1981). In: *Aquatic Toxicology and Hazard Assessment: Fourth Conference, ASTM STP 737*, D.R. Branson and K.L. Dickson (eds.), American Society for Testing and Materials, Philadelphia, p. 348-356.

Sample purity: 98%
Temperature: 15°C
Test parameter: EC50
Effect: Reduction in light output

Concentration: 0.020 mg/l
Exposure time: 15 min
Comment: Test solutions were made up freshly just before use and were shielded from light.

Bibliographical reference: King, E.F. (1984). In: *Toxicity Screening Procedures Using Bacterial Systems*, D. Liu and B.J. Dutka (eds.), Marcel Dekker, New York, p. 175-194.

Sample purity: Reagent grade quality
Temperature: 15 ± 0.3°C
Test parameter: EC50
Effect: Reduction in light output

Concentration: 0.020 mg/l
Exposure times: 5 and 10 min

Concentration: 0.022 mg/l
Exposure time: 20 min

Comment: Mean of four assays. EC50 values were calculated from nominal concentrations.

Bibliographical reference: Devillers, J., Steiman, R., Seigle-Murandi, F., Prévot, P., André, C., and Benoit-Guyod, J.L. (1990). *Tox. Assess.* **5**, 405-416.

2,1,3-BENZOTHIADIAZOLE

CAS RN: 273-13-2
Synonym: Piazthiole

Sample purity: 98%
Temperature: 15°C
Test parameter: EC50
Effect: Reduction in light output

Concentration: 18.8 mg/l
Exposure time: 5 min

Concentration: 19.7 mg/l
Exposure time: 15 min

Concentration: 23.7 mg/l
Exposure time: 30 min

Comment: Mean of three assays. Methanol (<10%) was used to prepare the stock solutions. EC50 values were calculated from nominal concentrations.

Bibliographical reference: Kaiser, K.L.E., and Palabrica, V.S. (1992). National Water Research Institute, Burlington, Ontario, Canada, unpublished results.

BENZOTHIAZOLE

CAS RN: 95-16-9

Sample purity: >99%
Temperature: 15°C
Test parameter: EC50
Effect: Reduction in light output

Concentration: 1.82 mg/l
Exposure time: 5 min

Concentration: 1.91 mg/l
Exposure time: 15 min

Concentration: 2.09 mg/l
Exposure time: 30 min

Comment: Mean of four assays. Methanol (<10%) was used to prepare the stock solutions. EC50 values were calculated from nominal concentrations.

Bibliographical reference: Kaiser, K.L.E., and Palabrica, V.S. (1991). *Water Poll. Res. J. Canada* **26**, 361-431.

BENZOYL CHLORIDE

CAS RN: 98-88-4

Sample purity: 99%
Temperature: 15°C

Test parameter: EC50
Effect: Reduction in light output

Concentration: 10.4 mg/l
Exposure time: 5 min

Concentration: 11.7 mg/l
Exposure time: 15 min

Concentration: 12.2 mg/l*
Exposure time: 30 min

Comment: Mean of three assays. Methanol (<10%) was used to prepare the stock solutions. EC50 values were calculated from nominal concentrations.

Bibliographical references: Kaiser, K.L.E., and Palabrica, V.S. (1991). *Water Poll. Res. J. Canada* **26**, 361-431.
* Kaiser, K.L.E., Palabrica, V.S., and Ribo, J.M. (1987). In: *QSAR in Environmental Toxicology - II*, K.L.E. Kaiser (ed.), D. Reidel Publishing Company, Dordrecht, p. 153-168.

BENZOYL CYANIDE

CAS RN: 613-90-1

Sample purity: 98%
Temperature: 15°C
Test parameter: EC50
Effect: Reduction in light output

Concentration: 6.28 mg/l
Exposure time: 5 min

Concentration: 4.99 mg/l
Exposure time: 15 min

Concentration: 3.29 mg/l*
Exposure time: 30 min

Comment: Mean of three assays. Methanol (<10%) was used to prepare the stock solutions. EC50 values were calculated from nominal concentrations.

Bibliographical references: Kaiser, K.L.E., and Palabrica, V.S. (1991). *Water Poll. Res. J. Canada* **26**, 361-431.
* Kaiser, K.L.E., Palabrica, V.S., and Ribo, J.M. (1987). In: *QSAR in Environmental Toxicology - II*, K.L.E. Kaiser (ed.), D. Reidel Publishing Company, Dordrecht, p. 153-168.

4-BENZOYLPYRIDINE

CAS RN: 14548-46-0

Sample purity: 98%
Temperature: 15°C
Test parameter: EC50
Effect: Reduction in light output

Concentration: 27.1 mg/l
Exposure time: 5 min

Concentration: 29.0 mg/l
Exposure time: 15 min

Concentration: 38.3 mg/l
Exposure time: 30 min

Comment: Mean of four assays. Methanol (<10%) was used to prepare the stock solutions. EC50 values were calculated from nominal concentrations.

Bibliographical reference: Kaiser, K.L.E., and Palabrica, V.S. (1991). *Water Poll. Res. J. Canada* **26**, 361-431.

BENZYL ACETATE

CAS RN: 140-11-4

Sample purity: >99%
Temperature: 15°C
Test parameter: EC50
Effect: Reduction in light output

Concentration: 3.29 mg/l
Exposure time: 5 min

Concentration: 3.86 mg/l
Exposure time: 15 min

Concentration: 4.54 mg/l
Exposure time: 30 min

Comment: Mean of three assays. Methanol (<10%) was used to prepare the stock solutions. EC50 values were calculated from nominal concentrations.

Bibliographical reference: Kaiser, K.L.E., and Palabrica, V.S. (1991). *Water Poll. Res. J. Canada* **26**, 361-431.

BENZYL ALCOHOL

CAS RN: 100-51-6

Temperature: 15°C
Test parameter: EC50
Effect: Reduction in light output
Concentration: 50 mg/l
Exposure time: 5 min
Comment: Toxicity value was calculated from nominal concentrations.

Bibliographical reference: Ferard, J.F., Vasseur, P., Danoux, L., and Larbaigt, G. (1983). *Rev. Fr. Sci. Eau* **2**, 221-237.

Sample purity: 99%
Temperature: 15°C
Test parameter: EC50
Effect: Reduction in light output

Concentration: 63.7 mg/l
Exposure times: 5 and 15 min

Concentration: 71.4 mg/l*
Exposure time: 30 min

Comment: Mean of three assays. Methanol (<10%) was used to prepare the stock solutions. EC50 values were calculated from nominal concentrations.

Bibliographical references: Kaiser, K.L.E., and Palabrica, V.S.

(1991). *Water Poll. Res. J. Canada* **26**, 361-431.
* Kaiser, K.L.E., Palabrica, V.S., and Ribo, J.M. (1987). In: *QSAR in Environmental Toxicology - II*, K.L.E. Kaiser (ed.), D. Reidel Publishing Company, Dordrecht, p. 153-168.

BENZYLAMINE

CAS RN: 100-46-9

Sample purity: 99%
Temperature: 15°C
Test parameter: EC50
Effect: Reduction in light output

Concentration: 21.4 mg/l
Exposure time: 5 min

Concentration: 17.0 mg/l
Exposure time: 15 min

Concentration: 17.0 mg/l*
Exposure time: 30 min

Comment: Mean of three assays. Methanol (<10%) was used to prepare the stock solutions. EC50 values were calculated from nominal concentrations.

Bibliographical references: Kaiser, K.L.E., and Palabrica, V.S. (1991). *Water Poll. Res. J. Canada* **26**, 361-431.
* Kaiser, K.L.E., Palabrica, V.S., and Ribo, J.M. (1987). In: *QSAR in Environmental Toxicology - II*, K.L.E. Kaiser (ed.), D. Reidel Publishing Company, Dordrecht, p. 153-168.

BENZYL CHLORIDE

CAS RN: 100-44-7

Sample purity: 97%
Temperature: 15°C
Test parameter: EC50
Effect: Reduction in light output

Concentration: 1.92 mg/l
Exposure time: 5 min

Concentration: 2.25 mg/l
Exposure time: 15 min

Concentration: 2.97 mg/l*
Exposure time: 30 min

Comment: Mean of three assays. Methanol (<10%) was used to prepare the stock solutions. EC50 values were calculated from nominal concentrations.

Bibliographical references: Kaiser, K.L.E., and Palabrica, V.S. (1991). *Water Poll. Res. J. Canada* **26**, 361-431.
* Kaiser, K.L.E., Palabrica, V.S., and Ribo, J.M. (1987). In: *QSAR in Environmental Toxicology - II*, K.L.E. Kaiser (ed.), D. Reidel Publishing Company, Dordrecht, p. 153-168.

BENZYL CYANIDE

CAS RN: 140-29-4
Synonym: Phenylacetonitrile

Sample purity: 99%
Temperature: 15°C
Test parameter: EC50
Effect: Reduction in light output

Concentration: 1.51 mg/l
Exposure time: 5 min

Concentration: 1.20 mg/l
Exposure time: 15 min

Concentration: 1.35 mg/l*
Exposure time: 30 min

Comment: Mean of three assays. Methanol (<10%) was used to prepare the stock solutions. EC50 values were calculated from nominal concentrations.

Bibliographical references: Kaiser, K.L.E., and Palabrica, V.S. (1991). *Water Poll. Res. J. Canada* **26**, 361-431.

* Kaiser, K.L.E., Palabrica, V.S., and Ribo, J.M. (1987). In: *QSAR in Environmental Toxicology - II*, K.L.E. Kaiser (ed.), D. Reidel Publishing Company, Dordrecht, p. 153-168.

BENZYL ISOTHIOCYANATE

CAS RN: 622-78-6

Sample purity: 97%
Temperature: 15°C
Test parameter: EC50
Effect: Reduction in light output

Concentration: 0.010 mg/l
Exposure times: 5 and 15 min

Concentration: 0.010 mg/l*
Exposure time: 30 min

Comment: Mean of three assays. Methanol (<10%) was used to prepare the stock solutions. EC50 values were calculated from nominal concentrations.

Bibliographical references: Kaiser, K.L.E., and Palabrica, V.S. (1991). *Water Poll. Res. J. Canada* **26**, 361-431.
* Kaiser, K.L.E., Palabrica, V.S., and Ribo, J.M. (1987). In: *QSAR in Environmental Toxicology - II*, K.L.E. Kaiser (ed.), D. Reidel Publishing Company, Dordrecht, p. 153-168.

BENZYL MERCAPTAN

CAS RN: 100-53-8

Sample purity: 99%
Temperature: 15°C
Test parameter: EC50
Effect: Reduction in light output

Concentration: 0.96 mg/l
Exposure time: 5 min

Concentration: 0.99 mg/l

Exposure time: 15 min

Concentration: 1.21 mg/l*
Exposure time: 30 min

Comment: Mean of three assays. Methanol (<10%) was used to prepare the stock solutions. EC50 values were calculated from nominal concentrations.

Bibliographical references: Kaiser, K.L.E., and Palabrica, V.S. (1991). *Water Poll. Res. J. Canada* **26**, 361-431.
* Kaiser, K.L.E., Palabrica, V.S., and Ribo, J.M. (1987). In: *QSAR in Environmental Toxicology - II*, K.L.E. Kaiser (ed.), D. Reidel Publishing Company, Dordrecht, p. 153-168.

4-(BENZYLOXY)PHENOL

CAS RN: 103-16-2
Synonym: Hydroquinone monobenzyl ether

Sample purity: 98%
Temperature: 15°C
Test parameter: EC50
Effect: Reduction in light output

Concentration: 1.63 mg/l
Exposure time: 5 min

Concentration: 1.87 mg/l
Exposure time: 15 min

Concentration: 2.15 mg/l
Exposure time: 30 min

Comment: Mean of five assays. Methanol (<10%) was used to prepare the stock solutions. EC50 values were calculated from nominal concentrations.

Bibliographical reference: Kaiser, K.L.E., and Palabrica, V.S. (1991). *Water Poll. Res. J. Canada* **26**, 361-431.

4-BENZYLPHENOL

CAS RN: 101-53-1
Synonym: 4-Hydroxydiphenylmethane

Sample purity: 96%
Temperature: 15°C
Test parameter: EC50
Effect: Reduction in light output

Concentration: 0.26 mg/l
Exposure times: 5 and 15 min

Concentration: 0.25 mg/l*
Exposure time: 30 min

Comment: Mean of three assays. Methanol (<10%) was used to prepare the stock solutions. EC50 values were calculated from nominal concentrations.

Bibliographical references: Kaiser, K.L.E., and Palabrica, V.S. (1991). *Water Poll. Res. J. Canada* **26**, 361-431.
* Kaiser, K.L.E., Palabrica, V.S., and Ribo, J.M. (1987). In: *QSAR in Environmental Toxicology - II*, K.L.E. Kaiser (ed.), D. Reidel Publishing Company, Dordrecht, p. 153-168.

2-BENZYLPYRIDINE

CAS RN: 101-82-6

Sample purity: 98%
Temperature: 15°C
Test parameter: EC50
Effect: Reduction in light output

Concentration: 22.3 mg/l
Exposure time: 5 min

Concentration: 21.8 mg/l
Exposure time: 15 min

Concentration: 20.8 mg/l
Exposure time: 30 min

Comment: Mean of four assays. Methanol (<10%) was used to prepare the stock solutions. EC50 values were calculated from nominal concentrations.

Bibliographical reference: Kaiser, K.L.E., and Palabrica, V.S. (1991). *Water Poll. Res. J. Canada* **26**, 361-431.

3-BENZYLPYRIDINE

CAS RN: 620-95-1

Sample purity: 97%
Temperature: 15°C
Test parameter: EC50
Effect: Reduction in light output

Concentration: 1.47 mg/l
Exposure time: 5 min

Concentration: 1.65 mg/l
Exposure time: 15 min

Concentration: 1.90 mg/l
Exposure time: 30 min

Comment: Mean of three assays. Methanol (<10%) was used to prepare the stock solutions. EC50 values were calculated from nominal concentrations.

Bibliographical reference: Kaiser, K.L.E., and Palabrica, V.S. (1991). *Water Poll. Res. J. Canada* **26**, 361-431.

4-BENZYLPYRIDINE

CAS RN: 2116-65-6

Sample purity: >99%
Temperature: 15°C
Test parameter: EC50
Effect: Reduction in light output

Concentration: 3.70 mg/l

Exposure times: 5 and 15 min

Concentration: 4.25 mg/l
Exposure time: 30 min

Comment: Mean of three assays. Methanol (<10%) was used to prepare the stock solutions. EC50 values were calculated from nominal concentrations.

Bibliographical reference: Kaiser, K.L.E., and Palabrica, V.S. (1991). *Water Poll. Res. J. Canada* **26**, 361-431.

BENZYLTHIOACETONITRILE

CAS RN: 17377-30-9

Temperature: 15°C
Test parameter: EC50
Effect: Reduction in light output

Concentration: 1.92 mg/l
Exposure time: 5 min

Concentration: 2.10 mg/l
Exposure time: 15 min

Concentration: 2.53 mg/l
Exposure time: 30 min

Comment: Mean of three assays. Methanol (<10%) was used to prepare the stock solutions. EC50 values were calculated from nominal concentrations.

Bibliographical reference: Kaiser, K.L.E., and Palabrica, V.S. (1991). *Water Poll. Res. J. Canada* **26**, 361-431.

BENZYL THIOCYANATE

CAS RN: 3012-37-1

Sample purity: 97%
Temperature: 15°C

Test parameter: EC50
Effect: Reduction in light output

Concentration: 0.54 mg/l
Exposure time: 5 min

Concentration: 0.51 mg/l
Exposure time: 15 min

Concentration: 0.45 mg/l*
Exposure time: 30 min

Comment: Mean of two assays. Methanol (<10%) was used to prepare the stock solutions. EC50 values were calculated from nominal concentrations.

Bibliographical references: Kaiser, K.L.E., and Palabrica, V.S. (1991). *Water Poll. Res. J. Canada* **26**, 361-431.
* Kaiser, K.L.E., Palabrica, V.S., and Ribo, J.M. (1987). In: *QSAR in Environmental Toxicology - II*, K.L.E. Kaiser (ed.), D. Reidel Publishing Company, Dordrecht, p. 153-168.

BERYLLIUM SULFATE TETRAHYDRATE

CAS RN: 7787-56-6

Temperature: 15 ± 0.1°C
Test parameter: EC50
Effect: Reduction in light output
Concentration: 252.5 mg/l
Exposure time: 5 min
Comment: Test was performed on *Photobacterium phosphoreum* NZ11D obtained from the Scripps Institute of Oceanography (La Jolla, CA).

Bibliographical reference: McFeters, G.A., Bond, P.J., Olson, S.B., and Tchan, Y.T. (1983). *Water Res.* **17**, 1757-1762.

4,4'-BIPHENOL

CAS RN: 92-88-6
Synonym: 4,4'-Dihydroxybiphenyl

Sample purity: >99%
Temperature: 15°C
Test parameter: EC50
Effect: Reduction in light output

Concentration: 2.00 mg/l
Exposure time: 5 min

Concentration: 2.29 mg/l
Exposure time: 15 min

Concentration: 2.57 mg/l
Exposure time: 30 min

Comment: Mean of three assays. Methanol (<10%) was used to prepare the stock solutions. EC50 values were calculated from nominal concentrations.

Bibliographical reference: Kaiser, K.L.E., and Palabrica, V.S. (1991). *Water Poll. Res. J. Canada* **26**, 361-431.

BIPHENYL

CAS RN: 92-52-4

Temperature: 15°C
Test parameter: EC50
Effect: Reduction in light output
Concentration: 1.90 mg/l
Exposure time: 30 min
Comment: Mean of three assays. Methanol (<10%) was used to prepare the stock solutions. EC50 values were calculated from nominal concentrations.

Bibliographical reference: Kaiser, K.L.E., Palabrica, V.S., and Ribo, J.M. (1987). In: *QSAR in Environmental Toxicology - II*, K.L.E. Kaiser (ed.), D. Reidel Publishing Company, Dordrecht, p. 153-168.

Sample purity: 99%
Temperature: 15°C
Test parameter: EC50
Effect: Reduction in light output

Concentration: 3.20 mg/l
Exposure time: 5 min

Concentration: 3.30 mg/l
Exposure time: 15 min

Comment: Chemical was prepared in an initial solution of 5% methanol.

Bibliographical reference: Cronin, M.T.D. (1993). Liverpool John Moores University, UK, private communication.

4-BIPHENYLACETIC ACID

CAS RN: 5728-52-9

Sample purity: 98%
Temperature: 15°C
Test parameter: EC50
Effect: Reduction in light output

Concentration: 21.7 mg/l
Exposure time: 5 min

Concentration: 20.7 mg/l
Exposure times: 15 and 30 min

Comment: Mean of three assays. Methanol (<10%) was used to prepare the stock solutions. EC50 values were calculated from nominal concentrations.

Bibliographical reference: Kaiser, K.L.E., and Palabrica, V.S. (1991). *Water Poll. Res. J. Canada* **26**, 361-431.

4-BIPHENYLCARBONITRILE

CAS RN: 2920-38-9
Synonym: 4-Phenylbenzonitrile

Sample purity: 99%
Temperature: 15°C
Test parameter: EC50

Effect: Reduction in light output

Concentration: 0.19 mg/l
Exposure times: 5 and 15 min

Concentration: 0.18 mg/l*
Exposure time: 30 min

Comment: Mean of three assays. Methanol (<10%) was used to prepare the stock solutions. EC50 values were calculated from nominal concentrations.

Bibliographical references: Kaiser, K.L.E., and Palabrica, V.S. (1991). *Water Poll. Res. J. Canada* **26**, 361-431.
* Kaiser, K.L.E., Palabrica, V.S., and Ribo, J.M. (1987). In: *QSAR in Environmental Toxicology - II*, K.L.E. Kaiser (ed.), D. Reidel Publishing Company, Dordrecht, p. 153-168.

4-BIPHENYLCARBOXYLIC ACID

CAS RN: 92-92-2
Synonym: 4-Phenylbenzoic acid

Sample purity: 95%
Temperature: 15°C
Test parameter: EC50
Effect: Reduction in light output

Concentration: 18.5 mg/l
Exposure time: 5 min

Concentration: 13.7 mg/l
Exposure time: 15 min

Concentration: 13.4 mg/l
Exposure time: 30 min

Comment: Mean of three assays. Methanol (<10%) was used to prepare the stock solutions. EC50 values were calculated from nominal concentrations.

Bibliographical reference: Kaiser, K.L.E., and Palabrica, V.S. (1991). *Water Poll. Res. J. Canada* **26**, 361-431.

4-BIPHENYLMETHANOL

CAS RN: 3597-91-9
Synonym: 4-Phenylbenzyl alcohol

Temperature: 15°C
Test parameter: EC50
Effect: Reduction in light output

Concentration: 2.79 mg/l
Exposure time: 5 min

Concentration: 2.49 mg/l
Exposure time: 15 min

Concentration: 2.60 mg/l*
Exposure time: 30 min

Comment: Mean of three assays. Methanol (<10%) was used to prepare the stock solutions. EC50 values were calculated from nominal concentrations.

Bibliographical references: Kaiser, K.L.E., and Palabrica, V.S. (1991). *Water Poll. Res. J. Canada* **26**, 361-431.
* Kaiser, K.L.E., Palabrica, V.S., and Ribo, J.M. (1987). In: *QSAR in Environmental Toxicology - II*, K.L.E. Kaiser (ed.), D. Reidel Publishing Company, Dordrecht, p. 153-168.

BIS(2-CARBOBUTOXYETHYL)TINBIS(THIOGLYCOLIC ACID ISOOCTYL ESTER)

CAS RN: 63397-60-4

Test parameter: EC50
Effect: Reduction in light output
Concentration: >0.59 mg/l
Exposure time: 30 min

Bibliographical reference: Steinhäuser, K.G., Amann, W., Späth, A., and Polenz, A. (1985). *Vom Wasser* **65**, 203-214.

BIS(TRI-*n*-BUTYLTIN) OXIDE

CAS RN: 56-35-9

Test parameter: EC50
Effect: Reduction in light output
Concentration: 0.0011 mg/l
Exposure time: 30 min

Bibliographical reference: Steinhäuser, K.G., Amann, W., Späth, A., and Polenz, A. (1985). *Vom Wasser* **65**, 203-214.

3,5-BIS(TRIFLUOROMETHYL)BENZYL ALCOHOL

CAS RN: 32707-89-4

Sample purity: 98%
Temperature: 15°C
Test parameter: EC50
Effect: Reduction in light output

Concentration: 59.9 mg/l
Exposure time: 5 min

Concentration: 51.0 mg/l
Exposure time: 15 min

Concentration: 50.0 mg/l
Exposure time: 30 min

Comment: Mean of three assays. Methanol (<10%) was used to prepare the stock solutions. EC50 values were calculated from nominal concentrations.

Bibliographical reference: Kaiser, K.L.E., and Palabrica, V.S. (1991). *Water Poll. Res. J. Canada* **26**, 361-431.

3,5-BIS(TRIFLUOROMETHYL)NITROBENZENE

CAS RN: 328-75-6

Sample purity: >98%

Temperature: 15°C
Test parameter: EC50
Effect: Reduction in light output

Concentration: 49.4 mg/l
Exposure time: 5 min

Concentration: 38.3 mg/l
Exposure time: 15 min

Concentration: 26.5 mg/l
Exposure time: 30 min

Comment: Mean of four assays. Methanol (<10%) was used to prepare the stock solutions. EC50 values were calculated from nominal concentrations.

Bibliographical reference: Kaiser, K.L.E., and Palabrica, V.S. (1991). *Water Poll. Res. J. Canada* **26**, 361-431.

3,5-BIS(TRIFLUOROMETHYL)PHENOL

CAS RN: 349-58-6

Sample purity: 95%
Temperature: 15°C
Test parameter: EC50
Effect: Reduction in light output

Concentration: 0.90 mg/l
Exposure time: 5 min

Concentration: 0.84 mg/l
Exposure time: 15 min

Concentration: 0.82 mg/l
Exposure time: 30 min

Comment: Mean of four assays. Methanol (<10%) was used to prepare the stock solutions. EC50 values were calculated from nominal concentrations.

Bibliographical reference: Kaiser, K.L.E., and Palabrica, V.S. (1991). *Water Poll. Res. J. Canada* **26**, 361-431.

(-)-BORNYL ACETATE

CAS RN: 5655-61-8

Sample purity: 97%
Temperature: 15°C
Test parameter: EC50
Effect: Reduction in light output

Concentration: 1.03 mg/l
Exposure time: 5 min

Concentration: 0.42 mg/l
Exposure time: 15 min

Concentration: 0.29 mg/l
Exposure time: 30 min

Comment: Mean of three assays. Methanol (<10%) was used to prepare the stock solutions. EC50 values were calculated from nominal concentrations.

Bibliographical reference: Kaiser, K.L.E., and Palabrica, V.S. (1991). *Water Poll. Res. J. Canada* **26**, 361-431.

BROMACIL

CAS RN: 314-40-9
Synonym: 5-Bromo-6-methyl-3-(1-methylpropyl)-2,4(1*H*,3*H*)-pyrimidinedione

Sample purity: Analytical reference standard grade
Temperature: 15 ± 0.1°C
Test parameter: EC50
Effect: Reduction in light output
Concentration: 6.65 mg/l
Exposure time: 5 min
Comment: Test was performed on *Photobacterium phosphoreum* NZ11D obtained from the Scripps Institute of Oceanography (La Jolla, CA).

Bibliographical reference: McFeters, G.A., Bond, P.J., Olson, S.B., and Tchan, Y.T. (1983). *Water Res.* **17**, 1757-1762.

4'-BROMOACETOPHENONE

CAS RN: 99-90-1

Sample purity: 98%
Temperature: 15°C
Test parameter: EC50
Effect: Reduction in light output

Concentration: 6.29 mg/l
Exposure time: 5 min

Concentration: 6.01 mg/l
Exposure time: 15 min

Concentration: 6.01 mg/l*
Exposure time: 30 min

Comment: Mean of three assays. Methanol (<10%) was used to prepare the stock solutions. EC50 values were calculated from nominal concentrations.

Bibliographical references: Kaiser, K.L.E., and Palabrica, V.S. (1991). *Water Poll. Res. J. Canada* **26**, 361-431.
* Kaiser, K.L.E., and Gough, K.M. (1989). In: *Aquatic Toxicology and Environmental Fate: Eleventh Volume, ASTM STP 1007*, G.W. Suter and M.A. Lewis (eds.), American Society for Testing and Materials, Philadelphia, p. 424-441.

4-BROMOANISOLE

CAS RN: 104-92-7

Sample purity: 99%
Temperature: 15°C
Test parameter: EC50
Effect: Reduction in light output

Concentration: 2.25 mg/l
Exposure time: 5 min

Concentration: 2.41 mg/l
Exposure time: 15 min

Concentration: 2.58 mg/l*
Exposure time: 30 min

Comment: Mean of three assays. Methanol (<10%) was used to prepare the stock solutions. EC50 values were calculated from nominal concentrations.

Bibliographical references: Kaiser, K.L.E., and Palabrica, V.S. (1991). *Water Poll. Res. J. Canada* **26**, 361-431.
* Kaiser, K.L.E., and Gough, K.M. (1989). In: *Aquatic Toxicology and Environmental Fate: Eleventh Volume, ASTM STP 1007*, G.W. Suter and M.A. Lewis (eds.), American Society for Testing and Materials, Philadelphia, p. 424-441.

BROMOBENZENE

CAS RN: 108-86-1

Temperature: 15°C
Test parameter: EC50
Effect: Reduction in light output
Concentration: 9.46 mg/l
Exposure time: 30 min
Comment: Mean of three assays. Methanol (<10%) was used to prepare the stock solutions. EC50 values were calculated from nominal concentrations.

Bibliographical reference: Kaiser, K.L.E., Palabrica, V.S., and Ribo, J.M. (1987). In: *QSAR in Environmental Toxicology - II*, K.L.E. Kaiser (ed.), D. Reidel Publishing Company, Dordrecht, p. 153-168.

4-BROMOBENZENESULFONIC ACID HYDRAZIDE

CAS RN: 2297-64-5

Temperature: 15°C
Test parameter: EC50
Effect: Reduction in light output

Concentration: 26.3 mg/l
Exposure time: 5 min

Concentration: 23.4 mg/l
Exposure time: 15 min

Concentration: 26.9 mg/l
Exposure time: 30 min

Comment: Mean of three assays. Methanol (<10%) was used to prepare the stock solutions. EC50 values were calculated from nominal concentrations.

Bibliographical reference: Kaiser, K.L.E., and Palabrica, V.S. (1991). *Water Poll. Res. J. Canada* **26**, 361-431.

4-BROMOBENZONITRILE

CAS RN: 623-00-7

Sample purity: 98%
Temperature: 15°C
Test parameter: EC50
Effect: Reduction in light output

Concentration: 8.13 mg/l
Exposure time: 5 min

Concentration: 8.32 mg/l
Exposure time: 15 min

Concentration: 8.71 mg/l*
Exposure time: 30 min

Comment: Mean of three assays. Methanol (<10%) was used to prepare the stock solutions. EC50 values were calculated from nominal concentrations.

Bibliographical references: Kaiser, K.L.E., and Palabrica, V.S. (1991). *Water Poll. Res. J. Canada* **26**, 361-431.
* Kaiser, K.L.E., and Gough, K.M. (1989). In: *Aquatic Toxicology and Environmental Fate: Eleventh Volume, ASTM STP 1007*, G.W. Suter and M.A. Lewis (eds.), American Society for Testing and Materials, Philadelphia, p. 424-441.

4-BROMOBENZOTRIFLUORIDE

CAS RN: 402-43-7
Synonym: 4-Bromo-α,α,α-trifluorotoluene

Sample purity: 99%
Temperature: 15°C
Test parameter: EC50
Effect: Reduction in light output

Concentration: 3.65 mg/l
Exposure time: 5 min

Concentration: 5.15 mg/l
Exposure time: 15 min

Concentration: 6.34 mg/l*
Exposure time: 30 min

Comment: Mean of three assays. Methanol (<10%) was used to prepare the stock solutions. EC50 values were calculated from nominal concentrations.

Bibliographical references: Kaiser, K.L.E., and Palabrica, V.S. (1991). *Water Poll. Res. J. Canada* **26**, 361-431.
* Kaiser, K.L.E., and Gough, K.M. (1989). In: *Aquatic Toxicology and Environmental Fate: Eleventh Volume, ASTM STP 1007*, G.W. Suter and M.A. Lewis (eds.), American Society for Testing and Materials, Philadelphia, p. 424-441.

4-BROMOCHLOROBENZENE

CAS RN: 106-39-8

Sample purity: 99+%
Temperature: 15°C
Test parameter: EC50
Effect: Reduction in light output

Concentration: 4.59 mg/l
Exposure time: 5 min

Concentration: 5.52 mg/l
Exposure time: 15 min

Concentration: 6.49 mg/l*
Exposure time: 30 min

Comment: Mean of three assays. Methanol (<10%) was used to prepare the stock solutions. EC50 values were calculated from nominal concentrations.

Bibliographical references: Kaiser, K.L.E., and Palabrica, V.S. (1991). *Water Poll. Res. J. Canada* **26**, 361-431.
* Kaiser, K.L.E. (1987). In: *QSAR in Environmental Toxicology - II*, K.L.E. Kaiser (ed.), D. Reidel Publishing Company, Dordrecht, p. 169-188.

2-BROMO-4-CHLORO-5-NITROTOLUENE

CAS RN: 40371-64-0

Sample purity: 99%
Temperature: 15°C
Test parameter: EC50
Effect: Reduction in light output

Concentration: 15.1 mg/l
Exposure time: 5 min

Concentration: 17.7 mg/l
Exposure time: 15 min

Concentration: 22.3 mg/l
Exposure time: 30 min

Comment: Mean of four assays. Methanol (<10%) was used to prepare the stock solutions. EC50 values were calculated from nominal concentrations.

Bibliographical reference: Kaiser, K.L.E., and Palabrica, V.S. (1991). *Water Poll. Res. J. Canada* **26**, 361-431.

2-BROMO-2',5'-DIMETHOXYACETOPHENONE

CAS RN: 1204-21-3

Sample purity: 97%
Temperature: 15°C
Test parameter: EC50
Effect: Reduction in light output

Concentration: 0.13 mg/l
Exposure time: 5 min

Concentration: 0.068 mg/l
Exposure time: 15 min

Concentration: 0.047 mg/l
Exposure time: 30 min

Comment: Mean of three assays. Methanol (<10%) was used to prepare the stock solutions. EC50 values were calculated from nominal concentrations.

Bibliographical reference: Kaiser, K.L.E., and Palabricą, V.S. (1991). *Water Poll. Res. J. Canada* **26**, 361-431.

2-BROMO-4,6-DINITROANILINE

CAS RN: 1817-73-8

Sample purity: 94%
Temperature: 15°C
Test parameter: EC50
Effect: Reduction in light output

Concentration: 29.4 mg/l
Exposure time: 5 min

Concentration: 23.9 mg/l
Exposure time: 15 min

Concentration: 24.5 mg/l
Exposure time: 30 min

Comment: Mean of four assays. Methanol (<10%) was used to prepare the stock solutions. EC50 values were calculated from nominal concentrations.

Bibliographical reference: Kaiser, K.L.E., and Palabrica, V.S.

(1992). National Water Research Institute, Burlington, Ontario, Canada, unpublished results.

(2-BROMOETHYL)BENZENE

CAS RN: 103-63-9
Synonym: Phenethyl bromide

Sample purity: 98%
Temperature: 15°C
Test parameter: EC50
Effect: Reduction in light output

Concentration: 0.66 mg/l
Exposure time: 5 min

Concentration: 0.70 mg/l
Exposure time: 15 min

Concentration: 0.79 mg/l
Exposure time: 30 min

Comment: Mean of three assays. Methanol (<10%) was used to prepare the stock solutions. EC50 values were calculated from nominal concentrations.

Bibliographical reference: Kaiser, K.L.E., and Palabrica, V.S. (1991). *Water Poll. Res. J. Canada* **26**, 361-431.

1-BROMO-4-FLUOROBENZENE

CAS RN: 460-00-4

Sample purity: 99%
Temperature: 15°C
Test parameter: EC50
Effect: Reduction in light output

Concentration: 23.1 mg/l
Exposure time: 5 min

Concentration: 24.2 mg/l

Exposure time: 15 min

Concentration: 29.0 mg/l*
Exposure time: 30 min

Comment: Mean of three assays. Methanol (<10%) was used to prepare the stock solutions. EC50 values were calculated from nominal concentrations.

Bibliographical references: Kaiser, K.L.E., and Palabrica, V.S. (1991). *Water Poll. Res. J. Canada* **26**, 361-431.
* Kaiser, K.L.E., and Gough, K.M. (1989). In: *Aquatic Toxicology and Environmental Fate: Eleventh Volume, ASTM STP 1007*, G.W. Suter and M.A. Lewis (eds.), American Society for Testing and Materials, Philadelphia, p. 424-441.

1-BROMO-4-IODOBENZENE

CAS RN: 589-87-7

Sample purity: 98%
Temperature: 15°C
Test parameter: EC50
Effect: Reduction in light output

Concentration: 1.21 mg/l
Exposure time: 5 min

Concentration: 1.42 mg/l
Exposure time: 15 min

Concentration: 1.59 mg/l
Exposure time: 30 min

Comment: Mean of three assays. Methanol (<10%) was used to prepare the stock solutions. EC50 values were calculated from nominal concentrations.

Bibliographical reference: Kaiser, K.L.E., and Palabrica, V.S. (1991). *Water Poll. Res. J. Canada* **26**, 361-431.

3-BROMO-4-METHYLANILINE

CAS RN: 7745-91-7

Sample purity: 98%
Temperature: 15°C
Test parameter: EC50
Effect: Reduction in light output

Concentration: 0.33 mg/l
Exposure time: 5 min

Concentration: 0.35 mg/l
Exposure time: 15 min

Concentration: 0.41 mg/l
Exposure time: 30 min

Comment: Mean of four assays. Methanol (<10%) was used to prepare the stock solutions. EC50 values were calculated from nominal concentrations.

Bibliographical reference: Kaiser, K.L.E., and Palabrica, V.S. (1991). *Water Poll. Res. J. Canada* **26**, 361-431.

4-(BROMOMETHYL)BENZONITRILE

CAS RN: 17201-43-3

Sample purity: 98%
Temperature: 15°C
Test parameter: EC50
Effect: Reduction in light output

Concentration: 1.03 mg/l
Exposure time: 5 min

Concentration: 0.57 mg/l
Exposure time: 15 min

Concentration: 0.33 mg/l
Exposure time: 30 min

Comment: Mean of three assays. Methanol (<10%) was used to

prepare the stock solutions. EC50 values were calculated from nominal concentrations.

Bibliographical reference: Kaiser, K.L.E., and Palabrica, V.S. (1991). *Water Poll. Res. J. Canada* **26**, 361-431.

1-BROMO-2-NITROBENZENE

CAS RN: 577-19-5

Sample purity: 98%
Temperature: 15°C
Test parameter: EC50
Effect: Reduction in light output

Concentration: 1.57 mg/l
Exposure times: 5 and 15 min

Concentration: 1.60 mg/l
Exposure time: 30 min

Comment: Mean of three assays. Methanol (<10%) was used to prepare the stock solutions. EC50 values were calculated from nominal concentrations.

Bibliographical reference: Kaiser, K.L.E., and Palabrica, V.S. (1991). *Water Poll. Res. J. Canada* **26**, 361-431.

1-BROMO-3-NITROBENZENE

CAS RN: 585-79-5

Sample purity: 99%
Temperature: 15°C
Test parameter: EC50
Effect: Reduction in light output

Concentration: 3.59 mg/l
Exposure time: 5 min

Concentration: 4.12 mg/l
Exposure time: 15 min

Concentration: 4.96 mg/l
Exposure time: 30 min

Comment: Mean of three assays. Methanol (<10%) was used to prepare the stock solutions. EC50 values were calculated from nominal concentrations.

Bibliographical reference: Kaiser, K.L.E., and Palabrica, V.S. (1992). National Water Research Institute, Burlington, Ontario, Canada, unpublished results.

1-BROMO-4-NITROBENZENE

CAS RN: 586-78-7

Temperature: 15°C
Test parameter: EC50
Effect: Reduction in light output

Concentration: 14.0 mg/l
Exposure time: 5 min

Concentration: 15.3 mg/l
Exposure time: 15 min

Concentration: 16.8 mg/l*
Exposure time: 30 min

Comment: Mean of four assays. Methanol (<10%) was used to prepare the stock solutions. EC50 values were calculated from nominal concentrations.

Bibliographical references: Kaiser, K.L.E., and Palabrica, V.S. (1991). *Water Poll. Res. J. Canada* **26**, 361-431.
* Kaiser, K.L.E. (1987). In: *QSAR in Environmental Toxicology - II*, K.L.E. Kaiser (ed.), D. Reidel Publishing Company, Dordrecht, p. 169-188.

2-BROMO-5-NITRO-BENZOTRIFLUORIDE

CAS RN: 367-67-9
Synonym: 1-Bromo-4-nitro-2-(trifluoromethyl)benzene

Sample purity: 90%
Temperature: 15°C
Test parameter: EC50
Effect: Reduction in light output

Concentration: 4.28 mg/l
Exposure time: 5 min

Concentration: 4.69 mg/l
Exposure time: 15 min

Concentration: 5.39 mg/l
Exposure time: 30 min

Comment: Mean of three assays. Methanol (<10%) was used to prepare the stock solutions. EC50 values were calculated from nominal concentrations.

Bibliographical reference: Kaiser, K.L.E., and Palabrica, V.S. (1991). *Water Poll. Res. J. Canada* **26**, 361-431.

4-BROMO-3-NITRO-BENZOTRIFLUORIDE

CAS RN: 349-03-1
Synonym: 4-Bromo-α,α,α-trifluoro-3-nitrotoluene

Sample purity: 97%
Temperature: 15°C
Test parameter: EC50
Effect: Reduction in light output

Concentration: 13.2 mg/l
Exposure time: 5 min

Concentration: 13.8 mg/l
Exposure time: 15 min

Concentration: 14.8 mg/l
Exposure time: 30 min

Comment: Mean of three assays. Methanol (<10%) was used to prepare the stock solutions. EC50 values were calculated from nominal concentrations.

Bibliographical reference: Kaiser, K.L.E., and Palabrica, V.S. (1991). *Water Poll. Res. J. Canada* **26**, 361-431.

5-BROMO-2-NITRO-BENZOTRIFLUORIDE

CAS RN: 344-38-7
Synonym: 4-Bromo-1-nitro-2-(trifluoromethyl)benzene

Sample purity: 98%
Temperature: 15°C
Test parameter: EC50
Effect: Reduction in light output

Concentration: 6.63 mg/l
Exposure time: 5 min

Concentration: 7.79 mg/l
Exposure time: 15 min

Concentration: 10.3 mg/l
Exposure time: 30 min

Comment: Mean of four assays. Methanol (<10%) was used to prepare the stock solutions. EC50 values were calculated from nominal concentrations.

Bibliographical reference: Kaiser, K.L.E., and Palabrica, V.S. (1991). *Water Poll. Res. J. Canada* **26**, 361-431.

2-BROMO-2-NITRO-1,3-DIHYDROXYPROPANE

CAS RN: 52-51-7

Sample purity: 98%
Temperature: 15°C
Test parameter: EC50
Effect: Reduction in light output

Concentration: 0.91 mg/l
Exposure time: 5 min

Concentration: 0.50 mg/l

Exposure time: 15 min

Concentration: 0.41 mg/l
Exposure time: 30 min

Comment: Mean of three assays. Methanol (<10%) was used to prepare the stock solutions. EC50 values were calculated from nominal concentrations.

Bibliographical reference: Kaiser, K.L.E., and Palabrica, V.S. (1992). National Water Research Institute, Burlington, Ontario, Canada, unpublished results.

4-BROMO-2-NITROPHENOL

CAS RN: 7693-52-9

Sample purity: 99%
Temperature: 15°C
Test parameter: EC50
Effect: Reduction in light output

Concentration: 11.4 mg/l
Exposure time: 5 min

Concentration: 10.9 mg/l
Exposure time: 15 min

Concentration: 10.7 mg/l
Exposure time: 30 min

Comment: Mean of four assays. Methanol (<10%) was used to prepare the stock solutions. EC50 values were calculated from nominal concentrations.

Bibliographical reference: Kaiser, K.L.E., and Palabrica, V.S. (1991). *Water Poll. Res. J. Canada* **26**, 361-431.

2-BROMO-4-NITROTOLUENE

CAS RN: 7745-93-9

Sample purity: 98%
Temperature: 15°C
Test parameter: EC50
Effect: Reduction in light output

Concentration: 1.36 mg/l
Exposure time: 5 min

Concentration: 1.60 mg/l
Exposure time: 15 min

Concentration: 2.06 mg/l
Exposure time: 30 min

Comment: Mean of three assays. Methanol (<10%) was used to prepare the stock solutions. EC50 values were calculated from nominal concentrations.

Bibliographical reference: Kaiser, K.L.E., and Palabrica, V.S. (1992). National Water Research Institute, Burlington, Ontario, Canada, unpublished results.

2-BROMO-5-NITROTOLUENE

CAS RN: 7149-70-4

Sample purity: 98%
Temperature: 15°C
Test parameter: EC50
Effect: Reduction in light output

Concentration: 1.43 mg/l
Exposure time: 5 min

Concentration: 1.72 mg/l
Exposure time: 15 min

Concentration: 2.21 mg/l
Exposure time: 30 min

Comment: Mean of three assays. Methanol (<10%) was used to prepare the stock solutions. EC50 values were calculated from nominal concentrations.

Bibliographical reference: Kaiser, K.L.E., and Palabrica, V.S. (1992). National Water Research Institute, Burlington, Ontario, Canada, unpublished results.

2-BROMO-6-NITROTOLUENE

CAS RN: 55289-35-5

Sample purity: 98%
Temperature: 15°C
Test parameter: EC50
Effect: Reduction in light output

Concentration: 0.160 mg/l
Exposure time: 5 min

Concentration: 0.168 mg/l
Exposure time: 15 min

Concentration: 0.184 mg/l
Exposure time: 30 min

Comment: Mean of three assays. Methanol (<10%) was used to prepare the stock solutions. EC50 values were calculated from nominal concentrations.

Bibliographical reference: Kaiser, K.L.E., and Palabrica, V.S. (1992). National Water Research Institute, Burlington, Ontario, Canada, unpublished results.

2-BROMOPHENOL

CAS RN: 95-56-7

Temperature: 15°C
Test parameter: EC50
Effect: Reduction in light output

Concentration: 21.2 mg/l
Exposure time: 5 min

Concentration: 20.6 mg/l

Exposure time: 30 min

Bibliographical reference: Speece, R. (1987). Drexel University, Philadelphia, USA, private communication.

3-BROMOPHENOL

CAS RN: 591-20-8

Temperature: 15°C
Test parameter: EC50
Effect: Reduction in light output

Concentration: 3.90 mg/l
Exposure time: 5 min

Concentration: 3.55 mg/l
Exposure time: 30 min

Bibliographical reference: Speece, R. (1987). Drexel University, Philadelphia, USA, private communication.

4-BROMOPHENOL

CAS RN: 106-41-2

Temperature: 15°C
Test parameter: EC50
Effect: Reduction in light output

Concentration: 0.41 mg/l
Exposure time: 5 min

Concentration: 0.47 mg/l
Exposure time: 30 min

Bibliographical reference: Speece, R. (1987). Drexel University, Philadelphia, USA, private communication.

4-BROMOPHENYLACETONITRILE

CAS RN: 16532-79-9
Synonym: 4-Bromobenzyl cyanide

Sample purity: 97%
Temperature: 15°C
Test parameter: EC50
Effect: Reduction in light output

Concentration: 0.14 mg/l
Exposure time: 5 min

Concentration: 0.15 mg/l
Exposure time: 15 min

Concentration: 0.19 mg/l
Exposure time: 30 min

Comment: Mean of three assays. Methanol (<10%) was used to prepare the stock solutions. EC50 values were calculated from nominal concentrations.

Bibliographical reference: Kaiser, K.L.E., and Palabrica, V.S. (1991). *Water Poll. Res. J. Canada* **26**, 361-431.

4-BROMOPYRIDINE HYDROCHLORIDE

CAS RN: 19524-06-2

Temperature: 15°C
Test parameter: EC50
Effect: Reduction in light output

Concentration: 21.3 mg/l
Exposure time: 5 min

Concentration: 22.8 mg/l
Exposure time: 15 min

Concentration: 23.4 mg/l
Exposure time: 30 min

Comment: Mean of three assays. Methanol (<10%) was used to

prepare the stock solutions. EC50 values were calculated from nominal concentrations.

Bibliographical reference: Kaiser, K.L.E., and Palabrica, V.S. (1991). *Water Poll. Res. J. Canada* **26**, 361-431.

5-BROMOSALICYLALDEHYDE

CAS RN: 1761-61-1

Sample purity: 99%
Temperature: 15°C
Test parameter: EC50
Effect: Reduction in light output

Concentration: 9.00 mg/l
Exposure time: 5 min

Concentration: 6.80 mg/l*
Exposure time: 15 min

Comment: Chemical was prepared in an initial solution of 3% methanol.

Bibliographical references: Cronin, M.T.D., Dearden, J.C., and Dobbs, A.J. (1991). *Sci. Total Environ.* **109/110**, 431-439.
* Cronin, M.T.D. (1990). Thesis, Liverpool Polytechnic, Liverpool, UK.

4-BROMOTOLUENE

CAS RN: 106-38-7

Sample purity: 98%
Temperature: 15°C
Test parameter: EC50
Effect: Reduction in light output

Concentration: 1.46 mg/l
Exposure time: 5 min

Concentration: 1.83 mg/l

Exposure time: 15 min

Concentration: 2.53 mg/l
Exposure time: 30 min

Comment: Mean of three assays. Methanol (<10%) was used to prepare the stock solutions. EC50 values were calculated from nominal concentrations.

Bibliographical reference: Kaiser, K.L.E., and Palabrica, V.S. (1991). *Water Poll. Res. J. Canada* **26**, 361-431.

BUFENCARB

CAS RN: 8065-36-9
Synonym: 3-(1-Methylbutyl)phenyl methylcarbamate mixture with 3-(1-ethylpropyl)phenyl methylcarbamate

Sample purity: Analytical reference standard grade
Temperature: 15 ± 0.1°C
Test parameter: EC50
Effect: Reduction in light output
Concentration: 0.256 mg/l
Exposure time: 5 min
Comment: Test was performed on *Photobacterium phosphoreum* NZ11D obtained from the Scripps Institute of Oceanography (La Jolla, CA).

Bibliographical reference: McFeters, G.A., Bond, P.J., Olson, S.B., and Tchan, Y.T. (1983). *Water Res.* **17**, 1757-1762.

BUTANAL

CAS RN: 123-72-8
Synonym: Butyraldehyde

Test parameter: EC50
Effect: Reduction in light output
Concentration: 16.4 mg/l
Exposure time: 5 min
Comment: Concentrations in the test were measured.

Bibliographical reference: Curtis, C., Lima, A., Lozano, S.J., and Veith, G.D. (1982). In: *Aquatic Toxicology and Hazard Assessment: Fifth Conference, ASTM STP 766*, J.G. Pearson, R.B. Foster, and W.E. Bishop (eds.), American Society for Testing and Materials, Philadelphia, p. 170-178.

Sample purity: 99%
Temperature: 15°C
Test parameter: EC50
Effect: Reduction in light output

Concentration: 268 mg/l
Exposure time: 5 min

Concentration: 185 mg/l*
Exposure time: 15 min

Comment: Chemical was prepared in an initial solution of 3% methanol.

Bibliographical references: Cronin, M.T.D., Dearden, J.C., and Dobbs, A.J. (1991). *Sci. Total Environ.* **109/110**, 431-439.
* Cronin, M.T.D. (1990). Thesis, Liverpool Polytechnic, Liverpool, UK.

Sample purity: 99%
Temperature: 15°C
Test parameter: EC50
Effect: Reduction in light output

Concentration: 153 ± 5.83 mg/l
Exposure time: 5 min

Concentration: 107 ± 6.05 mg/l
Exposure time: 15 min

Concentration: 99 ± 0.79 mg/l
Exposure time: 25 min

Comment: Mean of five assays. The values were converted to mg/l from the original data expressed in µM and a rounded molecular weight of 72 given by the authors. Phenol solution was used for quality control/quality assurance. The 5-min EC50 value was 18.2 mg/l. The EC50 value at 15 min was 20.7 mg/l with a relative error of <5%.

Bibliographical reference: Chou, C.C., and Que Hee, S.S. (1992). *Ecotoxicol. Environ. Safety* **23**, 355-363.

1-BUTANOL

CAS RN: 71-36-3
Synonyms: Butyl alcohol; *n*-Butanol

Temperature: 15°C
Test parameter: EC50
Effect: Reduction in light output
Concentration: 3300 mg/l
Exposure time: 5 min

Bibliographical reference: Bulich, A.A., Greene, M.W., and Isenberg, D.L. (1981). In: *Aquatic Toxicology and Hazard Assessment: Fourth Conference, ASTM STP 737*, D.R. Branson and K.L. Dickson, (eds.), American Society for Testing and Materials, Philadelphia, p. 338-347.

Temperature: 15 ± 0.1°C
Test parameter: EC50
Effect: Reduction in light output
Concentration: 44000 mg/l
Exposure time: 5 min

Bibliographical reference: Chang, J.C., Taylor, P.B., and Leach, F.R. (1981). *Bull. Environ. Contam. Toxicol.* **26**, 150-156.

Test parameter: EC50
Effect: Reduction in light output
Concentration: 2300 mg/l
Exposure time: 5 min
Comment: The EC50 value was calculated from nominal concentrations.

Bibliographical reference: Curtis, C., Lima, A., Lozano, S.J., and Veith, G.D. (1982). In: *Aquatic Toxicology and Hazard Assessment: Fifth Conference, ASTM STP 766*, J.G. Pearson, R.B. Foster, and W.E. Bishop (eds.), American Society for Testing and Materials, Philadelphia, p. 170-178.

Temperature: 15°C
Test parameter: EC50
Effect: Reduction in light output
Concentration: 2818 mg/l
Exposure time: 15 min

Bibliographical reference: Hermens, J., Busser, F., Leeuwangh, P., and Musch, A. (1985). *Ecotoxicol. Environ. Safety* **9**, 17-25.

Temperature: 15 ± 0.1°C
Test parameter: EC50
Effect: Reduction in light output

Concentration: 3370 mg/l
Exposure time: 5 min

Concentration: 3690 mg/l
Exposure time: 15 min

Concentration: 3710 mg/l
Exposure time: 30 min

Comment: The pH was not adjusted.

Bibliographical reference: Tarkpea, M., Hansson, M., and Samuelsson, B. (1986). *Ecotoxicol. Environ. Safety* **11**, 127-143.

Temperature: 15°C
Test parameter: EC50
Effect: Reduction in light output

Concentration: 2056 mg/l
Exposure time: 5 min

Concentration: 2186 mg/l
Exposure time: 30 min

Bibliographical reference: Speece, R. (1987). Drexel University, Philadelphia, USA, private communication.

2-BUTANOL

CAS RN: 78-92-2
Synonym: *sec*-Butyl alcohol

Temperature: 15°C
Test parameter: EC50
Effect: Reduction in light output
Concentration: 173 mg/l
Exposure time: 5 min

Bibliographical reference: Microtox® Application Notes (1982). Beckman Instruments, Inc., Carlsbad, California.

2-BUTANONE

CAS RN: 78-93-3
Synonym: Methyl ethyl ketone

Test parameter: EC50
Effect: Reduction in light output
Concentrations: 4350 mg/l
5750 mg/l
Exposure time: 5 min
Comment: Toxicity values were calculated from nominal concentrations.

Bibliographical reference: Curtis, C., Lima, A., Lozano, S.J., and Veith, G.D. (1982). In: *Aquatic Toxicology and Hazard Assessment: Fifth Conference, ASTM STP 766*, J.G. Pearson, R.B. Foster, and W.E. Bishop (eds.), American Society for Testing and Materials, Philadelphia, p. 170-178.

Temperature: 15°C
Test parameter: EC50
Effect: Reduction in light output

Concentration: 3426 mg/l
Exposure time: 5 min

Concentration: 3403 mg/l
Exposure time: 30 min

Bibliographical reference: Speece, R. (1987). Drexel University, Philadelphia, USA, private communication.

2-BUTANONE OXIME

CAS RN: 96-29-7

Test parameter: EC50
Effect: Reduction in light output
Concentration: 950 mg/l
Exposure time: 5 min
Comment: The EC50 value was calculated from nominal concentrations.

Bibliographical reference: Curtis, C., Lima, A., Lozano, S.J., and Veith, G.D. (1982). In: *Aquatic Toxicology and Hazard Assessment: Fifth Conference, ASTM STP 766*, J.G. Pearson, R.B. Foster, and W.E. Bishop (eds.), American Society for Testing and Materials, Philadelphia, p. 170-178.

BUTYL ACETATE

CAS RN: 123-86-4

Sample purity: 99+%
Temperature: 15°C
Test parameter: EC50
Effect: Reduction in light output

Concentration: 100 mg/l
Exposure time: 5 min

Concentration: 130 mg/l*
Exposure time: 15 min

Bibliographical references: Cronin, M.T.D., Dearden, J.C., and Dobbs, A.J. (1991). *Sci. Total Environ.* **109/110**, 431-439.
* Cronin, M.T.D. (1993). Liverpool John Moores University, UK, private communication.

Sample purity: 99%

Temperature: 15°C
Test parameter: EC50
Effect: Reduction in light output

Concentration: 70.0 mg/l
Exposure time: 5 min

Concentration: 82.2 mg/l
Exposure time: 15 min

Concentration: 98.9 mg/l
Exposure time: 30 min

Comment: Mean of three assays. Methanol (<10%) was used to prepare the stock solutions. EC50 values were calculated from nominal concentrations.

Bibliographical reference: Kaiser, K.L.E., and Palabrica, V.S. (1991). *Water Poll. Res. J. Canada* **26**, 361-431.

tert-BUTYL ACETATE

CAS RN: 540-88-5

Sample purity: 99+%
Temperature: 15°C
Test parameter: EC50
Effect: Reduction in light output

Concentration: 6.38 mg/l
Exposure time: 5 min

Concentration: 8.04 mg/l
Exposure time: 15 min

Concentration: 11.1 mg/l
Exposure time: 30 min

Comment: Mean of four assays. Methanol (<10%) was used to prepare the stock solutions. EC50 values were calculated from nominal concentrations.

Bibliographical reference: Kaiser, K.L.E., and Palabrica, V.S. (1992). National Water Research Institute, Burlington, Ontario, Canada, unpublished results.

BUTYL ACRYLATE

CAS RN: 141-32-2

Temperature: 15°C
Test parameter: EC50
Effect: Reduction in light output

Concentration: 37 mg/l
Exposure time: 5 min

Concentration: 35 mg/l
Exposure time: 15 min

Concentration: 31 mg/l
Exposure time: 30 min

Comment: Chemical was diluted to give five sample concentrations in duplicate.

Bibliographical reference: Benson, W.H., and Stackhouse, R.A. (1986). *Drug Chem. Toxicol.* **9**, 275-283.

n-BUTYLAMINE

CAS RN: 109-73-9

Sample purity: >99%
Temperature: 15°C
Test parameter: EC50
Effect: Reduction in light output

Concentration: 41.1 mg/l
Exposure time: 5 min

Concentration: 28.5 mg/l
Exposure time: 15 min

Concentration: 24.8 mg/l
Exposure time: 30 min

Comment: Mean of three assays. Methanol (<10%) was used to prepare the stock solutions. EC50 values were calculated from nominal concentrations.

Bibliographical reference: Kaiser, K.L.E., and Palabrica, V.S. (1992). National Water Research Institute, Burlington, Ontario, Canada, unpublished results.

BUTYLATE

CAS RN: 2008-41-5

Sample purity: 98%
Temperature: 15°C
Test parameter: EC50
Effect: Reduction in light output

Concentration: 15.0 mg/l
Exposure time: 5 min

Concentration: 16.1 mg/l
Exposure time: 15 min

Concentration: 18.1 mg/l
Exposure time: 30 min

Comment: Mean of four assays. Methanol (<10%) was used to prepare the stock solutions. EC50 values were calculated from nominal concentrations.

Bibliographical reference: Kaiser, K.L.E., and Palabrica, V.S. (1991). *Water Poll. Res. J. Canada* **26**, 361-431.

n-BUTYL ETHER

CAS RN: 142-96-1

Test parameter: EC50
Effect: Reduction in light output
Concentration: 63.0 mg/l
Exposure time: 5 min
Comment: Concentrations in the test were measured.

Bibliographical reference: Curtis, C., Lima, A., Lozano, S.J., and Veith, G.D. (1982). In: *Aquatic Toxicology and Hazard Assessment: Fifth Conference, ASTM STP 766*, J.G. Pearson, R.B. Foster, and W.E. Bishop (eds.), American Society for Testing and Materials, Philadelphia, p. 170-178.

BUTYL METHACRYLATE

CAS RN: 97-88-1

Temperature: 15°C
Test parameter: EC50
Effect: Reduction in light output

Concentration: 37 mg/l
Exposure time: 5 min

Concentration: 49 mg/l
Exposure time: 15 min

Concentration: 55 mg/l
Exposure time: 30 min

Comment: Chemical was diluted to give five sample concentrations in duplicate.

Bibliographical reference: Benson, W.H., and Stackhouse, R.A. (1986). *Drug Chem. Toxicol.* **9**, 275-283.

tert-BUTYL METHYL ETHER

CAS RN: 1634-04-4

Sample purity: >99%
Temperature: 15°C
Test parameter: EC50
Effect: Reduction in light output

Concentration: 8.23 mg/l
Exposure time: 5 min

Concentration: 9.67 mg/l

Exposure time: 15 min

Concentration: 11.4 mg/l
Exposure time: 30 min

Comment: Mean of three assays. Methanol (<10%) was used to prepare the stock solutions. EC50 values were calculated from nominal concentrations.

Bibliographical reference: Kaiser, K.L.E., and Palabrica, V.S. (1991). *Water Poll. Res. J. Canada* **26**, 361-431.

4-*tert*-BUTYLPHENOL

CAS RN: 98-54-4

Test parameter: EC50
Effect: Reduction in light output
Concentration: 0.21 mg/l
Exposure time: 5 min
Comment: Concentrations in the test were measured.

Bibliographical reference: Curtis, C., Lima, A., Lozano, S.J., and Veith, G.D. (1982). In: *Aquatic Toxicology and Hazard Assessment: Fifth Conference, ASTM STP 766*, J.G. Pearson, R.B. Foster, and W.E. Bishop (eds.), American Society for Testing and Materials, Philadelphia, p. 170-178.

BUTYLTHIOSTANNOUS ACID

CAS RN: 26410-42-4

Test parameter: EC50
Effect: Reduction in light output
Concentration: >0.20 mg/l
Exposure time: 30 min

Bibliographical reference: Steinhäuser, K.G., Amann, W., Späth, A., and Polenz, A. (1985). *Vom Wasser* **65**, 203-214.

2-BUTYNE

CAS RN: 503-17-3
Synonym: Crotonylene

Sample purity: 99%
Temperature: 15°C
Test parameter: EC50
Effect: Reduction in light output

Concentration: 81.9 mg/l
Exposure time: 5 min

Concentration: 69.7 mg/l
Exposure time: 15 min

Concentration: 48.2 mg/l
Exposure time: 30 min

Comment: Mean of three assays. Methanol (<10%) was used to prepare the stock solutions. EC50 values were calculated from nominal concentrations.

Bibliographical reference: Kaiser, K.L.E., and Palabrica, V.S. (1991). *Water Poll. Res. J. Canada* **26**, 361-431.

BUTYRIC ACID

CAS RN: 107-92-6

Sample purity: 99+%
Temperature: 15°C
Test parameter: EC50
Effect: Reduction in light output

Concentration: 16.9 ± 0.52 mg/l
Exposure time: 5 min

Concentration: 16.9 ± 0.48 mg/l
Exposure time: 15 min

Concentration: 17.2 ± 0.62 mg/l
Exposure time: 25 min

Comment: Mean of six assays. The values were converted to mg/l from the original data expressed in μM and a rounded molecular weight of 88 given by the authors. Phenol solution was used for quality control/quality assurance. The 5-min EC50 value was 18.2 mg/l. The EC50 value at 15 min was 20.7 mg/l with a relative error of <5%.

Bibliographical reference: Chou, C.C., and Que Hee, S.S. (1992). *Ecotoxicol. Environ. Safety* **23**, 355-363.

CADMIUM ACETATE DIHYDRATE

CAS RN: 5743-04-4

Sample purity: Analytical grade
Temperature: 15°C
Test parameter: EC50
Effect: Reduction in light output

Concentration: 154 mg/l Cd^{++}
Exposure time: 5 min

Concentration: 70.8 mg/l Cd^{++}
Exposure time: 10 min

Concentration: 41.4 mg/l Cd^{++}
Exposure time: 15 min

Concentration: 27.2 mg/l Cd^{++}
Exposure time: 20 min

Concentration: 14.9 mg/l Cd^{++}
Exposure time: 30 min

Comment: Results derived from the average of two replicates.

Bibliographical reference: Qureshi, A.A., Coleman, R.N., and Paran, J.H. (1984). In: *Toxicity Screening Procedures Using Bacterial Systems*, D. Liu and B.J. Dutka (eds.), Marcel Dekker, New York, p. 1-22.

Temperature: 15°C
Test parameter: EC50
Effect: Reduction in light output

Result:

	15 min	30 min
Microbics strain	13.9 ± 7.8 mg/l Cd^{++}	5.3 ± 4.1 mg/l Cd^{++}
	5.9 ± 3.3 mg/l Cd^{++}	2.2 ± 1.7 mg/l Cd^{++}
Dr. Lange strain	23.6 ± 3.3 mg/l Cd^{++}	9.15 ± 3.0 mg/l Cd^{++}
	10.0 ± 1.4 mg/l Cd^{++}	3.9 ± 1.3 mg/l Cd^{++}

Comment: EC50 values were calculated from nominal concentrations.

Bibliographical reference: Vasseur, P. (1992). Centre des Sciences de l'Environnement, Metz, France, private communication.

CADMIUM CHLORIDE

CAS RN: 10108-64-2

Sample purity: Reagent grade
Test parameter: EC50
Effect: Reduction in light output

Concentration: 98 mg/l Cd^{++} (72-130)
Exposure time: 5 min

Concentration: 17 mg/l Cd^{++} (16-18)
Exposure time: 15 min

Concentration: 5.4 mg/l Cd^{++} (4.5-6.1)
Exposure time: 30 min

Bibliographical reference: Elnabarawy, M.T., Robideau, R.R., and Beach, S.A. (1988). *Tox. Assess.* **3**, 361-370.

CADMIUM NITRATE TETRAHYDRATE

CAS RN: 10022-68-1

Temperature: 15°C
Test parameter: EC50
Effect: Reduction in light output
Concentration: 23.6 ± 3.9 mg/l Cd^{++}
Exposure time: 10 min
Comment: Mean of four assays. The EC50 values were calculated from

nominal concentrations.

Bibliographical reference: Ferard, J.F., Vasseur, P., Danoux, L., and Larbaigt, G. (1983). *Rev. Fr. Sci. Eau* **2**, 221-237.

Temperature: 20°C
Test parameter: EC50
Effect: Reduction in light output
Concentration: 12.9 ± 3.5 mg/l Cd^{++}
Exposure time: 10 min
Comment: Mean of five assays. The EC50 values were calculated from nominal concentrations.

Bibliographical reference: Ferard, J.F., Vasseur, P., Danoux, L., and Larbaigt, G. (1983). *Rev. Fr. Sci. Eau* **2**, 221-237.

Temperature: 20°C
Test parameter: EC50
Effect: Reduction in light output
Concentration: 2.46 ± 0.57 mg/l Cd^{++}
Exposure time: 30 min
Comment: Toxicity values were calculated from nominal concentrations.

Bibliographical reference: Vasseur, P., Bois, F., Ferard, J.F., Rast, C., and Larbaigt, G. (1986). *Tox. Assess.* **1**, 283-300.

Temperature: 20°C
Test parameter: EC50
Effect: Reduction in light output

Concentration: 1.78 mg/l Cd^{++}
Exposure time: 30 min
Comment: pH = 6.

Concentration: 1.92 mg/l Cd^{++}
Exposure time: 30 min
Comment: pH = 6.5.

Concentration: 1.74 mg/l Cd^{++}
Exposure time: 30 min
Comment: pH = 7.

Concentration: 1.62 mg/l Cd^{++}
Exposure time: 30 min
Comment: pH = 7.5.

Bibliographical reference: Vasseur, P., Bois, F., Ferard, J.F., Rast, C., and Larbaigt, G. (1986). *Tox. Assess.* **1**, 283-300.

CAFFEINE

CAS RN: 58-08-2

Temperature: 15°C
Test parameter: EC50
Effect: Reduction in light output
Concentrations: 603 mg/l
707 mg/l
Exposure time: 5 min

Bibliographical reference: King, E.F., and Painter, H.A. (1981). In: *Les Tests de Toxicité Aiguë en Milieu Aquatique. Acute Aquatic Ecotoxicological Tests*, H. Leclerc and D. Dive (eds.), INSERM 106, Paris, p. 143-153.

CAPTAFOL

CAS RN: 2425-06-1
Synonyms: 3a,4,7,7a-Tetrahydro-*N*-(1,1,2,2-tetrachloroethanesulfenyl)phthalimide; *N*-(1,1,2,2-Tetrachloroethylthio)cyclohex-4-ene-1,2-dicarboximide; 3a,4,7,7a-Tetrahydro-2-[(1,1,2,2-tetrachloroethyl)thio]-1*H*-isoindole-1,3(2*H*)-dione

Temperature: 15 ± 0.1°C
Test parameter: EC50
Effect: Reduction in light output
Concentration: 7 mg/l
Exposure time: 5 min

Bibliographical reference: Chang, J.C., Taylor, P.B., and Leach, F.R. (1981). *Bull. Environ. Contam. Toxicol.* **26**, 150-156.

CARBARYL

CAS RN: 63-25-2
Synonym: 1-Naphthalenyl methylcarbamate

Temperature: 15 ± 0.1°C
Test parameter: EC50
Effect: Reduction in light output
Concentration: 2 mg/l
Exposure time: 5 min

Bibliographical reference: Chang, J.C., Taylor, P.B., and Leach, F.R. (1981). *Bull. Environ. Contam. Toxicol.* **26**, 150-156.

Sample purity: 98%
Temperature: 15°C
Test parameter: EC50
Effect: Reduction in light output
Concentration: 0.63 ± 0.15 mg/l
Exposure time: 30 min
Comment: The assay was run in triplicate (pH = 6.2-6.6) using different bacterial reagents. The toxicity values were calculated from nominal concentrations. Carbaryl displayed no interactive effects with copper ($CuSO_4.5H_2O$) even at relatively high doses (0.72 mg/l carbaryl and 0.27 mg/l Cu^{2+}).

Bibliographical reference: Vasseur, P., Dive, D., Sokar, Z., and Bonnemain, H. (1988). *Chemosphere* **17**, 767-782.

Temperature: 15 ± 0.1°C
Test parameter: EC50
Effect: Reduction in light output
Concentration: 5.0 mg/l
Exposure time: 5 min

Bibliographical reference: Somasundaram, L., Coats, J.R., Racke, K.D., and Stahr, H.M. (1990). *Bull. Environ. Contam. Toxicol.* **44**, 254-259.

CARBAZOLE

CAS RN: 86-74-8

Sample purity: 99%
Temperature: 15°C
Test parameter: EC50
Effect: Reduction in light output

Concentration: 13.6 mg/l
Exposure time: 5 min

Concentration: 10.6 mg/l
Exposure time: 15 min

Concentration: 11.6 mg/l
Exposure time: 30 min

Comment: Mean of four assays. Methanol (<10%) was used to prepare the stock solutions. EC50 values were calculated from nominal concentrations.

Bibliographical reference: Kaiser, K.L.E., and Palabrica, V.S. (1991). *Water Poll. Res. J. Canada* **26**, 361-431.

CARBOFURAN

CAS RN: 1563-66-2
Synonym: 2,3-Dihydro-2,2-dimethyl-7-benzofuranyl methylcarbamate

Temperature: 15 ± 0.1°C
Test parameter: EC50
Effect: Reduction in light output
Concentration: 20.5 mg/l
Exposure time: 5 min

Bibliographical reference: Somasundaram, L., Coats, J.R., Racke, K.D., and Stahr, H.M. (1990). *Bull. Environ. Contam. Toxicol.* **44**, 254-259.

CARBOFURAN PHENOL

CAS RN: 1563-38-8

Temperature: 15 ± 0.1°C
Test parameter: EC50

Effect: Reduction in light output
Concentration: 60.9 mg/l
Exposure time: 5 min

Bibliographical reference: Somasundaram, L., Coats, J.R., Racke, K.D., and Stahr, H.M. (1990). *Bull. Environ. Contam. Toxicol.* **44**, 254-259.

CARBON DISULFIDE

CAS RN: 75-15-0

Sample purity: ≥99%
Test parameter: EC50
Effect: Reduction in light output
Concentration: 341 mg/l (260-448)
Exposure time: 15 min

Bibliographical reference: van Leeuwen, C.J., Maas-Diepeveen, J.L., Niebeek, G., Vergouw, W.H.A., Griffioen, P.S., and Luijken, M.W. (1985). *Aquat. Toxicol.* **7**, 145-164.

CARBON TETRACHLORIDE

CAS RN: 56-23-5
Synonym: Tetrachloromethane

Temperature: 15°C
Test parameter: EC50
Effect: Reduction in light output
Concentration: 5.6 mg/l
Exposure time: 5 min

Bibliographical reference: Microtox® Application Notes (1982). Beckman Instruments, Inc., Carlsbad, California.

Temperature: 15°C
Test parameter: EC50
Effect: Reduction in light output
Concentration: 34 mg/l
Exposure time: 10 min

Comment: Mean of six assays. Test solutions were analyzed using gas chromatography.

Bibliographical reference: Bazin, C., Chambon, P., Bonnefille, M., and Larbaigt, G. (1987). *Sci. Eau* **6**, 403-413.

4-CARBOXYBENZENESULFONAMIDE

CAS RN: 138-41-0

Sample purity: 99%
Temperature: 15°C
Test parameter: EC50
Effect: Reduction in light output

Concentration: 156 mg/l
Exposure time: 5 min

Concentration: 153 mg/l
Exposure time: 15 min

Concentration: 164 mg/l
Exposure time: 30 min

Comment: Mean of three assays. Methanol (<10%) was used to prepare the stock solutions. EC50 values were calculated from nominal concentrations.

Bibliographical reference: Kaiser, K.L.E., and Palabrica, V.S. (1991). *Water Poll. Res. J. Canada* **26**, 361-431.

CATECHOL

CAS RN: 120-80-9

Temperature: 15°C
Test parameter: EC50
Effect: Reduction in light output
Concentration: 32 mg/l
Exposure time: 5 min

Bibliographical reference: Lebsack, M.E., Anderson, A.D.,

DeGraeve, G.M., and Bergman, H.L. (1981). In: *Aquatic Toxicology and Hazard Assessment: Fourth Conference*, *ASTM STP 737*, D.R. Branson and K.L. Dickson (eds.), American Society for Testing and Materials, Philadelphia, p. 348-356.

Temperature: 15°C
Test parameter: EC50
Effect: Reduction in light output

Concentration: 32.0 mg/l
Exposure time: 5 min

Concentration: 29.7 mg/l
Exposure time: 30 min

Bibliographical reference: Speece, R. (1987). Drexel University, Philadelphia, USA, private communication.

CEQUARTYL A

CAS RN: 139-07-1

Test parameter: EC50
Effect: Reduction in light output

Concentration: 1.62 mg/l
Exposure time: 5 min

Concentration: 1.43 mg/l
Exposure time: 10 min

Bibliographical reference: Herschke, B., and Lhotellier, D. (1983). *Eau Industrie Nuisances* **75**, 68-72.

CEQUARTYL O

Test parameter: EC50
Effect: Reduction in light output

Concentrations: 1.82 mg/l
1.43 mg/l

1.70 mg/l
Exposure time: 5 min

Concentrations: 1.44 mg/l
0.79 mg/l
1.18 mg/l
Exposure time: 10 min

Bibliographical reference: Herschke, B., and Lhotellier, D. (1983). *Eau Industrie Nuisances* **75**, 68-72.

CETYLTRIMETHYLAMMONIUM BROMIDE

CAS RN: 57-09-0
Synonym: Cetrimonium bromide

Sample purity: Analytical grade
Temperature: 15°C
Test parameter: EC50
Effect: Reduction in light output
Concentration: 9.84 mg/l
Exposure time: 5 min

Bibliographical reference: Beaubien, A., Lapierre, L., Bouchard, A., and Jolicoeur, C. (1986). *Tox. Assess.* **1**, 187-200.

CETYLTRIMETHYLAMMONIUM CHLORIDE

CAS RN: 112-02-7
Synonym: Hexadecyltrimethylammonium chloride

Temperature: 15°C
Test parameter: EC50
Effect: Reduction in light output
Concentration: 0.80 mg/l
Exposure time: 5 min

Bibliographical reference: King, E.F., and Painter, H.A. (1981). In: *Les Tests de Toxicité Aiguë en Milieu Aquatique. Acute Aquatic Ecotoxicological Tests*, H. Leclerc and D. Dive (eds.), INSERM 106, Paris, p. 143-153.

Temperature: 15°C
Test parameter: EC50
Effect: Reduction in light output

Concentration: 1.35 mg/l
Exposure time: 5 min

Concentration: 0.98 mg/l
Exposure time: 10 min

Concentration: 0.86 mg/l
Exposure time: 15 min

Bibliographical reference: Dutka, B.J., Nyholm, N., and Petersen, J. (1983). *Water Res.* **17**, 1363-1368.

2-CHLOROACETAMIDE

CAS RN: 79-07-2

Sample purity: 98%
Temperature: 15°C
Test parameter: EC50
Effect: Reduction in light output

Concentration: 10.3 mg/l
Exposure time: 5 min

Concentration: 19.5 mg/l
Exposure time: 15 min

Concentration: 31.7 mg/l
Exposure time: 30 min

Comment: Mean of three assays. Methanol (<10%) was used to prepare the stock solutions. EC50 values were calculated from nominal concentrations.

Bibliographical reference: Kaiser, K.L.E., and Palabrica, V.S. (1991). *Water Poll. Res. J. Canada* **26**, 361-431.

CHLOROACETONE

CAS RN: 78-95-5
Synonym: Chloro-2-propanone

Sample purity: 90%
Temperature: 15°C
Test parameter: EC50
Effect: Reduction in light output

Concentration: 27.3 mg/l
Exposure time: 5 min

Concentration: 9.91 mg/l
Exposure time: 15 min

Concentration: 5.84 mg/l
Exposure time: 30 min

Comment: Mean of three assays. Methanol (<10%) was used to prepare the stock solutions. EC50 values were calculated from nominal concentrations.

Bibliographical reference: Kaiser, K.L.E., and Palabrica, V.S. (1991). *Water Poll. Res. J. Canada* **26**, 361-431.

CHLOROACETONITRILE

CAS RN: 107-14-2

Sample purity: 98+%
Temperature: 15°C
Test parameter: EC50
Effect: Reduction in light output

Concentration: 720 mg/l
Exposure time: 5 min

Concentration: 210 mg/l
Exposure time: 15 min

Bibliographical reference: Cronin, M.T.D. (1993). Liverpool John Moores University, UK, private communication.

4'-CHLOROACETOPHENONE

CAS RN: 99-91-2
Synonym: 1-(4-Chlorophenyl)ethanone

Sample purity: 97%
Temperature: 15°C
Test parameter: EC50
Effect: Reduction in light output

Concentration: 6.75 mg/l
Exposure time: 5 min

Concentration: 7.07 mg/l
Exposure time: 15 min

Concentration: 6.91 mg/l*
Exposure time: 30 min

Comment: Mean of three assays. Methanol (<10%) was used to prepare the stock solutions. EC50 values were calculated from nominal concentrations.

Bibliographical references: Kaiser, K.L.E., and Palabrica, V.S. (1991). *Water Poll. Res. J. Canada* **26**, 361-431.
* Kaiser, K.L.E., Ribo, J.M., and Zaruk, B.M. (1985). *Water Poll. Res. J. Canada* **20**, 36-43.

4-CHLORO-*N*-ACETYLANILINE

CAS RN: 539-03-7

Temperature: 15°C
Test parameter: EC50
Effect: Reduction in light output

Concentration: 38.0 mg/l
Exposure time: 5 min

Concentration: 44.6 mg/l
Exposure time: 15 min

Concentration: 48.9 mg/l*
Exposure time: 30 min

Comment: Mean of three assays. Methanol (<10%) was used to prepare the stock solutions. EC50 values were calculated from nominal concentrations.

Bibliographical references: Kaiser, K.L.E., and Palabrica, V.S. (1991). *Water Poll. Res. J. Canada* **26**, 361-431.
* Kaiser, K.L.E., Ribo, J.M., and Zaruk, B.M. (1985). *Water Poll. Res. J. Canada* **20**, 36-43.

2-CHLOROANILINE

CAS RN: 95-51-2

Test parameter: EC50
Effect: Reduction in light output
Concentrations: 16 mg/l
17 mg/l
Exposure time: 5 min

Bibliographical reference: King, E.F., and Painter, H.A. (1981). In: *Les Tests de Toxicité Aiguë en Milieu Aquatique. Acute Aquatic Ecotoxicological Tests*, H. Leclerc and D. Dive (eds.), INSERM 106, Paris; p. 143-153.

Temperature: 15°C
Test parameter: EC50
Effect: Reduction in light output

Concentration: 14.3 mg/l
Exposure time: 5 min

Concentration: 15.0 mg/l
Exposure time: 15 min

Concentration: 15.7 mg/l
Exposure time: 30 min

Comment: Mean of three assays. Methanol (<10%) was used to prepare the stock solutions. EC50 values were calculated from nominal concentrations.

Bibliographical reference: Ribo, J.M., and Kaiser, K.L.E. (1984). In: *QSAR in Environmental Toxicology*, K.L.E. Kaiser (ed.), D. Reidel

Publishing Company, Dordrecht, p. 319-336.

Temperature: 20°C
Test parameter: EC50
Effect: Reduction in light output
Concentration: 36.5 mg/l
Exposure time: 15 min
Comment: Chemical dissolved in DMSO (1%). The EC50 value was calculated from nominal concentrations.

Bibliographical reference: Vasseur, P. (1992). Centre des Sciences de l'Environnement, Metz, France, private communication.

3-CHLOROANILINE

CAS RN: 108-42-9

Temperature: 15°C
Test parameter: EC50
Effect: Reduction in light output

Concentration: 12.5 mg/l
Exposure time: 5 min

Concentration: 13.4 mg/l
Exposure time: 15 min

Concentration: 14.0 mg/l
Exposure time: 30 min

Comment: Mean of three assays. Methanol (<10%) was used to prepare the stock solutions. EC50 values were calculated from nominal concentrations.

Bibliographical reference: Ribo, J.M., and Kaiser, K.L.E. (1984). In: *QSAR in Environmental Toxicology*, K.L.E. Kaiser (ed.), D. Reidel Publishing Company, Dordrecht, p. 319-336.

Temperature: 20°C
Test parameter: EC50
Effect: Reduction in light output
Concentration: 39.5 mg/l

Exposure time: 15 min

Comment: Chemical dissolved in DMSO (1%). The EC50 value was calculated from nominal concentrations.

Bibliographical reference: Vasseur, P. (1992). Centre des Sciences de l'Environnement, Metz, France, private communication.

4-CHLOROANILINE

CAS RN: 106-47-8

Temperature: 15°C
Test parameter: EC50
Effect: Reduction in light output

Concentration: 3.20 mg/l
Exposure time: 5 min

Concentration: 3.77 mg/l
Exposure time: 15 min

Concentration: 5.08 mg/l
Exposure time: 30 min

Comment: Mean of three assays. Methanol (<10%) was used to prepare the stock solutions. EC50 values were calculated from nominal concentrations.

Bibliographical reference: Ribo, J.M., and Kaiser, K.L.E. (1984). In: *QSAR in Environmental Toxicology*, K.L.E. Kaiser (ed.), D. Reidel Publishing Company, Dordrecht, p. 319-336.

Temperature: 20°C
Test parameter: EC50
Effect: Reduction in light output
Concentration: 21 mg/l
Exposure time: 15 min
Comment: The EC50 value was calculated from nominal concentrations.

Bibliographical reference: Vasseur, P. (1992). Centre des Sciences de l'Environnement, Metz, France, private communication.

4-CHLOROANISOLE

CAS RN: 623-12-1

Temperature: 15°C
Test parameter: EC50
Effect: Reduction in light output

Concentration: 3.19 mg/l
Exposure time: 5 min

Concentration: 3.50 mg/l
Exposure time: 15 min

Concentration: 3.58 mg/l*
Exposure time: 30 min

Comment: Methanol (<10%) was used to prepare the stock solution. EC50 values were calculated from nominal concentrations.

Bibliographical references: Kaiser, K.L.E., and Palabrica, V.S. (1991). *Water Poll. Res. J. Canada* **26**, 361-431.
* Kaiser, K.L.E., Ribo, J.M., and Zaruk, B.M. (1985). *Water Poll. Res. J. Canada* **20**, 36-43.

4-CHLOROBENZALDEHYDE

CAS RN: 104-88-1

Temperature: 15°C
Test parameter: EC50
Effect: Reduction in light output

Concentration: 10.2 mg/l
Exposure time: 5 min

Concentration: 10.9 mg/l
Exposure time: 15 min

Concentration: 10.4 mg/l*
Exposure time: 30 min

Comment: Mean of three assays. Methanol (<10%) was used to prepare the stock solutions. EC50 values were calculated from nominal

concentrations.

Bibliographical references: Kaiser, K.L.E., and Palabrica, V.S. (1991). *Water Poll. Res. J. Canada* **26**, 361-431.
* Kaiser, K.L.E., Ribo, J.M., and Zaruk, B.M. (1985). *Water Poll. Res. J. Canada* **20**, 36-43.

CHLOROBENZENE

CAS RN: 108-90-7

Temperature: 15°C
Test parameter: EC50
Effect: Reduction in light output
Concentration: 14.6 ± 0.9 mg/l
Exposure time: 5 min
Comment: Mean of four assays. Toxicity values were calculated from nominal concentrations.

Bibliographical reference: Ferard, J.F., Vasseur, P., Danoux, L., and Larbaigt, G. (1983). *Rev. Fr. Sci. Eau* **2**, 221-237.

Temperature: 15°C
Test parameter: EC50
Effect: Reduction in light output
Concentration: 18.0 mg/l
Exposure time: 10 min
Comment: Mean of two assays. Toxicity values were calculated from nominal concentrations.

Bibliographical reference: Ferard, J.F., Vasseur, P., Danoux, L., and Larbaigt, G. (1983). *Rev. Fr. Sci. Eau* **2**, 221-237.

Temperature: 15°C
Test parameter: EC50
Effect: Reduction in light output

Concentration: 9.36 mg/l
Exposure time: 5 min

Concentration: 11.5 mg/l
Exposure time: 15 min

Concentration: 11.3 mg/l
Exposure time: 30 min

Comment: Mean of three assays. Methanol (<10%) was used to prepare the stock solutions. EC50 values were calculated from nominal concentrations.

Bibliographical reference: Ribo, J.M., and Kaiser, K.L.E. (1983). *Chemosphere* **12**, 1421-1442.

Temperature: 15°C
Test parameter: EC50
Effect: Reduction in light output
Concentration: 14.8 mg/l
Exposure time: 15 min

Bibliographical reference: Hermens, J., Busser, F., Leeuwangh, P., and Musch, A. (1985). *Ecotoxicol. Environ. Safety* **9**, 17-25.

Temperature: 15°C
Test parameter: EC50
Effect: Reduction in light output
Concentration: 20 mg/l
Exposure time: 10 min
Comment: Mean of three assays. Test solutions were analyzed using gas chromatography.

Bibliographical reference: Bazin, C., Chambon, P., Bonnefille, M., and Larbaigt, G. (1987). *Sci. Eau* **6**, 403-413.

4-CHLOROBENZENESULFONAMIDE

CAS RN: 98-64-6

Temperature: 15°C
Test parameter: EC50
Effect: Reduction in light output

Concentration: 69.6 mg/l
Exposure time: 5 min

Concentration: 83.7 mg/l

Exposure time: 15 min

Concentration: 101 mg/l*
Exposure time: 30 min

Comment: Mean of three assays. Methanol (<10%) was used to prepare the stock solutions. EC50 values were calculated from nominal concentrations.

Bibliographical references: Kaiser, K.L.E., and Palabrica, V.S. (1991). *Water Poll. Res. J. Canada* **26**, 361-431.
* Kaiser, K.L.E., Ribo, J.M., and Zaruk, B.M. (1985). *Water Poll. Res. J. Canada* **20**, 36-43.

4-CHLOROBENZOIC ACID

CAS RN: 74-11-3

Temperature: 15°C
Test parameter: EC50
Effect: Reduction in light output

Concentration: 5.68 mg/l
Exposure time: 5 min

Concentration: 6.23 mg/l
Exposure time: 15 min

Concentration: 6.68 mg/l*
Exposure time: 30 min

Comment: Mean of three assays. Methanol (<10%) was used to prepare the stock solutions. EC50 values were calculated from nominal concentrations.

Bibliographical references: Kaiser, K.L.E., and Palabrica, V.S. (1991). *Water Poll. Res. J. Canada* **26**, 361-431.
* Kaiser, K.L.E., Ribo, J.M., and Zaruk, B.M. (1985). *Water Poll. Res. J. Canada* **20**, 36-43.

4-CHLOROBENZOIC ACID AMIDE

CAS RN: 619-56-7

Temperature: 15°C
Test parameter: EC50
Effect: Reduction in light output

Concentration: 42.9 mg/l
Exposure times: 5 and 15 min

Concentration: 43.8 mg/l*
Exposure time: 30 min

Comment: Mean of three assays. Methanol (<10%) was used to prepare the stock solutions. EC50 values were calculated from nominal concentrations.

Bibliographical references: Kaiser, K.L.E., and Palabrica, V.S. (1991). *Water Poll. Res. J. Canada* **26**, 361-431.
* Kaiser, K.L.E., Ribo, J.M., and Zaruk, B.M. (1985). *Water Poll. Res. J. Canada* **20**, 36-43.

4-CHLOROBENZOIC ACID HYDRAZIDE

CAS RN: 536-40-3

Temperature: 15°C
Test parameter: EC50
Effect: Reduction in light output

Concentration: 59.2 mg/l
Exposure time: 5 min

Concentration: 64.9 mg/l
Exposure time: 15 min

Concentration: 60.5 mg/l*
Exposure time: 30 min

Comment: Mean of three assays. Methanol (<10%) was used to prepare the stock solutions. EC50 values were calculated from nominal concentrations.

Bibliographical references: Kaiser, K.L.E., and Palabrica, V.S. (1991). *Water Poll. Res. J. Canada* **26**, 361-431.
* Kaiser, K.L.E., Ribo, J.M., and Zaruk, B.M. (1985). *Water Poll. Res. J. Canada* **20**, 36-43.

4-CHLOROBENZONITRILE

CAS RN: 623-03-0

Temperature: 15°C
Test parameter: EC50
Effect: Reduction in light output

Concentration: 4.06 mg/l
Exposure time: 5 min

Concentration: 3.88 mg/l
Exposure time: 15 min

Concentration: 4.45 mg/l*
Exposure time: 30 min

Comment: Mean of four assays. Methanol (<10%) was used to prepare the stock solutions. EC50 values were calculated from nominal concentrations.

Bibliographical references: Kaiser, K.L.E., and Palabrica, V.S. (1991). *Water Poll. Res. J. Canada* **26**, 361-431.
* Kaiser, K.L.E., Ribo, J.M., and Zaruk, B.M. (1985). *Water Poll. Res. J. Canada* **20**, 36-43.

4-CHLOROBENZOPHENONE

CAS RN: 134-85-0

Temperature: 15°C
Test parameter: EC50
Effect: Reduction in light output

Concentration: 1.06 mg/l
Exposure time: 5 min

Concentration: 1.22 mg/l
Exposure time: 15 min

Concentration: 1.40 mg/l*
Exposure time: 30 min

Comment: Mean of three assays. Methanol (<10%) was used to prepare the stock solutions. EC50 values were calculated from nominal concentrations.

Bibliographical references: Kaiser, K.L.E., and Palabrica, V.S. (1991). *Water Poll. Res. J. Canada* **26**, 361-431.
* Kaiser, K.L.E., Ribo, J.M., and Zaruk, B.M. (1985). *Water Poll. Res. J. Canada* **20**, 36-43.

4-CHLOROBENZOTRIFLUORIDE

CAS RN: 98-56-6
Synonym: 4-Chloro-α,α,α-trifluorotoluene

Temperature: 20°C

Test parameter: EC20
Effect: Reduction in light output
Concentration: 0.78 mg/l (0.69-0.88)
Exposure time: 5 min

Test parameter: EC50
Effect: Reduction in light output
Concentration: 2.78 mg/l (2.61-2.96)
Exposure time: 5 min

Bibliographical reference: Casseri, N.A., Ying, W, and Sojka, S.A. (1983). Proceedings 38th Industrial Waste Conference, USA, May 1983, p. 867-878.

Temperature: 20°C

Test parameter: EC20
Effect: Reduction in light output
Concentration: 0.97 mg/l
Exposure time: 15 min

Test parameter: EC50
Effect: Reduction in light output
Concentration: 3.57 mg/l
Exposure time: 15 min

Bibliographical reference: Casseri, N.A., Ying, W, and Sojka, S.A. (1983). Proceedings 38th Industrial Waste Conference, USA, May 1983, p. 867-878.

Temperature: 20°C
Test parameter: EC50
Effect: Reduction in light output

Concentration: 2.75 mg/l (2.58-2.93)
Exposure time: 5 min

Concentration: 3.54 mg/l (3.36-3.73)
Exposure time: 15 min

Bibliographical reference: Casseri, N.A. (1985). Occidental Chemical Corporation, Grand Island, New York, USA, private communication.

Temperature: 15°C
Test parameter: EC50
Effect: Reduction in light output

Concentration: 11.1 mg/l
Exposure time: 5 min

Concentration: 13.4 mg/l
Exposure time: 15 min

Concentration: 14.3 mg/l*
Exposure time: 30 min

Comment: Mean of three assays. Methanol (<10%) was used to prepare the stock solutions. EC50 values were calculated from nominal concentrations.

Bibliographical references: Kaiser, K.L.E., and Palabrica, V.S. (1991). *Water Poll. Res. J. Canada* **26**, 361-431.
* Kaiser, K.L.E., Ribo, J.M., and Zaruk, B.M. (1985). *Water Poll. Res. J. Canada* **20**, 36-43.

4-CHLOROBENZOYL CHLORIDE

CAS RN: 122-01-0

Temperature: 15°C
Test parameter: EC50
Effect: Reduction in light output

Concentration: 4.82 mg/l
Exposure time: 5 min

Concentration: 5.80 mg/l
Exposure time: 15 min

Concentration: 5.80 mg/l*
Exposure time: 30 min

Comment: Mean of three assays. Methanol (<10%) was used to prepare the stock solutions. EC50 values were calculated from nominal concentrations.

Bibliographical references: Kaiser, K.L.E., and Palabrica, V.S. (1991). *Water Poll. Res. J. Canada* **26**, 361-431.
* Kaiser, K.L.E., Ribo, J.M., and Zaruk, B.M. (1985). *Water Poll. Res. J. Canada* **20**, 36-43.

4-CHLOROBENZYL ALCOHOL

CAS RN: 873-76-7

Temperature: 15°C
Test parameter: EC50
Effect: Reduction in light output

Concentration: 10.3 mg/l
Exposure time: 5 min

Concentration: 11.6 mg/l
Exposure time: 15 min

Concentration: 10.6 mg/l*
Exposure time: 30 min

Comment: Mean of three assays. Methanol (<10%) was used to

prepare the stock solutions. EC50 values were calculated from nominal concentrations.

Bibliographical references: Kaiser, K.L.E., and Palabrica, V.S. (1991). *Water Poll. Res. J. Canada* **26**, 361-431.
* Kaiser, K.L.E., Ribo, J.M., and Zaruk, B.M. (1985). *Water Poll. Res. J. Canada* **20**, 36-43.

4-CHLOROBENZYLAMINE

CAS RN: 104-86-9

Temperature: 15°C
Test parameter: EC50
Effect: Reduction in light output

Concentration: 14.2 mg/l
Exposure time: 5 min

Concentration: 17.8 mg/l
Exposure time: 15 min

Concentration: 24.6 mg/l*
Exposure time: 30 min

Comment: Mean of three assays. Methanol (<10%) was used to prepare the stock solutions. EC50 values were calculated from nominal concentrations.

Bibliographical references: Kaiser, K.L.E., and Palabrica, V.S. (1991). *Water Poll. Res. J. Canada* **26**, 361-431.
* Kaiser, K.L.E., Ribo, J.M., and Zaruk, B.M. (1985). *Water Poll. Res. J. Canada* **20**, 36-43.

3-CHLOROBENZYL CHLORIDE

CAS RN: 620-20-2

Sample purity: 98%
Temperature: 15°C
Test parameter: EC50
Effect: Reduction in light output

Concentration: 0.67 mg/l
Exposure time: 5 min

Concentration: 0.75 mg/l
Exposure time: 15 min

Concentration: 0.85 mg/l
Exposure time: 30 min

Comment: Mean of four assays. Methanol (<10%) was used to prepare the stock solutions. EC50 values were calculated from nominal concentrations.

Bibliographical reference: Kaiser, K.L.E., and Palabrica, V.S. (1991). *Water Poll. Res. J. Canada* **26**, 361-431.

4-CHLOROBENZYL CHLORIDE

CAS RN: 104-83-6

Temperature: 15°C
Test parameter: EC50
Effect: Reduction in light output

Concentration: 0.51 mg/l
Exposure time: 5 min

Concentration: 0.60 mg/l
Exposure time: 15 min

Concentration: 0.64 mg/l*
Exposure time: 30 min

Comment: Mean of three assays. Methanol (<10%) was used to prepare the stock solutions. EC50 values were calculated from nominal concentrations.

Bibliographical references: Kaiser, K.L.E., and Palabrica, V.S. (1991). *Water Poll. Res. J. Canada* **26**, 361-431.
* Kaiser, K.L.E., Ribo, J.M., and Zaruk, B.M. (1985). *Water Poll. Res. J. Canada* **20**, 36-43.

3-CHLOROBENZYL CYANIDE

CAS RN: 1529-41-5
Synonym: (3-Chlorophenyl)acetonitrile

Sample purity: 99%
Temperature: 15°C
Test parameter: EC50
Effect: Reduction in light output

Concentration: 1.29 mg/l
Exposure time: 5 min

Concentration: 1.26 mg/l
Exposure time: 15 min

Concentration: 1.18 mg/l
Exposure time: 30 min

Comment: Mean of three assays. Methanol (<10%) was used to prepare the stock solutions. EC50 values were calculated from nominal concentrations.

Bibliographical reference: Kaiser, K.L.E., and Palabrica, V.S. (1991). *Water Poll. Res. J. Canada* **26**, 361-431.

4-CHLOROBENZYL CYANIDE

CAS RN: 140-53-4
Synonym: (4-Chlorophenyl)acetonitrile

Temperature: 15°C
Test parameter: EC50
Effect: Reduction in light output

Concentration: 0.30 mg/l
Exposure time: 5 min

Concentration: 0.31 mg/l
Exposure time: 15 min

Concentration: 0.30 mg/l*
Exposure time: 30 min

Comment: Recrystallized. Mean of three assays. Methanol (<10%) was used to prepare the stock solutions. EC50 values were calculated from nominal concentrations.

Bibliographical references: Kaiser, K.L.E., and Palabrica, V.S. (1991). *Water Poll. Res. J. Canada* **26**, 361-431.
* Kaiser, K.L.E., Ribo, J.M., and Zaruk, B.M. (1985). *Water Poll. Res. J. Canada* **20**, 36-43.

4-CHLOROBENZYL MERCAPTAN

CAS RN: 6258-66-8
Synonym: 4-Chloro-α-toluenethiol

Sample purity: 98%
Temperature: 15°C
Test parameter: EC50
Effect: Reduction in light output

Concentration: 0.54 mg/l
Exposure time: 5 min

Concentration: 0.43 mg/l
Exposure time: 15 min

Concentration: 0.46 mg/l*
Exposure time: 30 min

Comment: Mean of two assays. Methanol (<10%) was used to prepare the stock solutions. EC50 values were calculated from nominal concentrations.

Bibliographical references: Kaiser, K.L.E., and Palabrica, V.S. (1991). *Water Poll. Res. J. Canada* **26**, 361-431.
* Kaiser, K.L.E., Ribo, J.M., and Zaruk, B.M. (1985). *Water Poll. Res. J. Canada* **20**, 36-43.

1-CHLOROBUTANE

CAS RN: 109-69-3
Synonym: Butyl chloride

Temperature: 15°C
Test parameter: EC50
Effect: Reduction in light output

Concentration: 485 mg/l
Exposure time: 5 min

Concentration: 732 mg/l
Exposure time: 30 min

Bibliographical reference: Speece, R. (1987). Drexel University, Philadelphia, USA, private communication.

trans-4-CHLOROCINNAMIC ACID

CAS RN: 1615-02-7
Synonym: 3-(4-Chlorophenyl)-2-propenoic acid

Temperature: 15°C
Test parameter: EC50
Effect: Reduction in light output

Concentration: 42.8 mg/l
Exposure times: 5 and 15 min

Concentration: 40.9 mg/l*
Exposure time: 30 min

Comment: Mean of three assays. Methanol (<10%) was used to prepare the stock solutions. EC50 values were calculated from nominal concentrations.

Bibliographical references: Kaiser, K.L.E., and Palabrica, V.S. (1991). *Water Poll. Res. J. Canada* **26**, 361-431.
* Kaiser, K.L.E., Ribo, J.M., and Zaruk, B.M. (1985). *Water Poll. Res. J. Canada* **20**, 36-43.

1-CHLORODECANE

CAS RN: 1002-69-3
Synonym: Decyl chloride

Temperature: 15°C
Test parameter: EC50
Effect: Reduction in light output

Concentration: 109 mg/l
Exposure time: 5 min

Concentration: 148 mg/l
Exposure time: 30 min

Bibliographical reference: Speece, R. (1987). Drexel University, Philadelphia, USA, private communication.

4-CHLORO-2,6-DINITROANILINE

CAS RN: 5388-62-5

Sample purity: 98%
Temperature: 15°C
Test parameter: EC50
Effect: Reduction in light output

Concentration: 6.27 mg/l
Exposure time: 5 min

Concentration: 4.05 mg/l
Exposure time: 15 min

Concentration: 3.78 mg/l
Exposure time: 30 min

Comment: Mean of three assays. Methanol (<10%) was used to prepare the stock solutions. EC50 values were calculated from nominal concentrations.

Bibliographical reference: Kaiser, K.L.E., and Palabrica, V.S. (1991). *Water Poll. Res. J. Canada* **26**, 361-431.

6-CHLORO-2,4-DINITROANILINE

CAS RN: 3531-19-9

Sample purity: 97%
Temperature: 15°C
Test parameter: EC50
Effect: Reduction in light output
Concentration: 22.8 mg/l
Exposure time: 5 min

Concentration: 19.8 mg/l
Exposure time: 15 min

Concentration: 20.3 mg/l
Exposure time: 30 min

Comment: Mean of three assays. Methanol (<10%) was used to prepare the stock solutions. EC50 values were calculated from nominal concentrations.

Bibliographical reference: Kaiser, K.L.E., and Palabrica, V.S. (1991). *Water Poll. Res. J. Canada* **26**, 361-431.

1-CHLORO-2,4-DINITROBENZENE

CAS RN: 97-00-7

Sample purity: >99%
Temperature: 15°C
Test parameter: EC50
Effect: Reduction in light output

Concentration: 12.8 mg/l
Exposure time: 5 min

Concentration: 5.71 mg/l
Exposure time: 15 min

Concentration: 3.52 mg/l
Exposure time: 30 min

Comment: Mean of three assays. Methanol (<10%) was used to prepare the stock solutions. EC50 values were calculated from nominal concentrations.

Bibliographical reference: Kaiser, K.L.E., and Palabrica, V.S. (1991). *Water Poll. Res. J. Canada* **26**, 361-431.

1-CHLORO-3,4-DINITROBENZENE

CAS RN: 610-40-2

Sample purity: 90%
Temperature: 15°C
Test parameter: EC50
Effect: Reduction in light output

Concentration: 2.44 mg/l
Exposure time: 5 min

Concentration: 0.88 mg/l
Exposure time: 15 min

Concentration: 0.52 mg/l
Exposure time: 30 min

Comment: Mean of three assays. Methanol (<10%) was used to prepare the stock solutions. EC50 values were calculated from nominal concentrations.

Bibliographical reference: Kaiser, K.L.E., and Palabrica, V.S. (1991). *Water Poll. Res. J. Canada* **26**, 361-431.

2-CHLORO-3,5-DINITROBENZOTRIFLUORIDE

CAS RN: 392-95-0

Sample purity: 97%
Temperature: 15°C
Test parameter: EC50
Effect: Reduction in light output

Concentration: 3.49 mg/l
Exposure time: 5 min

Concentration: 0.78 mg/l
Exposure time: 15 min

Concentration: 0.45 mg/l
Exposure time: 30 min

Comment: Mean of three assays. Methanol (<10%) was used to

prepare the stock solutions. EC50 values were calculated from nominal concentrations.

Bibliographical reference: Kaiser, K.L.E., and Palabrica, V.S. (1992). National Water Research Institute, Burlington, Ontario, Canada, unpublished results.

4-CHLORO-3,5-DINITROBENZOTRIFLUORIDE

CAS RN: 393-75-9

Sample purity: 97%
Temperature: 15°C
Test parameter: EC50
Effect: Reduction in light output

Concentration: 2.83 mg/l
Exposure time: 5 min

Concentration: 0.80 mg/l
Exposure time: 15 min

Concentration: 0.50 mg/l
Exposure time: 30 min

Comment: Mean of three assays. Methanol (<10%) was used to prepare the stock solutions. EC50 values were calculated from nominal concentrations.

Bibliographical reference: Kaiser, K.L.E., and Palabrica, V.S. (1991). *Water Poll. Res. J. Canada* **26**, 361-431.

2-CHLORO-3,5-DINITROPYRIDINE

CAS RN: 2578-45-2

Sample purity: 99%
Temperature: 15°C
Test parameter: EC50
Effect: Reduction in light output

Concentration: 0.13 mg/l

Exposure time: 5 min

Concentration: 0.047 mg/l
Exposure time: 15 min

Concentration: 0.035 mg/l
Exposure time: 30 min

Comment: Mean of three assays. Methanol (<10%) was used to prepare the stock solutions. EC50 values were calculated from nominal concentrations.

Bibliographical reference: Kaiser, K.L.E., and Palabrica, V.S. (1991). *Water Poll. Res. J. Canada* **26**, 361-431.

2-CHLOROETHANOL

CAS RN: 107-07-3
Synonym: Ethylene chlorohydrin

Test parameter: EC50
Effect: Reduction in light output
Concentration: 13400 mg/l
Exposure time: 5 min
Comment: The EC50 value was calculated from nominal concentrations.

Bibliographical reference: Curtis, C., Lima, A., Lozano, S.J., and Veith, G.D. (1982). In: *Aquatic Toxicology and Hazard Assessment: Fifth Conference, ASTM STP 766*, J.G. Pearson, R.B. Foster, and W.E. Bishop (eds.), American Society for Testing and Materials, Philadelphia, p. 170-178.

Temperature: 15°C
Test parameter: EC50
Effect: Reduction in light output
Concentration: 390.8 mg/l (326.9-467.1)
Exposure time: 15 min
Comment: Single batches of Microtox® reagent were used for less than 2 h before being discarded. Four concentrations were tested. Concentrations were unmeasured.

Bibliographical reference: Nacci, D., Jackim, E., and Walsh, R.

(1986). *Environ. Toxicol. Chem.* **5**, 521-525.

Temperature: 15°C
Test parameter: EC50
Effect: Reduction in light output

Concentration: 13130 mg/l
Exposure time: 5 min

Concentration: 12800 mg/l
Exposure time: 15 min

Concentration: 12380 mg/l
Exposure time: 30 min

Comment: Chemical was diluted to give five sample concentrations in duplicate.

Bibliographical reference: Benson, W.H., and Stackhouse, R.A. (1986). *Drug Chem. Toxicol.* **9**, 275-283.

2-CHLORO-6-FLUOROBENZALDEHYDE

CAS RN: 387-45-1

Sample purity: 95%
Temperature: 15°C
Test parameter: EC50
Effect: Reduction in light output

Concentration: 28 mg/l
Exposure time: 5 min

Concentration: 21 mg/l*
Exposure time: 15 min

Comment: Chemical was prepared in an initial solution of 3% methanol.

Bibliographical references: Cronin, M.T.D., Dearden, J.C., and Dobbs, A.J. (1991). *Sci. Total Environ.* **109/110**, 431-439.
* Cronin, M.T.D. (1993). Liverpool John Moores University, UK, private communication.

1-CHLORO-4-FLUOROBENZENE

CAS RN: 352-33-0

Temperature: 15°C
Test parameter: EC50
Effect: Reduction in light output

Concentration: 99.0 mg/l
Exposure time: 5 min

Concentration: 119 mg/l
Exposure time: 15 min

Concentration: 137 mg/l*
Exposure time: 30 min

Comment: Mean of three assays. Methanol (<10%) was used to prepare the stock solutions. EC50 values were calculated from nominal concentrations.

Bibliographical references: Kaiser, K.L.E., and Palabrica, V.S. (1991). *Water Poll. Res. J. Canada* **26**, 361-431.
* Kaiser, K.L.E., Ribo, J.M., and Zaruk, B.M. (1985). *Water Poll. Res. J. Canada* **20**, 36-43.

4-CHLORO-2-FLUORO-5-NITROTOLUENE

CAS RN: 18349-11-6

Sample purity: 98%
Temperature: 15°C
Test parameter: EC50
Effect: Reduction in light output

Concentration: 20.8 mg/l
Exposure time: 5 min

Concentration: 25.0 mg/l
Exposure time: 15 min

Concentration: 30.7 mg/l
Exposure time: 30 min

Comment: Mean of three assays. Methanol (<10%) was used to prepare the stock solutions. EC50 values were calculated from nominal concentrations.

Bibliographical reference: Kaiser, K.L.E., and Palabrica, V.S. (1991). *Water Poll. Res. J. Canada* **26**, 361-431.

CHLOROFORM

CAS RN: 67-66-3
Synonym: Trichloromethane

Temperature: 15 ± 0.3°C
Test parameter: EC50
Effect: Reduction in light output
Concentrations: 441 mg/l
429 mg/l
Exposure time: 5 min
Comment: Test solutions were analyzed using gas chromatography.

Bibliographical reference: Qureshi, A.A., Flood, K.W., Thompson, S.R., Janhurst, S.M., Inniss, C.S., and Rokosh, D.A. (1982). In: *Aquatic Toxicology and Hazard Assessment: Fifth Conference, ASTM STP 766*, J.G. Pearson, R.B. Foster, and W.E. Bishop (eds.), American Society for Testing and Materials, Philadelphia, p. 179-195.

Temperature: 15°C
Test parameter: EC50
Effect: Reduction in light output
Concentration: 2464 mg/l
Exposure time: 10 min
Comment: Mean of 10 assays. Test solutions were analyzed using gas chromatography.

Bibliographical reference: Bazin, C., Chambon, P., Bonnefille, M., and Larbaigt, G. (1987). *Sci. Eau* **6**, 403-413.

Sample purity: Reagent grade
Test parameter: EC50
Effect: Reduction in light output

Concentration: 520 mg/l (490-560)

Exposure time: 5 min

Concentration: 670 mg/l (610-740)
Exposure times: 15 and 30 min

Bibliographical reference: Elnabarawy, M.T., Robideau, R.R., and Beach, S.A. (1988). *Tox. Assess.* **3**, 361-370.

Temperature: 15°C
Test parameter: EC50
Effect: Reduction in light output
Concentration: 736 mg/l
Exposure time: 5 min

Bibliographical reference: Kahru, A. (1993). *ATLA* **21**, 210-215.

1-CHLOROHEXANE

CAS RN: 544-10-5

Temperature: 15°C
Test parameter: EC50
Effect: Reduction in light output

Concentration: 203 mg/l*
Exposure time: 5 min

Concentration: 256 mg/l*
Exposure time: 30 min

* EC50 values from linear regression fit to data.

Bibliographical reference: Speece, R. (1987). Drexel University, Philadelphia, USA, private communication.

CHLOROHYDROQUINONE

CAS RN: 615-67-8

Sample purity: 95%
Temperature: 15°C

Test parameter: EC50
Effect: Reduction in light output

Concentration: 11.2 mg/l
Exposure time: 5 min

Concentration: 7.25 mg/l
Exposure time: 15 min

Concentration: 5.76 mg/l
Exposure time: 30 min

Comment: Mean of three assays. Methanol (<10%) was used to prepare the stock solutions. EC50 values were calculated from nominal concentrations.

Bibliographical reference: Kaiser, K.L.E., and Palabrica, V.S. (1991). *Water Poll. Res. J. Canada* **26**, 361-431.

1-CHLORO-4-IODOBENZENE

CAS RN: 637-87-6

Temperature: 15°C
Test parameter: EC50
Effect: Reduction in light output

Concentration: 2.17 mg/l
Exposure time: 5 min

Concentration: 1.98 mg/l
Exposure time: 15 min

Concentration: 1.65 mg/l*
Exposure time: 30 min

Comment: Mean of three assays. Methanol (<10%) was used to prepare the stock solutions. EC50 values were calculated from nominal concentrations.

Bibliographical references: Kaiser, K.L.E., and Palabrica, V.S. (1991). *Water Poll. Res. J. Canada* **26**, 361-431.
* Kaiser, K.L.E., Ribo, J.M., and Zaruk, B.M. (1985). *Water Poll. Res. J. Canada* **20**, 36-43.

2-CHLORO-4-METHYLANILINE

CAS RN: 615-65-6
Synonym: 2-Chloro-*p*-toluidine

Sample purity: 98%
Temperature: 15°C
Test parameter: EC50
Effect: Reduction in light output

Concentration: 5.3 mg/l
Exposure time: 5 min

Concentration: 6.1 mg/l*
Exposure time: 15 min

Comment: Chemical was prepared in an initial solution of 2% methanol.

Bibliographical references: Cronin, M.T.D., Dearden, J.C., and Dobbs, A.J. (1991). *Sci. Total Environ.* **109/110**, 431-439.
* Cronin, M.T.D. (1990). Thesis, Liverpool Polytechnic, Liverpool, UK.

4-CHLORO-*N*-METHYLANILINE

CAS RN: 932-96-7

Temperature: 15°C
Test parameter: EC50
Effect: Reduction in light output

Concentration: 0.91 mg/l
Exposure time: 5 min

Concentration: 1.00 mg/l
Exposure time: 15 min

Concentration: 1.03 mg/l*
Exposure time: 30 min

Comment: Mean of three assays. Methanol (<10%) was used to prepare the stock solutions. EC50 values were calculated from nominal concentrations.

Bibliographical references: Kaiser, K.L.E., and Palabrica, V.S. (1991). *Water Poll. Res. J. Canada* **26**, 361-431.
* Kaiser, K.L.E., Ribo, J.M., and Zaruk, B.M. (1985). *Water Poll. Res. J. Canada* **20**, 36-43.

4-(CHLOROMETHYL)BENZOIC ACID

CAS RN: 1642-81-5

Sample purity: 99%
Temperature: 15°C
Test parameter: EC50
Effect: Reduction in light output

Concentration: 44.9 mg/l
Exposure time: 5 min

Concentration: 40.9 mg/l
Exposure time: 15 min

Concentration: 33.3 mg/l
Exposure time: 30 min

Comment: Mean of three assays. Methanol (<10%) was used to prepare the stock solutions. EC50 values were calculated from nominal concentrations.

Bibliographical reference: Kaiser, K.L.E., and Palabrica, V.S. (1991). *Water Poll. Res. J. Canada* **26**, 361-431.

4-(CHLOROMETHYL)BENZOYL CHLORIDE

CAS RN: 876-08-4

Sample purity: 98%
Temperature: 15°C
Test parameter: EC50
Effect: Reduction in light output

Concentration: 1.40 mg/l
Exposure time: 5 min

Concentration: 1.47 mg/l
Exposure time: 15 min

Concentration: 1.93 mg/l
Exposure time: 30 min

Comment: Mean of three assays. Methanol (<10%) was used to prepare the stock solutions. EC50 values were calculated from nominal concentrations.

Bibliographical reference: Kaiser, K.L.E., and Palabrica, V.S. (1991). *Water Poll. Res. J. Canada* **26**, 361-431.

5-CHLORO-2-METHYL-4-NITROANILINE

CAS RN: 13852-51-2

Sample purity: 40% in water (adjusted to 100%)
Temperature: 15°C
Test parameter: EC50
Effect: Reduction in light output

Concentration: 0.78 mg/l
Exposure time: 5 min

Concentration: 0.91 mg/l
Exposure time: 15 min

Concentration: 1.12 mg/l
Exposure time: 30 min

Comment: Mean of three assays. Methanol (<10%) was used to prepare the stock solutions. EC50 values were calculated from nominal concentrations.

Bibliographical reference: Kaiser, K.L.E., and Palabrica, V.S. (1992). National Water Research Institute, Burlington, Ontario, Canada, unpublished results.

4-CHLORO-3-METHYLPHENOL

CAS RN: 59-50-7

Sample purity: 99%
Temperature: 15°C
Test parameter: EC50
Effect: Reduction in light output

Concentration: 0.27 mg/l
Exposure time: 5 min

Concentration: 0.28 mg/l
Exposure time: 15 min

Concentration: 0.34 mg/l
Exposure time: 30 min

Comment: Mean of three assays. Methanol (<10%) was used to prepare the stock solutions. EC50 values were calculated from nominal concentrations.

Bibliographical reference: Kaiser, K.L.E., and Palabrica, V.S. (1991). *Water Poll. Res. J. Canada* **26**, 361-431.

4-CHLORO-1'-METHYLPHENYLACETONITRILE

CAS RN: 2184-88-5

Temperature: 15°C
Test parameter: EC50
Effect: Reduction in light output

Concentration: 0.18 mg/l
Exposure time: 5 min

Concentration: 0.17 mg/l
Exposure times: 15 and 30 min

Comment: Mean of three assays. Methanol (<10%) was used to prepare the stock solutions. EC50 values were calculated from nominal concentrations.

Bibliographical reference: Kaiser, K.L.E., and Palabrica, V.S. (1991). *Water Poll. Res. J. Canada* **26**, 361-431.

1-CHLORO-2-METHYLPROPENE

CAS RN: 513-37-1

Temperature: 15°C
Test parameter: EC50
Effect: Reduction in light output

Concentration: 453 mg/l*
Exposure time: 5 min

Concentration: 490 mg/l*
Exposure time: 30 min

* EC50 values from linear regression fit to data.

Bibliographical reference: Speece, R. (1987). Drexel University, Philadelphia, USA, private communication.

3-CHLORO-2-METHYLPROPENE

CAS RN: 563-47-3

Sample purity: 98%
Temperature: 15°C
Test parameter: EC50
Effect: Reduction in light output

Concentration: 102 mg/l
Exposure time: 5 min

Concentration: 128 mg/l
Exposure time: 15 min

Concentration: 154 mg/l
Exposure time: 30 min

Comment: Mean of three assays. Methanol (<10%) was used to prepare the stock solutions. EC50 values were calculated from nominal concentrations.

Bibliographical reference: Kaiser, K.L.E., and Palabrica, V.S. (1991). *Water Poll. Res. J. Canada* **26**, 361-431.

4-CHLORO-1-NAPHTHOL

CAS RN: 604-44-4

Sample purity: 97%
Temperature: 15°C
Test parameter: EC50
Effect: Reduction in light output

Concentration: 0.59 mg/l
Exposure times: 5 and 15 min

Concentration: 0.58 mg/l
Exposure time: 30 min

Comment: Mean of four assays. Methanol (<10%) was used to prepare the stock solutions. EC50 values were calculated from nominal concentrations.

Bibliographical reference: Kaiser, K.L.E., and Palabrica, V.S. (1991). *Water Poll. Res. J. Canada* **26**, 361-431.

2-CHLORO-4-NITROANILINE

CAS RN: 121-87-9

Sample purity: 99%
Temperature: 15°C
Test parameter: EC50
Effect: Reduction in light output

Concentration: 3.00 mg/l
Exposure time: 5 min

Concentration: 3.21 mg/l
Exposure time: 15 min

Concentration: 3.69 mg/l
Exposure time: 30 min

Comment: Mean of three assays. Methanol (<10%) was used to prepare the stock solutions. EC50 values were calculated from nominal concentrations.

Bibliographical reference: Kaiser, K.L.E., and Palabrica, V.S. (1992). National Water Research Institute, Burlington, Ontario, Canada, unpublished results.

2-CHLORO-5-NITROANILINE

CAS RN: 6283-25-6

Sample purity: 98%
Temperature: 15°C
Test parameter: EC50
Effect: Reduction in light output

Concentrations: 16.9 mg/l
18.1 mg/l
Exposure time: 5 min

Concentrations: 18.1 mg/l
19.4 mg/l
Exposure time: 15 min

Concentrations: 19.8 mg/l
26.7 mg/l
Exposure time: 30 min

Comment: The two series of tests were performed in triplicate. Methanol (<10%) was used to prepare the stock solutions. EC50 values were calculated from nominal concentrations.

Bibliographical reference: Kaiser, K.L.E., and Palabrica, V.S. (1991). *Water Poll. Res. J. Canada* **26**, 361-431.

4-CHLORO-2-NITROANILINE

CAS RN: 89-63-4

Temperature: 15°C
Test parameter: EC50
Effect: Reduction in light output
Concentration: 19.8 mg/l
Exposure time: 30 min
Comment: Mean of three assays. Methanol (<10%) was used to

prepare the stock solutions. EC50 values were calculated from nominal concentrations.

Bibliographical reference: Kaiser, K.L.E., and Palabrica, V.S. (1991). *Water Poll. Res. J. Canada* **26**, 361-431.

4-CHLORO-3-NITROANILINE

CAS RN: 635-22-3

Sample purity: 97%
Temperature: 15°C
Test parameter: EC50
Effect: Reduction in light output

Concentration: 5.33 mg/l
Exposure time: 5 min

Concentration: 5.09 mg/l
Exposure time: 15 min

Concentration: 5.46 mg/l
Exposure time: 30 min

Comment: Mean of four assays. Methanol (<10%) was used to prepare the stock solutions. EC50 values were calculated from nominal concentrations.

Bibliographical reference: Kaiser, K.L.E., and Palabrica, V.S. (1992). National Water Research Institute, Burlington, Ontario, Canada, unpublished results.

5-CHLORO-2-NITROANILINE

CAS RN: 1635-61-6

Sample purity: 97%
Temperature: 15°C
Test parameter: EC50
Effect: Reduction in light output

Concentration: 5.09 mg/l

Exposure time: 5 min

Concentration: 5.46 mg/l
Exposure time: 15 min

Concentration: 5.85 mg/l
Exposure time: 30 min

Comment: Mean of three assays. Methanol (<10%) was used to prepare the stock solutions. EC50 values were calculated from nominal concentrations.

Bibliographical reference: Kaiser, K.L.E., and Palabrica, V.S. (1992). National Water Research Institute, Burlington, Ontario, Canada, unpublished results.

4-CHLORO-3-NITROANISOLE

CAS RN: 10298-80-3

Sample purity: 98%
Temperature: 15°C
Test parameter: EC50
Effect: Reduction in light output

Concentration: 3.92 mg/l
Exposure time: 5 min

Concentration: 4.30 mg/l
Exposure time: 15 min

Concentration: 4.50 mg/l
Exposure time: 30 min

Comment: Mean of three assays. Methanol (<10%) was used to prepare the stock solutions. EC50 values were calculated from nominal concentrations.

Bibliographical reference: Kaiser, K.L.E., and Palabrica, V.S. (1991). *Water Poll. Res. J. Canada* **26**, 361-431.

2-CHLORONITROBENZENE

CAS RN: 88-73-3

Test parameter: EC50
Effect: Reduction in light output
Concentration: 4.54 mg/l
Exposure time: 15 min
Comment: The concentrations of the compound increased geometrically with a factor of 3.2. The concentrations were calculated on the basis of added amounts of material.

Bibliographical reference: Deneer, J.W., van Leeuwen, C.J., Seinen, W., Maas-Diepeveen, J.L., and Hermens, J.L.M. (1989). *Aquat. Toxicol.* **15**, 83-98.

Temperature: 15°C
Test parameter: EC50
Effect: Reduction in light output

Concentration: 4.05 mg/l
Exposure time: 5 min

Concentration: 4.24 mg/l
Exposure time: 15 min

Concentration: 4.34 mg/l*
Exposure time: 30 min

Comment: Mean of three assays. Methanol (<10%) was used to prepare the stock solutions. EC50 values were calculated from nominal concentrations.

Bibliographical references: Kaiser, K.L.E., and Palabrica, V.S. (1991). *Water Poll. Res. J. Canada* **26**, 361-431.
* Kaiser, K.L.E., and Ribo, J.M. (1985). In: *QSAR in Toxicology and Xenobiochemistry*, M. Tichy (ed.), Elsevier, Amsterdam, p. 27-38.

Sample purity: 99+%
Temperature: 15°C
Test parameter: EC50
Effect: Reduction in light output

Concentration: 4.10 mg/l

Exposure time: 5 min

Concentration: 4.40 mg/l
Exposure time: 15 min

Bibliographical reference: Cronin, M.T.D. (1993). Liverpool John Moores University, UK, private communication.

3-CHLORONITROBENZENE

CAS RN: 121-73-3

Test parameter: EC50
Effect: Reduction in light output
Concentration: 13.1 mg/l
Exposure time: 15 min
Comment: The concentrations of the compound increased geometrically with a factor of 3.2. The concentrations were calculated on the basis of added amounts of material.

Bibliographical reference: Deneer, J.W., van Leeuwen, C.J., Seinen, W., Maas-Diepeveen, J.L., and Hermens, J.L.M. (1989). *Aquat. Toxicol.* **15**, 83-98.

Temperature: 15°C
Test parameter: EC50
Effect: Reduction in light output

Concentration: 15.0 mg/l
Exposure time: 5 min

Concentration: 17.3 mg/l
Exposure time: 15 min

Concentration: 19.8 mg/l*
Exposure time: 30 min

Comment: Mean of three assays. Methanol (<10%) was used to prepare the stock solutions. EC50 values were calculated from nominal concentrations.

Bibliographical references: Kaiser, K.L.E., and Palabrica, V.S. (1991). *Water Poll. Res. J. Canada* **26**, 361-431.

* Kaiser, K.L.E., and Ribo, J.M. (1985). In: *QSAR in Toxicology and Xenobiochemistry*, M. Tichy (ed.), Elsevier, Amsterdam, p. 27-38.

Temperature: 15°C
Test parameter: EC50
Effect: Reduction in light output
Concentration: 11 mg/l
Exposure times: 5 and 15 min
Comment: Chemical was prepared in an initial solution of 2% methanol.

Bibliographical reference: Cronin, M.T.D. (1993). Liverpool John Moores University, UK, private communication.

4-CHLORONITROBENZENE

CAS RN: 100-00-5

Test parameter: EC50
Effect: Reduction in light output
Concentration: 33.7 mg/l
Exposure time: 15 min
Comment: The concentrations of the compound increased geometrically with a factor of 3.2. The concentrations were calculated on the basis of added amounts of material.

Bibliographical reference: Deneer, J.W., van Leeuwen, C.J., Seinen, W., Maas-Diepeveen, J.L., and Hermens, J.L.M. (1989). *Aquat. Toxicol.* **15**, 83-98.

Temperature: 15°C
Test parameter: EC50
Effect: Reduction in light output

Concentration: 20.8 mg/l
Exposure time: 5 min

Concentration: 21.3 mg/l
Exposure time: 15 min

Concentration: 23.8 mg/l*
Exposure time: 30 min

Comment: Mean of three assays. Methanol (<10%) was used to prepare the stock solutions. EC50 values were calculated from nominal concentrations.

Bibliographical references: Kaiser, K.L.E., and Palabrica, V.S. (1991). *Water Poll. Res. J. Canada* **26**, 361-431.
* Kaiser, K.L.E., and Ribo, J.M. (1985). In: *QSAR in Toxicology and Xenobiochemistry*, M. Tichy (ed.), Elsevier, Amsterdam, p. 27-38.

Temperature: 15°C
Test parameter: EC50
Effect: Reduction in light output

Concentration: 24 mg/l
Exposure time: 5 min

Concentration: 26 mg/l
Exposure time: 15 min

Comment: Chemical was prepared in an initial solution of 2% methanol.

Bibliographical reference: Cronin, M.T.D. (1993). Liverpool John Moores University, UK, private communication.

4-CHLORO-3-NITROBENZOIC ACID

CAS RN: 96-99-1

Sample purity: 99%
Temperature: 15°C
Test parameter: EC50
Effect: Reduction in light output

Concentration: 90.0 mg/l
Exposure time: 5 min

Concentration: 86.0 mg/l
Exposure times: 15 and 30 min

Comment: Mean of three assays. Methanol (<10%) was used to prepare the stock solutions. EC50 values were calculated from nominal concentrations.

Bibliographical reference: Kaiser, K.L.E., and Palabrica, V.S. (1991). *Water Poll. Res. J. Canada* **26**, 361-431.

4-CHLORO-3-NITROPHENOL

CAS RN: 610-78-6

Sample purity: 99%
Temperature: 15°C
Test parameter: EC50
Effect: Reduction in light output

Concentration: 4.16 mg/l
Exposure time: 5 min

Concentration: 3.80 mg/l
Exposure times: 15 and 30 min

Comment: Mean of three assays. Methanol (<10%) was used to prepare the stock solutions. EC50 values were calculated from nominal concentrations.

Bibliographical reference: Kaiser, K.L.E., and Palabrica, V.S. (1991). *Water Poll. Res. J. Canada* **26**, 361-431.

2-CHLORO-3-NITROPYRIDINE

CAS RN: 5470-18-8

Sample purity: 99%
Temperature: 15°C
Test parameter: EC50
Effect: Reduction in light output

Concentration: 39.8 mg/l
Exposure time: 5 min

Concentration: 33.9 mg/l
Exposure time: 15 min

Concentration: 34.7 mg/l

Exposure time: 30 min

Comment: Mean of four assays. Methanol (<10%) was used to prepare the stock solutions. EC50 values were calculated from nominal concentrations.

Bibliographical reference: Kaiser, K.L.E., and Palabrica, V.S. (1991). *Water Poll. Res. J. Canada* **26**, 361-431.

2-CHLORO-4-NITROTOLUENE

CAS RN: 121-86-8

Sample purity: 98%
Temperature: 15°C
Test parameter: EC50
Effect: Reduction in light output

Concentration: 3.3 mg/l
Exposure time: 5 min

Concentration: 3.9 mg/l
Exposure time: 15 min

Comment: Chemical was prepared in an initial solution of 2% methanol.

Bibliographical reference: Cronin, M.T.D. (1993). Liverpool John Moores University, UK, private communication.

2-CHLORO-6-NITROTOLUENE

CAS RN: 83-42-1

Test parameter: EC50
Effect: Reduction in light output
Concentration: 0.88 mg/l
Exposure time: 15 min
Comment: The concentrations of the compound increased geometrically with a factor of 3.2. The concentrations were calculated on the basis of added amounts of material.

Bibliographical reference: Deneer, J.W., van Leeuwen, C.J., Seinen, W., Maas-Diepeveen, J.L., and Hermens, J.L.M. (1989). *Aquat. Toxicol.* **15**, 83-98.

Sample purity: 98%
Temperature: 15°C
Test parameter: EC50
Effect: Reduction in light output

Concentration: 0.72 mg/l
Exposure time: 5 min

Concentration: 0.79 mg/l
Exposure time: 15 min

Comment: Chemical was prepared in an initial solution of 2% methanol.

Bibliographical reference: Cronin, M.T.D. (1993). Liverpool John Moores University, UK, private communication.

4-CHLORO-2-NITROTOLUENE

CAS RN: 89-59-8

Test parameter: EC50
Effect: Reduction in light output
Concentration: 4.84 mg/l
Exposure time: 15 min
Comment: The concentrations of the compound increased geometrically with a factor of 3.2. The concentrations were calculated on the basis of added amounts of material.

Bibliographical reference: Deneer, J.W., van Leeuwen, C.J., Seinen, W., Maas-Diepeveen, J.L., and Hermens, J.L.M. (1989). *Aquat. Toxicol.* **15**, 83-98.

4-CHLORO-3-NITROTOLUENE

CAS RN: 89-60-1

Temperature: 15°C
Test parameter: EC50
Effect: Reduction in light output

Concentration: 5.10 mg/l
Exposure time: 5 min

Concentration: 5.30 mg/l*
Exposure time: 15 min

Comment: Chemical was prepared in an initial solution of 2% methanol.

Bibliographical references: Cronin, M.T.D., Dearden, J.C., and Dobbs, A.J. (1991). *Sci. Total Environ.* **109/110**, 431-439.
* Cronin, M.T.D. (1990). Thesis, Liverpool Polytechnic, Liverpool, UK.

1-CHLOROOCTANE

CAS RN: 111-85-3

Temperature: 15°C
Test parameter: EC50
Effect: Reduction in light output

Concentration: 291 mg/l*
Exposure time: 5 min

Concentration: 398 mg/l*
Exposure time: 30 min

* EC50 values from linear regression fit to data.

Bibliographical reference: Speece, R. (1987). Drexel University, Philadelphia, USA, private communication.

1-CHLOROPENTANE

CAS RN: 543-59-9
Synonyms: Amyl chloride; Pentyl chloride

Temperature: 15°C
Test parameter: EC50
Effect: Reduction in light output

Concentration: 227 mg/l
Exposure time: 5 min

Concentration: 245 mg/l
Exposure time: 30 min

Bibliographical reference: Speece, R. (1987). Drexel University, Philadelphia, USA, private communication.

5-CHLORO-1-PENTYNE

CAS RN: 14267-92-6

Temperature: 15°C
Test parameter: EC50
Effect: Reduction in light output

Concentration: 23.2 mg/l
Exposure time: 5 min

Concentration: 16.1 mg/l*
Exposure time: 30 min

* EC50 value from linear regression fit to data.

Bibliographical reference: Speece, R. (1987). Drexel University, Philadelphia, USA, private communication.

4-CHLOROPHENETHYL ALCOHOL

CAS RN: 1875-88-3

Sample purity: 99%
Temperature: 15°C
Test parameter: EC50
Effect: Reduction in light output

Concentration: 1.46 mg/l

Exposure time: 5 min

Concentration: 1.68 mg/l
Exposure time: 15 min

Concentration: 1.88 mg/l*
Exposure time: 30 min

Comment: Mean of three assays. Methanol (<10%) was used to prepare the stock solutions. EC50 values were calculated from nominal concentrations.

Bibliographical references: Kaiser, K.L.E., and Palabrica, V.S. (1991). *Water Poll. Res. J. Canada* **26**, 361-431.
* Kaiser, K.L.E. (1987). In: *QSAR in Environmental Toxicology - II*, K.L.E. Kaiser (ed.), D. Reidel Publishing Company, Dordrecht, p. 169-188.

2-CHLOROPHENOL

CAS RN: 95-57-8

Test parameter: EC50
Effect: Reduction in light output
Concentration: 22.1 mg/l
Exposure time: 5 min
Comment: The EC50 value was calculated from nominal concentrations.

Bibliographical reference: Curtis, C., Lima, A., Lozano, S.J., and Veith, G.D. (1982). In: *Aquatic Toxicology and Hazard Assessment: Fifth Conference, ASTM STP 766*, J.G. Pearson, R.B. Foster, and W.E. Bishop (eds.), American Society for Testing and Materials, Philadelphia, p. 170-178.

Temperature: 15°C
Test parameter: EC50
Effect: Reduction in light output

Concentration: 37.1 mg/l
Exposure time: 5 min

Concentration: 39.7 mg/l

Exposure time: 15 min

Concentration: 33.8 mg/l
Exposure time: 30 min

Comment: Mean of three assays. Methanol (<10%) was used to prepare the stock solutions. EC50 values were calculated from nominal concentrations.

Bibliographical reference: Ribo, J.M., and Kaiser, K.L.E. (1983). *Chemosphere* **12**, 1421-1442.

Temperature: 15°C
Test parameter: EC50
Effect: Reduction in light output
Concentration: 34 mg/l
Exposure time: 10 min
Comment: Mean of four assays. Test solutions were analyzed using gas chromatography.

Bibliographical reference: Bazin, C., Chambon, P., Bonnefille, M., and Larbaigt, G. (1987). *Sci. Eau* **6**, 403-413.

Temperature: 15°C
Test parameter: EC50
Effect: Reduction in light output

Concentration: 18.1 mg/l
Exposure time: 5 min

Concentration: 21.1 mg/l
Exposure time: 30 min

Bibliographical reference: Speece, R. (1987). Drexel University, Philadelphia, USA, private communication.

Test parameter: EC50
Effect: Reduction in light output

Concentration: 27.8 mg/l
Exposure time: 5 min

Concentration: 28.5 mg/l

Exposure time: 15 min

Bibliographical reference: Ribo, J.M., and Rogers, F. (1990). *Tox. Assess.* **5**, 135-152.

3-CHLOROPHENOL

CAS RN: 108-43-0

Temperature: 15°C
Test parameter: EC50
Effect: Reduction in light output

Concentration: 9.98 mg/l
Exposure time: 5 min

Concentration: 13.2 mg/l
Exposure time: 15 min

Concentration: 14.1 mg/l
Exposure time: 30 min

Comment: Mean of three assays. Methanol (<10%) was used to prepare the stock solutions. EC50 values were calculated from nominal concentrations.

Bibliographical reference: Ribo, J.M., and Kaiser, K.L.E. (1983). *Chemosphere* **12**, 1421-1442.

Temperature: 15°C
Test parameter: EC50
Effect: Reduction in light output

Concentration: 6.13 mg/l
Exposure time: 5 min

Concentration: 6.97 mg/l
Exposure time: 30 min

Bibliographical reference: Speece, R. (1987). Drexel University, Philadelphia, USA, private communication.

Test parameter: EC50
Effect: Reduction in light output

Concentration: 16.6 mg/l
Exposure time: 5 min

Concentration: 19.0 mg/l
Exposure time: 15 min

Bibliographical reference: Ribo, J.M., and Rogers, F. (1990). *Tox. Assess.* **5**, 135-152.

4-CHLOROPHENOL

CAS RN: 106-48-9

Temperature: 15°C
Test parameter: EC50
Effect: Reduction in light output

Concentration: 8.49 mg/l
Exposure time: 5 min

Concentration: 9.10 mg/l
Exposure time: 15 min

Concentration: 8.30 mg/l
Exposure time: 30 min

Comment: Mean of three assays. Methanol (<10%) was used to prepare the stock solutions. EC50 values were calculated from nominal concentrations.

Bibliographical reference: Ribo, J.M., and Kaiser, K.L.E. (1983). *Chemosphere* **12**, 1421-1442.

Temperature: 15°C
Test parameter: EC50
Effect: Reduction in light output

Concentration: 0.96 mg/l
Exposure time: 5 min

Concentration: 1.07 mg/l
Exposure time: 30 min

Bibliographical reference: Speece, R. (1987). Drexel University, Philadelphia, USA, private communication.

4-CHLOROPHENOXYACETIC ACID

CAS RN: 122-88-3

Sample purity: 98%
Temperature: 15°C
Test parameter: EC50
Effect: Reduction in light output

Concentration: 148 mg/l
Exposure time: 5 min

Concentration: 120 mg/l
Exposure time: 15 min

Concentration: 97.9 mg/l*
Exposure time: 30 min

Comment: Mean of three assays. Methanol (<10%) was used to prepare the stock solutions. EC50 values were calculated from nominal concentrations.

Bibliographical references: Kaiser, K.L.E., and Palabrica, V.S. (1991). *Water Poll. Res. J. Canada* **26**, 361-431.
* Kaiser, K.L.E., Ribo, J.M., and Zaruk, B.M. (1985). *Water Poll. Res. J. Canada* **20**, 36-43.

4-CHLOROPHENYLACETIC ACID

CAS RN: 1878-66-6

Temperature: 15°C
Test parameter: EC50
Effect: Reduction in light output

Concentration: 64.9 mg/l

Exposure time: 5 min

Concentration: 69.5 mg/l
Exposure time: 15 min

Concentration: 79.8 mg/l*
Exposure time: 30 min

Comment: Mean of three assays. Methanol (<10%) was used to prepare the stock solutions. EC50 values were calculated from nominal concentrations.

Bibliographical references: Kaiser, K.L.E., and Palabrica, V.S. (1991). *Water Poll. Res. J. Canada* **26**, 361-431.
* Kaiser, K.L.E., Ribo, J.M., and Zaruk, B.M. (1985). *Water Poll. Res. J. Canada* **20**, 36-43.

4'-CHLOROPHENYLACETONE

CAS RN: 5586-88-9

Temperature: 15°C
Test parameter: EC50
Effect: Reduction in light output

Concentration: 5.33 mg/l
Exposure time: 5 min

Concentration: 5.85 mg/l
Exposure time: 15 min

Concentration: 5.58 mg/l*
Exposure time: 30 min

Comment: Mean of three assays. Methanol (<10%) was used to prepare the stock solutions. EC50 values were calculated from nominal concentrations.

Bibliographical references: Kaiser, K.L.E., and Palabrica, V.S. (1991). *Water Poll. Res. J. Canada* **26**, 361-431.
* Kaiser, K.L.E., Ribo, J.M., and Zaruk, B.M. (1985). *Water Poll. Res. J. Canada* **20**, 36-43.

DL-4-CHLOROPHENYLALANINE

CAS RN: 7424-00-2

Temperature: 15°C
Test parameter: EC50
Effect: Reduction in light output
Concentration: 934 mg/l
Exposure time: 30 min
Comment: Mean of three assays. Methanol (<10%) was used to prepare the stock solutions. EC50 values were calculated from nominal concentrations.

Bibliographical reference: Kaiser, K.L.E., Ribo, J.M., and Zaruk, B.M. (1985). *Water Poll. Res. J. Canada* **20**, 36-43.

Sample purity: 98%
Temperature: 15°C
Test parameter: EC50
Effect: Reduction in light output

Concentration: 112 mg/l
Exposure time: 5 min

Concentration: 115 mg/l
Exposure time: 15 min

Concentration: 115 mg/l*
Exposure time: 30 min

Comment: Mean of three assays. Methanol (<10%) was used to prepare the stock solutions. EC50 values were calculated from nominal concentrations.

Bibliographical references: Kaiser, K.L.E., and Palabrica, V.S. (1991). *Water Poll. Res. J. Canada* **26**, 361-431.
* Kaiser, K.L.E. (1987). In: *QSAR in Environmental Toxicology - II*, K.L.E. Kaiser (ed.), D. Reidel Publishing Company, Dordrecht, p. 169-188.

4-CHLORO-1,2-PHENYLENEDIAMINE

CAS RN: 95-83-0

Sample purity: 96%
Temperature: 15°C
Test parameter: EC50
Effect: Reduction in light output

Concentration: 1.60 mg/l
Exposure time: 5 min

Concentration: 1.80 mg/l
Exposure time: 15 min

Concentration: 2.85 mg/l
Exposure time: 30 min

Comment: Mean of three assays. Methanol (<10%) was used to prepare the stock solutions. EC50 values were calculated from nominal concentrations.

Bibliographical reference: Kaiser, K.L.E., and Palabrica, V.S. (1991). *Water Poll. Res. J. Canada* **26**, 361-431.

2-(4-CHLOROPHENYL)ETHYLAMINE

CAS RN: 156-41-2

Temperature: 15°C
Test parameter: EC50
Effect: Reduction in light output

Concentration: 30.3 mg/l
Exposure time: 5 min

Concentration: 26.4 mg/l
Exposure time: 15 min

Concentration: 26.4 mg/l*
Exposure time: 30 min

Comment: Methanol (<10%) was used to prepare the stock solution. EC50 values were calculated from nominal concentrations.

Bibliographical references: Kaiser, K.L.E., and Palabrica, V.S. (1991). *Water Poll. Res. J. Canada* **26**, 361-431.
* Kaiser, K.L.E., Ribo, J.M., and Zaruk, B.M. (1985). *Water Poll. Res.*

J. Canada **20**, 36-43.

4-CHLOROPHENYLHYDRAZINE HYDROCHLORIDE

CAS RN: 1073-70-7

Sample purity: 98%
Temperature: 15°C
Test parameter: EC50
Effect: Reduction in light output

Concentration: 2.06 mg/l
Exposure time: 5 min

Concentration: 2.53 mg/l
Exposure time: 15 min

Concentration: 3.57 mg/l
Exposure time: 30 min

Comment: Mean of four assays. Methanol (<10%) was used to prepare the stock solutions. EC50 values were calculated from nominal concentrations.

Bibliographical reference: Kaiser, K.L.E., and Palabrica, V.S. (1991). *Water Poll. Res. J. Canada* **26**, 361-431.

4-CHLOROPHENYL ISOCYANATE

CAS RN: 104-12-1

Temperature: 15°C
Test parameter: EC50
Effect: Reduction in light output

Concentration: 2.67 mg/l
Exposure time: 5 min

Concentration: 2.49 mg/l
Exposure time: 15 min

Concentration: 2.32 mg/l*

Exposure time: 30 min

Comment: Methanol (<10%) was used to prepare the stock solution. EC50 values were calculated from nominal concentrations.

Bibliographical references: Kaiser, K.L.E., and Palabrica, V.S. (1991). *Water Poll. Res. J. Canada* **26**, 361-431.
* Kaiser, K.L.E., Ribo, J.M., and Zaruk, B.M. (1985). *Water Poll. Res. J. Canada* **20**, 36-43.

4-CHLOROPHENYL ISOTHIOCYANATE

CAS RN: 2131-55-7

Temperature: 15°C
Test parameter: EC50
Effect: Reduction in light output

Concentration: 0.50 mg/l
Exposure time: 5 min

Concentration: 0.45 mg/l
Exposure time: 15 min

Concentration: 0.36 mg/l*
Exposure time: 30 min

Comment: Mean of three assays. Methanol (<10%) was used to prepare the stock solutions. EC50 values were calculated from nominal concentrations.

Bibliographical references: Kaiser, K.L.E., and Palabrica, V.S. (1991). *Water Poll. Res. J. Canada* **26**, 361-431.
* Kaiser, K.L.E., Ribo, J.M., and Zaruk, B.M. (1985). *Water Poll. Res. J. Canada* **20**, 36-43.

4-CHLOROPHENYL PHENYL SULFONE

CAS RN: 80-00-2

Sample purity: 97+%
Temperature: 15°C

Test parameter: EC50
Effect: Reduction in light output

Concentration: 7.46 mg/l
Exposure time: 5 min

Concentration: 8.56 mg/l
Exposure time: 15 min

Concentration: 9.39 mg/l*
Exposure time: 30 min

Comment: Mean of three assays. Methanol (<10%) was used to prepare the stock solutions. EC50 values were calculated from nominal concentrations.

Bibliographical references: Kaiser, K.L.E., and Palabrica, V.S. (1991). *Water Poll. Res. J. Canada* **26**, 361-431.
* Kaiser, K.L.E., Ribo, J.M., and Zaruk, B.M. (1985). *Water Poll. Res. J. Canada* **20**, 36-43.

4-CHLOROPHENYL SELENINIC ACID

CAS RN: 20753-53-1

Temperature: 15°C
Test parameter: EC50
Effect: Reduction in light output

Concentration: 75.9 mg/l
Exposure time: 5 min

Concentration: 57.6 mg/l
Exposure time: 15 min

Concentration: 44.7 mg/l*
Exposure time: 30 min

Comment: Mean of three assays. Methanol (<10%) was used to prepare the stock solutions. EC50 values were calculated from nominal concentrations.

Bibliographical references: Kaiser, K.L.E., and Palabrica, V.S. (1991). *Water Poll. Res. J. Canada* **26**, 361-431.

* Kaiser, K.L.E., Ribo, J.M., and Zaruk, B.M. (1985). *Water Poll. Res. J. Canada* **20**, 36-43.

1-CHLOROPROPANE

CAS RN: 540-54-5

Temperature: 15°C
Test parameter: EC50
Effect: Reduction in light output

Concentration: 831 mg/l
Exposure time: 5 min

Concentration: 1333 mg/l
Exposure time: 30 min

Bibliographical reference: Speece, R. (1987). Drexel University, Philadelphia, USA, private communication.

2-CHLOROPROPANE

CAS RN: 75-29-6
Synonym: Isopropyl chloride

Temperature: 15°C
Test parameter: EC50
Effect: Reduction in light output

Concentration: 210 mg/l
Exposure time: 5 min

Concentration: 406 mg/l
Exposure time: 30 min

Bibliographical reference: Speece, R. (1987). Drexel University, Philadelphia, USA, private communication.

3-CHLORO-1,2-PROPANEDIOL

CAS RN: 96-24-2

Temperature: 15°C
Test parameter: EC50
Effect: Reduction in light output

Concentration: 5649 mg/l
Exposure time: 5 min

Concentration: 4209 mg/l
Exposure time: 30 min

Bibliographical reference: Speece, R. (1987). Drexel University, Philadelphia, USA, private communication.

2-CHLOROPROPIONIC ACID

CAS RN: 598-78-7

Temperature: 15°C
Test parameter: EC50
Effect: Reduction in light output

Concentration: 6408 mg/l
Exposure time: 5 min

Concentration: 4039 mg/l
Exposure time: 30 min

Bibliographical reference: Speece, R. (1987). Drexel University, Philadelphia, USA, private communication.

Sample purity: 98%
Temperature: 15°C
Test parameter: EC50
Effect: Reduction in light output

Concentration: 86.2 mg/l
Exposure time: 5 min

Concentration: 75.1 mg/l

Exposure time: 15 min

Concentration: 66.9 mg/l
Exposure time: 30 min

Comment: Mean of three assays. Methanol (<10%) was used to prepare the stock solutions. EC50 values were calculated from nominal concentrations.

Bibliographical reference: Kaiser, K.L.E., and Palabrica, V.S. (1991). *Water Poll. Res. J. Canada* **26**, 361-431.

4'-CHLOROPROPIOPHENONE

CAS RN: 6285-05-8

Sample purity: 97%
Temperature: 15°C
Test parameter: EC50
Effect: Reduction in light output

Concentration: 4.75 mg/l
Exposure time: 5 min

Concentration: 5.21 mg/l
Exposure time: 15 min

Concentration: 5.33 mg/l*
Exposure time: 30 min

Comment: Mean of three assays. Methanol (<10%) was used to prepare the stock solutions. EC50 values were calculated from nominal concentrations.

Bibliographical references: Kaiser, K.L.E., and Palabrica, V.S. (1991). *Water Poll. Res. J. Canada* **26**, 361-431.
* Kaiser, K.L.E. (1987). In: *QSAR in Environmental Toxicology - II*, K.L.E. Kaiser (ed.), D. Reidel Publishing Company, Dordrecht, p. 169-188.

3-CHLORO-1-PROPYNE

CAS RN: 624-65-7
Synonym: Propargyl chloride

Temperature: 15°C
Test parameter: EC50
Effect: Reduction in light output

Concentration: 79.2 mg/l
Exposure time: 5 min

Concentration: 24.4 mg/l
Exposure time: 30 min

Bibliographical reference: Speece, R. (1987). Drexel University, Philadelphia, USA, private communication.

2-CHLOROPYRIDINE

CAS RN: 109-09-1

Temperature: 15°C
Test parameter: EC50
Effect: Reduction in light output

Concentration: 70.0 mg/l
Exposure time: 5 min

Concentration: 71.6 mg/l
Exposure time: 15 min

Concentration: 71.6 mg/l*
Exposure time: 30 min

Comment: Mean of three assays. Methanol (<10%) was used to prepare the stock solutions. EC50 values were calculated from nominal concentrations.

Bibliographical references: Kaiser, K.L.E., and Palabrica, V.S. (1991). *Water Poll. Res. J. Canada* **26**, 361-431.
* Kaiser, K.L.E., and Ribo, J.M. (1985). In: *QSAR in Toxicology and Xenobiochemistry*, M. Tichy (ed.), Elsevier, Amsterdam, p. 27-38.

3-CHLOROPYRIDINE

CAS RN: 626-60-8

Temperature: 15°C
Test parameter: EC50
Effect: Reduction in light output

Concentration: 66.9 mg/l
Exposure time: 5 min

Concentration: 54.3 mg/l
Exposure time: 15 min

Concentration: 70.0 mg/l*
Exposure time: 30 min

Comment: Mean of three assays. Methanol (<10%) was used to prepare the stock solutions. EC50 values were calculated from nominal concentrations.

Bibliographical references: Kaiser, K.L.E., and Palabrica, V.S. (1991). *Water Poll. Res. J. Canada* **26**, 361-431.
* Kaiser, K.L.E., and Ribo, J.M. (1985). In: *QSAR in Toxicology and Xenobiochemistry*, M. Tichy (ed.), Elsevier, Amsterdam, p. 27-38.

5-CHLORO-2-PYRIDINOL

CAS RN: 4214-79-3

Sample purity: 99%
Temperature: 15°C
Test parameter: EC50
Effect: Reduction in light output

Concentration: 374 mg/l
Exposure time: 5 min

Concentration: 349 mg/l
Exposure time: 15 min

Concentration: 382 mg/l
Exposure time: 30 min

Comment: Mean of three assays. Methanol (<10%) was used to prepare the stock solutions. EC50 values were calculated from nominal concentrations.

Bibliographical reference: Kaiser, K.L.E., and Palabrica, V.S. (1991). *Water Poll. Res. J. Canada* **26**, 361-431.

4-CHLOROSTYRENE

CAS RN: 1073-67-2

Sample purity: 99%
Temperature: 15°C
Test parameter: EC50
Effect: Reduction in light output

Concentration: 1.13 mg/l
Exposure time: 5 min

Concentration: 1.39 mg/l
Exposure time: 15 min

Concentration: 1.67 mg/l*
Exposure time: 30 min

Comment: Mean of three assays. Methanol (<10%) was used to prepare the stock solutions. EC50 values were calculated from nominal concentrations.

Bibliographical references: Kaiser, K.L.E., and Palabrica, V.S. (1991). *Water Poll. Res. J. Canada* **26**, 361-431.
* Kaiser, K.L.E. (1987). In: *QSAR in Environmental Toxicology - II*, K.L.E. Kaiser (ed.), D. Reidel Publishing Company, Dordrecht, p. 169-188.

4-CHLOROTHIOPHENOL

CAS RN: 106-54-7

Temperature: 15°C
Test parameter: EC50
Effect: Reduction in light output

Concentration: 0.91 mg/l
Exposure time: 5 min

Concentration: 0.65 mg/l
Exposure time: 15 min

Concentration: 0.55 mg/l*
Exposure time: 30 min

Comment: Mean of two assays. Methanol (<10%) was used to prepare the stock solutions. EC50 values were calculated from nominal concentrations.

Bibliographical references: Kaiser, K.L.E., and Palabrica, V.S. (1991). *Water Poll. Res. J. Canada* **26**, 361-431.
* Kaiser, K.L.E., Ribo, J.M., and Zaruk, B.M. (1985). *Water Poll. Res. J. Canada* **20**, 36-43.

2-CHLOROTOLUENE

CAS RN: 95-49-8

Temperature: 20°C

Test parameter: EC20
Effect: Reduction in light output
Concentrations: 2.19 mg/l
2.27 mg/l
Exposure time: 5 min

Test parameter: EC50
Effect: Reduction in light output
Concentrations: 5.95 mg/l
6.13 mg/l
Exposure time: 5 min

Bibliographical reference: Casseri, N.A., Ying, W, and Sojka, S.A. (1983). Proceedings 38th Industrial Waste Conference, USA, May 1983, p. 867-878.

Temperature: 20°C

Test parameter: EC20

Effect: Reduction in light output
Concentrations: 2.35 mg/l
2.29 mg/l
Exposure time: 15 min

Test parameter: EC50
Effect: Reduction in light output
Concentrations: 6.15 mg/l
6.23 mg/l
Exposure time: 15 min

Bibliographical reference: Casseri, N.A., Ying, W, and Sojka, S.A. (1983). Proceedings 38th Industrial Waste Conference, USA, May 1983, p. 867-878.

Temperature: 20°C
Test parameter: EC50
Effect: Reduction in light output

Concentration: 5.98 mg/l (5.57-6.41)
Exposure time: 5 min

Concentration: 6.12 mg/l (5.62-6.68)
Exposure time: 15 min

Bibliographical reference: Casseri, N.A. (1985). Occidental Chemical Corporation, Grand Island, New York, USA, private communication.

Sample purity: 99%
Temperature: 15°C
Test parameter: EC50
Effect: Reduction in light output

Concentration: 2.90 mg/l
Exposure time: 5 min

Concentration: 3.33 mg/l
Exposure time: 15 min

Concentration: 4.70 mg/l
Exposure time: 30 min

Comment: Mean of three assays. Methanol (<10%) was used to

prepare the stock solutions. EC50 values were calculated from nominal concentrations.

Bibliographical reference: Kaiser, K.L.E., and Palabrica, V.S. (1991). *Water Poll. Res. J. Canada* **26**, 361-431.

3-CHLOROTOLUENE

CAS RN: 108-41-8

Temperature: 15°C
Test parameter: EC50
Effect: Reduction in light output

Concentration: 2.9 mg/l
Exposure time: 5 min

Concentration: 3.2 mg/l
Exposure time: 15 min

Bibliographical reference: Cronin, M.T.D. (1993). Liverpool John Moores University, UK, private communication.

4-CHLOROTOLUENE

CAS RN: 106-43-4

Temperature: 20°C

Test parameter: EC20
Effect: Reduction in light output
Concentrations: 2.12 mg/l
2.17 mg/l
Exposure time: 5 min

Test parameter: EC50
Effect: Reduction in light output
Concentrations: 6.23 mg/l
6.53 mg/l
Exposure time: 5 min

Bibliographical reference: Casseri, N.A., Ying, W, and Sojka, S.A.

(1983). Proceedings 38th Industrial Waste Conference, USA, May 1983, p. 867-878.

Temperature: 20°C

Test parameter: EC20
Effect: Reduction in light output
Concentrations: 2.44 mg/l
2.65 mg/l
Exposure time: 15 min

Test parameter: EC50
Effect: Reduction in light output
Concentrations: 7.17 mg/l
7.64 mg/l
Exposure time: 15 min

Bibliographical reference: Casseri, N.A., Ying, W, and Sojka, S.A. (1983). Proceedings 38th Industrial Waste Conference, USA, May 1983, p. 867-878.

Temperature: 20°C
Test parameter: EC50
Effect: Reduction in light output

Concentration: 6.31 mg/l (6.01-6.63)
Exposure time: 5 min

Concentration: 7.32 mg/l (6.91-7.76)
Exposure time: 15 min

Bibliographical reference: Casseri, N.A. (1985). Occidental Chemical Corporation, Grand Island, New York, USA, private communication.

Temperature: 15°C
Test parameter: EC50
Effect: Reduction in light output

Concentration: 4.92 mg/l
Exposure time: 5 min

Concentration: 5.79 mg/l

Exposure time: 15 min

Concentration: 6.49 mg/l*
Exposure time: 30 min

Comment: Mean of three assays. Methanol (<10%) was used to prepare the stock solutions. EC50 values were calculated from nominal concentrations.

Bibliographical references: Kaiser, K.L.E., and Palabrica, V.S. (1991). *Water Poll. Res. J. Canada* **26**, 361-431.
* Kaiser, K.L.E., Ribo, J.M., and Zaruk, B.M. (1985). *Water Poll. Res. J. Canada* **20**, 36-43.

4-CHLORO-1',1',1'-TRICHLOROTOLUENE

CAS RN: 5216-25-1

Temperature: 15°C
Test parameter: EC50
Effect: Reduction in light output

Concentration: 6.33 mg/l
Exposure time: 5 min

Concentration: 7.27 mg/l
Exposure time: 15 min

Concentration: 10.8 mg/l*
Exposure time: 30 min

Comment: Mean of three assays. Methanol (<10%) was used to prepare the stock solutions. EC50 values were calculated from nominal concentrations.

Bibliographical references: Kaiser, K.L.E., and Palabrica, V.S. (1991). *Water Poll. Res. J. Canada* **26**, 361-431.
* Kaiser, K.L.E., Ribo, J.M., and Zaruk, B.M. (1985). *Water Poll. Res. J. Canada* **20**, 36-43.

2-(2-CHLORO-1,1,2-TRIFLUOROETHYLTHIO)ANILINE

CAS RN: 81029-02-9

Sample purity: 98+%
Temperature: 15°C
Test parameter: EC50
Effect: Reduction in light output

Concentration: 4.10 mg/l
Exposure time: 5 min

Concentration: 4.40 mg/l
Exposure time: 15 min

Concentration: 4.50 mg/l
Exposure time: 30 min

Comment: Mean of three assays. Methanol (<10%) was used to prepare the stock solutions. EC50 values were calculated from nominal concentrations.

Bibliographical reference: Kaiser, K.L.E., and Palabrica, V.S. (1991). *Water Poll. Res. J. Canada* **26**, 361-431.

2-CHLORO-5-(TRIFLUOROMETHYL)PYRIDINE

CAS RN: 52334-81-3

Sample purity: 99%
Temperature: 15°C
Test parameter: EC50
Effect: Reduction in light output

Concentration: 30.1 mg/l
Exposure time: 5 min

Concentration: 37.9 mg/l
Exposure time: 15 min

Concentration: 45.6 mg/l
Exposure time: 30 min

Comment: Mean of three assays. Methanol (<10%) was used to

prepare the stock solutions. EC50 values were calculated from nominal concentrations.

Bibliographical reference: Kaiser, K.L.E., and Palabrica, V.S. (1991). *Water Poll. Res. J. Canada* **26**, 361-431.

4-CHLORO-*m*-XYLENOL

CAS RN: 88-04-0
Synonyms: 4-Chloro-3,5-dimethylphenol; 4-Chloro-3,5-xylenol

Temperature: 15°C
Test parameter: EC50
Effect: Reduction in light output
Concentrations: 1.60 mg/l
2.30 mg/l
Exposure time: 5 min

Bibliographical reference: King, E.F., and Painter, H.A. (1981). In: *Les Tests de Toxicité Aiguë en Milieu Aquatique. Acute Aquatic Ecotoxicological Tests*, H. Leclerc and D. Dive (eds.), INSERM 106, Paris, p. 143-153.

Sample purity: 99%
Temperature: 15°C
Test parameter: EC50
Effect: Reduction in light output
Concentration: 8.0 mg/l
Exposure time: 15 min
Comment: To dissolve the chemical, 1M sodium hydroxide was added and since the solution could not be neutralized to pH 7 without precipitate it was used at approximately pH 8.

Bibliographical reference: King, E.F. (1984). In: *Toxicity Screening Procedures Using Bacterial Systems*, D. Liu and B.J. Dutka (eds.), Marcel Dekker, New York, p. 175-194.

CHLORPYRIFOS

CAS RN: 2921-88-2
Synonym: *O,O*-Diethyl *O*-3,5,6-trichloro-2-pyridyl phosphorothioate

Temperature: 15 ± 0.1°C
Test parameter: EC50
Effect: Reduction in light output
Concentration: 46.3 mg/l
Exposure time: 5 min
Comment: Chemical was first dissolved in a solvent and then serially diluted with the diluent. The concentration of solvent did not exceed 8% (v/v) in the test samples.

Bibliographical reference: Somasundaram, L., Coats, J.R., Racke, K.D., and Stahr, H.M. (1990). *Bull. Environ. Contam. Toxicol.* **44**, 254-259.

CHROMIUM POTASSIUM SULFATE DODECAHYDRATE

CAS RN: 7788-99-0

Sample purity: Analytical grade
Temperature: 15°C
Test parameter: EC50
Effect: Reduction in light output

Concentration: 10.7 mg/l Cr^{3+}
Exposure time: 5 min

Concentration: 12.6 mg/l Cr^{3+}
Exposure time: 10 min

Concentration: 15.3 mg/l Cr^{3+}
Exposure time: 15 min

Concentration: 15.8 mg/l Cr^{3+}
Exposure time: 20 min

Concentration: 16.0 mg/l Cr^{3+}
Exposure time: 30 min

Comment: Results derived from the average of two replicates.

Bibliographical reference: Qureshi, A.A., Coleman, R.N., and Paran, J.H. (1984). In: *Toxicity Screening Procedures Using Bacterial Systems*, D. Liu and B.J. Dutka (eds.), Marcel Dekker, New York, p. 1-22.

trans-CINNAMIC ACID

CAS RN: 140-10-3

Sample purity: 99+%
Temperature: 15°C
Test parameter: EC50
Effect: Reduction in light output

Concentration: 66.2 mg/l
Exposure times: 5 and 15 min

Concentration: 64.7 mg/l*
Exposure time: 30 min

Comment: Mean of four assays. Methanol (<10%) was used to prepare the stock solutions. EC50 values were calculated from nominal concentrations.

Bibliographical references: Kaiser, K.L.E., and Palabrica, V.S. (1991). *Water Poll. Res. J. Canada* **26**, 361-431.
* Kaiser, K.L.E., Palabrica, V.S., and Ribo, J.M. (1987). In: *QSAR in Environmental Toxicology - II*, K.L.E. Kaiser (ed.), D. Reidel Publishing Company, Dordrecht, p. 153-168.

CITRIC ACID

CAS RN: 77-92-9

Temperature: 15°C
Test parameter: EC50
Effect: Reduction in light output
Concentration: 14 mg/l
Exposure time: 15 min

Bibliographical reference: Bulich, A.A., Tung, K.K., and Scheibner, G. (1990). *J. Biolumin. Chemilumin.* **5**, 71-77.

CITRININ

CAS RN: 518-75-2

Temperature: 15°C
Test parameter: EC50
Effect: Reduction in light output

Concentration: 27.74 mg/l
Exposure time: 5 min

Concentration: 20.46 mg/l
Exposure time: 10 min

Concentration: 17.07 mg/l
Exposure time: 15 min

Concentration: 14.91 mg/l
Exposure time: 20 min

Comment: Freshly reconstituted bacterial suspensions. Chemical was dissolved in methanol.

Bibliographical reference: Yates, I.E., and Porter, J.K. (1982). *Appl. Environ. Microbiol.* **44**, 1072-1075.

Temperature: 15°C
Test parameter: EC20
Effect: Reduction in light output

Concentration: 11.08 mg/l
Exposure time: 5 min

Concentration: 7.00 mg/l
Exposure time: 20 min

Comment: Freshly reconstituted bacterial suspensions. Chemical was dissolved in methanol.

Bibliographical reference: Yates, I.E., and Porter, J.K. (1982). *Appl. Environ. Microbiol.* **44**, 1072-1075.

Temperature: 15°C
Test parameter: EC50
Effect: Reduction in light output

Concentration: 30.70 mg/l
Exposure time: 5 min

Concentration: 19.99 mg/l
Exposure time: 10 min

Concentration: 16.60 mg/l
Exposure time: 15 min

Concentration: 14.46 mg/l
Exposure time: 20 min

Comment: Performed on bacterial suspensions maintained at 3°C for 5 h after reconstitution. Chemical was dissolved in methanol.

Bibliographical reference: Yates, I.E., and Porter, J.K. (1982). *Appl. Environ. Microbiol.* **44**, 1072-1075.

COBALT CHLORIDE HEXAHYDRATE

CAS RN: 7791-13-1

Sample purity: Reagent grade
Test parameter: EC50
Effect: Reduction in light output

Concentration: 160 mg/l Co^{++} (140-180)
Exposure time: 5 min

Concentration: 16 mg/l Co^{++} (14-18)
Exposure time: 15 min

Concentration: 2.8 mg/l Co^{++} (2.3-3.4)
Exposure time: 30 min

Bibliographical reference: Elnabarawy, M.T., Robideau, R.R., and Beach, S.A. (1988). *Tox. Assess.* **3**, 361-370.

COPPER(II) CHLORIDE DIHYDRATE

CAS RN: 10125-13-0

Temperature: 15 ± 0.1°C
Test parameter: EC50
Effect: Reduction in light output

Concentration: 1.29 mg/l Cu^{++}
Exposure time: 5 min

Concentration: 0.27 mg/l Cu^{++}
Exposure time: 15 min

Concentration: 0.16 mg/l Cu^{++}
Exposure time: 30 min

Comment: The pH was not adjusted.

Bibliographical reference: Tarkpea, M., Hansson, M., and Samuelsson, B. (1986). *Ecotoxicol. Environ. Safety* **11**, 127-143.

COPPER SULFATE PENTAHYDRATE

CAS RN: 7758-99-8

Temperature: 15°C
Test parameter: EC50
Effect: Reduction in light output
Concentration: 0.80 ± 0.08 mg/l Cu^{++}
Exposure time: 10 min
Comment: Mean of four assays. Toxicity values were calculated from nominal concentrations.

Bibliographical reference: Ferard, J.F., Vasseur, P., Danoux, L., and Larbaigt, G. (1983). *Rev. Fr. Sci. Eau* **2**, 221-237.

Temperature: 20°C
Test parameter: EC50
Effect: Reduction in light output
Concentration: 0.58 ± 0.07 mg/l Cu^{++}
Exposure time: 10 min
Comment: Mean of four assays. Toxicity values were calculated from nominal concentrations.

Bibliographical reference: Ferard, J.F., Vasseur, P., Danoux, L., and Larbaigt, G. (1983). *Rev. Fr. Sci. Eau* **2**, 221-237.

Sample purity: Analytical grade
Temperature: 15°C

Test parameter: EC50
Effect: Reduction in light output

Concentrations: 2.46 mg/l Cu^{++} (time after reconstitution = 0.5 h)
0.72 mg/l Cu^{++} (time after reconstitution = 4 h)
Exposure time: 5 min

Concentrations: 0.30 mg/l Cu^{++} (time after reconstitution = 0.5 h)
0.27 mg/l Cu^{++} (time after reconstitution = 4 h)
Exposure time: 15 min

Bibliographical reference: Qureshi, A.A., Coleman, R.N., and Paran, J.H. (1984). In: *Toxicity Screening Procedures Using Bacterial Systems*, D. Liu and B.J. Dutka (eds.), Marcel Dekker, New York, p. 1-22.

Sample purity: Analytical grade
Temperature: 15°C
Test parameter: EC50
Effect: Reduction in light output

Concentration: 1.84 mg/l Cu^{++}
Exposure time: 5 min

Concentration: 1.98 mg/l Cu^{++}
Exposure time: 10 min

Concentration: 0.28 mg/l Cu^{++}
Exposure time: 15 min

Concentration: 0.18 mg/l Cu^{++}
Exposure time: 20 min

Concentration: 0.13 mg/l Cu^{++}
Exposure time: 30 min

Comment: Results derived from the average of two replicates.

Bibliographical reference: Qureshi, A.A., Coleman, R.N., and Paran, J.H. (1984). In: *Toxicity Screening Procedures Using Bacterial Systems*, D. Liu and B.J. Dutka (eds.), Marcel Dekker, New York, p. 1-22.

Temperature: 15°C

Test parameter: EC50
Effect: Reduction in light output
Concentration: 0.2 ± 0.03 mg/l Cu^{++}
Exposure time: 30 min
Comment: The assay was run in triplicate (pH = 6.2-6.6) using different bacterial reagents. The toxicity values were calculated from nominal concentrations.

Bibliographical reference: Vasseur, P., Dive, D., Sokar, Z., and Bonnemain, H. (1988). *Chemosphere* **17**, 767-782.

Sample purity: Reagent grade
Test parameter: EC50
Effect: Reduction in light output

Concentration: 1.3 mg/l Cu^{++} (1.1-1.6)
Exposure time: 5 min

Concentration: 0.25 mg/l Cu^{++} (0.24-0.26)
Exposure time: 15 min

Concentration: <0.25 mg/l Cu^{++}
Exposure time: 30 min

Bibliographical reference: Elnabarawy, M.T., Robideau, R.R., and Beach, S.A. (1988). *Tox. Assess.* **3**, 361-370.

Temperature: 15°C
Test parameter: EC50
Effect: Reduction in light output
Concentration: 1.33 mg/l Cu^{++}
Exposure time: 5 min
Comment: The test was performed in duplicate. The pH of the chemical solutions was not adjusted.

Bibliographical reference: Awong, J., Bitton, G., Koopman, B., and Morel, J.L. (1989). *Bull. Environ. Contam. Toxicol.* **43**, 118-122.

COTININE

CAS RN: 486-56-6

Sample purity: 98% (Aldrich Chemical Company, Milwaukee, WI)
Test parameter: EC50
Effect: Reduction in light output
Result:

Chemical	toxicity (10^{-4} mol/l)		
	5 min	15 min	25 min
Cotinine	10.6 ± 1.05	11.6 ± 1.25	11.8 ± 1.35
Nicotine*	7.77 ± 1.54	7.31 ± 1.83	7.39 ± 1.46
Nicotine/Cotinine = 0.4	15.9 ± 3.72	14.9 ± 2.37	15.3 ± 2.50
Nicotine/Cotinine = 1.0	14.1 ± 3.26	13.4 ± 2.15	13.5 ± 1.77
Nicotine/Cotinine = 1.6	14.1 ± 2.94	12.1 ± 2.44	12.0 ± 2.24

* purity 98% (Eastman Kodak Company, Rochester, NY)

Bibliographical reference: Chou, C.C., and Que Hee, S.S. (1993). *J. Biolumin. Chemilumin.* **8**, 39-48.

COUMARIN

CAS RN: 91-64-5

Temperature: 15°C
Test parameter: EC50
Effect: Reduction in light output

Concentration: 11.6 mg/l
Exposure time: 5 min

Concentration: 11.9 mg/l
Exposure time: 15 min

Concentration: 12.4 mg/l
Exposure time: 30 min

Comment: Mean of four assays. Methanol (<10%) was used to prepare the stock solutions. EC50 values were calculated from nominal concentrations.

Bibliographical reference: Kaiser, K.L.E., and Palabrica, V.S. (1991). *Water Poll. Res. J. Canada* **26**, 361-431.

o-CRESOL

CAS RN: 95-48-7
Synonym: 2-Methylphenol

Temperature: 15°C
Test parameter: EC50
Effect: Reduction in light output
Concentration: 32 mg/l
Exposure time: 5 min

Bibliographical reference: Lebsack, M.E., Anderson, A.D., DeGraeve, G.M., and Bergman, H.L. (1981). In: *Aquatic Toxicology and Hazard Assessment: Fourth Conference*, *ASTM STP 737*, D.R. Branson and K.L. Dickson (eds.), American Society for Testing and Materials, Philadelphia, p. 348-356.

Sample purity: >98%
Temperature: 15 ± 0.1°C
Test parameter: EC50
Effect: Reduction in light output

Concentration: 20.7 mg/l
Exposure time: 5 min

Concentration: 15.4 mg/l
Exposure time: 15 min

Comment: Test was performed in duplicate. Toxicity values were based on nominal test concentrations.

Bibliographical reference: de Zwart, D., and Slooff, W. (1983). *Aquat. Toxicol.* **4**, 129-138.

Sample purity: >98%
Temperature: 15 ± 0.1°C
Test parameter: EC10
Effect: Reduction in light output

Concentration: 7.1 mg/l
Exposure time: 5 min

Concentration: 7.2 mg/l
Exposure time: 15 min

Comment: Test was performed in duplicate. Toxicity values were based on nominal test concentrations.

Bibliographical reference: de Zwart, D., and Slooff, W. (1983). *Aquat. Toxicol.* **4**, 129-138.

Sample purity: 99%
Temperature: 15°C
Test parameter: EC50
Effect: Reduction in light output

Concentration: 22.6 mg/l
Exposure time: 5 min

Concentration: 25.9 mg/l
Exposure time: 15 min

Concentration: 26.5 mg/l
Exposure time: 30 min

Comment: Mean of four assays. Methanol (<10%) was used to prepare the stock solutions. EC50 values were calculated from nominal concentrations.

Bibliographical reference: Kaiser, K.L.E., and Palabrica, V.S. (1991). *Water Poll. Res. J. Canada* **26**, 361-431.

m-CRESOL

CAS RN: 108-39-4
Synonym: 3-Methylphenol

Temperature: 15°C
Test parameter: EC50
Effect: Reduction in light output
Concentration: 8.20 mg/l
Exposure time: 5 min

Bibliographical reference: Lebsack, M.E., Anderson, A.D., DeGraeve, G.M., and Bergman, H.L. (1981). In: *Aquatic Toxicology and Hazard Assessment: Fourth Conference, ASTM STP 737*, D.R. Branson and K.L. Dickson (eds.), American Society for Testing and Materials, Philadelphia, p. 348-356.

Temperature: 15 ± 0.1°C
Test parameter: EC50
Effect: Reduction in light output
Concentration: 11 mg/l
Exposure time: 5 min

Bibliographical reference: Chang, J.C., Taylor, P.B., and Leach, F.R. (1981). *Bull. Environ. Contam. Toxicol.* **26**, 150-156.

Sample purity: 99+%
Temperature: 15°C
Test parameter: EC50
Effect: Reduction in light output

Concentration: 6.82 mg/l
Exposure time: 5 min

Concentration: 7.48 mg/l
Exposure time: 15 min

Concentration: 7.83 mg/l
Exposure time: 30 min

Comment: Mean of three assays. Methanol (<10%) was used to prepare the stock solutions. EC50 values were calculated from nominal concentrations.

Bibliographical reference: Kaiser, K.L.E., and Palabrica, V.S. (1991). *Water Poll. Res. J. Canada* **26**, 361-431.

p-CRESOL

CAS RN: 106-44-5
Synonym: 4-Methylphenol

Temperature: 15°C
Test parameter: EC50
Effect: Reduction in light output
Concentration: 1.5 mg/l
Exposure time: 5 min

Bibliographical reference: Bulich, A.A., Greene, M.W., and Isenberg, D.L. (1981). In: *Aquatic Toxicology and Hazard Assessment:*

Fourth Conference, ASTM STP 737, D.R. Branson and K.L. Dickson (eds.), American Society for Testing and Materials, Philadelphia, p. 338-347.

Temperature: 15°C
Test parameter: EC50
Effect: Reduction in light output
Concentration: 1.30 mg/l
Exposure time: 5 min

Bibliographical reference: Lebsack, M.E., Anderson, A.D., DeGraeve, G.M., and Bergman, H.L. (1981). In: *Aquatic Toxicology and Hazard Assessment: Fourth Conference, ASTM STP 737*, D.R. Branson and K.L. Dickson (eds.), American Society for Testing and Materials, Philadelphia, p. 348-356.

Temperature: 15°C
Test parameter: EC50
Effect: Reduction in light output

Concentration: 2.06 mg/l
Exposure time: 5 min

Concentration: 2.31 mg/l
Exposure time: 15 min

Concentration: 2.37 mg/l
Exposure time: 30 min

Comment: Mean of three assays. Methanol (<10%) was used to prepare the stock solutions. EC50 values were calculated from nominal concentrations.

Bibliographical reference: Ribo, J.M., and Kaiser, K.L.E. (1983). *Chemosphere* **12**, 1421-1442.

Temperature: 15°C
Test parameter: EC50
Effect: Reduction in light output
Concentration: 1.4 mg/l
Exposure time: 5 min

Bibliographical reference: Kahru, A. (1993). *ATLA* **21**, 210-215.

CUMENE

CAS RN: 98-82-8
Synonym: Isopropylbenzene

Sample purity: 99%
Temperature: 15°C
Test parameter: EC50
Effect: Reduction in light output

Concentration: 0.89 mg/l
Exposure time: 5 min

Concentration: 1.10 mg/l
Exposure time: 15 min

Concentration: 1.48 mg/l
Exposure time: 30 min

Comment: Mean of three assays. Methanol (<10%) was used to prepare the stock solutions. EC50 values were calculated from nominal concentrations.

Bibliographical reference: Kaiser, K.L.E., and Palabrica, V.S. (1991). *Water Poll. Res. J. Canada* **26**, 361-431.

4-CYANOBENZALDEHYDE

CAS RN: 105-07-7

Sample purity: 99%
Temperature: 15°C
Test parameter: EC50
Effect: Reduction in light output

Concentration: 14.1 mg/l
Exposure time: 5 min

Concentration: 13.7 mg/l
Exposure time: 15 min

Concentration: 12.5 mg/l*
Exposure time: 30 min

Comment: Mean of three assays. Methanol (<10%) was used to prepare the stock solutions. EC50 values were calculated from nominal concentrations.

Bibliographical references: Kaiser, K.L.E., and Palabrica, V.S. (1991). *Water Poll. Res. J. Canada* **26**, 361-431.
* Kaiser, K.L.E., and Gough, K.M. (1989). In: *Aquatic Toxicology and Environmental Fate: Eleventh Volume, ASTM STP 1007*, G.W. Suter and M.A. Lewis (eds.), American Society for Testing and Materials, Philadelphia, p. 424-441.

4-CYANOBENZOIC ACID

CAS RN: 619-65-8

Sample purity: 99%
Temperature: 15°C
Test parameter: EC50
Effect: Reduction in light output

Concentration: 73.7 mg/l
Exposure time: 5 min

Concentration: 80.9 mg/l
Exposure time: 15 min

Concentration: 92.8 mg/l
Exposure time: 30 min

Comment: Mean of three assays. Methanol (<10%) was used to prepare the stock solutions. EC50 values were calculated from nominal concentrations.

Bibliographical reference: Kaiser, K.L.E., and Palabrica, V.S. (1991). *Water Poll. Res. J. Canada* **26**, 361-431.

4-CYANONITROBENZENE

CAS RN: 619-72-7
Synonym: 4-Nitrobenzonitrile

Temperature: 15°C

Test parameter: EC50
Effect: Reduction in light output
Concentration: 2.40 mg/l
Exposure time: 30 min
Comment: Mean of three assays. Methanol (<10%) was used to prepare the stock solutions. EC50 values were calculated from nominal concentrations.

Bibliographical reference: Kaiser, K.L.E. (1987). In: *QSAR in Environmental Toxicology - II*, K.L.E. Kaiser (ed.), D. Reidel Publishing Company, Dordrecht, p. 169-188.

2-CYANOPYRIDINE

CAS RN: 100-70-9

Sample purity: 99%
Temperature: 15°C
Test parameter: EC50
Effect: Reduction in light output

Concentration: 314 mg/l
Exposure time: 5 min

Concentration: 161 mg/l
Exposure time: 15 min

Concentration: 88.6 mg/l
Exposure time: 30 min

Comment: Mean of four assays. Methanol (<10%) was used to prepare the stock solutions. EC50 values were calculated from nominal concentrations.

Bibliographical reference: Kaiser, K.L.E., and Palabrica, V.S. (1991). *Water Poll. Res. J. Canada* **26**, 361-431.

3-CYANOPYRIDINE

CAS RN: 100-54-9

Sample purity: 98%

Temperature: 15°C
Test parameter: EC50
Effect: Reduction in light output

Concentration: 642 mg/l
Exposure times: 5 and 15 min

Concentration: 657 mg/l
Exposure time: 30 min

Comment: Mean of three assays. Methanol (<10%) was used to prepare the stock solutions. EC50 values were calculated from nominal concentrations.

Bibliographical reference: Kaiser, K.L.E., and Palabrica, V.S. (1991). *Water Poll. Res. J. Canada* **26**, 361-431.

4-CYANOPYRIDINE

CAS RN: 100-48-1

Sample purity: 98%
Temperature: 15°C
Test parameter: EC50
Effect: Reduction in light output

Concentrations: 454 mg/l
487 mg/l
Exposure time: 5 min

Concentrations: 414 mg/l
434 mg/l
Exposure time: 15 min

Concentrations: 396 mg/l
487 mg/l
Exposure time: 30 min

Comment: The two series of tests were performed in triplicate. Methanol (<10%) was used to prepare the stock solutions. EC50 values were calculated from nominal concentrations.

Bibliographical reference: Kaiser, K.L.E., and Palabrica, V.S. (1991). *Water Poll. Res. J. Canada* **26**, 361-431.

4-CYANOPYRIDINE *N*-OXIDE

CAS RN: 14906-59-3
Synonym: Isonicotinonitrile 1-oxide

Sample purity: 96%
Temperature: 15°C
Test parameter: EC50
Effect: Reduction in light output

Concentration: 288 mg/l
Exposure time: 5 min

Concentration: 316 mg/l
Exposure time: 15 min

Concentration: 339 mg/l
Exposure time: 30 min

Comment: Mean of three assays. Methanol (<10%) was used to prepare the stock solutions. EC50 values were calculated from nominal concentrations.

Bibliographical reference: Kaiser, K.L.E., and Palabrica, V.S. (1991). *Water Poll. Res. J. Canada* **26**, 361-431.

CYCLOHEXANE

CAS RN: 110-82-7

Temperature: 15°C
Test parameter: EC50
Effect: Reduction in light output

Concentration: 85.5 mg/l
Exposure time: 5 min

Concentration: 93 mg/l
Exposure time: 10 min

Comment: Mean of two assays. Chemical was dissolved in DMSO (0.1%) (P. Vasseur, private communication). Toxicity values were calculated from nominal concentrations.

Bibliographical reference: Ferard, J.F., Vasseur, P., Danoux, L., and Larbaigt, G. (1983). *Rev. Fr. Sci. Eau* **2**, 221-237.

1,4-CYCLOHEXANEDIOL

CAS RN: 556-48-9

Sample purity: 99%
Temperature: 15°C
Test parameter: EC50
Effect: Reduction in light output

Concentration: 3850 mg/l
Exposure time: 5 min

Concentration: 3940 mg/l
Exposure time: 15 min

Concentration: 4030 mg/l
Exposure time: 30 min

Comment: Mean of three assays. Methanol (<10%) was used to prepare the stock solutions. EC50 values were calculated from nominal concentrations.

Bibliographical reference: Kaiser, K.L.E., and Palabrica, V.S. (1991). *Water Poll. Res. J. Canada* **26**, 361-431.

CYCLOHEXANOL

CAS RN: 108-93-0

Test parameter: EC50
Effect: Reduction in light output
Concentration: 115 mg/l
Exposure time: 5 min
Comment: The EC50 value was calculated from nominal concentrations.

Bibliographical reference: Curtis, C., Lima, A., Lozano, S.J., and Veith, G.D. (1982). In: *Aquatic Toxicology and Hazard Assessment: Fifth Conference, ASTM STP 766*, J.G. Pearson, R.B. Foster, and W.E.

Bishop (eds.), American Society for Testing and Materials, Philadelphia, p. 170-178.

Temperature: 15°C
Test parameter: EC50
Effect: Reduction in light output

Concentration: 83 mg/l
Exposure time: 5 min

Concentration: 42.5 mg/l
Exposure time: 10 min

Comment: Mean of two assays. Toxicity values were calculated from nominal concentrations.

Bibliographical reference: Ferard, J.F., Vasseur, P., Danoux, L., and Larbaigt, G. (1983). *Rev. Fr. Sci. Eau* **2**, 221-237.

CYCLOHEXANONE

CAS RN: 108-94-1

Test parameter: EC50
Effect: Reduction in light output
Concentration: 18.5 mg/l
Exposure time: 5 min
Comment: The EC50 value was calculated from nominal concentrations.

Bibliographical reference: Curtis, C., Lima, A., Lozano, S.J., and Veith, G.D. (1982). In: *Aquatic Toxicology and Hazard Assessment: Fifth Conference, ASTM STP 766*, J.G. Pearson, R.B. Foster, and W.E. Bishop (eds.), American Society for Testing and Materials, Philadelphia, p. 170-178.

Temperature: 15°C
Test parameter: EC50
Effect: Reduction in light output

Concentration: 25 mg/l
Exposure time: 5 min

Concentration: 21.3 mg/l
Exposure time: 10 min

Comment: Mean of two assays. Toxicity values were calculated from nominal concentrations.

Bibliographical reference: Ferard, J.F., Vasseur, P., Danoux, L., and Larbaigt, G. (1983). *Rev. Fr. Sci. Eau* **2**, 221-237.

CYCLOHEXANONE OXIME

CAS RN: 100-64-1

Sample purity: 97%
Temperature: 15°C
Test parameter: EC50
Effect: Reduction in light output

Concentration: 19.7 mg/l
Exposure time: 5 min

Concentration: 22.1 mg/l
Exposure time: 15 min

Concentration: 25.9 mg/l
Exposure time: 30 min

Comment: Mean of four assays. Methanol (<10%) was used to prepare the stock solutions. EC50 values were calculated from nominal concentrations.

Bibliographical reference: Kaiser, K.L.E., and Palabrica, V.S. (1992). National Water Research Institute, Burlington, Ontario, Canada, unpublished results.

CYCLOHEXYLAMINE

CAS RN: 108-91-8

Temperature: 15 ± 0.2°C
Test parameter: EC50
Effect: Reduction in light output

Concentration: 120 mg/l
Exposure time: 30 min
Comment: The test was run in duplicate with a minimum of five serial concentration dilutions. Samples were pH adjusted to 7 ± 0.5 pH units.

Bibliographical reference: Indorato, A.M., Snyder, K.B., and Usinowicz, P.J. (1984). In: *Toxicity Screening Procedures Using Bacterial Systems*, D. Liu and B.J. Dutka (eds.), Marcel Dekker, New York, p. 37-53.

CYHEXATIN

CAS RN: 13121-70-5

Test parameter: EC50
Effect: Reduction in light output
Concentration: 0.383 mg/l
Exposure time: 30 min

Bibliographical reference: Steinhäuser, K.G., Amann, W., Späth, A., and Polenz, A. (1985). *Vom Wasser* **65**, 203-214.

CYPERMETHRIN

CAS RN: 52315-07-8

Sample purity: 74%
Temperature: 15°C
Test parameter: EC50
Effect: Reduction in light output

Concentration: 5.36 mg/l
Exposure time: 5 min

Concentration: 6.60 mg/l
Exposure time: 15 min

Concentration: 8.50 mg/l
Exposure time: 30 min

Comment: Mean of three assays. Methanol (<10%) was used to prepare the stock solutions. EC50 values were calculated from nominal

concentrations.

Bibliographical reference: Kaiser, K.L.E., and Palabrica, V.S. (1991). *Water Poll. Res. J. Canada* **26**, 361-431.

2,4-D

CAS RN: 94-75-7
Synonym: 2,4-Dichlorophenoxyacetic acid

Sample purity: Analytical reference standard grade
Temperature: 15 ± 0.1°C
Test parameter: EC50
Effect: Reduction in light output
Concentration: 61.6 mg/l
Exposure time: 5 min
Comment: Test was performed on *Photobacterium phosphoreum* NZ11D obtained from the Scripps Institute of Oceanography (La Jolla, CA).

Bibliographical reference: McFeters, G.A., Bond, P.J., Olson, S.B., and Tchan, Y.T. (1983). *Water Res.* **17**, 1757-1762.

Sample purity: 94%
Test parameter: EC50
Effect: Reduction in light output

Concentration: 112 mg/l
Exposure time: 5 min

Concentration: 107 mg/l
Exposure time: 15 min

Concentration: 128 mg/l
Exposure time: 30 min

Bibliographical reference: Miller, W.E., Peterson, S.A., Greene, J.C., and Callahan, C.A. (1985). *J. Environ. Qual.* **14**, 569-574.

Temperature: 15 ± 0.1°C
Test parameter: EC50
Effect: Reduction in light output

Concentration: 100.7 mg/l
Exposure time: 5 min

Bibliographical reference: Somasundaram, L., Coats, J.R., Racke, K.D., and Stahr, H.M. (1990). *Bull. Environ. Contam. Toxicol.* **44**, 254-259.

Temperature: 15°C
Test parameter: EC50
Effect: Reduction in light output
Concentration: 5.74 ± 0.41 mg/l
Exposure time: 15 min

Bibliographical reference: Rychert, R., and Mortimer, M. (1991). *Environ. Toxicol. Water Qual.* **6**, 415-421.

p,p'-DDT

CAS RN: 50-29-3
Synonym: Dichloro diphenyl trichloroethane

Temperature: 15°C
Test parameter: EC50
Effect: Reduction in light output
Concentration: Not toxic at 10 µg/l
Exposure time: 10 min
Comment: Result of three assays. Chemical dissolved in 0.1 ml/l acetone. Test solutions were analyzed using gas chromatography.

Bibliographical reference: Bazin, C. (1985). DEA Ecologie Fondamentale et Appliquée des Eaux Continentales. University Lyon I, France.

DECANOIC ACID

CAS RN: 334-48-5
Synonym: Capric acid

Sample purity: 98%
Temperature: 15°C
Test parameter: EC50

Effect: Reduction in light output

Concentration: 11.2 ± 1.27 mg/l
Exposure time: 5 min

Concentration: 9.31 ± 0.93 mg/l
Exposure time: 15 min

Concentration: 9.0 ± 0.89 mg/l
Exposure time: 25 min

Comment: Mean of six assays. The values were converted to mg/l from the original data expressed in μM and a rounded molecular weight of 172 given by the authors. The EC50 values were determined in 0.45% (v/v) methanol in 2% (w/w) sodium chloride solution. The 5-min EC50 value for methanol was 43000 mg/l. Phenol solution was used for quality control/quality assurance. The 5-min EC50 value was 18.2 mg/l. The EC50 value at 15 min was 20.7 mg/l with a relative error of <5%.

Bibliographical reference: Chou, C.C., and Que Hee, S.S. (1992). *Ecotoxicol. Environ. Safety* **23**, 355-363.

1-DECANOL

CAS RN: 112-30-1

Temperature: 15°C
Test parameter: EC50
Effect: Reduction in light output

Concentration: 1.31 mg/l
Exposure time: 5 min

Concentration: 1.47 mg/l
Exposure time: 30 min

Bibliographical reference: Speece, R. (1987). Drexel University, Philadelphia, USA, private communication.

2-DECANOL

CAS RN: 1120-06-5

Test parameter: EC50
Effect: Reduction in light output
Concentration: 1.16 mg/l
Exposure time: 5 min
Comment: The EC50 value was calculated from nominal concentrations.

Bibliographical reference: Curtis, C., Lima, A., Lozano, S.J., and Veith, G.D. (1982). In: *Aquatic Toxicology and Hazard Assessment: Fifth Conference, ASTM STP 766*, J.G. Pearson, R.B. Foster, and W.E. Bishop (eds.), American Society for Testing and Materials, Philadelphia, p. 170-178.

2-DECANONE

CAS RN: 693-54-9

Test parameter: EC50
Effect: Reduction in light output
Concentrations: 6.1 mg/l
9.7 mg/l
Exposure time: 5 min
Comment: Concentrations in the tests were measured.

Bibliographical reference: Curtis, C., Lima, A., Lozano, S.J., and Veith, G.D. (1982). In: *Aquatic Toxicology and Hazard Assessment: Fifth Conference, ASTM STP 766*, J.G. Pearson, R.B. Foster, and W.E. Bishop (eds.), American Society for Testing and Materials, Philadelphia, p. 170-178.

DECYL ALDEHYDE

CAS RN: 112-31-2

Sample purity: 98%
Temperature: 15°C
Test parameter: EC50
Effect: Reduction in light output

Concentration: 4.71 ± 2.03 mg/l
Exposure time: 5 min

Concentration: 3.59 ± 0.97 mg/l
Exposure time: 15 min

Concentration: 2.90 ± 0.66 mg/l
Exposure time: 25 min

Comment: Mean of five assays. The values were converted to mg/l from the original data expressed in μM and a rounded molecular weight of 156 given by the authors. The EC50 values were determined in 0.45% (v/v) methanol in 2% (w/w) sodium chloride solution. The 5-min EC50 value for methanol was 43000 mg/l. Phenol solution was used for quality control/quality assurance. The 5-min EC50 value was 18.2 mg/l. The EC50 value at 15 min was 20.7 mg/l with a relative error of <5%.

Bibliographical reference: Chou, C.C., and Que Hee, S.S. (1992). *Ecotoxicol. Environ. Safety* **23**, 355-363.

DECYL SULFATE SODIUM SALT

CAS RN: 142-87-0

Sample purity: 99%
Temperature: 15°C
Test parameter: EC50
Effect: Reduction in light output

Concentration: 18.9 mg/l
Exposure time: 5 min

Concentration: 9.90 mg/l
Exposure time: 15 min

Concentration: 7.01 mg/l
Exposure time: 30 min

Comment: Mean of three assays. Methanol (<10%) was used to prepare the stock solutions. EC50 values were calculated from nominal concentrations.

Bibliographical reference: Kaiser, K.L.E., and Palabrica, V.S.

(1991). *Water Poll. Res. J. Canada* **26**, 361-431.

1,4-DIACETYLBENZENE

CAS RN: 1009-61-6

Sample purity: 98%
Temperature: 15°C
Test parameter: EC50
Effect: Reduction in light output
Concentration: 2.95 mg/l
Exposure time: 30 min
Comment: Mean of three assays. Methanol (<10%) was used to prepare the stock solutions. EC50 values were calculated from nominal concentrations.

Bibliographical reference: Ribo, J.M., and Kaiser, K.L.E. National Water Research Institute, Burlington, Ontario, Canada, unpublished results (value later published by Kaiser, K.L.E., and Palabrica, V.S. (1991). *Water Poll. Res. J. Canada* **26**, 361-431).

DIALLYL DIBUTYLTIN

CAS RN: 15336-98-8

Temperature: 15°C
Test parameter: EC50
Effect: Reduction in light output

Concentration: 2.10 ± 0.36 mg/l
Exposure time: 5 min

Concentration: 1.29 ± 0.38 mg/l
Exposure time: 15 min

Comment: Chemical was dissolved in ethanol (~0.05%).

Bibliographical reference: Dooley, C.A., and Kenis, P. (1987). In: *Oceans '87, International Organotin Symposium*, Department of Fisheries and Oceans, Canada, and the Society for Underwater Technology, William MacNab & Son, Halifax, p. 1517-1524.

3,3'-DIAMINOBENZIDINE

CAS RN: 91-95-2
Synonym: 3,3',4,4'-Biphenyltetramine

Sample purity: 99%
Temperature: 15°C
Test parameter: EC50
Effect: Reduction in light output

Concentration: 24.6 mg/l
Exposure time: 5 min

Concentration: 25.2 mg/l
Exposure time: 15 min

Concentration: 25.8 mg/l
Exposure time: 30 min

Comment: Mean of three assays. Methanol (<10%) was used to prepare the stock solutions. EC50 values were calculated from nominal concentrations.

Bibliographical reference: Kaiser, K.L.E., and Palabrica, V.S. (1992). National Water Research Institute, Burlington, Ontario, Canada, unpublished results.

3,4-DIAMINOBENZOIC ACID

CAS RN: 619-05-6

Sample purity: 97%
Temperature: 15°C
Test parameter: EC50
Effect: Reduction in light output

Concentration: 121 mg/l
Exposure time: 5 min

Concentration: 115 mg/l
Exposure time: 15 min

Concentration: 105 mg/l
Exposure time: 30 min

Comment: Mean of three assays. Methanol (<10%) was used to prepare the stock solutions. EC50 values were calculated from nominal concentrations.

Bibliographical reference: Kaiser, K.L.E., and Palabrica, V.S. (1991). *Water Poll. Res. J. Canada* **26**, 361-431.

3,5-DIAMINOBENZOIC ACID

CAS RN: 535-87-5

Sample purity: 98%
Temperature: 15°C
Test parameter: EC50
Effect: Reduction in light output

Concentration: 27.7 mg/l
Exposure time: 5 min

Concentration: 28.3 mg/l
Exposure time: 15 min

Concentration: 31.8 mg/l
Exposure time: 30 min

Comment: Mean of three assays. Methanol (<10%) was used to prepare the stock solutions. EC50 values were calculated from nominal concentrations.

Bibliographical reference: Kaiser, K.L.E., and Palabrica, V.S. (1991). *Water Poll. Res. J. Canada* **26**, 361-431.

4,4'-DIAMINOBIBENZYL

CAS RN: 621-95-4

Sample purity: 97%
Temperature: 15°C
Test parameter: EC50
Effect: Reduction in light output

Concentration: 10.9 mg/l

Exposure time: 5 min

Concentration: 11.4 mg/l
Exposure time: 15 min

Concentration: 10.4 mg/l
Exposure time: 30 min

Comment: Mean of three assays. Methanol (<10%) was used to prepare the stock solutions. EC50 values were calculated from nominal concentrations.

Bibliographical reference: Kaiser, K.L.E., and Palabrica, V.S. (1991). *Water Poll. Res. J. Canada* **26**, 361-431.

4,4'-DIAMINOBIBENZYLMETHANE

CAS RN: 101-77-9

Sample purity: 99%
Temperature: 15°C
Test parameter: EC50
Effect: Reduction in light output

Concentration: 5.99 mg/l
Exposure time: 5 min

Concentration: 6.57 mg/l
Exposure times: 15 and 30 min

Comment: Mean of three assays. Methanol (<10%) was used to prepare the stock solutions. EC50 values were calculated from nominal concentrations.

Bibliographical reference: Kaiser, K.L.E., and Palabrica, V.S. (1991). *Water Poll. Res. J. Canada* **26**, 361-431.

4,4'-DIAMINODIPHENYL ETHER

CAS RN: 101-80-4

Sample purity: 98%

Temperature: 15°C
Test parameter: EC50
Effect: Reduction in light output

Concentration: 3.73 mg/l
Exposure time: 5 min

Concentration: 3.56 mg/l
Exposure time: 15 min

Concentration: 3.17 mg/l*
Exposure time: 30 min

Comment: Mean of three assays. Methanol (<10%) was used to prepare the stock solutions. EC50 values were calculated from nominal concentrations.

Bibliographical references: Kaiser, K.L.E., and Palabrica, V.S. (1991). *Water Poll. Res. J. Canada* **26**, 361-431.
* Kaiser, K.L.E. (1987). In: *QSAR in Environmental Toxicology - II*, K.L.E. Kaiser (ed.), D. Reidel Publishing Company, Dordrecht, p. 169-188.

4,4'-DIAMINODIPHENYL SULFONE

CAS RN: 80-08-0

Sample purity: 99%
Temperature: 15°C
Test parameter: EC50
Effect: Reduction in light output

Concentration: 119 mg/l
Exposure time: 5 min

Concentration: 108 mg/l
Exposure time: 15 min

Concentration: 98.9 mg/l
Exposure time: 30 min

Comment: Mean of three assays. Methanol (<10%) was used to prepare the stock solutions. EC50 values were calculated from nominal concentrations.

Bibliographical reference: Kaiser, K.L.E., and Palabrica, V.S. (1991). *Water Poll. Res. J. Canada* **26**, 361-431.

2,7-DIAMINOFLUORENE

CAS RN: 525-64-4

Sample purity: 97%
Temperature: 15°C
Test parameter: EC50
Effect: Reduction in light output

Concentration: 31.1 mg/l
Exposure time: 5 min

Concentration: 30.4 mg/l
Exposure time: 15 min

Concentration: 29.0 mg/l
Exposure time: 30 min

Comment: Mean of three assays. Methanol (<10%) was used to prepare the stock solutions. EC50 values were calculated from nominal concentrations.

Bibliographical reference: Kaiser, K.L.E., and Palabrica, V.S. (1991). *Water Poll. Res. J. Canada* **26**, 361-431.

1,5-DIAMINONAPHTHALENE

CAS RN: 2243-62-1

Sample purity: 97%
Temperature: 15°C
Test parameter: EC50
Effect: Reduction in light output

Concentration: 45.6 mg/l
Exposure time: 5 min

Concentration: 37.9 mg/l
Exposure time: 15 min

Concentration: 36.2 mg/l
Exposure time: 30 min

Comment: Mean of three assays. Methanol (<10%) was used to prepare the stock solutions. EC50 values were calculated from nominal concentrations.

Bibliographical reference: Kaiser, K.L.E., and Palabrica, V.S. (1991). *Water Poll. Res. J. Canada* **26**, 361-431.

1,8-DIAMINONAPHTHALENE

CAS RN: 479-27-6

Sample purity: 97%
Temperature: 15°C
Test parameter: EC50
Effect: Reduction in light output

Concentration: 14.8 mg/l
Exposure time: 5 min

Concentration: 16.6 mg/l
Exposure time: 15 min

Concentration: 17.8 mg/l
Exposure time: 30 min

Comment: Mean of four assays. Methanol (<10%) was used to prepare the stock solutions. EC50 values were calculated from nominal concentrations.

Bibliographical reference: Kaiser, K.L.E., and Palabrica, V.S. (1991). *Water Poll. Res. J. Canada* **26**, 361-431.

2,3-DIAMINONAPHTHALENE

CAS RN: 771-97-1

Sample purity: 97%
Temperature: 15°C
Test parameter: EC50

Effect: Reduction in light output

Concentration: 13.5 mg/l
Exposure time: 5 min

Concentration: 12.6 mg/l
Exposure time: 15 min

Concentration: 12.0 mg/l
Exposure time: 30 min

Comment: Mean of four assays. Methanol (<10%) was used to prepare the stock solutions. EC50 values were calculated from nominal concentrations.

Bibliographical reference: Kaiser, K.L.E., and Palabrica, V.S. (1991). *Water Poll. Res. J. Canada* **26**, 361-431.

1,2-DIAMINOPROPANE

CAS RN: 78-90-0
Synonyms: 1,2-Propanediamine; Propylenediamine

Sample purity: 99+%
Temperature: 15°C
Test parameter: EC50
Effect: Reduction in light output

Concentration: 25.0 mg/l
Exposure time: 5 min

Concentration: 22.0 mg/l*
Exposure time: 15 min

Bibliographical references: Cronin, M.T.D., Dearden, J.C., and Dobbs, A.J. (1991). *Sci. Total Environ.* **109/110**, 431-439.
* Cronin, M.T.D. (1990). Thesis, Liverpool Polytechnic, Liverpool, UK.

2,3-DIAMINOPYRIDINE

CAS RN: 452-58-4

Sample purity: 98%
Temperature: 15°C
Test parameter: EC50
Effect: Reduction in light output

Concentration: 689 mg/l
Exposure time: 5 min

Concentration: 628 mg/l
Exposure time: 15 min

Concentration: 600 mg/l
Exposure time: 30 min

Comment: Mean of three assays. Methanol (<10%) was used to prepare the stock solutions. EC50 values were calculated from nominal concentrations.

Bibliographical reference: Kaiser, K.L.E., and Palabrica, V.S. (1991). *Water Poll. Res. J. Canada* **26**, 361-431.

2,6-DIAMINOPYRIDINE

CAS RN: 141-86-6

Sample purity: >99%
Temperature: 15°C
Test parameter: EC50
Effect: Reduction in light output

Concentration: 560 mg/l
Exposure time: 5 min

Concentration: 522 mg/l
Exposure times: 15 and 30 min

Comment: Mean of three assays. Methanol (<10%) was used to prepare the stock solutions. EC50 values were calculated from nominal concentrations.

Bibliographical reference: Kaiser, K.L.E., and Palabrica, V.S. (1991). *Water Poll. Res. J. Canada* **26**, 361-431.

3,4-DIAMINOPYRIDINE

CAS RN: 54-96-6

Sample purity: 98%
Temperature: 15°C
Test parameter: EC50
Effect: Reduction in light output

Concentration: 37.0 mg/l
Exposure time: 5 min

Concentration: 32.2 mg/l
Exposure time: 15 min

Concentration: 33.7 mg/l
Exposure time: 30 min

Comment: Mean of three assays. Methanol (<10%) was used to prepare the stock solutions. EC50 values were calculated from nominal concentrations.

Bibliographical reference: Kaiser, K.L.E., and Palabrica, V.S. (1991). *Water Poll. Res. J. Canada* **26**, 361-431.

2,3-DIAMINOTOLUENE

CAS RN: 2687-25-4

Sample purity: 97%
Temperature: 15°C
Test parameter: EC50
Effect: Reduction in light output

Concentration: 28.0 mg/l
Exposure time: 5 min

Concentration: 19.8 mg/l
Exposure time: 15 min

Concentration: 17.7 mg/l
Exposure time: 30 min

Comment: Mean of four assays. Methanol (<10%) was used to prepare the stock solutions. EC50 values were calculated from nominal concentrations.

Bibliographical reference: Kaiser, K.L.E., and Palabrica, V.S. (1991). *Water Poll. Res. J. Canada* **26**, 361-431.

2,4-DIAMINOTOLUENE

CAS RN: 95-80-7

Temperature: 15 ± 0.2°C
Test parameter: EC50
Effect: Reduction in light output

Concentration: 73 mg/l
Exposure time: 5 min

Concentration: 86 mg/l
Exposure time: 30 min

Comment: The test was run in duplicate with a minimum of five serial concentration dilutions. Samples were pH adjusted to 7 ± 0.5 pH units.

Bibliographical reference: Indorato, A.M., Snyder, K.B., and Usinowicz, P.J. (1984). In: *Toxicity Screening Procedures Using Bacterial Systems*, D. Liu and B.J. Dutka (eds.), Marcel Dekker, New York, p. 37-53.

Sample purity: 98%
Temperature: 15°C
Test parameter: EC50
Effect: Reduction in light output

Concentration: 106 mg/l
Exposure time: 5 min

Concentration: 97.0 mg/l
Exposure time: 15 min

Concentration: 102 mg/l
Exposure time: 30 min

Comment: Mean of three assays. Methanol (<10%) was used to

prepare the stock solutions. EC50 values were calculated from nominal concentrations.

Bibliographical reference: Kaiser, K.L.E., and Palabrica, V.S. (1991). *Water Poll. Res. J. Canada* **26**, 361-431.

3,4-DIAMINOTOLUENE

CAS RN: 496-72-0

Sample purity: 97%
Temperature: 15°C
Test parameter: EC50
Effect: Reduction in light output

Concentration: 40.5 mg/l
Exposure time: 5 min

Concentration: 28.0 mg/l
Exposure time: 15 min

Concentration: 24.9 mg/l
Exposure time: 30 min

Comment: Mean of three assays. Methanol (<10%) was used to prepare the stock solutions. EC50 values were calculated from nominal concentrations.

Bibliographical reference: Kaiser, K.L.E., and Palabrica, V.S. (1991). *Water Poll. Res. J. Canada* **26**, 361-431.

DIAZINON

CAS RN: 333-41-5
Synonym: *O*,*O*-Diethyl *O*-(2-isopropyl-6-methyl-4-pyrimidinyl) phosphorothioate

Temperature: 15 ± 0.1°C
Test parameter: EC50
Effect: Reduction in light output
Concentration: 1.7 mg/l
Exposure time: 5 min

Bibliographical reference: Chang, J.C., Taylor, P.B., and Leach, F.R. (1981). *Bull. Environ. Contam. Toxicol.* **26**, 150-156.

Test parameter: EC50
Effect: Reduction in light output
Concentration: 9.8 mg/l
Exposure time: 5 min
Comment: Concentrations in the test were measured.

Bibliographical reference: Curtis, C., Lima, A., Lozano, S.J., and Veith, G.D. (1982). In: *Aquatic Toxicology and Hazard Assessment: Fifth Conference, ASTM STP 766*, J.G. Pearson, R.B. Foster, and W.E. Bishop (eds.), American Society for Testing and Materials, Philadelphia, p. 170-178.

Temperature: 15 ± 0.1°C
Test parameter: EC50
Effect: Reduction in light output
Concentration: 10.3 mg/l
Exposure time: 5 min

Bibliographical reference: Somasundaram, L., Coats, J.R., Racke, K.D., and Stahr, H.M. (1990). *Bull. Environ. Contam. Toxicol.* **44**, 254-259.

DIBENZOFURAN

CAS RN: 132-64-9

Sample purity: 99+%
Temperature: 15°C
Test parameter: EC50
Effect: Reduction in light output

Concentration: 0.79 mg/l
Exposure time: 5 min

Concentration: 0.86 mg/l
Exposure time: 15 min

Concentration: 1.09 mg/l
Exposure time: 30 min

Comment: Mean of three assays. Methanol (<10%) was used to prepare the stock solutions. EC50 values were calculated from nominal concentrations.

Bibliographical reference: Kaiser, K.L.E., and Palabrica, V.S. (1991). *Water Poll. Res. J. Canada* **26**, 361-431.

DIBENZOTHIOPHENE

CAS RN: 132-65-0

Sample purity: 95%
Temperature: 15°C
Test parameter: EC50
Effect: Reduction in light output

Concentration: 0.10 mg/l
Exposure time: 5 min

Concentration: 0.11 mg/l
Exposure time: 15 min

Concentration: 0.12 mg/l
Exposure time: 30 min

Comment: Mean of three assays. Methanol (<10%) was used to prepare the stock solutions. EC50 values were calculated from nominal concentrations.

Bibliographical reference: Kaiser, K.L.E., and Palabrica, V.S. (1991). *Water Poll. Res. J. Canada* **26**, 361-431.

1,4-DIBROMOBENZENE

CAS RN: 106-37-6

Sample purity: 99%
Temperature: 15°C
Test parameter: EC50
Effect: Reduction in light output
Concentration: 2.84 mg/l
Exposure time: 30 min

Comment: Mean of three assays. Methanol (<10%) was used to prepare the stock solutions. EC50 values were calculated from nominal concentrations.

Bibliographical reference: Kaiser, K.L.E., and Palabrica, V.S. (1991). *Water Poll. Res. J. Canada* **26**, 361-431.

1,2-DIBROMOETHANE

CAS RN: 106-93-4
Synonym: Ethylene dibromide

Temperature: 15°C

Test parameter: EC20
Effect: Reduction in light output
Concentration: 308 mg/l
Exposure time: 5 min

Test parameter: EC50
Effect: Reduction in light output
Concentration: 735 mg/l
Exposure time: 5 min

Comment: Toxicity values were calculated from nominal concentrations.

Bibliographical reference: Reteuna, C. (1988). Thesis, University of Metz, Metz, France.

2,5-DIBROMONITROBENZENE

CAS RN: 3460-18-2

Temperature: 15°C
Test parameter: EC50
Effect: Reduction in light output

Concentration: 2.50 mg/l
Exposure time: 5 min

Concentration: 2.80 mg/l

Exposure time: 15 min

Concentration: 3.50 mg/l
Exposure time: 30 min

Comment: Mean of four assays. Methanol (<10%) was used to prepare the stock solutions. EC50 values were calculated from nominal concentrations.

Bibliographical reference: Kaiser, K.L.E., and Palabrica, V.S. (1993). National Water Research Institute, Burlington, Ontario, Canada, unpublished results.

2,6-DIBROMO-4-NITROPHENOL

CAS RN: 99-28-5

Sample purity: 98%
Temperature: 15°C
Test parameter: EC50
Effect: Reduction in light output

Concentration: 5.66 mg/l
Exposure time: 5 min

Concentration: 4.71 mg/l
Exposure times: 15 and 30 min

Comment: Mean of four assays. Methanol (<10%) was used to prepare the stock solutions. EC50 values were calculated from nominal concentrations.

Bibliographical reference: Kaiser, K.L.E., and Palabrica, V.S. (1992). National Water Research Institute, Burlington, Ontario, Canada, unpublished results.

2,6-DIBROMOPHENOL

CAS RN: 608-33-3

Temperature: 15°C
Test parameter: EC50

Effect: Reduction in light output

Concentration: 17.2 mg/l
Exposure time: 5 min

Concentration: 15.1 mg/l
Exposure time: 30 min

Bibliographical reference: Speece, R. (1987). Drexel University, Philadelphia, USA, private communication.

2,3-DIBROMOPROPANOL

CAS RN: 96-13-9

Test parameter: EC50
Effect: Reduction in light output
Concentration: 320 mg/l
Exposure time: 5 min
Comment: The EC50 value was calculated from nominal concentrations.

Bibliographical reference: Curtis, C., Lima, A., Lozano, S.J., and Veith, G.D. (1982). In: *Aquatic Toxicology and Hazard Assessment: Fifth Conference, ASTM STP 766*, J.G. Pearson, R.B. Foster, and W.E. Bishop (eds.), American Society for Testing and Materials, Philadelphia, p. 170-178.

DIBUT-3-ENYLTIN DIBROMIDE

Temperature: 15°C
Test parameter: EC50
Effect: Reduction in light output

Concentration: 1.99 ± 0.27 mg/l
Exposure time: 5 min

Concentration: 1.10 ± 0.09 mg/l
Exposure time: 15 min

Comment: Chemical was dissolved in ethanol (~0.05%).

Bibliographical reference: Dooley, C.A., and Kenis, P. (1987). In: *Oceans '87, International Organotin Symposium*, Department of Fisheries and Oceans, Canada, and the Society for Underwater Technology, William MacNab & Son, Halifax, p. 1517-1524.

DIBUTYL ADIPATE

CAS RN: 105-99-7

Sample purity: 96%
Temperature: 15°C
Test parameter: EC50
Effect: Reduction in light output

Concentration: 3.0 mg/l
Exposure time: 5 min

Concentration: 3.1 mg/l*
Exposure time: 15 min

Comment: Chemical was prepared in an initial solution of 3% methanol.

Bibliographical references: Cronin, M.T.D., Dearden, J.C., and Dobbs, A.J. (1991). *Sci. Total Environ.* **109/110**, 431-439.
* Cronin, M.T.D. (1993). Liverpool John Moores University, UK, private communication.

2,6-DI-*tert*-BUTYL-1,4-BENZOQUINONE

CAS RN: 719-22-2

Sample purity: 98%
Temperature: 15°C
Test parameter: EC50
Effect: Reduction in light output

Concentration: 11.0 mg/l
Exposure time: 5 min

Concentration: 8.00 mg/l
Exposure time: 15 min

Concentration: 7.30 mg/l
Exposure time: 30 min

Comment: Mean of three assays. Methanol (<10%) was used to prepare the stock solutions. EC50 values were calculated from nominal concentrations.

Bibliographical reference: Kaiser, K.L.E., and Palabrica, V.S. (1991). *Water Poll. Res. J. Canada* **26**, 361-431.

2,5-DI-*tert*-BUTYLHYDROQUINONE

CAS RN: 88-58-4

Sample purity: 97%
Temperature: 15°C
Test parameter: EC50
Effect: Reduction in light output

Concentration: 4.24 mg/l
Exposure time: 5 min

Concentration: 3.86 mg/l
Exposure time: 15 min

Concentration: 4.65 mg/l
Exposure time: 30 min

Comment: Mean of three assays. Methanol (<10%) was used to prepare the stock solutions. EC50 values were calculated from nominal concentrations.

Bibliographical reference: Kaiser, K.L.E., and Palabrica, V.S. (1991). *Water Poll. Res. J. Canada* **26**, 361-431.

3,5-DI-*tert*-BUTYL-4-HYDROXYANISOLE

CAS RN: 489-01-0

Sample purity: 97%
Temperature: 15°C
Test parameter: EC50

Effect: Reduction in light output

Concentration: 56.7 mg/l
Exposure time: 5 min

Concentration: 41.1 mg/l
Exposure times: 15 and 30 min

Comment: Mean of three assays. Methanol (<10%) was used to prepare the stock solutions. EC50 values were calculated from nominal concentrations.

Bibliographical reference: Kaiser, K.L.E., and Palabrica, V.S. (1991). *Water Poll. Res. J. Canada* **26**, 361-431.

2,6-DI-*tert*-BUTYL-4-METHYLPHENOL

CAS RN: 128-37-0
Synonyms: BHT; Butylated hydroxytoluene

Sample purity: >99%
Temperature: 15°C
Test parameter: EC50
Effect: Reduction in light output

Concentration: 7.82 mg/l
Exposure time: 5 min

Concentration: 8.57 mg/l
Exposure time: 15 min

Concentration: 8.98 mg/l
Exposure time: 30 min

Comment: Mean of three assays. Methanol (<10%) was used to prepare the stock solutions. EC50 values were calculated from nominal concentrations.

Bibliographical reference: Kaiser, K.L.E., and Palabrica, V.S. (1991). *Water Poll. Res. J. Canada* **26**, 361-431.

DIBUTYL PHTHALATE

CAS RN: 84-74-2

Temperature: 15 ± 0.1°C
Test parameter: EC50
Effect: Reduction in light output

Concentration: 10.9 mg/l
Exposure time: 5 min

Concentration: 11.1 mg/l
Exposure time: 15 min

Concentration: 10.9 mg/l
Exposure time: 30 min

Comment: The pH was not adjusted.

Bibliographical reference: Tarkpea, M., Hansson, M., and Samuelsson, B. (1986). *Ecotoxicol. Environ. Safety* **11**, 127-143.

Temperature: 15°C
Test parameter: EC50
Effect: Reduction in light output

Concentration: 23 mg/l
Exposure time: 5 min

Concentration: 26 mg/l
Exposure time: 15 min

Concentration: 23 mg/l
Exposure time: 30 min

Comment: Chemical was diluted to give five sample concentrations in duplicate. It was dissolved in 2% ethanol prior to addition of NaCl.

Bibliographical reference: Benson, W.H., and Stackhouse, R.A. (1986). *Drug Chem. Toxicol.* **9**, 275-283.

N,N'-DIBUTYLTHIOUREA

CAS RN: 109-46-6

Sample purity: ≥97%
Test parameter: EC50
Effect: Reduction in light output
Concentration: 7.84 mg/l
Exposure time: 15 min
Comment: EC50 value converted to mg/l from the original data expressed in log(1/µmol l^{-1}) and a rounded molecular weight of 188 given by the authors.

Bibliographical reference: Govers, H., Ruepert, C., Stevens, T., and van Leeuwen, C.J. (1986). *Chemosphere* **15**, 383-393.

DIBUTYLTIN DICHLORIDE

CAS RN: 683-18-1

Test parameter: EC50
Effect: Reduction in light output
Concentration: 0.217 mg/l
Exposure time: 30 min

Bibliographical reference: Steinhäuser, K.G., Amann, W., Späth, A., and Polenz, A. (1985). *Vom Wasser* **65**, 203-214.

Temperature: 15°C
Test parameter: EC50
Effect: Reduction in light output

Concentration: 0.64 ± 0.21 mg/l
Exposure time: 5 min

Concentration: 0.33 ± 0.07 mg/l
Exposure time: 15 min

Comment: Chemical was dissolved in ethanol (~0.05%).

Bibliographical reference: Dooley, C.A., and Kenis, P. (1987). In: *Oceans '87, International Organotin Symposium*, Department of Fisheries and Oceans, Canada, and the Society for Underwater

Technology, William MacNab & Son, Halifax, p. 1517-1524.

DIBUTYLTIN DILAURATE

CAS RN: 77-58-7

Test parameter: EC50
Effect: Reduction in light output
Concentration: 0.570 mg/l
Exposure time: 30 min

Bibliographical reference: Steinhäuser, K.G., Amann, W., Späth, A., and Polenz, A. (1985). *Vom Wasser* **65**, 203-214.

1,3-DICHLOROACETONE

CAS RN: 534-07-6

Sample purity: >95%
Temperature: 15°C
Test parameter: EC50
Effect: Reduction in light output

Concentration: 1.53 mg/l
Exposure time: 5 min

Concentration: 0.43 mg/l
Exposure time: 15 min

Concentration: 0.20 mg/l
Exposure time: 30 min

Comment: Mean of four assays. Methanol (<10%) was used to prepare the stock solutions. EC50 values were calculated from nominal concentrations.

Bibliographical reference: Kaiser, K.L.E., and Palabrica, V.S. (1991). *Water Poll. Res. J. Canada* **26**, 361-431.

2,3-DICHLOROANILINE

CAS RN: 608-27-5

Temperature: 15°C
Test parameter: EC50
Effect: Reduction in light output

Concentration: 2.51 mg/l
Exposure time: 5 min

Concentration: 2.75 mg/l
Exposure times: 15 and 30 min

Comment: Mean of three assays. Methanol (<10%) was used to prepare the stock solutions. EC50 values were calculated from nominal concentrations.

Bibliographical reference: Ribo, J.M., and Kaiser, K.L.E. (1984). In: *QSAR in Environmental Toxicology*, K.L.E. Kaiser (ed.), D. Reidel Publishing Company, Dordrecht, p. 319-336.

2,4-DICHLOROANILINE

CAS RN: 554-00-7

Temperature: 15°C
Test parameter: EC50
Effect: Reduction in light output

Concentration: 3.98 mg/l
Exposure time: 5 min

Concentration: 4.57 mg/l
Exposure time: 15 min

Concentration: 4.67 mg/l
Exposure time: 30 min

Comment: Mean of three assays. Methanol (<10%) was used to prepare the stock solutions. EC50 values were calculated from nominal concentrations.

Bibliographical reference: Ribo, J.M., and Kaiser, K.L.E. (1984).

In: *QSAR in Environmental Toxicology*, K.L.E. Kaiser (ed.), D. Reidel Publishing Company, Dordrecht, p. 319-336.

2,5-DICHLOROANILINE

CAS RN: 95-82-9

Temperature: 15°C
Test parameter: EC50
Effect: Reduction in light output

Concentration: 3.39 mg/l
Exposure time: 5 min

Concentration: 3.63 mg/l
Exposure time: 15 min

Concentration: 3.80 mg/l
Exposure time: 30 min

Comment: Mean of three assays. Methanol (<10%) was used to prepare the stock solutions. EC50 values were calculated from nominal concentrations.

Bibliographical reference: Ribo, J.M., and Kaiser, K.L.E. (1984). In: *QSAR in Environmental Toxicology*, K.L.E. Kaiser (ed.), D. Reidel Publishing Company, Dordrecht, p. 319-336.

2,6-DICHLOROANILINE

CAS RN: 608-31-1

Temperature: 15°C
Test parameter: EC50
Effect: Reduction in light output

Concentration: 1.48 mg/l
Exposure time: 5 min

Concentration: 1.70 mg/l
Exposure time: 15 min

Concentration: 1.74 mg/l
Exposure time: 30 min

Comment: Mean of three assays. Methanol (<10%) was used to prepare the stock solutions. EC50 values were calculated from nominal concentrations.

Bibliographical reference: Ribo, J.M., and Kaiser, K.L.E. (1984). In: *QSAR in Environmental Toxicology*, K.L.E. Kaiser (ed.), D. Reidel Publishing Company, Dordrecht, p. 319-336.

3,4-DICHLOROANILINE

CAS RN: 95-76-1

Temperature: 15°C
Test parameter: EC50
Effect: Reduction in light output

Concentration: 0.45 mg/l
Exposure time: 5 min

Concentration: 0.56 mg/l
Exposure time: 15 min

Concentration: 0.65 mg/l
Exposure time: 30 min

Comment: Mean of three assays. Methanol (<10%) was used to prepare the stock solutions. EC50 values were calculated from nominal concentrations.

Bibliographical reference: Ribo, J.M., and Kaiser, K.L.E. (1984). In: *QSAR in Environmental Toxicology*, K.L.E. Kaiser (ed.), D. Reidel Publishing Company, Dordrecht, p. 319-336.

3,5-DICHLOROANILINE

CAS RN: 626-43-7

Temperature: 15°C
Test parameter: EC50

Effect: Reduction in light output

Concentration: 9.54 mg/l
Exposure time: 5 min

Concentration: 10.7 mg/l
Exposure time: 15 min

Concentration: 10.5 mg/l
Exposure time: 30 min

Comment: Mean of three assays. Methanol (<10%) was used to prepare the stock solutions. EC50 values were calculated from nominal concentrations.

Bibliographical reference: Ribo, J.M., and Kaiser, K.L.E. (1984). In: *QSAR in Environmental Toxicology*, K.L.E. Kaiser (ed.), D. Reidel Publishing Company, Dordrecht, p. 319-336.

2,4-DICHLOROBENZALDEHYDE

CAS RN: 874-42-0

Temperature: 15°C
Test parameter: EC50
Effect: Reduction in light output

Concentration: 5.4 mg/l
Exposure time: 5 min

Concentration: 5.2 mg/l
Exposure time: 15 min

Bibliographical reference: Cronin, M.T.D. (1993). Liverpool John Moores University, UK, private communication.

1,2-DICHLOROBENZENE

CAS RN: 95-50-1

Temperature: 15 ± 0.1°C
Test parameter: EC50

Effect: Reduction in light output
Concentration: 10.25 mg/l
Exposure time: 5 min
Comment: Test was performed on *Photobacterium phosphoreum* NZ11D obtained from the Scripps Institute of Oceanography (La Jolla, CA).

Bibliographical reference: McFeters, G.A., Bond, P.J., Olson, S.B., and Tchan, Y.T. (1983). *Water Res.* **17**, 1757-1762.

Temperature: 15°C
Test parameter: EC50
Effect: Reduction in light output

Concentration: 2.74 mg/l
Exposure time: 5 min

Concentration: 3.14 mg/l
Exposure time: 15 min

Concentration: 4.05 mg/l
Exposure time: 30 min

Comment: Mean of three assays. Methanol (<10%) was used to prepare the stock solutions. EC50 values were calculated from nominal concentrations.

Bibliographical reference: Ribo, J.M., and Kaiser, K.L.E. (1983). *Chemosphere* **12**, 1421-1442.

Sample purity: 99+%
Temperature: 15°C
Test parameter: EC50
Effect: Reduction in light output

Concentration: 4.76 mg/l
Exposure time: 5 min

Concentration: 4.98 mg/l
Exposure time: 15 min

Concentration: 5.99 mg/l
Exposure time: 30 min

Comment: Mean of three assays. Methanol (<10%) was used to prepare the stock solutions. EC50 values were calculated from nominal concentrations.

Bibliographical reference: Kaiser, K.L.E., and Palabrica, V.S. (1991). *Water Poll. Res. J. Canada* **26**, 361-431.

1,3-DICHLOROBENZENE

CAS RN: 541-73-1

Temperature: 15°C
Test parameter: EC50
Effect: Reduction in light output

Concentration: 3.07 mg/l
Exposure time: 5 min

Concentration: 4.14 mg/l
Exposure time: 15 min

Concentration: 5.10 mg/l
Exposure time: 30 min

Comment: Mean of three assays. Methanol (<10%) was used to prepare the stock solutions. EC50 values were calculated from nominal concentrations.

Bibliographical reference: Ribo, J.M., and Kaiser, K.L.E. (1983). *Chemosphere* **12**, 1421-1442.

Temperature: 15°C
Test parameter: EC50
Effect: Reduction in light output
Concentration: 3.29 mg/l
Exposure time: 15 min

Bibliographical reference: Hermens, J., Busser, F., Leeuwangh, P., and Musch, A. (1985). *Ecotoxicol. Environ. Safety* **9**, 17-25.

Temperature: 15°C
Test parameter: EC50

Effect: Reduction in light output
Concentration: 4.20 mg/l
Exposure time: 10 min
Comment: Mean of three assays. Test solutions were analyzed using gas chromatography.

Bibliographical reference: Bazin, C., Chambon, P., Bonnefille, M., and Larbaigt, G. (1987). *Sci. Eau* **6**, 403-413.

Sample purity: 98%
Temperature: 15°C
Test parameter: EC50
Effect: Reduction in light output

Concentration: 3.29 mg/l
Exposure time: 5 min

Concentration: 3.96 mg/l
Exposure time: 15 min

Concentration: 5.10 mg/l
Exposure time: 30 min

Comment: Mean of three assays. Methanol (<10%) was used to prepare the stock solutions. EC50 values were calculated from nominal concentrations.

Bibliographical reference: Kaiser, K.L.E., and Palabrica, V.S. (1991). *Water Poll. Res. J. Canada* **26**, 361-431.

1,4-DICHLOROBENZENE

CAS RN: 106-46-7

Temperature: 15°C
Test parameter: EC50
Effect: Reduction in light output

Concentration: 4.34 mg/l
Exposure time: 5 min

Concentration: 4.87 mg/l
Exposure time: 15 min

Concentration: 5.34 mg/l
Exposure time: 30 min

Comment: Mean of three assays. Methanol (<10%) was used to prepare the stock solutions. EC50 values were calculated from nominal concentrations.

Bibliographical reference: Ribo, J.M., and Kaiser, K.L.E. (1983). *Chemosphere* **12**, 1421-1442.

3,3'-DICHLOROBENZIDINE

CAS RN: 91-94-1

Test parameter: EC50
Effect: Reduction in light output
Concentration: 0.06 mg/l
Exposure time: 15 min
Comment: Test was performed in duplicate at pH = 6.7.

Bibliographical references: Dutka, B.J., and Kwan, K.K. (1981). *Bull. Environ. Contam. Toxicol.* **27**, 753-757.
Dutka, B.J., and Kwan, K.K. (1984). In: *Toxicity Screening Procedures Using Bacterial Systems*, D. Liu and B.J. Dutka (eds.), Marcel Dekker, New York, p. 125-138.

3,5-DICHLOROBENZONITRILE

CAS RN: 6575-00-4

Sample purity: 97%
Temperature: 15°C
Test parameter: EC50
Effect: Reduction in light output

Concentration: 18.9 mg/l
Exposure time: 5 min

Concentration: 21.7 mg/l
Exposure time: 15 min

Concentration: 24.9 mg/l

Exposure time: 30 min

Comment: Mean of three assays. Methanol (<10%) was used to prepare the stock solutions. EC50 values were calculated from nominal concentrations.

Bibliographical reference: Kaiser, K.L.E., and Palabrica, V.S. (1991). *Water Poll. Res. J. Canada* **26**, 361-431.

3,4-DICHLOROBENZOTRIFLUORIDE

CAS RN: 328-84-7
Synonym: 3,4-Dichloro-α,α,α-trifluorotoluene

Temperature: 20°C

Test parameter: EC20
Effect: Reduction in light output
Concentration: 0.47 mg/l (0.44-0.50)
Exposure time: 5 min

Test parameter: EC50
Effect: Reduction in light output
Concentration: 1.57 mg/l (1.50-1.65)
Exposure time: 5 min

Bibliographical reference: Casseri, N.A., Ying, W, and Sojka, S.A. (1983). Proceedings 38th Industrial Waste Conference, USA, May 1983, p. 867-878.

Temperature: 20°C
Test parameter: EC50
Effect: Reduction in light output
Concentration: 1.56 mg/l (1.48-1.64)
Exposure time: 5 min

Bibliographical reference: Casseri, N.A. (1985). Occidental Chemical Corporation, Grand Island, New York, USA, private communication.

1,2-DICHLOROBUTANE

CAS RN: 616-21-7

Temperature: 15°C
Test parameter: EC50
Effect: Reduction in light output

Concentration: 61.5 mg/l*
Exposure time: 5 min

Concentration: 89.1 mg/l*
Exposure time: 30 min

* EC50 values from linear regression fit to data.

Bibliographical reference: Speece, R. (1987). Drexel University, Philadelphia, USA, private communication.

5,5'-DICHLORO-2,2'-DIHYDROXYDIPHENYLMETHANE

CAS RN: 97-23-4

Temperature: 15°C
Test parameter: EC50
Effect: Reduction in light output
Concentrations: 0.02 mg/l
0.06 mg/l
Exposure time: 5 min

Bibliographical reference: King, E.F., and Painter, H.A. (1981). In: *Les Tests de Toxicité Aiguë en Milieu Aquatique. Acute Aquatic Ecotoxicological Tests*, H. Leclerc and D. Dive (eds.), INSERM 106, Paris, p. 143-153.

Sample purity: 95%
Temperature: 15°C
Test parameter: EC50
Effect: Reduction in light output
Concentration: 0.055 mg/l
Exposure time: 15 min

Comment: To dissolve the chemical, 1M sodium hydroxide was added and since the solution could not be neutralized to pH 7 without precipitate it was used at approximately pH 8.

Bibliographical reference: King, E.F. (1984). In: *Toxicity Screening Procedures Using Bacterial Systems*, D. Liu and B.J. Dutka (eds.), Marcel Dekker, New York, p. 175-194.

1,2-DICHLORO-4,5-DINITROBENZENE

CAS RN: 6306-39-4

Sample purity: 98%
Temperature: 15°C
Test parameter: EC50
Effect: Reduction in light output

Concentration: 0.99 mg/l
Exposure time: 5 min

Concentration: 0.33 mg/l
Exposure time: 15 min

Concentration: 0.20 mg/l
Exposure time: 30 min

Comment: Mean of three assays. Methanol (<10%) was used to prepare the stock solutions. EC50 values were calculated from nominal concentrations.

Bibliographical reference: Kaiser, K.L.E., and Palabrica, V.S. (1991). *Water Poll. Res. J. Canada* **26**, 361-431.

1,3-DICHLORO-4,6-DINITROBENZENE

CAS RN: 3698-83-7

Sample purity: 97%
Temperature: 15°C
Test parameter: EC50
Effect: Reduction in light output

Concentration: 2.72 mg/l
Exposure time: 5 min

Concentration: 1.01 mg/l
Exposure time: 15 min

Concentration: 0.52 mg/l
Exposure time: 30 min

Comment: Mean of three assays. Methanol (<10%) was used to prepare the stock solutions. EC50 values were calculated from nominal concentrations.

Bibliographical reference: Kaiser, K.L.E., and Palabrica, V.S. (1992). National Water Research Institute, Burlington, Ontario, Canada, unpublished results.

1,5-DICHLORO-2,3-DINITROBENZENE

CAS RN: 28689-08-9

Sample purity: 98%
Temperature: 15°C
Test parameter: EC50
Effect: Reduction in light output

Concentration: 0.78 mg/l
Exposure time: 5 min

Concentration: 0.24 mg/l
Exposure time: 15 min

Concentration: 0.16 mg/l
Exposure time: 30 min

Comment: Mean of three assays. Methanol (<10%) was used to prepare the stock solutions. EC50 values were calculated from nominal concentrations.

Bibliographical reference: Kaiser, K.L.E., and Palabrica, V.S. (1991). *Water Poll. Res. J. Canada* **26**, 361-431.

1,1-DICHLOROETHANE

CAS RN: 75-34-3

Temperature: 15°C
Test parameter: EC50
Effect: Reduction in light output

Concentration: 269 mg/l
Exposure time: 5 min

Concentration: 347 mg/l
Exposure time: 30 min

Bibliographical reference: Speece, R. (1987). Drexel University, Philadelphia, USA, private communication.

1,2-DICHLOROETHANE

CAS RN: 107-06-2
Synonym: Ethylene dichloride

Temperature: 15 ± 0.3°C
Test parameter: EC50
Effect: Reduction in light output
Concentrations: 153 mg/l
162 mg/l
Exposure time: 5 min
Comment: Test solutions were analyzed using gas chromatography.

Bibliographical reference: Qureshi, A.A., Flood, K.W., Thompson, S.R., Janhurst, S.M., Inniss, C.S., and Rokosh, D.A. (1982). In: *Aquatic Toxicology and Hazard Assessment: Fifth Conference, ASTM STP 766*, J.G. Pearson, R.B. Foster, and W.E. Bishop (eds.), American Society for Testing and Materials, Philadelphia, p. 179-195.

Temperature: 15°C
Test parameter: EC50
Effect: Reduction in light output
Concentration: 1110 mg/l
Exposure time: 15 min

Bibliographical reference: Hermens, J., Busser, F., Leeuwangh, P., and Musch, A. (1985). *Ecotoxicol. Environ. Safety* **9**, 17-25.

Temperature: 15°C
Test parameter: EC50
Effect: Reduction in light output
Concentration: 2063 mg/l
Exposure time: 10 min
Comment: Mean of six assays. Test solutions were analyzed using gas chromatography.

Bibliographical reference: Bazin, C., Chambon, P., Bonnefille, M., and Larbaigt, G. (1987). *Sci. Eau* **6**, 403-413.

Temperature: 15°C
Test parameter: EC50
Effect: Reduction in light output

Concentration: 696 mg/l
Exposure time: 5 min

Concentration: 918 mg/l
Exposure time: 30 min

Bibliographical reference: Speece, R. (1987). Drexel University, Philadelphia, USA, private communication.

cis-1,2-DICHLOROETHYLENE

CAS RN: 156-59-2

Temperature: 15°C
Test parameter: EC50
Effect: Reduction in light output

Concentration: 721 mg/l*
Exposure time: 5 min

Concentration: 905 mg/l*
Exposure time: 30 min

* EC50 values from linear regression fit to data.

Bibliographical reference: Speece, R. (1987). Drexel University, Philadelphia, USA, private communication.

trans-1,2-DICHLOROETHYLENE

CAS RN: 156-60-5

Temperature: 15°C
Test parameter: EC50
Effect: Reduction in light output

Concentration: 1142 mg/l*
Exposure time: 5 min

Concentration: 1546 mg/l*
Exposure time: 30 min

* EC50 values from linear regression fit to data.

Bibliographical reference: Speece, R. (1987). Drexel University, Philadelphia, USA, private communication.

DICHLOROMETHANE

CAS RN: 75-09-2
Synonym: Methylene chloride

Test parameter: EC50
Effect: Reduction in light output
Concentrations: >1000 mg/l
1000 mg/l*
Exposure time: 15 min
Comment: Test was performed in duplicate at pH = 6.7.

Bibliographical references: Dutka, B.J., and Kwan, K.K. (1981). *Bull. Environ. Contam. Toxicol.* **27**, 753-757.
* Dutka, B.J., and Kwan, K.K. (1984). In: *Toxicity Screening Procedures Using Bacterial Systems*, D. Liu and B.J. Dutka (eds.), Marcel Dekker, New York, p. 125-138.

Temperature: 15°C

Test parameter: EC50
Effect: Reduction in light output
Concentration: 2878 mg/l
Exposure time: 15 min

Bibliographical reference: Hermens, J., Busser, F., Leeuwangh, P., and Musch, A. (1985). *Ecotoxicol. Environ. Safety* **9**, 17-25.

Temperature: 15°C
Test parameter: EC50
Effect: Reduction in light output
Concentration: 2.15 ul/ml
Exposure time: 15 min
Comment: Four concentrations of chemical were tested in duplicate.

Bibliographical reference: Schiewe, M.H., Hawk, E.G., Actor, D.I., and Krahn, M.M. (1985). *Can. J. Fish. Aquat. Sci.* **42**, 1244-1248.

1,2-DICHLORO-2-METHYLPROPANE

CAS RN: 594-37-6

Temperature: 15°C
Test parameter: EC50
Effect: Reduction in light output

Concentration: 5.73 mg/l
Exposure time: 5 min

Concentration: 6.53 mg/l
Exposure time: 30 min

Bibliographical reference: Speece, R. (1987). Drexel University, Philadelphia, USA, private communication.

2,4-DICHLORO-1-NAPHTHOL

CAS RN: 2050-76-2

Sample purity: 98%
Temperature: 15°C

Test parameter: EC50
Effect: Reduction in light output

Concentration: 0.18 mg/l
Exposure time: 5 min

Concentration: 0.17 mg/l
Exposure times: 15 and 30 min

Comment: Mean of three assays. Methanol (<10%) was used to prepare the stock solutions. EC50 values were calculated from nominal concentrations.

Bibliographical reference: Kaiser, K.L.E., and Palabrica, V.S. (1991). *Water Poll. Res. J. Canada* **26**, 361-431.

2,4-DICHLORO-6-NITROANILINE

CAS RN: 2683-43-4

Sample purity: 98%
Temperature: 15°C
Test parameter: EC50
Effect: Reduction in light output

Concentration: 7.87 mg/l
Exposure times: 5 and 15 min

Concentration: 8.43 mg/l
Exposure time: 30 min

Comment: Mean of four assays. Methanol (<10%) was used to prepare the stock solutions. EC50 values were calculated from nominal concentrations.

Bibliographical reference: Kaiser, K.L.E., and Palabrica, V.S. (1991). *Water Poll. Res. J. Canada* **26**, 361-431.

2,6-DICHLORO-4-NITROANILINE

CAS RN: 99-30-9

Sample purity: 98%
Temperature: 15°C
Test parameter: EC50
Effect: Reduction in light output

Concentration: 3.52 mg/l
Exposure time: 5 min

Concentration: 3.28 mg/l
Exposure time: 15 min

Concentration: 3.68 mg/l
Exposure time: 30 min

Comment: Mean of four assays. Methanol (<10%) was used to prepare the stock solutions. EC50 values were calculated from nominal concentrations.

Bibliographical reference: Kaiser, K.L.E., and Palabrica, V.S. (1991). *Water Poll. Res. J. Canada* **26**, 361-431.

4,5-DICHLORO-2-NITROANILINE

CAS RN: 6641-64-1

Sample purity: 96%
Temperature: 15°C
Test parameter: EC50
Effect: Reduction in light output

Concentration: 8.24 mg/l
Exposure times: 5 and 15 min

Concentration: 8.83 mg/l
Exposure time: 30 min

Comment: Mean of four assays. Methanol (<10%) was used to prepare the stock solutions. EC50 values were calculated from nominal concentrations.

Bibliographical reference: Kaiser, K.L.E., and Palabrica, V.S. (1991). *Water Poll. Res. J. Canada* **26**, 361-431.

2,3-DICHLORONITROBENZENE

CAS RN: 3209-22-1

Test parameter: EC50
Effect: Reduction in light output
Concentration: 1.49 mg/l
Exposure time: 15 min
Comment: The concentrations of the compound increased geometrically with a factor of 3.2. The concentrations were calculated on the basis of added amounts of material.

Bibliographical reference: Deneer, J.W., van Leeuwen, C.J., Seinen, W., Maas-Diepeveen, J.L., and Hermens, J.L.M. (1989). *Aquat. Toxicol.* **15**, 83-98.

Temperature: 15°C
Test parameter: EC50
Effect: Reduction in light output

Concentration: 1.30 mg/l
Exposure time: 5 min

Concentration: 1.39 mg/l
Exposure time: 15 min

Concentration: 1.46 mg/l*
Exposure time: 30 min

Comment: Mean of three assays. Methanol (<10%) was used to prepare the stock solutions. EC50 values were calculated from nominal concentrations.

Bibliographical references: Kaiser, K.L.E., and Palabrica, V.S. (1991). *Water Poll. Res. J. Canada* **26**, 361-431.
* Kaiser, K.L.E., and Ribo, J.M. (1985). In: *QSAR in Toxicology and Xenobiochemistry*, M. Tichy (ed.), Elsevier, Amsterdam, p. 27-38.

2,4-DICHLORONITROBENZENE

CAS RN: 611-06-3

Test parameter: EC50

Effect: Reduction in light output
Concentration: 1.71 mg/l
Exposure time: 15 min
Comment: The concentrations of the compound increased geometrically with a factor of 3.2. The concentrations were calculated on the basis of added amounts of material.

Bibliographical reference: Deneer, J.W., van Leeuwen, C.J., Seinen, W., Maas-Diepeveen, J.L., and Hermens, J.L.M. (1989). *Aquat. Toxicol.* **15**, 83-98.

Temperature: 15°C
Test parameter: EC50
Effect: Reduction in light output

Concentration: 2.91 mg/l
Exposure time: 5 min

Concentration: 3.04 mg/l
Exposure time: 15 min

Concentration: 3.19 mg/l*
Exposure time: 30 min

Comment: Mean of three assays. Methanol (<10%) was used to prepare the stock solutions. EC50 values were calculated from nominal concentrations.

Bibliographical references: Kaiser, K.L.E., and Palabrica, V.S. (1991). *Water Poll. Res. J. Canada* **26**, 361-431.
* Kaiser, K.L.E., and Ribo, J.M. (1985). In: *QSAR in Toxicology and Xenobiochemistry*, M. Tichy (ed.), Elsevier, Amsterdam, p. 27-38.

2,5-DICHLORONITROBENZENE

CAS RN: 89-61-2

Test parameter: EC50
Effect: Reduction in light output
Concentration: 8.38 mg/l
Exposure time: 15 min

Comment: The concentrations of the compound increased geometrically with a factor of 3.2. The concentrations were calculated on the basis of added amounts of material.

Bibliographical reference: Deneer, J.W., van Leeuwen, C.J., Seinen, W., Maas-Diepeveen, J.L., and Hermens, J.L.M. (1989). *Aquat. Toxicol.* **15**, 83-98.

Temperature: 15°C
Test parameter: EC50
Effect: Reduction in light output

Concentration: 7.82 mg/l
Exposure time: 5 min

Concentration: 8.38 mg/l
Exposure time: 15 min

Concentration: 8.78 mg/l*
Exposure time: 30 min

Comment: Mean of three assays. Methanol (<10%) was used to prepare the stock solutions. EC50 values were calculated from nominal concentrations.

Bibliographical references: Kaiser, K.L.E., and Palabrica, V.S. (1991). *Water Poll. Res. J. Canada* **26**, 361-431.
* Kaiser, K.L.E., and Ribo, J.M. (1985). In: *QSAR in Toxicology and Xenobiochemistry*, M. Tichy (ed.), Elsevier, Amsterdam, p. 27-38.

3,4-DICHLORONITROBENZENE

CAS RN: 99-54-7

Temperature: 15°C
Test parameter: EC50
Effect: Reduction in light output

Concentration: 9.19 mg/l
Exposure time: 5 min

Concentration: 10.1 mg/l
Exposure time: 15 min

Concentration: 10.1 mg/l*
Exposure time: 30 min

Comment: Mean of three assays. Methanol (<10%) was used to prepare the stock solutions. EC50 values were calculated from nominal concentrations.

Bibliographical references: Kaiser, K.L.E., and Palabrica, V.S. (1991). *Water Poll. Res. J. Canada* **26**, 361-431.
* Kaiser, K.L.E., and Ribo, J.M. (1985). In: *QSAR in Toxicology and Xenobiochemistry*, M. Tichy (ed.), Elsevier, Amsterdam, p. 27-38.

3,5-DICHLORONITROBENZENE

CAS RN: 618-62-2

Test parameter: EC50
Effect: Reduction in light output
Concentration: 17.9 mg/l
Exposure time: 15 min
Comment: The concentrations of the compound increased geometrically with a factor of 3.2. The concentrations were calculated on the basis of added amounts of material.

Bibliographical reference: Deneer, J.W., van Leeuwen, C.J., Seinen, W., Maas-Diepeveen, J.L., and Hermens, J.L.M. (1989). *Aquat. Toxicol.* **15**, 83-98.

Temperature: 15°C
Test parameter: EC50
Effect: Reduction in light output

Concentration: 22.6 mg/l
Exposure time: 5 min

Concentration: 19.2 mg/l
Exposure time: 15 min

Concentration: 17.1 mg/l*
Exposure time: 30 min

Comment: Mean of three assays. Methanol (<10%) was used to prepare the stock solutions. EC50 values were calculated from nominal concentrations.

Bibliographical references: Kaiser, K.L.E., and Palabrica, V.S. (1991). *Water Poll. Res. J. Canada* **26**, 361-431.
* Kaiser, K.L.E., and Ribo, J.M. (1985). In: *QSAR in Toxicology and Xenobiochemistry*, M. Tichy (ed.), Elsevier, Amsterdam, p. 27-38.

1,5-DICHLOROPENTANE

CAS RN: 628-76-2

Temperature: 15°C
Test parameter: EC50
Effect: Reduction in light output

Concentration: 15.7 mg/l
Exposure time: 5 min

Concentration: 20.7 mg/l
Exposure time: 30 min

Bibliographical reference: Speece, R. (1987). Drexel University, Philadelphia, USA, private communication.

2,3-DICHLOROPHENOL

CAS RN: 576-24-9

Temperature: 15°C
Test parameter: EC50
Effect: Reduction in light output

Concentration: 4.29 mg/l
Exposure time: 5 min

Concentration: 4.81 mg/l
Exposure time: 15 min

Concentration: 4.92 mg/l
Exposure time: 30 min

Comment: Mean of three assays. Methanol (<10%) was used to prepare the stock solutions. EC50 values were calculated from nominal concentrations.

Bibliographical reference: Ribo, J.M., and Kaiser, K.L.E. (1983). *Chemosphere* **12**, 1421-1442.

Temperature: 15°C
Test parameter: EC50
Effect: Reduction in light output

Concentration: 4.32 mg/l
Exposure time: 5 min

Concentration: 5.26 mg/l
Exposure time: 30 min

Bibliographical reference: Speece, R. (1987). Drexel University, Philadelphia, USA, private communication.

2,4-DICHLOROPHENOL

CAS RN: 120-83-2

Test parameter: EC50
Effect: Reduction in light output
Concentration: 3.63 mg/l
Exposure time: 5 min
Comment: The EC50 value was calculated from nominal concentrations.

Bibliographical reference: Curtis, C., Lima, A., Lozano, S.J., and Veith, G.D. (1982). In: *Aquatic Toxicology and Hazard Assessment: Fifth Conference, ASTM STP 766*, J.G. Pearson, R.B. Foster, and W.E. Bishop (eds.), American Society for Testing and Materials, Philadelphia, p. 170-178.

Temperature: 15°C
Test parameter: EC50
Effect: Reduction in light output

Concentration: 4.70 mg/l

Exposure time: 5 min

Concentration: 5.04 mg/l
Exposure time: 15 min

Concentration: 5.52 mg/l
Exposure time: 30 min

Comment: Mean of three assays. Methanol (<10%) was used to prepare the stock solutions. EC50 values were calculated from nominal concentrations.

Bibliographical reference: Ribo, J.M., and Kaiser, K.L.E. (1983). *Chemosphere* **12**, 1421-1442.

Temperature: 15°C
Test parameter: EC50
Effect: Reduction in light output
Concentration: 6 mg/l
Exposure time: 10 min
Comment: Mean of three assays. Test solutions were analyzed using gas chromatography.

Bibliographical reference: Bazin, C., Chambon, P., Bonnefille, M., and Larbaigt, G. (1987). *Sci. Eau* **6**, 403-413.

Temperature: 15°C
Test parameter: EC50
Effect: Reduction in light output

Concentration: 2.32 mg/l
Exposure time: 5 min

Concentration: 2.27 mg/l
Exposure time: 30 min

Bibliographical reference: Speece, R. (1987). Drexel University, Philadelphia, USA, private communication.

Temperature: 15 ± 0.1°C
Test parameter: EC50
Effect: Reduction in light output
Concentration: 5.0 mg/l

Exposure time: 5 min

Bibliographical reference: Somasundaram, L., Coats, J.R., Racke, K.D., and Stahr, H.M. (1990). *Bull. Environ. Contam. Toxicol.* **44**, 254-259.

Sample purity: 95%
Temperature: 15°C
Test parameter: EC50
Effect: Reduction in light output

Concentration: 1.10 mg/l
Exposure time: 5 min

Concentration: 1.18 mg/l
Exposure time: 15 min

Concentration: 1.24 mg/l
Exposure time: 30 min

Comment: Mean of three assays. Methanol (<10%) was used to prepare the stock solutions. EC50 values were calculated from nominal concentrations.

Bibliographical reference: Kaiser, K.L.E., and Palabrica, V.S. (1991). *Water Poll. Res. J. Canada* **26**, 361-431.

Temperature: 15°C
Test parameter: EC50
Effect: Reduction in light output
Concentration: 3.5 mg/l
Exposure time: 5 min

Bibliographical reference: Kahru, A. (1993). *ATLA* **21**, 210-215.

2,5-DICHLOROPHENOL

CAS RN: 583-78-8

Temperature: 15°C
Test parameter: EC50
Effect: Reduction in light output

Concentration: 8.36 mg/l
Exposure time: 5 min

Concentration: 9.60 mg/l
Exposure time: 15 min

Concentration: 9.38 mg/l
Exposure time: 30 min

Comment: Mean of three assays. Methanol (<10%) was used to prepare the stock solutions. EC50 values were calculated from nominal concentrations.

Bibliographical reference: Ribo, J.M., and Kaiser, K.L.E. (1983). *Chemosphere* **12**, 1421-1442.

Temperature: 15°C
Test parameter: EC50
Effect: Reduction in light output

Concentration: 9.07 mg/l
Exposure time: 5 min

Concentration: 10.5 mg/l
Exposure time: 30 min

Bibliographical reference: Speece, R. (1987). Drexel University, Philadelphia, USA, private communication.

Test parameter: EC50
Effect: Reduction in light output

Concentration: 12.1 mg/l
Exposure time: 5 min

Concentration: 12.2 mg/l
Exposure time: 15 min

Bibliographical reference: Ribo, J.M., and Rogers, F. (1990). *Tox. Assess.* **5**, 135-152.

2,6-DICHLOROPHENOL

CAS RN: 87-65-0

Temperature: 15°C
Test parameter: EC50
Effect: Reduction in light output

Concentration: 9.60 mg/l
Exposure time: 5 min

Concentration: 13.6 mg/l
Exposure time: 15 min

Concentration: 13.2 mg/l
Exposure time: 30 min

Comment: Mean of three assays. Methanol (<10%) was used to prepare the stock solutions. EC50 values were calculated from nominal concentrations.

Bibliographical reference: Ribo, J.M., and Kaiser, K.L.E. (1983). *Chemosphere* **12**, 1421-1442.

Temperature: 15°C
Test parameter: EC50
Effect: Reduction in light output

Concentration: 26.0 mg/l
Exposure time: 5 min

Concentration: 30.1 mg/l
Exposure time: 30 min

Bibliographical reference: Speece, R. (1987). Drexel University, Philadelphia, USA, private communication.

3,4-DICHLOROPHENOL

CAS RN: 95-77-2

Temperature: 15°C
Test parameter: EC50

Effect: Reduction in light output

Concentration: 1.27 mg/l
Exposure time: 5 min

Concentration: 1.67 mg/l
Exposure time: 15 min

Concentration: 1.63 mg/l
Exposure time: 30 min

Comment: Mean of three assays. Methanol (<10%) was used to prepare the stock solutions. EC50 values were calculated from nominal concentrations.

Bibliographical reference: Ribo, J.M., and Kaiser, K.L.E. (1983). *Chemosphere* **12**, 1421-1442.

Temperature: 15°C
Test parameter: EC50
Effect: Reduction in light output

Concentration: 0.51 mg/l
Exposure time: 5 min

Concentration: 0.64 mg/l
Exposure time: 30 min

Bibliographical reference: Speece, R. (1987). Drexel University, Philadelphia, USA, private communication.

Temperature: 15°C
Test parameter: EC50
Effect: Reduction in light output
Concentration: 2.6 mg/l
Exposure time: 15 min
Comment: pH = 7.2.

Bibliographical reference: Neilson, A.H., Allard, A.S., Fischer, S., Malmberg, M., and Viktor, T. (1990). *Ecotoxicol. Environ. Safety* **20**, 82-97.

3,5-DICHLOROPHENOL

CAS RN: 591-35-5

Temperature: 15°C
Test parameter: EC50
Effect: Reduction in light output
Concentrations: 2.8 mg/l
3.9 mg/l
Exposure time: 5 min

Bibliographical reference: King, E.F., and Painter, H.A. (1981). In: *Les Tests de Toxicité Aiguë en Milieu Aquatique. Acute Aquatic Ecotoxicological Tests*, H. Leclerc and D. Dive (eds.), INSERM 106, Paris, p. 143-153.

Temperature: 15°C
Test parameter: EC50
Effect: Reduction in light output

Concentration: 3.2 mg/l
Exposure time: 5 min

Concentration: 3.0 mg/l
Exposure time: 10 min

Concentration: 2.9 mg/l
Exposure time: 15 min

Comment: Chemical was tested at pH 6.7.

Bibliographical references: Dutka, B.J., and Kwan, K.K. (1982). *Environ. Pollut. Ser. A* **29**, 125-134.
Dutka, B.J., Nyholm, N., and Petersen, J. (1983). *Water Res.* **17**, 1363-1368.

Temperature: 15°C
Test parameter: EC50
Effect: Reduction in light output

Concentration: 3.91 mg/l
Exposure time: 5 min

Concentration: 3.18 mg/l

Exposure time: 15 min

Concentration: 2.77 mg/l
Exposure time: 30 min

Comment: Mean of three assays. Methanol (<10%) was used to prepare the stock solutions. EC50 values were calculated from nominal concentrations.

Bibliographical reference: Ribo, J.M., and Kaiser, K.L.E. (1983). *Chemosphere* **12**, 1421-1442.

Temperature: 15°C
Test parameter: EC50
Effect: Reduction in light output
Concentration: 4.5 ± 0.5 mg/l
Exposure time: 5 min
Comment: Mean of three assays. Chemical dissolved in DMSO (0.1%). Toxicity values were calculated from nominal concentrations.

Bibliographical reference: Ferard, J.F., Vasseur, P., Danoux, L., and Larbaigt, G. (1983). *Rev. Fr. Sci. Eau* **2**, 221-237.

Sample purity: 99%
Temperature: 15°C
Test parameter: EC50
Effect: Reduction in light output
Concentration: 3.20 mg/l
Exposure time: 15 min

Bibliographical reference: King, E.F. (1984). In: *Toxicity Screening Procedures Using Bacterial Systems*, D. Liu and B.J. Dutka (eds.), Marcel Dekker, New York, p. 175-194.

Temperature: 15°C
Test parameter: EC50
Effect: Reduction in light output

Concentration: 4.47 mg/l
Exposure time: 5 min

Concentration: 4.24 mg/l
Exposure time: 10 min

Concentration: 4.15 mg/l
Exposure time: 15 min

Comment: Toxicity values were calculated from nominal concentrations.

Bibliographical reference: Reteuna, C., Vasseur, P., Cabridenc, R., and Lepailleur, H. (1986). *Tox. Assess.* **1**, 159-168.

Temperature: 15°C
Test parameter: EC50
Effect: Reduction in light output

Concentration: 7.77 mg/l
Exposure time: 5 min

Concentration: 8.12 mg/l
Exposure time: 30 min

Bibliographical reference: Speece, R. (1987). Drexel University, Philadelphia, USA, private communication.

Sample purity: Reagent grade
Test parameter: EC50
Effect: Reduction in light output

Concentration: 4.4 mg/l (4.3-4.5)
Exposure time: 5 min

Concentration: 3.7 mg/l (3.4-4.0)
Exposure time: 15 min

Concentration: 3.9 mg/l (3.7-4.1)
Exposure time: 30 min

Bibliographical reference: Elnabarawy, M.T., Robideau, R.R., and Beach, S.A. (1988). *Tox. Assess.* **3**, 361-370.

2,4-DICHLOROPHENYLACETIC ACID

CAS RN: 19719-28-9

Sample purity: 99%
Temperature: 15°C
Test parameter: EC50
Effect: Reduction in light output

Concentration: 93.7 mg/l
Exposure time: 5 min

Concentration: 89.5 mg/l
Exposure time: 15 min

Concentration: 85.5 mg/l
Exposure time: 30 min

Comment: Mean of three assays. Methanol (<10%) was used to prepare the stock solutions. EC50 values were calculated from nominal concentrations.

Bibliographical reference: Kaiser, K.L.E., and Palabrica, V.S. (1991). *Water Poll. Res. J. Canada* **26**, 361-431.

2,4-DICHLOROPHENYLACETONITRILE

CAS RN: 6306-60-1

Sample purity: 98%
Temperature: 15°C
Test parameter: EC50
Effect: Reduction in light output

Concentration: 0.78 mg/l
Exposure time: 5 min

Concentration: 0.83 mg/l
Exposure time: 15 min

Concentration: 1.00 mg/l
Exposure time: 30 min

Comment: Mean of three assays. Methanol (<10%) was used to prepare the stock solutions. EC50 values were calculated from nominal concentrations.

Bibliographical reference: Kaiser, K.L.E., and Palabrica, V.S. (1991). *Water Poll. Res. J. Canada* **26**, 361-431.

2,4-DICHLOROPHENYL ISOCYANATE

CAS RN: 2612-57-9

Sample purity: 99%
Temperature: 15°C
Test parameter: EC50
Effect: Reduction in light output

Concentration: 1.36 mg/l
Exposure time: 5 min

Concentration: 1.56 mg/l
Exposure time: 15 min

Concentration: 1.80 mg/l
Exposure time: 30 min

Comment: Mean of three assays. Methanol (<10%) was used to prepare the stock solutions. EC50 values were calculated from nominal concentrations.

Bibliographical reference: Kaiser, K.L.E., and Palabrica, V.S. (1991). *Water Poll. Res. J. Canada* **26**, 361-431.

3,4-DICHLOROPHENYL ISOCYANATE

CAS RN: 102-36-3

Sample purity: 97%
Temperature: 15°C
Test parameter: EC50
Effect: Reduction in light output

Concentration: 1.49 mg/l
Exposure time: 5 min

Concentration: 1.11 mg/l
Exposure time: 15 min

Concentration: 0.96 mg/l
Exposure time: 30 min

Comment: Mean of three assays. Methanol (<10%) was used to prepare the stock solutions. EC50 values were calculated from nominal concentrations.

Bibliographical reference: Kaiser, K.L.E., and Palabrica, V.S. (1991). *Water Poll. Res. J. Canada* **26**, 361-431.

1,2-DICHLOROPROPANE

CAS RN: 78-87-5

Temperature: 15°C
Test parameter: EC50
Effect: Reduction in light output

Concentration: 58.9 mg/l
Exposure time: 5 min

Concentration: 93.4 mg/l
Exposure time: 30 min

Bibliographical reference: Speece, R. (1987). Drexel University, Philadelphia, USA, private communication.

1,3-DICHLOROPROPANE

CAS RN: 142-28-9

Temperature: 15°C
Test parameter: EC50
Effect: Reduction in light output

Concentration: 71.3 mg/l
Exposure time: 5 min

Concentration: 122 mg/l
Exposure time: 30 min

Bibliographical reference: Speece, R. (1987). Drexel University, Philadelphia, USA, private communication.

1,3-DICHLORO-2-PROPANOL

CAS RN: 96-23-1

Sample purity: 95%
Temperature: 15°C
Test parameter: EC50
Effect: Reduction in light output

Concentration: 1900 mg/l
Exposure time: 5 min

Concentration: 1700 mg/l
Exposure time: 15 min

Bibliographical reference: Cronin, M.T.D. (1993). Liverpool John Moores University, UK, private communication.

1,3-DICHLOROPROPENE

CAS RN: 542-75-6

Temperature: 15°C
Test parameter: EC50
Effect: Reduction in light output

Concentration: 112 mg/l
Exposure time: 5 min

Concentration: 71.6 mg/l
Exposure time: 30 min

Bibliographical reference: Speece, R. (1987). Drexel University, Philadelphia, USA, private communication.

2,3-DICHLOROPYRIDINE

CAS RN: 2402-77-9

Temperature: 15°C
Test parameter: EC50
Effect: Reduction in light output

Concentration: 35.5 mg/l
Exposure time: 5 min

Concentration: 33.9 mg/l
Exposure time: 15 min

Concentration: 34.7 mg/l*
Exposure time: 30 min

Comment: Mean of three assays. Methanol (<10%) was used to prepare the stock solutions. EC50 values were calculated from nominal concentrations.

Bibliographical references: Kaiser, K.L.E., and Palabrica, V.S. (1991). *Water Poll. Res. J. Canada* **26**, 361-431.
* Kaiser, K.L.E., and Ribo, J.M. (1985). In: *QSAR in Toxicology and Xenobiochemistry*, M. Tichy (ed.), Elsevier, Amsterdam, p. 27-38.

2,5-DICHLOROPYRIDINE

CAS RN: 16110-09-1

Temperature: 15°C
Test parameter: EC50
Effect: Reduction in light output

Concentration: 81.3 mg/l
Exposure time: 5 min

Concentration: 83.2 mg/l
Exposure time: 15 min

Concentration: 74.2 mg/l*
Exposure time: 30 min

Comment: Mean of three assays. Methanol (<10%) was used to prepare the stock solutions. EC50 values were calculated from nominal concentrations.

Bibliographical references: Kaiser, K.L.E., and Palabrica, V.S. (1991). *Water Poll. Res. J. Canada* **26**, 361-431.
* Kaiser, K.L.E., and Ribo, J.M. (1985). In: *QSAR in Toxicology and Xenobiochemistry*, M. Tichy (ed.), Elsevier, Amsterdam, p. 27-38.

2,6-DICHLOROPYRIDINE

CAS RN: 2402-78-0

Temperature: 15°C
Test parameter: EC50
Effect: Reduction in light output

Concentration: 81.3 mg/l
Exposure time: 5 min

Concentration: 83.2 mg/l
Exposure time: 15 min

Concentration: 83.2 mg/l*
Exposure time: 30 min

Comment: Mean of three assays. Methanol (<10%) was used to prepare the stock solutions. EC50 values were calculated from nominal concentrations.

Bibliographical references: Kaiser, K.L.E., and Palabrica, V.S. (1991). *Water Poll. Res. J. Canada* **26**, 361-431.
* Kaiser, K.L.E., and Ribo, J.M. (1985). In: *QSAR in Toxicology and Xenobiochemistry*, M. Tichy (ed.), Elsevier, Amsterdam, p. 27-38.

3,5-DICHLOROPYRIDINE

CAS RN: 2457-47-8

Temperature: 15°C
Test parameter: EC50
Effect: Reduction in light output

Concentration: 74.2 mg/l
Exposure time: 5 min

Concentration: 72.5 mg/l
Exposure time: 15 min

Concentration: 70.8 mg/l*
Exposure time: 30 min

Comment: Mean of three assays. Methanol (<10%) was used to prepare the stock solutions. EC50 values were calculated from nominal concentrations.

Bibliographical references: Kaiser, K.L.E., and Palabrica, V.S. (1991). *Water Poll. Res. J. Canada* **26**, 361-431.
* Kaiser, K.L.E., and Ribo, J.M. (1985). In: *QSAR in Toxicology and Xenobiochemistry*, M. Tichy (ed.), Elsevier, Amsterdam, p. 27-38.

4,7-DICHLOROQUINOLINE

CAS RN: 86-98-6

Sample purity: 99%
Temperature: 15°C
Test parameter: EC50
Effect: Reduction in light output

Concentration: 3.29 mg/l
Exposure time: 5 min

Concentration: 3.21 mg/l
Exposure time: 15 min

Concentration: 3.07 mg/l
Exposure time: 30 min

Comment: Mean of three assays. Methanol (<10%) was used to prepare the stock solutions. EC50 values were calculated from nominal concentrations.

Bibliographical reference: Kaiser, K.L.E., and Palabrica, V.S. (1991). *Water Poll. Res. J. Canada* **26**, 361-431.

α,α-DICHLOROTOLUENE

CAS RN: 98-87-3

Sample purity: 99%
Temperature: 15°C
Test parameter: EC50
Effect: Reduction in light output

Concentration: 2.12 mg/l
Exposure time: 5 min

Concentration: 3.44 mg/l
Exposure time: 15 min

Concentration: 5.85 mg/l*
Exposure time: 30 min

Comment: Methanol (<10%) was used to prepare the stock solution. EC50 values were calculated from nominal concentrations.

Bibliographical references: Kaiser, K.L.E., and Palabrica, V.S. (1991). *Water Poll. Res. J. Canada* **26**, 361-431.
* Kaiser, K.L.E., Palabrica, V.S., and Ribo, J.M. (1987). In: *QSAR in Environmental Toxicology - II*, K.L.E. Kaiser (ed.), D. Reidel Publishing Company, Dordrecht, p. 153-168.

2,4-DICHLOROTOLUENE

CAS RN: 95-73-8

Sample purity: 99%
Temperature: 15°C
Test parameter: EC50
Effect: Reduction in light output

Concentration: 2.27 mg/l
Exposure time: 5 min

Concentration: 2.49 mg/l
Exposure time: 15 min

Concentration: 2.67 mg/l
Exposure time: 30 min

Comment: Mean of four assays. Methanol (<10%) was used to prepare the stock solutions. EC50 values were calculated from nominal concentrations.

Bibliographical reference: Kaiser, K.L.E., and Palabrica, V.S. (1991). *Water Poll. Res. J. Canada* **26**, 361-431.

3,4-DICHLOROTOLUENE

CAS RN: 95-75-0

Temperature: 15°C
Test parameter: EC50
Effect: Reduction in light output
Concentration: 1.40 mg/l
Exposure time: 15 min

Bibliographical reference: Hermens, J., Busser, F., Leeuwangh, P., and Musch, A. (1985). *Ecotoxicol. Environ. Safety* **9**, 17-25.

α,α'-DICHLORO-*p*-XYLENE

CAS RN: 623-25-6
Synonym: 1,4-Bis(chloromethyl)benzene

Temperature: 15°C
Test parameter: EC50
Effect: Reduction in light output

Concentration: 0.040 mg/l
Exposure time: 5 min

Concentration: 0.050 mg/l
Exposure times: 15 and 30 min

Comment: Mean of three assays. Methanol (<10%) was used to prepare the stock solutions. EC50 values were calculated from nominal concentrations.

Bibliographical reference: Kaiser, K.L.E., and Palabrica, V.S. (1991). *Water Poll. Res. J. Canada* **26**, 361-431.

1,4-DICYANOBUTANE

CAS RN: 111-69-3

Sample purity: 99%
Temperature: 15°C
Test parameter: EC50
Effect: Reduction in light output

Concentration: 393 mg/l
Exposure time: 5 min

Concentration: 402 mg/l
Exposure time: 15 min

Concentration: 411 mg/l
Exposure time: 30 min

Comment: Mean of four assays. Methanol (<10%) was used to prepare the stock solutions. EC50 values were calculated from nominal concentrations.

Bibliographical reference: Kaiser, K.L.E., and Palabrica, V.S. (1991). *Water Poll. Res. J. Canada* **26**, 361-431.

Sample purity: 99%
Temperature: 15°C
Test parameter: EC50
Effect: Reduction in light output
Concentration: 2720 mg/l
Exposure time: 5 min

Bibliographical reference: Cronin, M.T.D., Dearden, J.C., and Dobbs, A.J. (1991). *Sci. Total Environ.* **109/110**, 431-439.

1,6-DICYANOHEXANE

CAS RN: 629-40-3

Sample purity: 99%
Temperature: 15°C
Test parameter: EC50
Effect: Reduction in light output

Concentration: 16.0 mg/l
Exposure time: 5 min

Concentration: 18.0 mg/l*
Exposure time: 15 min

Bibliographical references: Cronin, M.T.D., Dearden, J.C., and Dobbs, A.J. (1991). *Sci. Total Environ.* **109/110**, 431-439.
* Cronin, M.T.D. (1993). Liverpool John Moores University, UK, private communication.

DIDT

CAS RN: 33813-20-6
Synonym: 5,6-Dihydro-3*H*-imidazo(2,1-c)-1,2,4-dithiazole-3-thione

Sample purity: ≥98%
Test parameter: EC50
Effect: Reduction in light output
Concentration: 0.03 mg/l (0.02-0.04)
Exposure time: 15 min

Bibliographical reference: van Leeuwen, C.J., Maas-Diepeveen, J.L., Niebeek, G., Vergouw, W.H.A., Griffioen, P.S., and Luijken, M.W. (1985). *Aquat. Toxicol.* **7**, 145-164.

DIETHANOLAMINE

CAS RN: 111-42-2

Temperature: 15°C
Test parameter: EC50
Effect: Reduction in light output
Concentration: 73 mg/l
Exposure time: 5 min
Comment: Value derived by extrapolation.

Bibliographical reference: Samak, Q.M., and Noiseux, R. (1981). *Can. Tech. Rep. Fish. Aquat. Sci.* **990**, 288-308.

1,4-DIETHOXYBENZENE

CAS RN: 122-95-2

Sample purity: >98%
Temperature: 15°C
Test parameter: EC50
Effect: Reduction in light output

Concentration: 8.14 mg/l
Exposure time: 5 min

Concentration: 8.72 mg/l
Exposure time: 15 min

Concentration: 11.8 mg/l
Exposure time: 30 min

Comment: Mean of three assays. Methanol (<10%) was used to prepare the stock solutions. EC50 values were calculated from nominal concentrations.

Bibliographical reference: Kaiser, K.L.E., and Palabrica, V.S. (1991). *Water Poll. Res. J. Canada* **26**, 361-431.

N,N-DIETHYLACETAMIDE

CAS RN: 685-91-6

Sample purity: 97%
Temperature: 15°C
Test parameter: EC50
Effect: Reduction in light output

Concentration: 155 mg/l
Exposure time: 5 min

Concentration: 170 mg/l
Exposure time: 15 min

Concentration: 152 mg/l
Exposure time: 30 min

Comment: Mean of four assays. Methanol (<10%) was used to prepare the stock solutions. EC50 values were calculated from nominal concentrations.

Bibliographical reference: Kaiser, K.L.E., and Palabrica, V.S. (1992). National Water Research Institute, Burlington, Ontario, Canada, unpublished results.

DIETHYL ADIPATE

CAS RN: 141-28-6

Temperature: 15°C
Test parameter: EC50
Effect: Reduction in light output

Concentration: 30.6 mg/l
Exposure time: 5 min

Concentration: 35.1 mg/l*
Exposure time: 15 min

Bibliographical references: Cronin, M.T.D., Dearden, J.C., and Dobbs, A.J. (1991). *Sci. Total Environ.* **109/110**, 431-439.
* Cronin, M.T.D. (1990). Thesis, Liverpool Polytechnic, Liverpool, UK.

DIETHYLAMINE

CAS RN: 109-89-7

Sample purity: ≥99%
Test parameter: EC50
Effect: Reduction in light output
Concentration: 21.8 mg/l (19.0-25.1)
Exposure time: 15 min

Bibliographical reference: van Leeuwen, C.J., Maas-Diepeveen, J.L., Niebeek, G., Vergouw, W.H.A., Griffioen, P.S., and Luijken, M.W. (1985). *Aquat. Toxicol.* **7**, 145-164.

Temperature: 15°C

Test parameter: EC10
Effect: Reduction in light output
Concentration: 15.2 mg/l
Exposure time: 5 min

Test parameter: EC50
Effect: Reduction in light output
Concentration: 23.4 mg/l
Exposure time: 5 min

Comment: Toxicity values were calculated from nominal concentrations.

Bibliographical reference: Reteuna, C. (1988). Thesis, University of Metz, Metz, France.

Sample purity: 98%
Temperature: 15°C
Test parameter: EC50
Effect: Reduction in light output

Concentration: 35.0 mg/l
Exposure time: 5 min

Concentration: 27.2 mg/l
Exposure time: 15 min

Concentration: 24.8 mg/l
Exposure time: 30 min

Comment: Mean of three assays. Methanol (<10%) was used to prepare the stock solutions. EC50 values were calculated from nominal concentrations.

Bibliographical reference: Kaiser, K.L.E., and Palabrica, V.S. (1992). National Water Research Institute, Burlington, Ontario, Canada, unpublished results.

N,N-DIETHYLANILINE

CAS RN: 91-66-7

Sample purity: 99%
Temperature: 15°C
Test parameter: EC50
Effect: Reduction in light output

Concentration: 6.50 mg/l
Exposure time: 5 min

Concentration: 7.70 mg/l*
Exposure time: 15 min

Comment: Chemical was prepared in an initial solution of 5% methanol.

Bibliographical references: Cronin, M.T.D., Dearden, J.C., and Dobbs, A.J. (1991). *Sci. Total Environ.* **109/110**, 431-439.
* Cronin, M.T.D. (1990). Thesis, Liverpool Polytechnic, Liverpool, UK.

DIETHYLENE GLYCOL

CAS RN: 111-46-6

Temperature: 15°C
Test parameter: EC50
Effect: Reduction in light output
Concentration: 8950 mg/l
Exposure time: 5 min

Bibliographical reference: Microtox® Application Notes (1982). Beckman Instruments, Inc., Carlsbad, California.

Temperature: 15°C
Test parameter: EC50
Effect: Reduction in light output
Concentration: 29228 mg/l
Exposure time: 15 min

Bibliographical reference: Hermens, J., Busser, F., Leeuwangh, P., and Musch, A. (1985). *Ecotoxicol. Environ. Safety* **9**, 17-25.

DIETHYL ETHER

CAS RN: 60-29-7

Temperature: 15°C
Test parameter: EC50
Effect: Reduction in light output
Concentration: 5623 mg/l
Exposure time: 15 min

Bibliographical reference: Hermens, J., Busser, F., Leeuwangh, P., and Musch, A. (1985). *Ecotoxicol. Environ. Safety* **9**, 17-25.

DI-2-ETHYLHEXYL PHTHALATE

CAS RN: 117-81-7

Temperature: 15°C
Test parameter: EC50
Effect: Reduction in light output
Concentration: 800 mg/l
Exposure times: 5, 15, and 30 min
Comment: Chemical was diluted to give five sample concentrations in duplicate.

Bibliographical reference: Benson, W.H., and Stackhouse, R.A. (1986). *Drug Chem. Toxicol.* **9**, 275-283.

DIETHYL PHTHALATE

CAS RN: 84-66-2

Temperature: 15°C
Test parameter: EC50
Effect: Reduction in light output

Concentration: 114 mg/l
Exposure times: 5 and 15 min

Concentration: 112 mg/l
Exposure time: 30 min

Comment: Chemical was diluted to give five sample concentrations in duplicate.

Bibliographical reference: Benson, W.H., and Stackhouse, R.A. (1986). *Drug Chem. Toxicol.* **9**, 275-283.

2,3-DIETHYLPYRAZINE

CAS RN: 15707-24-1

Sample purity: 98%
Temperature: 15°C
Test parameter: EC50
Effect: Reduction in light output

Concentration: 119 mg/l
Exposure time: 5 min

Concentration: 136 mg/l
Exposure time: 15 min

Concentration: 168 mg/l
Exposure time: 30 min

Comment: Mean of three assays. Methanol (<10%) was used to prepare the stock solutions. EC50 values were calculated from nominal concentrations.

Bibliographical reference: Kaiser, K.L.E., and Palabrica, V.S. (1991). *Water Poll. Res. J. Canada* **26**, 361-431.

DIETHYL SEBACATE

CAS RN: 110-40-7

Sample purity: 95%
Temperature: 15°C
Test parameter: EC50
Effect: Reduction in light output
Concentration: 0.60 mg/l
Exposure times: 5 and 15 min

Comment: Chemical was prepared in an initial solution of 3% methanol.

Bibliographical reference: Cronin, M.T.D. (1993). Liverpool John Moores University, UK, private communication.

N,N'-DIETHYLTHIOUREA

CAS RN: 105-55-5
Synonym: 1,3-Diethyl-2-thiourea

Sample purity: ≥97%
Test parameter: EC50
Effect: Reduction in light output
Concentration: 760 mg/l
Exposure time: 15 min
Comment: EC50 value converted to mg/l from the original data expressed in log(1/μmol l^{-1}) and a rounded molecular weight of 132 given by the authors.

Bibliographical reference: Govers, H., Ruepert, C., Stevens, T., and van Leeuwen, C.J. (1986). *Chemosphere* **15**, 383-393.

2,4-DIFLUOROANILINE

CAS RN: 367-25-9

Sample purity: 99%
Temperature: 15°C
Test parameter: EC50
Effect: Reduction in light output

Concentration: 97.9 mg/l
Exposure times: 5 and 15 min

Concentration: 93.5 mg/l
Exposure time: 30 min

Comment: Mean of three assays. Methanol (<10%) was used to prepare the stock solutions. EC50 values were calculated from nominal concentrations.

Bibliographical reference: Kaiser, K.L.E., and Palabrica, V.S. (1991). *Water Poll. Res. J. Canada* **26**, 361-431.

3,5-DIFLUOROANISOLE

CAS RN: 93343-10-3

Sample purity: 97%
Temperature: 15°C
Test parameter: EC50
Effect: Reduction in light output

Concentration: 61.5 mg/l
Exposure time: 5 min

Concentration: 73.9 mg/l
Exposure time: 15 min

Concentration: 93.1 mg/l
Exposure time: 30 min

Comment: Mean of four assays. Methanol (<10%) was used to prepare the stock solutions. EC50 values were calculated from nominal concentrations.

Bibliographical reference: Kaiser, K.L.E., and Palabrica, V.S. (1991). *Water Poll. Res. J. Canada* **26**, 361-431.

1,4-DIFLUOROBENZENE

CAS RN: 540-36-3

Sample purity: 99+%
Temperature: 15°C
Test parameter: EC50
Effect: Reduction in light output

Concentration: 169 mg/l
Exposure time: 5 min

Concentration: 117 mg/l
Exposure time: 15 min

Concentration: 104 mg/l*
Exposure time: 30 min

Comment: Mean of three assays. Methanol (<10%) was used to prepare the stock solutions. EC50 values were calculated from nominal concentrations.

Bibliographical references: Kaiser, K.L.E., and Palabrica, V.S. (1991). *Water Poll. Res. J. Canada* **26**, 361-431.
* Kaiser, K.L.E., and Gough, K.M. (1989). In: *Aquatic Toxicology and Environmental Fate: Eleventh Volume, ASTM STP 1007*, G.W. Suter and M.A. Lewis (eds.), American Society for Testing and Materials, Philadelphia, p. 424-441.

2,6-DIFLUOROBENZOIC ACID

CAS RN: 385-00-2

Sample purity: 98%
Temperature: 15°C
Test parameter: EC50
Effect: Reduction in light output

Concentration: 91.0 mg/l
Exposure time: 5 min

Concentration: 74.0 mg/l
Exposure time: 15 min

Concentration: 57.4 mg/l
Exposure time: 30 min

Comment: Mean of three assays. Methanol (<10%) was used to prepare the stock solutions. EC50 values were calculated from nominal concentrations.

Bibliographical reference: Kaiser, K.L.E., and Palabrica, V.S. (1991). *Water Poll. Res. J. Canada* **26**, 361-431.

2,6-DIFLUOROBENZONITRILE

CAS RN: 1897-52-5

Sample purity: 98%
Temperature: 15°C
Test parameter: EC50
Effect: Reduction in light output

Concentration: 55.4 mg/l
Exposure times: 5 and 15 min

Concentration: 58.0 mg/l
Exposure time: 30 min

Comment: Mean of three assays. Methanol (<10%) was used to prepare the stock solutions. EC50 values were calculated from nominal concentrations.

Bibliographical reference: Kaiser, K.L.E., and Palabrica, V.S. (1991). *Water Poll. Res. J. Canada* **26**, 361-431.

1,5-DIFLUORO-2,4-DINITROBENZENE

CAS RN: 327-92-4

Sample purity: 97%
Temperature: 15°C
Test parameter: EC50
Effect: Reduction in light output

Concentration: 0.28 mg/l
Exposure time: 5 min

Concentration: 0.11 mg/l
Exposure time: 15 min

Concentration: 0.06 mg/l
Exposure time: 30 min

Comment: Mean of three assays. Methanol (<10%) was used to prepare the stock solutions. EC50 values were calculated from nominal concentrations.

Bibliographical reference: Kaiser, K.L.E., and Palabrica, V.S. (1991). *Water Poll. Res. J. Canada* **26**, 361-431.

Sample purity: 99%
Temperature: 15°C
Test parameter: EC50
Effect: Reduction in light output

Concentration: 0.76 mg/l
Exposure time: 5 min

Concentration: 0.21 mg/l
Exposure time: 15 min

Concentration: 0.11 mg/l
Exposure time: 30 min

Comment: Mean of three assays. Methanol (<10%) was used to prepare the stock solutions. EC50 values were calculated from nominal concentrations.

Bibliographical reference: Kaiser, K.L.E., and Palabrica, V.S. (1991). *Water Poll. Res. J. Canada* **26**, 361-431.

4,5-DIFLUORO-2-NITROANILINE

CAS RN: 78056-39-0

Sample purity: 98%
Temperature: 15°C
Test parameter: EC50
Effect: Reduction in light output

Concentration: 34.7 mg/l
Exposure time: 5 min

Concentration: 24.0 mg/l
Exposure time: 15 min

Concentration: 18.2 mg/l
Exposure time: 30 min

Comment: Mean of three assays. Methanol (<10%) was used to prepare the stock solutions. EC50 values were calculated from nominal concentrations.

Bibliographical reference: Kaiser, K.L.E., and Palabrica, V.S. (1991). *Water Poll. Res. J. Canada* **26**, 361-431.

2,3-DIFLUOROPHENOL

CAS RN: 6418-38-8

Sample purity: 98%
Temperature: 15°C
Test parameter: EC50
Effect: Reduction in light output

Concentration: 88.0 mg/l
Exposure time: 5 min

Concentration: 80.2 mg/l
Exposure time: 15 min

Concentration: 76.6 mg/l
Exposure time: 30 min

Comment: Mean of three assays. Methanol (<10%) was used to prepare the stock solutions. EC50 values were calculated from nominal concentrations.

Bibliographical reference: Kaiser, K.L.E., and Palabrica, V.S. (1991). *Water Poll. Res. J. Canada* **26**, 361-431.

2,4-DIFLUOROPHENOL

CAS RN: 367-27-1

Sample purity: 99%
Temperature: 15°C
Test parameter: EC50
Effect: Reduction in light output

Concentration: 48.3 mg/l
Exposure time: 5 min

Concentration: 49.5 mg/l
Exposure time: 15 min

Concentration: 56.8 mg/l
Exposure time: 30 min

Comment: Mean of three assays. Methanol (<10%) was used to prepare the stock solutions. EC50 values were calculated from nominal concentrations.

Bibliographical reference: Kaiser, K.L.E., and Palabrica, V.S. (1991). *Water Poll. Res. J. Canada* **26**, 361-431.

3,4-DIFLUOROPHENOL

CAS RN: 2713-33-9

Sample purity: 99%
Temperature: 15°C
Test parameter: EC50
Effect: Reduction in light output

Concentration: 32.7 mg/l
Exposure time: 5 min

Concentration: 29.1 mg/l
Exposure times: 15 and 30 min

Comment: Mean of three assays. Methanol (<10%) was used to prepare the stock solutions. EC50 values were calculated from nominal concentrations.

Bibliographical reference: Kaiser, K.L.E., and Palabrica, V.S. (1991). *Water Poll. Res. J. Canada* **26**, 361-431.

3,5-DIFLUOROPHENOL

CAS RN: 2713-34-0

Sample purity: 99%
Temperature: 15°C
Test parameter: EC50
Effect: Reduction in light output

Concentration: 43.1 mg/l

Exposure time: 5 min

Concentration: 42.1 mg/l
Exposure time: 15 min

Concentration: 44.1 mg/l
Exposure time: 30 min

Comment: Mean of three assays. Methanol (<10%) was used to prepare the stock solutions. EC50 values were calculated from nominal concentrations.

Bibliographical reference: Kaiser, K.L.E., and Palabrica, V.S. (1991). *Water Poll. Res. J. Canada* **26**, 361-431.

2',5'-DIHYDROXYACETOPHENONE

CAS RN: 490-78-8

Sample purity: >99%
Temperature: 15°C
Test parameter: EC50
Effect: Reduction in light output

Concentration: 9.82 mg/l
Exposure time: 5 min

Concentration: 6.64 mg/l
Exposure time: 15 min

Concentration: 5.52 mg/l
Exposure time: 30 min

Comment: Mean of three assays. Methanol (<10%) was used to prepare the stock solutions. EC50 values were calculated from nominal concentrations.

Bibliographical reference: Kaiser, K.L.E., and Palabrica, V.S. (1991). *Water Poll. Res. J. Canada* **26**, 361-431.

3,5-DIHYDROXYBENZOIC ACID

CAS RN: 99-10-5

Sample purity: 97%
Temperature: 15°C
Test parameter: EC50
Effect: Reduction in light output

Concentration: 21.3 mg/l
Exposure time: 5 min

Concentration: 24.4 mg/l
Exposure time: 15 min

Concentration: 35.3 mg/l
Exposure time: 30 min

Comment: Mean of three assays. Methanol (<10%) was used to prepare the stock solutions. EC50 values were calculated from nominal concentrations.

Bibliographical reference: Kaiser, K.L.E., and Palabrica, V.S. (1991). *Water Poll. Res. J. Canada* **26**, 361-431.

1,5-DIHYDROXYNAPHTHALENE

CAS RN: 83-56-7

Sample purity: >97%
Temperature: 15°C
Test parameter: EC50
Effect: Reduction in light output

Concentration: 3.67 mg/l
Exposure time: 5 min

Concentration: 2.60 mg/l
Exposure time: 15 min

Concentration: 2.78 mg/l
Exposure time: 30 min

Comment: Mean of three assays. Methanol (<10%) was used to prepare the stock solutions. EC50 values were calculated from nominal concentrations.

Bibliographical reference: Kaiser, K.L.E., and Palabrica, V.S. (1991). *Water Poll. Res. J. Canada* **26**, 361-431.

2,3-DIHYDROXYPYRIDINE

CAS RN: 16867-04-2

Sample purity: 98%
Temperature: 15°C
Test parameter: EC50
Effect: Reduction in light output

Concentration: 292 mg/l
Exposure time: 5 min

Concentration: 267 mg/l
Exposure time: 15 min

Concentration: 249 mg/l
Exposure time: 30 min

Comment: Mean of three assays. Methanol (<10%) was used to prepare the stock solutions. EC50 values were calculated from nominal concentrations.

Bibliographical reference: Kaiser, K.L.E., and Palabrica, V.S. (1991). *Water Poll. Res. J. Canada* **26**, 361-431.

2,6-DIHYDROXYTOLUENE

CAS RN: 608-25-3
Synonym: 2-Methylresorcinol

Temperature: 15°C
Test parameter: EC50
Effect: Reduction in light output

Concentration: 80.2 mg/l

Exposure time: 5 min

Concentration: 68.2 mg/l
Exposure time: 15 min

Concentration: 62.2 mg/l
Exposure time: 30 min

Comment: Mean of three assays. Methanol (<10%) was used to prepare the stock solutions. EC50 values were calculated from nominal concentrations.

Bibliographical reference: Kaiser, K.L.E., and Palabrica, V.S. (1991). *Water Poll. Res. J. Canada* **26**, 361-431.

1,6-DIISOCYANATOHEXANE

CAS RN: 822-06-0
Synonym: Hexamethylene diisocyanate

Sample purity: 98%
Temperature: 15°C
Test parameter: EC50
Effect: Reduction in light output

Concentration: 53.2 mg/l
Exposure time: 5 min

Concentration: 25.5 mg/l
Exposure time: 15 min

Concentration: 15.7 mg/l
Exposure time: 30 min

Comment: Mean of three assays. Methanol (<10%) was used to prepare the stock solutions. EC50 values were calculated from nominal concentrations.

Bibliographical reference: Kaiser, K.L.E., and Palabrica, V.S. (1991). *Water Poll. Res. J. Canada* **26**, 361-431.

1,4-DIMETHOXYBENZENE

CAS RN: 150-78-7
Synonym: Hydroquinone dimethyl ether

Sample purity: 99%
Temperature: 15°C
Test parameter: EC50
Effect: Reduction in light output
Concentration: 4.17 mg/l
Exposure time: 30 min
Comment: Mean of three assays. Methanol (<10%) was used to prepare the stock solutions. EC50 values were calculated from nominal concentrations.

Bibliographical reference: Ribo, J.M., and Kaiser, K.L.E. National Water Research Institute, Burlington, Ontario, Canada, unpublished results (value later published by Kaiser, K.L.E., and Palabrica, V.S. (1991). *Water Poll. Res. J. Canada* **26**, 361-431).

3,3'-DIMETHOXYBENZIDINE

CAS RN: 119-90-4
Synonym: *o*-Dianisidine

Sample purity: 97%
Temperature: 15°C
Test parameter: EC50
Effect: Reduction in light output

Concentration: 84.7 mg/l
Exposure time: 5 min

Concentration: 88.7 mg/l
Exposure time: 15 min

Concentration: 101 mg/l
Exposure time: 30 min

Comment: Mean of three assays. Methanol (<10%) was used to prepare the stock solutions. EC50 values were calculated from nominal concentrations.

Bibliographical reference: Kaiser, K.L.E., and Palabrica, V.S. (1991). *Water Poll. Res. J. Canada* **26**, 361-431.

DIMETHOXYMETHANE

CAS RN: 109-87-5
Synonyms: Formaldehyde dimethyl acetal; Methylal

Sample purity: 99%
Temperature: 15°C
Test parameter: EC50
Effect: Reduction in light output

Concentration: 514 mg/l
Exposure time: 5 min

Concentration: 469 mg/l
Exposure times: 15 and 30 min

Comment: Mean of three assays. Methanol (<10%) was used to prepare the stock solutions. EC50 values were calculated from nominal concentrations.

Bibliographical reference: Kaiser, K.L.E., and Palabrica, V.S. (1992). National Water Research Institute, Burlington, Ontario, Canada, unpublished results.

1,2-DIMETHOXY-4-NITROBENZENE

CAS RN: 709-09-1
Synonym: 4-Nitroveratrole

Temperature: 15°C
Test parameter: EC50
Effect: Reduction in light output

Concentration: 46.0 mg/l
Exposure time: 5 min

Concentration: 52.8 mg/l
Exposure time: 15 min

Concentration: 63.5 mg/l
Exposure time: 30 min

Comment: Mean of four assays. Methanol (<10%) was used to prepare the stock solutions. EC50 values were calculated from nominal concentrations.

Bibliographical reference: Kaiser, K.L.E., and Palabrica, V.S. (1992). National Water Research Institute, Burlington, Ontario, Canada, unpublished results.

N,N-DIMETHYLACETAMIDE

CAS RN: 127-19-5

Temperature: 15°C
Test parameter: EC50
Effect: Reduction in light output

Concentration: 4815 mg/l
Exposure time: 5 min

Concentration: 2393 mg/l*
Exposure time: 30 min

* EC50 value from linear regression fit to data.

Bibliographical reference: Speece, R. (1987). Drexel University, Philadelphia, USA, private communication.

DIMETHYLAMINE

CAS RN: 124-40-3

Sample purity: ≥98%
Test parameter: EC50
Effect: Reduction in light output
Concentration: 26.8 mg/l (20.8-34.5)
Exposure time: 15 min
Comment: The dimethylamine used was a 40% solution in water.

Bibliographical reference: van Leeuwen, C.J., Maas-Diepeveen, J.L., Niebeek, G., Vergouw, W.H.A., Griffioen, P.S., and Luijken, M.W. (1985). *Aquat. Toxicol.* **7**, 145-164.

4-DIMETHYLAMINOAZOBENZENE

CAS RN: 60-11-7
Synonym: Methyl yellow

Temperature: 15°C
Test parameter: EC50
Effect: Reduction in light output

Concentration: 0.023 mg/l
Exposure time: 5 min

Concentration: 0.019 mg/l
Exposure times: 15 and 30 min

Comment: Mean of three assays. Methanol (<10%) was used to prepare the stock solutions. EC50 values were calculated from nominal concentrations.

Bibliographical reference: Kaiser, K.L.E. (1987). In: *QSAR in Environmental Toxicology - II*, K.L.E. Kaiser (ed.), D. Reidel Publishing Company, Dordrecht, p. 169-188.

4-(DIMETHYLAMINO)BENZALDEHYDE

CAS RN: 100-10-7

Sample purity: 99%
Temperature: 15°C
Test parameter: EC50
Effect: Reduction in light output

Concentration: 0.70 mg/l
Exposure time: 5 min

Concentration: 0.75 mg/l
Exposure time: 15 min

Concentration: 0.88 mg/l
Exposure time: 30 min

Comment: Mean of four assays. Methanol (<10%) was used to prepare the stock solutions. EC50 values were calculated from nominal concentrations.

Bibliographical reference: Kaiser, K.L.E., and Palabrica, V.S. (1991). *Water Poll. Res. J. Canada* **26**, 361-431.

4-(DIMETHYLAMINO)BENZONITRILE

CAS RN: 1197-19-9

Sample purity: 98%
Temperature: 15°C
Test parameter: EC50
Effect: Reduction in light output

Concentration: 0.18 mg/l
Exposure times: 5 and 15 min

Concentration: 0.17 mg/l*
Exposure time: 30 min

Comment: Mean of four assays. Methanol (<10%) was used to prepare the stock solutions. EC50 values were calculated from nominal concentrations.

Bibliographical references: Kaiser, K.L.E., and Palabrica, V.S. (1991). *Water Poll. Res. J. Canada* **26**, 361-431.
* Kaiser, K.L.E. (1987). In: *QSAR in Environmental Toxicology - II*, K.L.E. Kaiser (ed.), D. Reidel Publishing Company, Dordrecht, p. 169-188.

4-(DIMETHYLAMINO)BENZOPHENONE

CAS RN: 530-44-9

Sample purity: 98%
Temperature: 15°C
Test parameter: EC50

Effect: Reduction in light output
Concentration: 0.066 mg/l
Exposure time: 30 min
Comment: Methanol (<10%) was used to prepare the stock solution. EC50 value was calculated from nominal concentrations.

Bibliographical reference: Kaiser, K.L.E., Palabrica, V.S., and Ribo, J.M. (1987). In: *QSAR in Environmental Toxicology - II*, K.L.E. Kaiser (ed.), D. Reidel Publishing Company, Dordrecht, p. 153-168.

4-DIMETHYLAMINO-3-METHYL-2-BUTANONE

CAS RN: 22104-62-7

Test parameter: EC50
Effect: Reduction in light output
Concentration: 42.1 mg/l
Exposure time: 5 min
Comment: The EC50 value was calculated from nominal concentrations.

Bibliographical reference: Curtis, C., Lima, A., Lozano, S.J., and Veith, G.D. (1982). In: *Aquatic Toxicology and Hazard Assessment: Fifth Conference, ASTM STP 766*, J.G. Pearson, R.B. Foster, and W.E. Bishop (eds.), American Society for Testing and Materials, Philadelphia, p. 170-178.

4-(DIMETHYLAMINO)PHENETHYL ALCOHOL

CAS RN: 50438-75-0

Sample purity: 98%
Temperature: 15°C
Test parameter: EC50
Effect: Reduction in light output

Concentration: 8.67 mg/l
Exposure time: 5 min

Concentration: 8.87 mg/l
Exposure time: 15 min

Concentration: 8.28 mg/l*
Exposure time: 30 min

Comment: Mean of three assays. Methanol (<10%) was used to prepare the stock solutions. EC50 values were calculated from nominal concentrations.

Bibliographical references: Kaiser, K.L.E., and Palabrica, V.S. (1991). *Water Poll. Res. J. Canada* **26**, 361-431.
* Kaiser, K.L.E. (1987). In: *QSAR in Environmental Toxicology - II*, K.L.E. Kaiser (ed.), D. Reidel Publishing Company, Dordrecht, p. 169-188.

2-DIMETHYLAMINOPYRIDINE

CAS RN: 5683-33-0

Sample purity: 97%
Temperature: 15°C
Test parameter: EC50
Effect: Reduction in light output

Concentration: 99.3 mg/l
Exposure times: 5 and 15 min

Concentration: 109 mg/l
Exposure time: 30 min

Comment: Mean of three assays. Methanol (<10%) was used to prepare the stock solutions. EC50 values were calculated from nominal concentrations.

Bibliographical reference: Kaiser, K.L.E., and Palabrica, V.S. (1991). *Water Poll. Res. J. Canada* **26**, 361-431.

4-DIMETHYLAMINOPYRIDINE

CAS RN: 1122-58-3

Sample purity: 99%
Temperature: 15°C
Test parameter: EC50

Effect: Reduction in light output

Concentrations: 24.4 mg/l
26.7 mg/l
Exposure time: 5 min

Concentrations: 21.2 mg/l
24.9 mg/l
Exposure time: 15 min

Concentrations: 23.3 mg/l
26.7 mg/l
Exposure time: 30 min

Comment: The two series of tests were performed in triplicate. Methanol (<10%) was used to prepare the stock solutions. EC50 values were calculated from nominal concentrations.

Bibliographical reference: Kaiser, K.L.E., and Palabrica, V.S. (1991). *Water Poll. Res. J. Canada* **26**, 361-431.

N,N-DIMETHYLANILINE

CAS RN: 121-69-7

Sample purity: 99%
Temperature: 15°C
Test parameter: EC50
Effect: Reduction in light output

Concentration: 13.6 mg/l
Exposure times: 5 and 15 min

Concentration: 14.6 mg/l*
Exposure time: 30 min

Comment: Mean of three assays. Methanol (<10%) was used to prepare the stock solutions. EC50 values were calculated from nominal concentrations.

Bibliographical references: Kaiser, K.L.E., and Palabrica, V.S. (1991). *Water Poll. Res. J. Canada* **26**, 361-431.

* Kaiser, K.L.E., Palabrica, V.S., and Ribo, J.M. (1987). In: *QSAR in Environmental Toxicology - II*, K.L.E. Kaiser (ed.), D. Reidel Publishing Company, Dordrecht, p. 153-168.

2,3-DIMETHYLANILINE

CAS RN: 87-59-2
Synonym: 2,3-Xylidine

Sample purity: 99%
Temperature: 15°C
Test parameter: EC50
Effect: Reduction in light output

Concentration: 34.2 mg/l
Exposure time: 5 min

Concentration: 39.2 mg/l
Exposure time: 15 min

Concentration: 47.1 mg/l
Exposure time: 30 min

Comment: Mean of three assays. Methanol (<10%) was used to prepare the stock solutions. EC50 values were calculated from nominal concentrations.

Bibliographical reference: Kaiser, K.L.E., and Palabrica, V.S. (1991). *Water Poll. Res. J. Canada* **26**, 361-431.

2,4-DIMETHYLANILINE

CAS RN: 95-68-1
Synonym: 2,4-Xylidine

Sample purity: 99%
Temperature: 15°C
Test parameter: EC50
Effect: Reduction in light output

Concentration: 15.3 mg/l
Exposure time: 5 min

Concentration: 16.3 mg/l
Exposure time: 15 min

Concentration: 17.1 mg/l
Exposure time: 30 min

Comment: Mean of three assays. Methanol (<10%) was used to prepare the stock solutions. EC50 values were calculated from nominal concentrations.

Bibliographical reference: Kaiser, K.L.E., and Palabrica, V.S. (1991). *Water Poll. Res. J. Canada* **26**, 361-431.

2,5-DIMETHYLANILINE

CAS RN: 95-78-3
Synonym: 2,5-Xylidine

Sample purity: 99%
Temperature: 15°C
Test parameter: EC50
Effect: Reduction in light output

Concentration: 15.6 mg/l
Exposure time: 5 min

Concentration: 17.9 mg/l
Exposure time: 15 min

Concentration: 22.1 mg/l
Exposure time: 30 min

Comment: Mean of three assays. Methanol (<10%) was used to prepare the stock solutions. EC50 values were calculated from nominal concentrations.

Bibliographical reference: Kaiser, K.L.E., and Palabrica, V.S. (1991). *Water Poll. Res. J. Canada* **26**, 361-431.

2,6-DIMETHYLANILINE

CAS RN: 87-62-7

Synonym: 2,6-Xylidine

Sample purity: 99%
Temperature: 15°C
Test parameter: EC50
Effect: Reduction in light output

Concentration: 19.2 mg/l
Exposure time: 5 min

Concentration: 21.1 mg/l
Exposure time: 15 min

Concentration: 26.5 mg/l
Exposure time: 30 min

Comment: Mean of three assays. Methanol (<10%) was used to prepare the stock solutions. EC50 values were calculated from nominal concentrations.

Bibliographical reference: Kaiser, K.L.E., and Palabrica, V.S. (1991). *Water Poll. Res. J. Canada* **26**, 361-431.

3,4-DIMETHYLANILINE

CAS RN: 95-64-7
Synonym: 3,4-Xylidine

Sample purity: 98%
Temperature: 15°C
Test parameter: EC50
Effect: Reduction in light output

Concentration: 0.76 mg/l
Exposure time: 5 min

Concentration: 0.86 mg/l
Exposure time: 15 min

Concentration: 0.96 mg/l
Exposure time: 30 min

Comment: Mean of three assays. Methanol (<10%) was used to prepare the stock solutions. EC50 values were calculated from nominal concentrations.

Bibliographical reference: Kaiser, K.L.E., and Palabrica, V.S. (1991). *Water Poll. Res. J. Canada* **26**, 361-431.

3,5-DIMETHYLANILINE

CAS RN: 108-69-0
Synonym: 3,5-Xylidine

Sample purity: 98%
Temperature: 15°C
Test parameter: EC50
Effect: Reduction in light output

Concentration: 14.2 mg/l
Exposure time: 5 min

Concentration: 16.3 mg/l
Exposure time: 15 min

Concentration: 19.2 mg/l
Exposure time: 30 min

Comment: Mean of three assays. Methanol (<10%) was used to prepare the stock solutions. EC50 values were calculated from nominal concentrations.

Bibliographical reference: Kaiser, K.L.E., and Palabrica, V.S. (1991). *Water Poll. Res. J. Canada* **26**, 361-431.

3,3-DIMETHYL-2-BUTANONE

CAS RN: 75-97-8
Synonyms: Pinacolone; *tert*-Butyl methyl ketone

Sample purity: 95%
Temperature: 15°C
Test parameter: EC50
Effect: Reduction in light output

Concentration: 3.20 mg/l
Exposure time: 5 min

Concentration: 3.60 mg/l
Exposure time: 15 min

Bibliographical reference: Cronin, M.T.D. (1993). Liverpool John Moores University, UK, private communication.

N,N-DIMETHYLDODECYLAMINE

CAS RN: 112-18-5

Sample purity: 97%
Temperature: 15°C
Test parameter: EC50
Effect: Reduction in light output

Concentration: 0.27 mg/l
Exposure times: 5 and 15 min

Concentration: 0.33 mg/l
Exposure time: 30 min

Comment: Mean of three assays. Methanol (<10%) was used to prepare the stock solutions. EC50 values were calculated from nominal concentrations.

Bibliographical reference: Kaiser, K.L.E., and Palabrica, V.S. (1991). *Water Poll. Res. J. Canada* **26**, 361-431.

N,N-DIMETHYLFORMAMIDE

CAS RN: 68-12-2

Test parameter: EC50
Effect: Reduction in light output
Concentration: 20000 mg/l
Exposure time: 5 min
Comment: The EC50 value was calculated from nominal concentrations.

Bibliographical reference: Curtis, C., Lima, A., Lozano, S.J., and Veith, G.D. (1982). In: *Aquatic Toxicology and Hazard Assessment: Fifth Conference, ASTM STP 766*, J.G. Pearson, R.B. Foster, and W.E. Bishop (eds.), American Society for Testing and Materials, Philadelphia, p. 170-178.

DIMETHYL MALONATE

CAS RN: 108-59-8

Sample purity: 97%
Temperature: 15°C
Test parameter: EC50
Effect: Reduction in light output
Concentration: 9350 mg/l
Exposure time: 5 min

Bibliographical reference: Cronin, M.T.D., Dearden, J.C., and Dobbs, A.J. (1991). *Sci. Total Environ.* **109/110**, 431-439.

N,N-DIMETHYL-3-NITROANILINE

CAS RN: 619-31-8

Sample purity: 98%
Temperature: 15°C
Test parameter: EC50
Effect: Reduction in light output

Concentration: 1.86 mg/l
Exposure time: 5 min

Concentration: 2.04 mg/l
Exposure time: 15 min

Concentration: 2.40 mg/l
Exposure time: 30 min

Comment: Mean of three assays. Methanol (<10%) was used to prepare the stock solutions. EC50 values were calculated from nominal concentrations.

Bibliographical reference: Kaiser, K.L.E., and Palabrica, V.S. (1992). National Water Research Institute, Burlington, Ontario, Canada, unpublished results.

N,N-DIMETHYL-4-NITROANILINE

CAS RN: 100-23-2

Temperature: 15°C
Test parameter: EC50
Effect: Reduction in light output

Concentration: 6.62 mg/l
Exposure time: 5 min

Concentration: 6.17 mg/l
Exposure time: 15 min

Concentration: 6.03 mg/l*
Exposure time: 30 min

Comment: Mean of three assays. Methanol (<10%) was used to prepare the stock solutions. EC50 values were calculated from nominal concentrations.

Bibliographical references: Kaiser, K.L.E., and Palabrica, V.S. (1991). *Water Poll. Res. J. Canada* **26**, 361-431.
* Kaiser, K.L.E. (1987). In: *QSAR in Environmental Toxicology - II*, K.L.E. Kaiser (ed.), D. Reidel Publishing Company, Dordrecht, p. 169-188.

2,3-DIMETHYL-6-NITROANILINE

CAS RN: 59146-96-2

Sample purity: 98%
Temperature: 15°C
Test parameter: EC50
Effect: Reduction in light output

Concentration: 0.59 mg/l
Exposure time: 5 min

Concentration: 0.65 mg/l
Exposure time: 15 min

Concentration: 0.74 mg/l
Exposure time: 30 min

Comment: Mean of three assays. Methanol (<10%) was used to prepare the stock solutions. EC50 values were calculated from nominal concentrations.

Bibliographical reference: Kaiser, K.L.E., and Palabrica, V.S. (1992). National Water Research Institute, Burlington, Ontario, Canada, unpublished results.

2,3-DIMETHYLNITROBENZENE

CAS RN: 83-41-0

Test parameter: EC50
Effect: Reduction in light output
Concentration: 0.54 mg/l
Exposure time: 15 min
Comment: The concentrations of the compound increased geometrically with a factor of 3.2. The concentrations were calculated on the basis of added amounts of material.

Bibliographical reference: Deneer, J.W., van Leeuwen, C.J., Seinen, W., Maas-Diepeveen, J.L., and Hermens, J.L.M. (1989). *Aquat. Toxicol.* **15**, 83-98.

3,4-DIMETHYLNITROBENZENE

CAS RN: 99-51-4

Test parameter: EC50
Effect: Reduction in light output
Concentration: 2.14 mg/l
Exposure time: 15 min
Comment: The concentrations of the compound increased geometrically with a factor of 3.2. The concentrations were calculated on the basis of added amounts of material.

Bibliographical reference: Deneer, J.W., van Leeuwen, C.J., Seinen, W., Maas-Diepeveen, J.L., and Hermens, J.L.M. (1989). *Aquat. Toxicol.* **15**, 83-98.

DIMETHYL 4-NITROPHTHALATE

CAS RN: 610-22-0

Sample purity: 98%
Temperature: 15°C
Test parameter: EC50
Effect: Reduction in light output

Concentration: 37.0 mg/l
Exposure time: 5 min

Concentration: 26.8 mg/l
Exposure time: 15 min

Concentration: 24.5 mg/l
Exposure time: 30 min

Comment: Mean of four assays. Methanol (<10%) was used to prepare the stock solutions. EC50 values were calculated from nominal concentrations.

Bibliographical reference: Kaiser, K.L.E., and Palabrica, V.S. (1991). *Water Poll. Res. J. Canada* **26**, 361-431.

2,3-DIMETHYLPHENOL

CAS RN: 526-75-0

Sample purity: 97%
Temperature: 15°C
Test parameter: EC50
Effect: Reduction in light output

Concentration: 3.21 mg/l
Exposure time: 5 min

Concentration: 3.36 mg/l

Exposure times: 15 and 30 min

Comment: Mean of four assays. Methanol (<10%) was used to prepare the stock solutions. EC50 values were calculated from nominal concentrations.

Bibliographical reference: Kaiser, K.L.E., and Palabrica, V.S. (1991). *Water Poll. Res. J. Canada* **26**, 361-431.

2,4-DIMETHYLPHENOL

CAS RN: 105-67-9

Test parameter: EC50
Effect: Reduction in light output
Concentration: 4.4 mg/l
Exposure time: 5 min
Comment: The EC50 value was calculated from nominal concentrations.

Bibliographical reference: Curtis, C., Lima, A., Lozano, S.J., and Veith, G.D. (1982). In: *Aquatic Toxicology and Hazard Assessment: Fifth Conference, ASTM STP 766*, J.G. Pearson, R.B. Foster, and W.E. Bishop (eds.), American Society for Testing and Materials, Philadelphia, p. 170-178.

Sample purity: 97%
Temperature: 15°C
Test parameter: EC50
Effect: Reduction in light output

Concentration: 2.49 mg/l
Exposure time: 5 min

Concentration: 2.61 mg/l
Exposure time: 15 min

Concentration: 2.67 mg/l
Exposure time: 30 min

Comment: Mean of four assays. Methanol (<10%) was used to prepare the stock solutions. EC50 values were calculated from nominal concentrations.

Bibliographical reference: Kaiser, K.L.E., and Palabrica, V.S. (1991). *Water Poll. Res. J. Canada* **26**, 361-431.

2,5-DIMETHYLPHENOL

CAS RN: 95-87-4

Sample purity: 99+%
Temperature: 15°C
Test parameter: EC50
Effect: Reduction in light output

Concentration: 8.85 mg/l
Exposure time: 5 min

Concentration: 9.06 mg/l
Exposure times: 15 and 30 min

Comment: Mean of three assays. Methanol (<10%) was used to prepare the stock solutions. EC50 values were calculated from nominal concentrations.

Bibliographical reference: Kaiser, K.L.E., and Palabrica, V.S. (1991). *Water Poll. Res. J. Canada* **26**, 361-431.

2,6-DIMETHYLPHENOL

CAS RN: 576-26-1

Sample purity: 99+%
Temperature: 15°C
Test parameter: EC50
Effect: Reduction in light output

Concentration: 24.4 mg/l
Exposure time: 5 min

Concentration: 22.7 mg/l
Exposure times: 15 and 30 min

Comment: Mean of three assays. Methanol (<10%) was used to prepare the stock solutions. EC50 values were calculated from nominal concentrations.

Bibliographical reference: Kaiser, K.L.E., and Palabrica, V.S. (1991). *Water Poll. Res. J. Canada* **26**, 361-431.

3,4-DIMETHYLPHENOL

CAS RN: 95-65-8

Sample purity: 97%
Temperature: 15°C
Test parameter: EC50
Effect: Reduction in light output

Concentration: 0.44 mg/l
Exposure time: 5 min

Concentration: 0.49 mg/l
Exposure times: 15 and 30 min

Comment: Mean of three assays. Methanol (<10%) was used to prepare the stock solutions. EC50 values were calculated from nominal concentrations.

Bibliographical reference: Kaiser, K.L.E., and Palabrica, V.S. (1991). *Water Poll. Res. J. Canada* **26**, 361-431.

3,5-DIMETHYLPHENOL

CAS RN: 108-68-9

Sample purity: 99+%
Temperature: 15°C
Test parameter: EC50
Effect: Reduction in light output
Concentration: 12.5 mg/l
Exposure times: 5, 15, and 30 min
Comment: Mean of three assays. Methanol (<10%) was used to prepare the stock solutions. EC50 values were calculated from nominal concentrations.

Bibliographical reference: Kaiser, K.L.E., and Palabrica, V.S. (1991). *Water Poll. Res. J. Canada* **26**, 361-431.

DIMETHYL PHTHALATE

CAS RN: 131-11-3

Temperature: 15°C
Test parameter: EC50
Effect: Reduction in light output

Concentration: 16 mg/l
Exposure time: 5 min

Concentration: 18 mg/l
Exposure times: 15 and 30 min

Comment: Chemical was diluted to give five sample concentrations in duplicate.

Bibliographical reference: Benson, W.H., and Stackhouse, R.A. (1986). *Drug Chem. Toxicol.* **9**, 275-283.

2,4-DIMETHYLQUINOLINE

CAS RN: 1198-37-4

Sample purity: >97%
Temperature: 15 ± 0.3°C
Test parameter: EC50
Effect: Reduction in light output

Concentration: 30 mg/l
Exposure time: 5 min

Concentration: 22 mg/l
Exposure time: 15 min

Comment: Test was performed in duplicate at pH = 5.1. Dichloromethane was used as carrier solvent. Concentrations of the chemical were determined by high performance liquid chromatography.

Hexachloroethane was used for quality control/quality assurance (5-min EC50 = 0.31 mg/l; n = 16; sd = 0.08 mg/l).

Bibliographical reference: Birkholz, D.A., Coutts, R.T., Hrudey, S.E., Danell, R.W., and Lockhart, W.L. (1990). *Water Res.* **24**, 67-73.

Sample purity: 97%
Temperature: 15°C
Test parameter: EC50
Effect: Reduction in light output

Concentration: 18.9 mg/l
Exposure time: 5 min

Concentration: 17.2 mg/l
Exposure time: 15 min

Concentration: 15.0 mg/l
Exposure time: 30 min

Comment: Mean of three assays. Methanol (<10%) was used to prepare the stock solutions. EC50 values were calculated from nominal concentrations.

Bibliographical reference: Kaiser, K.L.E., and Palabrica, V.S. (1991). *Water Poll. Res. J. Canada* **26**, 361-431.

2,6-DIMETHYLQUINOLINE

CAS RN: 877-43-0

Sample purity: >97%
Temperature: 15 ± 0.3°C
Test parameter: EC50
Effect: Reduction in light output

Concentration: 5.7 mg/l
Exposure time: 5 min

Concentration: 6.3 mg/l
Exposure time: 15 min

Comment: Test was performed in duplicate at pH = 5.1. Dichloromethane was used as carrier solvent. Concentrations of the chemical were determined by high performance liquid chromatography. Hexachloroethane was used for quality control/quality assurance (5-min EC50 = 0.31 mg/l; n = 16; sd = 0.08 mg/l).

Bibliographical reference: Birkholz, D.A., Coutts, R.T., Hrudey, S.E., Danell, R.W., and Lockhart, W.L. (1990). *Water Res.* **24**, 67-73.

2,7-DIMETHYLQUINOLINE

CAS RN: 93-37-8

Sample purity: >97%
Temperature: 15 ± 0.3°C
Test parameter: EC50
Effect: Reduction in light output

Concentration: 9.6 mg/l
Exposure time: 5 min

Concentration: 11 mg/l
Exposure time: 15 min

Comment: Test was performed in duplicate at pH = 5.1. Dichloromethane was used as carrier solvent. Concentrations of the chemical were determined by high performance liquid chromatography. Hexachloroethane was used for quality control/quality assurance (5-min EC50 = 0.31 mg/l; n = 16; sd = 0.08 mg/l).

Bibliographical reference: Birkholz, D.A., Coutts, R.T., Hrudey, S.E., Danell, R.W., and Lockhart, W.L. (1990). *Water Res.* **24**, 67-73.

2,8-DIMETHYLQUINOLINE

CAS RN: 1463-17-8

Sample purity: >97%
Temperature: 15 ± 0.3°C
Test parameter: EC50
Effect: Reduction in light output
Concentration: 14 mg/l

Exposure times: 5 and 15 min
Comment: Test was performed in duplicate at pH = 5.1. Dichloromethane was used as carrier solvent. Concentrations of the chemical were determined by high performance liquid chromatography. Hexachloroethane was used for quality control/quality assurance (5-min EC50 = 0.31 mg/l; n = 16; sd = 0.08 mg/l).

Bibliographical reference: Birkholz, D.A., Coutts, R.T., Hrudey, S.E., Danell, R.W., and Lockhart, W.L. (1990). *Water Res.* **24**, 67-73.

3,5-DIMETHYLQUINOLINE

CAS RN: 20668-27-3

Sample purity: >97%
Temperature: 15 ± 0.3°C
Test parameter: EC50
Effect: Reduction in light output

Concentration: 1.0 mg/l
Exposure time: 5 min

Concentration: 1.1 mg/l
Exposure time: 15 min

Comment: Test was performed in duplicate at pH = 5.1. Dichloromethane was used as carrier solvent. Concentrations of the chemical were determined by high performance liquid chromatography. Hexachloroethane was used for quality control/quality assurance (5-min EC50 = 0.31 mg/l; n = 16; sd = 0.08 mg/l). The 5-min EC50 value for 3,5-dimethylquinoline at pH 7 was 1.2 mg/l.

Bibliographical reference: Birkholz, D.A., Coutts, R.T., Hrudey, S.E., Danell, R.W., and Lockhart, W.L. (1990). *Water Res.* **24**, 67-73.

3,6-DIMETHYLQUINOLINE

CAS RN: 20668-26-2

Sample purity: >97%
Temperature: 15 ± 0.3°C
Test parameter: EC50

Effect: Reduction in light output

Concentration: 0.30 mg/l
Exposure time: 5 min

Concentration: 0.36 mg/l
Exposure time: 15 min

Comment: Test was performed in duplicate at pH = 5.1. Dichloromethane was used as carrier solvent. Concentrations of the chemical were determined by high performance liquid chromatography. Hexachloroethane was used for quality control/quality assurance (5-min EC50 = 0.31 mg/l; n = 16; sd = 0.08 mg/l). The 5-min EC50 value for 3,6-dimethylquinoline at pH 7 was 0.36 mg/l.

Bibliographical reference: Birkholz, D.A., Coutts, R.T., Hrudey, S.E., Danell, R.W., and Lockhart, W.L. (1990). *Water Res.* **24**, 67-73.

3,7-DIMETHYLQUINOLINE

CAS RN: 20668-28-4

Sample purity: >97%
Temperature: 15 ± 0.3°C
Test parameter: EC50
Effect: Reduction in light output

Concentration: 4.5 mg/l
Exposure time: 5 min

Concentration: 3.8 mg/l
Exposure time: 15 min

Comment: Test was performed in duplicate at pH = 5.1. Dichloromethane was used as carrier solvent. Concentrations of the chemical were determined by high performance liquid chromatography. Hexachloroethane was used for quality control/quality assurance (5-min EC50 = 0.31 mg/l; n = 16; sd = 0.08 mg/l).

Bibliographical reference: Birkholz, D.A., Coutts, R.T., Hrudey, S.E., Danell, R.W., and Lockhart, W.L. (1990). *Water Res.* **24**, 67-73.

3,8-DIMETHYLQUINOLINE

CAS RN: 20668-29-5

Sample purity: 94%
Temperature: 15 ± 0.3°C
Test parameter: EC50
Effect: Reduction in light output

Concentration: 4.6 mg/l
Exposure time: 5 min

Concentration: 5.1 mg/l
Exposure time: 15 min

Comment: Test was performed in duplicate at pH = 5.1. Dichloromethane was used as carrier solvent. Concentrations of the chemical were determined by high performance liquid chromatography. Hexachloroethane was used for quality control/quality assurance (5-min EC50 = 0.31 mg/l; n = 16; sd = 0.08 mg/l).

Bibliographical reference: Birkholz, D.A., Coutts, R.T., Hrudey, S.E., Danell, R.W., and Lockhart, W.L. (1990). *Water Res.* **24**, 67-73.

4,6-DIMETHYLQUINOLINE

CAS RN: 826-77-7

Sample purity: >97%
Temperature: 15 ± 0.3°C
Test parameter: EC50
Effect: Reduction in light output

Concentration: 4.0 mg/l
Exposure time: 5 min

Concentration: 4.4 mg/l
Exposure time: 15 min

Comment: Test was performed in duplicate at pH = 5.1. Dichloromethane was used as carrier solvent. Concentrations of the chemical were determined by high performance liquid chromatography. Hexachloroethane was used for quality control/quality assurance (5-min EC50 = 0.31 mg/l; n = 16; sd = 0.08 mg/l).

Bibliographical reference: Birkholz, D.A., Coutts, R.T., Hrudey, S.E., Danell, R.W., and Lockhart, W.L. (1990). *Water Res.* **24**, 67-73.

4,8-DIMETHYLQUINOLINE

CAS RN: 13362-80-6

Sample purity: >97%
Temperature: 15 ± 0.3°C
Test parameter: EC50
Effect: Reduction in light output

Concentration: 1.7 mg/l
Exposure time: 5 min

Concentration: 2.0 mg/l
Exposure time: 15 min

Comment: Test was performed in duplicate at pH = 5.1. Dichloromethane was used as carrier solvent. Concentrations of the chemical were determined by high performance liquid chromatography. Hexachloroethane was used for quality control/quality assurance (5-min EC50 = 0.31 mg/l; n = 16; sd = 0.08 mg/l).

Bibliographical reference: Birkholz, D.A., Coutts, R.T., Hrudey, S.E., Danell, R.W., and Lockhart, W.L. (1990). *Water Res.* **24**, 67-73.

5,6-DIMETHYLQUINOLINE

CAS RN: 20668-30-8

Sample purity: >97%
Temperature: 15 ± 0.3°C
Test parameter: EC50
Effect: Reduction in light output

Concentration: 0.74 mg/l
Exposure time: 5 min

Concentration: 0.91 mg/l
Exposure time: 15 min

Comment: Test was performed in duplicate at pH = 5.1. Dichloromethane was used as carrier solvent. Concentrations of the chemical were determined by high performance liquid chromatography. Hexachloroethane was used for quality control/quality assurance (5-min EC50 = 0.31 mg/l; n = 16; sd = 0.08 mg/l). The 5-min EC50 value for 5,6-dimethylquinoline at pH 7 was 0.78 mg/l.

Bibliographical reference: Birkholz, D.A., Coutts, R.T., Hrudey, S.E., Danell, R.W., and Lockhart, W.L. (1990). *Water Res.* **24**, 67-73.

6,7-DIMETHYLQUINOLINE

CAS RN: 20668-33-1

Sample purity: >97%
Temperature: 15 ± 0.3°C
Test parameter: EC50
Effect: Reduction in light output

Concentration: 1.3 mg/l
Exposure time: 5 min

Concentration: 1.9 mg/l
Exposure time: 15 min

Comment: Test was performed in duplicate at pH = 5.1. Dichloromethane was used as carrier solvent. Concentrations of the chemical were determined by high performance liquid chromatography. Hexachloroethane was used for quality control/quality assurance (5-min EC50 = 0.31 mg/l; n = 16; sd = 0.08 mg/l).

Bibliographical reference: Birkholz, D.A., Coutts, R.T., Hrudey, S.E., Danell, R.W., and Lockhart, W.L. (1990). *Water Res.* **24**, 67-73.

6,8-DIMETHYLQUINOLINE

CAS RN: 2436-93-3

Sample purity: >97%
Temperature: 15 ± 0.3°C
Test parameter: EC50
Effect: Reduction in light output

Concentration: 2.2 mg/l
Exposure time: 5 min

Concentration: 2.4 mg/l
Exposure time: 15 min

Comment: Test was performed in duplicate at pH = 5.1. Dichloromethane was used as carrier solvent. Concentrations of the chemical were determined by high performance liquid chromatography. Hexachloroethane was used for quality control/quality assurance (5-min EC50 = 0.31 mg/l; n = 16; sd = 0.08 mg/l).

Bibliographical reference: Birkholz, D.A., Coutts, R.T., Hrudey, S.E., Danell, R.W., and Lockhart, W.L. (1990). *Water Res.* **24**, 67-73.

7,8-DIMETHYLQUINOLINE

CAS RN: 20668-35-3

Sample purity: 69%
Temperature: 15 ± 0.3°C
Test parameter: EC50
Effect: Reduction in light output

Concentration: 7.0 mg/l
Exposure time: 5 min

Concentration: 7.9 mg/l
Exposure time: 15 min

Comment: Test was performed in duplicate at pH = 5.1. Dichloromethane was used as carrier solvent. Concentrations of the chemical were determined by high performance liquid chromatography. Hexachloroethane was used for quality control/quality assurance (5-min EC50 = 0.31 mg/l; n = 16; sd = 0.08 mg/l).

Bibliographical reference: Birkholz, D.A., Coutts, R.T., Hrudey, S.E., Danell, R.W., and Lockhart, W.L. (1990). *Water Res.* **24**, 67-73.

2,3-DIMETHYLQUINOXALINE

CAS RN: 2379-55-7

Sample purity: 97%
Temperature: 15°C
Test parameter: EC50
Effect: Reduction in light output

Concentration: 173 mg/l
Exposure time: 5 min

Concentration: 123 mg/l
Exposure time: 15 min

Concentration: 77.5 mg/l
Exposure time: 30 min

Comment: Mean of three assays. Methanol (<10%) was used to prepare the stock solutions. EC50 values were calculated from nominal concentrations.

Bibliographical reference: Kaiser, K.L.E., and Palabrica, V.S. (1991). *Water Poll. Res. J. Canada* **26**, 361-431.

DIMETHYL SULFOXIDE

CAS RN: 67-68-5
Synonym: DMSO

Temperature: 15 ± 0.1°C
Test parameter: EC50
Effect: Reduction in light output
Concentration: 103400 mg/l
Exposure time: 5 min
Comment: Test was performed on *Photobacterium phosphoreum* NZ11D obtained from the Scripps Institute of Oceanography (La Jolla, CA).

Bibliographical reference: McFeters, G.A., Bond, P.J., Olson, S.B., and Tchan, Y.T. (1983). *Water Res.* **17**, 1757-1762.

Temperature: 15°C
Test parameter: EC50
Effect: Reduction in light output
Concentration: 58.39 µl/ml
Exposure time: 15 min

Comment: Four concentrations of chemical were tested in duplicate.

Bibliographical reference: Schiewe, M.H., Hawk, E.G., Actor, D.I., and Krahn, M.M. (1985). *Can. J. Fish. Aquat. Sci.* **42**, 1244-1248.

Effect: Toxicity (15-min EC50, µl/ml) of DMSO (A) and DMSO amended with 10% dichloromethane (B) after incubation (T in min) for up to 120 min at room temperature (R) and 110°C (H)
Result:

T	A-R	A-H	B-R	B-H
0	57.60	57.17	14.41	13.57
30	58.26	60.61	13.12	13.13
60	59.65	51.30	14.02	2.51
120	54.80	52.89	13.32	0.6163

Bibliographical reference: Schiewe, M.H., Hawk, E.G., Actor, D.I., and Krahn, M.M. (1985). *Can. J. Fish. Aquat. Sci.* **42**, 1244-1248.

Effect: Toxicity of DMSO and ethanol alone, and after solvent exchange
Result:

Chemical	15-min EC50
DMSO	56.43 µl/ml
Ethanol	33.41 µl/ml
Ethanol heated for 1 h	25.42 µl/ml
CH_2Cl_2 replaced by ethanol by heating 1 h	30.34 µl/ml
CH_2Cl_2 replaced by ethanol by heating 1 h; Ethanol replaced by DMSO by heating 1 h	46.46 µl/ml

Bibliographical reference: Schiewe, M.H., Hawk, E.G., Actor, D.I., and Krahn, M.M. (1985). *Can. J. Fish. Aquat. Sci.* **42**, 1244-1248.

Sample purity: ~99%
Temperature: 15 ± 0.1°C
Test parameter: EC50
Effect: Reduction in light output

Concentration: 87900 mg/l
Exposure time: 5 min

Concentration: 97500 mg/l

Exposure time: 15 min

Concentration: 98000 mg/l
Exposure time: 30 min

Comment: The pH was not adjusted.

Bibliographical reference: Tarkpea, M., Hansson, M., and Samuelsson, B. (1986). *Ecotoxicol. Environ. Safety* **11**, 127-143.

Temperature: 15°C
Test parameter: EC50
Effect: Reduction in light output
Concentration: 56.7 μl/ml (52.7-60.7)
Exposure time: 15 min
Comment: Mean of 15 bioassays.

Bibliographical reference: Jacobs, M.W., Delfino, J.J., and Bitton, G. (1992). *Environ. Toxicol. Chem.* **11**, 1137-1143.

Temperature: 15°C
Test parameter: EC50
Effect: Reduction in light output
Concentration: 77400 mg/l
Exposure time: 5 min

Bibliographical reference: Kahru, A. (1993). *ATLA* **21**, 210-215.

DIMETHYL TEREPHTHALATE

CAS RN: 120-61-6

Sample purity: 99%
Temperature: 15°C
Test parameter: EC50
Effect: Reduction in light output

Concentration: 22.8 mg/l
Exposure time: 5 min

Concentration: 20.3 mg/l
Exposure time: 15 min

Concentration: 18.5 mg/l
Exposure time: 30 min

Comment: Mean of three assays. Methanol (<10%) was used to prepare the stock solutions. EC50 values were calculated from nominal concentrations.

Bibliographical reference: Kaiser, K.L.E., and Palabrica, V.S. (1991). *Water Poll. Res. J. Canada* **26**, 361-431.

N,N-DIMETHYLTHIOUREA

CAS RN: 6972-05-0

Test parameter: EC50
Effect: Reduction in light output
Concentration: 1167 mg/l
Exposure time: 15 min
Comment: EC50 value converted to mg/l from the original data expressed in log(1/µmol l^{-1}) and a rounded molecular weight of 104 given by the authors.

Bibliographical reference: Govers, H., Ruepert, C., Stevens, T., and van Leeuwen, C.J. (1986). *Chemosphere* **15**, 383-393.

DIMETHYLTINBIS(THIOGLYCOLIC ACID ISOOCTYL ESTER)

CAS RN: 57583-35-4

Temperature: 15°C
Test parameter: EC50
Effect: Reduction in light output
Concentration: 0.100 mg/l
Exposure time: 30 min

Bibliographical reference: Steinhäuser, K.G., Amann, W., Späth, A., and Polenz, A. (1985). *Vom Wasser* **65**, 203-214.

5,7-DIMETHYL-*s*-TRIAZOLO[1,5-*a*]PYRIMIDINE

CAS RN: 7681-99-4

Sample purity: 98%
Temperature: 15°C
Test parameter: EC50
Effect: Reduction in light output

Concentration: 155 mg/l
Exposure time: 5 min

Concentration: 166 mg/l
Exposure time: 15 min

Concentration: 182 mg/l
Exposure time: 30 min

Comment: Mean of four assays. Methanol (<10%) was used to prepare the stock solutions. EC50 values were calculated from nominal concentrations.

Bibliographical reference: Kaiser, K.L.E., and Palabrica, V.S. (1991). *Water Poll. Res. J. Canada* **26**, 361-431.

2,4-DINITROANILINE

CAS RN: 97-02-9

Sample purity: 96%
Temperature: 15°C
Test parameter: EC50
Effect: Reduction in light output

Concentration: 59.3 mg/l
Exposure time: 5 min

Concentration: 50.4 mg/l
Exposure time: 15 min

Concentration: 48.2 mg/l
Exposure time: 30 min

Comment: Mean of four assays. Methanol (<10%) was used to prepare the stock solutions. EC50 values were calculated from nominal concentrations.

Bibliographical reference: Kaiser, K.L.E., and Palabrica, V.S. (1991). *Water Poll. Res. J. Canada* **26**, 361-431.

3,5-DINITROANILINE

CAS RN: 618-87-1

Sample purity: 97%
Temperature: 15°C
Test parameter: EC50
Effect: Reduction in light output

Concentration: 35.7 mg/l
Exposure time: 5 min

Concentration: 22.5 mg/l
Exposure time: 15 min

Concentration: 19.2 mg/l
Exposure time: 30 min

Comment: Mean of three assays. Methanol (<10%) was used to prepare the stock solutions. EC50 values were calculated from nominal concentrations.

Bibliographical reference: Kaiser, K.L.E., and Palabrica, V.S. (1991). *Water Poll. Res. J. Canada* **26**, 361-431.

3,5-DINITROBENZAMIDE

CAS RN: 121-81-3

Sample purity: 98+%
Temperature: 15°C
Test parameter: EC50
Effect: Reduction in light output

Concentration: 27.2 mg/l

Exposure time: 5 min

Concentration: 13.6 mg/l
Exposure time: 15 min

Concentration: 9.43 mg/l
Exposure time: 30 min

Comment: Mean of three assays. Methanol (<10%) was used to prepare the stock solutions. EC50 values were calculated from nominal concentrations.

Bibliographical reference: Kaiser, K.L.E., and Palabrica, V.S. (1992). National Water Research Institute, Burlington, Ontario, Canada, unpublished results.

1,2-DINITROBENZENE

CAS RN: 528-29-0

Test parameter: EC50
Effect: Reduction in light output
Concentration: 9.90 mg/l
Exposure time: 15 min
Comment: The concentrations of the compound increased geometrically with a factor of 3.2. The concentrations were calculated on the basis of added amounts of material.

Bibliographical reference: Deneer, J.W., van Leeuwen, C.J., Seinen, W., Maas-Diepeveen, J.L., and Hermens, J.L.M. (1989). *Aquat. Toxicol.* **15**, 83-98.

Sample purity: 99%
Temperature: 15°C
Test parameter: EC50
Effect: Reduction in light output

Concentration: 9.67 mg/l
Exposure time: 5 min

Concentration: 6.54 mg/l
Exposure time: 15 min

Concentration: 5.32 mg/l
Exposure time: 30 min

Comment: Mean of four assays. Methanol (<10%) was used to prepare the stock solutions. EC50 values were calculated from nominal concentrations.

Bibliographical reference: Kaiser, K.L.E., and Palabrica, V.S. (1991). *Water Poll. Res. J. Canada* **26**, 361-431.

1,3-DINITROBENZENE

CAS RN: 99-65-0

Test parameter: EC50
Effect: Reduction in light output
Concentration: 40.3 mg/l
Exposure time: 15 min
Comment: The concentrations of the compound increased geometrically with a factor of 3.2. The concentrations were calculated on the basis of added amounts of material.

Bibliographical reference: Deneer, J.W., van Leeuwen, C.J., Seinen, W., Maas-Diepeveen, J.L., and Hermens, J.L.M. (1989). *Aquat. Toxicol.* **15**, 83-98.

Sample purity: 97%
Temperature: 15°C
Test parameter: EC50
Effect: Reduction in light output

Concentration: 32.0 mg/l
Exposure times: 5 and 15 min

Concentration: 26.6 mg/l
Exposure time: 30 min

Comment: Mean of four assays. Methanol (<10%) was used to prepare the stock solutions. EC50 values were calculated from nominal concentrations.

Bibliographical reference: Kaiser, K.L.E., and Palabrica, V.S. (1991). *Water Poll. Res. J. Canada* **26**, 361-431.

1,4-DINITROBENZENE

CAS RN: 100-25-4

Test parameter: EC50
Effect: Reduction in light output
Concentration: 0.18 mg/l
Exposure time: 15 min
Comment: The concentrations of the compound increased geometrically with a factor of 3.2. The concentrations were calculated on the basis of added amounts of material.

Bibliographical reference: Deneer, J.W., van Leeuwen, C.J., Seinen, W., Maas-Diepeveen, J.L., and Hermens, J.L.M. (1989). *Aquat. Toxicol.* **15**, 83-98.

Temperature: 15°C
Test parameter: EC50
Effect: Reduction in light output

Concentration: 0.38 mg/l
Exposure time: 5 min

Concentration: 0.11 mg/l
Exposure time: 15 min

Concentration: 0.095 mg/l*
Exposure time: 30 min

Comment: Recrystallized from toluene/hexane. Mean of three assays. Methanol (<10%) was used to prepare the stock solutions. EC50 values were calculated from nominal concentrations.

Bibliographical references: Kaiser, K.L.E., and Palabrica, V.S. (1991). *Water Poll. Res. J. Canada* **26**, 361-431.
* Kaiser, K.L.E. (1987). In: *QSAR in Environmental Toxicology - II*, K.L.E. Kaiser (ed.), D. Reidel Publishing Company, Dordrecht, p. 169-188.

3,5-DINITROBENZOIC ACID

CAS RN: 99-34-3

Sample purity: >99%
Temperature: 15°C
Test parameter: EC50
Effect: Reduction in light output

Concentration: 280 mg/l
Exposure time: 5 min

Concentration: 244 mg/l
Exposure times: 15 and 30 min

Comment: Mean of four assays. Methanol (<10%) was used to prepare the stock solutions. EC50 values were calculated from nominal concentrations.

Bibliographical reference: Kaiser, K.L.E., and Palabrica, V.S. (1991). *Water Poll. Res. J. Canada* **26**, 361-431.

3,4-DINITROBENZONITRILE

CAS RN: 4248-33-3

Sample purity: 97%
Temperature: 15°C
Test parameter: EC50
Effect: Reduction in light output

Concentration: 0.34 mg/l
Exposure time: 5 min

Concentration: 0.17 mg/l
Exposure time: 15 min

Concentration: 0.15 mg/l
Exposure time: 30 min

Comment: Mean of three assays. Methanol (<10%) was used to prepare the stock solutions. EC50 values were calculated from nominal concentrations.

Bibliographical reference: Kaiser, K.L.E., and Palabrica, V.S. (1991). *Water Poll. Res. J. Canada* **26**, 361-431.

3,5-DINITROBENZONITRILE

CAS RN: 4110-35-4

Sample purity: 97%
Temperature: 15°C
Test parameter: EC50
Effect: Reduction in light output

Concentration: 5.08 mg/l
Exposure time: 5 min

Concentration: 1.31 mg/l
Exposure time: 15 min

Concentration: 0.70 mg/l
Exposure time: 30 min

Comment: Mean of three assays. Methanol (<10%) was used to prepare the stock solutions. EC50 values were calculated from nominal concentrations.

Bibliographical reference: Kaiser, K.L.E., and Palabrica, V.S. (1991). *Water Poll. Res. J. Canada* **26**, 361-431.

3,5-DINITROBENZOTRIFLUORIDE

CAS RN: 401-99-0

Sample purity: 99%
Temperature: 15°C
Test parameter: EC50
Effect: Reduction in light output

Concentration: 24.7 mg/l
Exposure time: 5 min

Concentration: 5.29 mg/l
Exposure time: 15 min

Concentration: 3.74 mg/l
Exposure time: 30 min

Comment: Mean of three assays. Methanol (<10%) was used to prepare the stock solutions. EC50 values were calculated from nominal concentrations.

Bibliographical reference: Kaiser, K.L.E., and Palabrica, V.S. (1991). *Water Poll. Res. J. Canada* **26**, 361-431.

3,5-DINITROBENZOYL CHLORIDE

CAS RN: 99-33-2

Temperature: 15°C
Test parameter: EC50
Effect: Reduction in light output

Concentration: 8.77 mg/l
Exposure time: 5 min

Concentration: 5.79 mg/l
Exposure time: 15 min

Concentration: 5.04 mg/l
Exposure time: 30 min

Comment: Mean of three assays. Methanol (<10%) was used to prepare the stock solutions. EC50 values were calculated from nominal concentrations.

Bibliographical reference: Kaiser, K.L.E., and Palabrica, V.S. (1992). National Water Research Institute, Burlington, Ontario, Canada, unpublished results.

3,4-DINITROBENZYL ALCOHOL

CAS RN: 79544-31-3

Sample purity: 99%
Temperature: 15°C
Test parameter: EC50
Effect: Reduction in light output

Concentration: 9.93 mg/l

Exposure time: 5 min

Concentration: 7.03 mg/l
Exposure time: 15 min

Concentration: 5.85 mg/l
Exposure time: 30 min

Comment: Mean of three assays. Methanol (<10%) was used to prepare the stock solutions. EC50 values were calculated from nominal concentrations.

Bibliographical reference: Kaiser, K.L.E., and Palabrica, V.S. (1991). *Water Poll. Res. J. Canada* **26**, 361-431.

3,5-DINITROBENZYL ALCOHOL

CAS RN: 71022-43-0

Sample purity: 98%
Temperature: 15°C
Test parameter: EC50
Effect: Reduction in light output

Concentration: 65.6 mg/l
Exposure time: 5 min

Concentration: 38.6 mg/l
Exposure time: 15 min

Concentration: 32.9 mg/l
Exposure time: 30 min

Comment: Mean of four assays. Methanol (<10%) was used to prepare the stock solutions. EC50 values were calculated from nominal concentrations.

Bibliographical reference: Kaiser, K.L.E., and Palabrica, V.S. (1991). *Water Poll. Res. J. Canada* **26**, 361-431.

2,6-DINITRO-4-METHYLANILINE

CAS RN: 6393-42-6

Sample purity: 99%
Temperature: 15°C
Test parameter: EC50
Effect: Reduction in light output

Concentration: 13.6 mg/l
Exposure time: 5 min

Concentration: 12.4 mg/l
Exposure time: 15 min

Concentration: 13.0 mg/l
Exposure time: 30 min

Comment: Mean of three assays. Methanol (<10%) was used to prepare the stock solutions. EC50 values were calculated from nominal concentrations.

Bibliographical reference: Kaiser, K.L.E., and Palabrica, V.S. (1991). *Water Poll. Res. J. Canada* **26**, 361-431.

2,6-DINITRO-4-METHYLPHENOL

CAS RN: 609-93-8
Synonym: 2,6-Dinitro-*p*-cresol

Sample purity: 90%
Temperature: 15°C
Test parameter: EC50
Effect: Reduction in light output

Concentration: 9.27 mg/l
Exposure time: 5 min

Concentration: 8.85 mg/l
Exposure time: 15 min

Concentration: 9.06 mg/l
Exposure time: 30 min

Comment: Mean of three assays. Methanol (<10%) was used to prepare the stock solutions. EC50 values were calculated from nominal concentrations.

Bibliographical reference: Kaiser, K.L.E., and Palabrica, V.S. (1991). *Water Poll. Res. J. Canada* **26**, 361-431.

4,6-DINITRO-2-METHYLPHENOL

CAS RN: 534-52-1
Synonym: 4,6-Dinitro-*o*-cresol

Test parameter: EC50
Effect: Reduction in light output
Concentration: 6.6 mg/l
Exposure time: 5 min
Comment: The EC50 value was calculated from nominal concentrations.

Bibliographical reference: Curtis, C., Lima, A., Lozano, S.J., and Veith, G.D. (1982). In: *Aquatic Toxicology and Hazard Assessment: Fifth Conference, ASTM STP 766*, J.G. Pearson, R.B. Foster, and W.E. Bishop (eds.), American Society for Testing and Materials, Philadelphia, p. 170-178.

Temperature: 15 ± 0.2°C
Test parameter: EC50
Effect: Reduction in light output

Concentration: 6.3 mg/l
Exposure time: 5 min

Concentration: 1.5 mg/l
Exposure time: 30 min

Comment: The test was run in duplicate with a minimum of five serial concentration dilutions. Samples were pH adjusted to 7 ± 0.5 pH units.

Bibliographical reference: Indorato, A.M., Snyder, K.B., and Usinowicz, P.J. (1984). In: *Toxicity Screening Procedures Using Bacterial Systems*, D. Liu and B.J. Dutka (eds.), Marcel Dekker, New York, p. 37-53.

Temperature: 15°C
Test parameter: EC50
Effect: Reduction in light output

Concentration: 7.19 mg/l
Exposure time: 5 min

Concentration: 5.85 mg/l
Exposure time: 15 min

Concentration: 6.12 mg/l
Exposure time: 30 min

Comment: Mean of three assays. Methanol (<10%) was used to prepare the stock solutions. EC50 values were calculated from nominal concentrations.

Bibliographical reference: Kaiser, K.L.E., and Palabrica, V.S. (1991). *Water Poll. Res. J. Canada* **26**, 361-431.

2,4-DINITROPHENOL

CAS RN: 51-28-5

Test parameter: EC50
Effect: Reduction in light output
Concentration: 15.8 mg/l
Exposure time: 5 min
Comment: The EC50 value was calculated from nominal concentrations.

Bibliographical reference: Curtis, C., Lima, A., Lozano, S.J., and Veith, G.D. (1982). In: *Aquatic Toxicology and Hazard Assessment: Fifth Conference, ASTM STP 766*, J.G. Pearson, R.B. Foster, and W.E. Bishop (eds.), American Society for Testing and Materials, Philadelphia, p. 170-178.

Temperature: 15°C
Test parameter: EC50
Effect: Reduction in light output
Concentration: 26.6 ± 5.7 mg/l
Exposure time: 5 min

Comment: Mean of three assays. Toxicity values were calculated from nominal concentrations.

Bibliographical reference: Ferard, J.F., Vasseur, P., Danoux, L., and Larbaigt, G. (1983). *Rev. Fr. Sci. Eau* **2**, 221-237.

Temperature: 15 ± 0.2°C
Test parameter: EC50
Effect: Reduction in light output
Concentration: 6.1 mg/l
Exposure times: 5 and 30 min
Comment: The test was run in duplicate with a minimum of five serial concentration dilutions. Samples were pH adjusted to 7 ± 0.5 pH units.

Bibliographical reference: Indorato, A.M., Snyder, K.B., and Usinowicz, P.J. (1984). In: *Toxicity Screening Procedures Using Bacterial Systems*, D. Liu and B.J. Dutka (eds.), Marcel Dekker, New York, p. 37-53.

Temperature: 15°C
Test parameter: EC50
Effect: Reduction in light output
Concentration: 45 mg/l
Exposure time: 15 min
Comment: pH = 7.2.

Bibliographical reference: Neilson, A.H., Allard, A.S., Fischer, S., Malmberg, M., and Viktor, T. (1990). *Ecotoxicol. Environ. Safety* **20**, 82-97.

Temperature: 15°C
Test parameter: EC50
Effect: Reduction in light output

Concentration: 10.1 mg/l
Exposure time: 5 min

Concentration: 9.66 mg/l
Exposure time: 15 min

Concentration: 10.6 mg/l
Exposure time: 30 min

Comment: Mean of four assays. Methanol (<10%) was used to prepare the stock solutions. EC50 values were calculated from nominal concentrations.

Bibliographical reference: Kaiser, K.L.E., and Palabrica, V.S. (1991). *Water Poll. Res. J. Canada* **26**, 361-431.

Temperature: 15°C
Test parameter: EC50
Effect: Reduction in light output
Concentration: 9.6 mg/l
Exposure time: 5 min

Bibliographical reference: Kahru, A. (1993). *ATLA* **21**, 210-215.

2,6-DINITROPHENOL

CAS RN: 573-56-8

Temperature: 15 ± 0.2°C
Test parameter: EC50
Effect: Reduction in light output
Concentration: 2.2 mg/l
Exposure time: 5 min
Comment: The test was run in duplicate with a minimum of five serial concentration dilutions. Samples were pH adjusted to 7 ± 0.5 pH units.

Bibliographical reference: Indorato, A.M., Snyder, K.B., and Usinowicz, P.J. (1984). In: *Toxicity Screening Procedures Using Bacterial Systems*, D. Liu and B.J. Dutka (eds.), Marcel Dekker, New York, p. 37-53.

Sample purity: 97%
Temperature: 15°C
Test parameter: EC50
Effect: Reduction in light output

Concentration: 11.6 mg/l
Exposure time: 5 min

Concentration: 12.2 mg/l
Exposure time: 15 min

Concentration: 12.4 mg/l
Exposure time: 30 min

Comment: Mean of three assays. Methanol (<10%) was used to prepare the stock solutions. EC50 values were calculated from nominal concentrations.

Bibliographical reference: Kaiser, K.L.E., and Palabrica, V.S. (1992). National Water Research Institute, Burlington, Ontario, Canada, unpublished results.

2,4-DINITROPHENYLACETIC ACID

CAS RN: 643-43-6

Sample purity: >98%
Temperature: 15°C
Test parameter: EC50
Effect: Reduction in light output

Concentration: 66.7 mg/l
Exposure time: 5 min

Concentration: 62.3 mg/l
Exposure time: 15 min

Concentration: 58.1 mg/l
Exposure time: 30 min

Comment: Mean of three assays. Methanol (<10%) was used to prepare the stock solutions. EC50 values were calculated from nominal concentrations.

Bibliographical reference: Kaiser, K.L.E., and Palabrica, V.S. (1991). *Water Poll. Res. J. Canada* **26**, 361-431.

2,4-DINITROPHENYLHYDRAZINE

CAS RN: 119-26-6

Sample purity: 70% in water, value adjusted to 100%
Temperature: 15°C

Test parameter: EC50
Effect: Reduction in light output

Concentration: 27.4 mg/l
Exposure time: 5 min

Concentration: 24.9 mg/l
Exposure time: 15 min

Concentration: 24.4 mg/l
Exposure time: 30 min

Comment: Mean of four assays. Methanol (<10%) was used to prepare the stock solutions. EC50 values were calculated from nominal concentrations.

Bibliographical reference: Kaiser, K.L.E., and Palabrica, V.S. (1992). National Water Research Institute, Burlington, Ontario, Canada, unpublished results.

3,5-DINITROSALICYLIC ACID

CAS RN: 609-99-4

Sample purity: 98%
Temperature: 15°C
Test parameter: EC50
Effect: Reduction in light output

Concentration: 169 mg/l
Exposure time: 5 min

Concentration: 161 mg/l
Exposure time: 15 min

Concentration: 165 mg/l
Exposure time: 30 min

Comment: Mean of three assays. Methanol (<10%) was used to prepare the stock solutions. EC50 values were calculated from nominal concentrations.

Bibliographical reference: Kaiser, K.L.E., and Palabrica, V.S. (1991). *Water Poll. Res. J. Canada* **26**, 361-431.

2,3-DINITROTOLUENE

CAS RN: 602-01-7

Test parameter: EC50
Effect: Reduction in light output
Concentration: 6.03 mg/l
Exposure time: 15 min
Comment: The concentrations of the compound increased geometrically with a factor of 3.2. The concentrations were calculated on the basis of added amounts of material.

Bibliographical reference: Deneer, J.W., van Leeuwen, C.J., Seinen, W., Maas-Diepeveen, J.L., and Hermens, J.L.M. (1989). *Aquat. Toxicol.* **15**, 83-98.

2,4-DINITROTOLUENE

CAS RN: 121-14-2

Temperature: 15 ± 0.2°C
Test parameter: EC50
Effect: Reduction in light output

Concentration: 33 mg/l
Exposure time: 5 min

Concentration: 21 mg/l
Exposure time: 30 min

Comment: The test was run in duplicate with a minimum of five serial concentration dilutions. Samples were pH adjusted to 7 ± 0.5 pH units.

Bibliographical reference: Indorato, A.M., Snyder, K.B., and Usinowicz, P.J. (1984). In: *Toxicity Screening Procedures Using Bacterial Systems*, D. Liu and B.J. Dutka (eds.), Marcel Dekker, New York, p. 37-53.

Test parameter: EC50
Effect: Reduction in light output
Concentration: 51.3 mg/l
Exposure time: 15 min

Comment: The concentrations of the compound increased geometrically with a factor of 3.2. The concentrations were calculated on the basis of added amounts of material.

Bibliographical reference: Deneer, J.W., van Leeuwen, C.J., Seinen, W., Maas-Diepeveen, J.L., and Hermens, J.L.M. (1989). *Aquat. Toxicol.* **15**, 83-98.

2,6-DINITROTOLUENE

CAS RN: 606-20-2

Test parameter: EC50
Effect: Reduction in light output
Concentration: 2.89 mg/l
Exposure time: 15 min
Comment: The concentrations of the compound increased geometrically with a factor of 3.2. The concentrations were calculated on the basis of added amounts of material.

Bibliographical reference: Deneer, J.W., van Leeuwen, C.J., Seinen, W., Maas-Diepeveen, J.L., and Hermens, J.L.M. (1989). *Aquat. Toxicol.* **15**, 83-98.

3,4-DINITROTOLUENE

CAS RN: 610-39-9

Test parameter: EC50
Effect: Reduction in light output
Concentration: 6.92 mg/l
Exposure time: 15 min
Comment: The concentrations of the compound increased geometrically with a factor of 3.2. The concentrations were calculated on the basis of added amounts of material.

Bibliographical reference: Deneer, J.W., van Leeuwen, C.J., Seinen, W., Maas-Diepeveen, J.L., and Hermens, J.L.M. (1989). *Aquat. Toxicol.* **15**, 83-98.

DIOCTYLTIN DICHLORIDE

CAS RN: 3542-36-7

Temperature: 15°C
Test parameter: EC50
Effect: Reduction in light output
Concentration: 0.0015 mg/l
Exposure time: 30 min

Bibliographical reference: Steinhäuser, K.G., Amann, W., Späth, A., and Polenz, A. (1985). *Vom Wasser* **65**, 203-214.

1,4-DIOXANE

CAS RN: 123-91-1

Sample purity: >99%
Temperature: 15°C
Test parameter: EC50
Effect: Reduction in light output

Concentration: 610 mg/l
Exposure time: 5 min

Concentration: 668 mg/l
Exposure time: 15 min

Concentration: 733 mg/l
Exposure time: 30 min

Comment: Mean of three assays. Methanol (<10%) was used to prepare the stock solutions. EC50 values were calculated from nominal concentrations.

Bibliographical reference: Kaiser, K.L.E., and Palabrica, V.S. (1991). *Water Poll. Res. J. Canada* **26**, 361-431.

DIPHENYLAMINE

CAS RN: 122-39-4

Sample purity: 99%
Temperature: 15°C
Test parameter: EC50
Effect: Reduction in light output

Concentration: 2.81 mg/l
Exposure time: 5 min

Concentration: 3.46 mg/l
Exposure time: 15 min

Concentration: 4.77 mg/l*
Exposure time: 30 min

Comment: Mean of three assays. Methanol (<10%) was used to prepare the stock solutions. EC50 values were calculated from nominal concentrations.

Bibliographical references: Kaiser, K.L.E., and Palabrica, V.S. (1991). *Water Poll. Res. J. Canada* **26**, 361-431.
* Kaiser, K.L.E., Palabrica, V.S., and Ribo, J.M. (1987). In: *QSAR in Environmental Toxicology - II*, K.L.E. Kaiser (ed.), D. Reidel Publishing Company, Dordrecht, p. 153-168.

DIPHENYL ETHER

CAS RN: 101-84-8

Sample purity: 99+%
Temperature: 15°C
Test parameter: EC50
Effect: Reduction in light output
Concentration: 3.64 mg/l
Exposure time: 30 min
Comment: Mean of three assays. Methanol (<10%) was used to prepare the stock solutions. EC50 values were calculated from nominal concentrations.

Bibliographical reference: Kaiser, K.L.E., Palabrica, V.S., and Ribo, J.M. (1987). In: *QSAR in Environmental Toxicology - II*, K.L.E. Kaiser (ed.), D. Reidel Publishing Company, Dordrecht, p. 153-168.

DIPHENYLMERCURY

CAS RN: 587-85-9

Sample purity: 99%
Temperature: 15°C
Test parameter: EC50
Effect: Reduction in light output

Concentration: 0.12 mg/l
Exposure time: 5 min

Concentration: 0.13 mg/l
Exposure time: 15 min

Concentration: 0.13 mg/l*
Exposure time: 30 min

Comment: Mean of four assays. Methanol (<10%) was used to prepare the stock solutions. EC50 values were calculated from nominal concentrations.

Bibliographical references: Kaiser, K.L.E., and Palabrica, V.S. (1991). *Water Poll. Res. J. Canada* **26**, 361-431.
* Kaiser, K.L.E., Palabrica, V.S., and Ribo, J.M. (1987). In: *QSAR in Environmental Toxicology - II*, K.L.E. Kaiser (ed.), D. Reidel Publishing Company, Dordrecht, p. 153-168.

DIPHENYLMETHANE

CAS RN: 101-81-5

Sample purity: 99%
Temperature: 15°C
Test parameter: EC50
Effect: Reduction in light output

Concentration: 1.43 mg/l
Exposure time: 5 min

Concentration: 2.27 mg/l
Exposure time: 15 min

Concentration: 3.52 mg/l*

Exposure time: 30 min

Comment: Mean of three assays. Methanol (<10%) was used to prepare the stock solutions. EC50 values were calculated from nominal concentrations.

Bibliographical references: Kaiser, K.L.E., and Palabrica, V.S. (1991). *Water Poll. Res. J. Canada* **26**, 361-431.
* Kaiser, K.L.E., Palabrica, V.S., and Ribo, J.M. (1987). In: *QSAR in Environmental Toxicology - II*, K.L.E. Kaiser (ed.), D. Reidel Publishing Company, Dordrecht, p. 153-168.

DIPHENYL PHOSPHATE

CAS RN: 838-85-7

Sample purity: 97%
Temperature: 15°C
Test parameter: EC50
Effect: Reduction in light output

Concentration: 435 mg/l
Exposure time: 5 min

Concentration: 415 mg/l
Exposure time: 15 min

Concentration: 425 mg/l*
Exposure time: 30 min

Comment: Mean of three assays. Methanol (<10%) was used to prepare the stock solutions. EC50 values were calculated from nominal concentrations.

Bibliographical references: Kaiser, K.L.E., and Palabrica, V.S. (1991). *Water Poll. Res. J. Canada* **26**, 361-431.
* Kaiser, K.L.E., Palabrica, V.S., and Ribo, J.M. (1987). In: *QSAR in Environmental Toxicology - II*, K.L.E. Kaiser (ed.), D. Reidel Publishing Company, Dordrecht, p. 153-168.

DIPHENYL PHTHALATE

CAS RN: 84-62-8

Sample purity: 99%
Temperature: 15°C
Test parameter: EC50
Effect: Reduction in light output

Concentration: 57.9 mg/l
Exposure times: 5 and 15 min

Concentration: 62.1 mg/l
Exposure time: 30 min

Comment: Mean of three assays. Methanol (<10%) was used to prepare the stock solutions. EC50 values were calculated from nominal concentrations.

Bibliographical reference: Kaiser, K.L.E., and Palabrica, V.S. (1992). National Water Research Institute, Burlington, Ontario, Canada, unpublished results.

N,N'-DIPHENYLTHIOUREA

CAS RN: 102-08-9

Sample purity: ≥97%
Test parameter: EC50
Effect: Reduction in light output
Concentration: 9.95 mg/l
Exposure time: 15 min
Comment: EC50 value converted to mg/l from the original data expressed in log(1/µmol l^{-1}) and a rounded molecular weight of 228 given by the authors.

Bibliographical reference: Govers, H., Ruepert, C., Stevens, T., and van Leeuwen, C.J. (1986). *Chemosphere* **15**, 383-393.

2,2'-DIPYRIDYL

CAS RN: 366-18-7

Sample purity: 99+%
Temperature: 15°C
Test parameter: EC50
Effect: Reduction in light output

Concentration: 127 mg/l
Exposure time: 5 min

Concentration: 113 mg/l
Exposure time: 15 min

Concentration: 101 mg/l
Exposure time: 30 min

Comment: Mean of three assays. Methanol (<10%) was used to prepare the stock solutions. EC50 values were calculated from nominal concentrations.

Bibliographical reference: Kaiser, K.L.E., and Palabrica, V.S. (1991). *Water Poll. Res. J. Canada* **26**, 361-431.

4,4'-DIPYRIDYL

CAS RN: 553-26-4

Sample purity: 98%
Temperature: 15°C
Test parameter: EC50
Effect: Reduction in light output

Concentration: 80.1 mg/l
Exposure time: 5 min

Concentration: 87.8 mg/l
Exposure time: 15 min

Concentration: 106 mg/l
Exposure time: 30 min

Comment: Mean of three assays. Methanol (<10%) was used to prepare the stock solutions. EC50 values were calculated from nominal concentrations.

Bibliographical reference: Kaiser, K.L.E., and Palabrica, V.S. (1991). *Water Poll. Res. J. Canada* **26**, 361-431.

2,2'-DIPYRIDYLAMINE

CAS RN: 1202-34-2

Sample purity: 98%
Temperature: 15°C
Test parameter: EC50
Effect: Reduction in light output

Concentration: 56.7 mg/l
Exposure time: 5 min

Concentration: 50.5 mg/l
Exposure time: 15 min

Concentration: 45.0 mg/l
Exposure time: 30 min

Comment: Mean of three assays. Methanol (<10%) was used to prepare the stock solutions. EC50 values were calculated from nominal concentrations.

Bibliographical reference: Kaiser, K.L.E., and Palabrica, V.S. (1991). *Water Poll. Res. J. Canada* **26**, 361-431.

DISULFINE BLUE

CAS RN: 129-17-9

Temperature: 15°C
Test parameter: EC50
Effect: Reduction in light output

Concentration: 20.1 mg/l
Exposure time: 5 min

Concentration: 18.8 mg/l
Exposure times: 15 and 30 min

Comment: Mean of three assays. Methanol (<10%) was used to prepare the stock solutions. EC50 values were calculated from nominal concentrations.

Bibliographical reference: Kaiser, K.L.E., and Palabrica, V.S. (1992). National Water Research Institute, Burlington, Ontario, Canada, unpublished results.

DISULFIRAM

CAS RN: 97-77-8
Synonym: Tetraethylthiuram disulfide

Sample purity: ≥97%
Test parameter: EC50
Effect: Reduction in light output
Concentration: 1.21 mg/l (0.91-1.61)
Exposure time: 15 min

Bibliographical reference: van Leeuwen, C.J., Maas-Diepeveen, J.L., Niebeek, G., Vergouw, W.H.A., Griffioen, P.S., and Luijken, M.W. (1985). *Aquat. Toxicol.* **7**, 145-164.

DIURON

CAS RN: 330-54-1
Synonym: 3-(3,4-Dichlorophenyl)-1,1-dimethylurea

Sample purity: Analytical reference standard grade
Temperature: 15 ± 0.1°C
Test parameter: EC50
Effect: Reduction in light output
Concentration: 16.38 mg/l
Exposure time: 5 min
Comment: Test was performed on *Photobacterium phosphoreum* NZ11D obtained from the Scripps Institute of Oceanography (La Jolla, CA).

Bibliographical reference: McFeters, G.A., Bond, P.J., Olson, S.B., and Tchan, Y.T. (1983). *Water Res.* **17**, 1757-1762.

1-DODECANOL

CAS RN: 112-53-8
Synonym: Lauryl alcohol

Temperature: 15°C
Test parameter: EC50
Effect: Reduction in light output

Concentration: 0.29 mg/l
Exposure time: 5 min

Concentration: 0.34 mg/l
Exposure time: 30 min

Bibliographical reference: Speece, R. (1987). Drexel University, Philadelphia, USA, private communication.

Sample purity: 98%
Temperature: 15°C
Test parameter: EC50
Effect: Reduction in light output

Concentration: 0.038 mg/l
Exposure time: 5 min

Concentration: 0.043 mg/l
Exposure time: 15 min

Concentration: 0.057 mg/l
Exposure time: 30 min

Comment: Mean of three assays. Methanol (<10%) was used to prepare the stock solutions. EC50 values were calculated from nominal concentrations.

Bibliographical reference: Kaiser, K.L.E., and Palabrica, V.S. (1991). *Water Poll. Res. J. Canada* **26**, 361-431.

DODECYLAMINE

CAS RN: 124-22-1

Sample purity: 99+%
Temperature: 15°C
Test parameter: EC50
Effect: Reduction in light output

Concentration: 0.93 mg/l
Exposure time: 5 min

Concentration: 0.91 mg/l
Exposure time: 15 min

Concentration: 0.93 mg/l
Exposure time: 30 min

Comment: Mean of three assays. Methanol (<10%) was used to prepare the stock solutions. EC50 values were calculated from nominal concentrations.

Bibliographical reference: Kaiser, K.L.E., and Palabrica, V.S. (1992). National Water Research Institute, Burlington, Ontario, Canada, unpublished results.

DODECYLBENZENESULFONIC ACID

CAS RN: 27176-87-0

Temperature: 15°C

Test parameter: EC20
Effect: Reduction in light output
Concentration: 12.7 mg/l
Exposure time: 15 min

Test parameter: EC50
Effect: Reduction in light output
Concentration: 29.2 mg/l
Exposure time: 15 min

Comment: Toxicity values were calculated from nominal concentrations.

Bibliographical reference: Reteuna, C. (1988). Thesis, University of Metz, Metz, France.

DUROQUINONE

CAS RN: 527-17-3
Synonym: Tetramethyl-1,4-benzoquinone

Sample purity: 97%
Temperature: 15°C
Test parameter: EC50
Effect: Reduction in light output

Concentration: 1.80 mg/l
Exposure time: 5 min

Concentration: 1.11 mg/l
Exposure time: 15 min

Concentration: 1.37 mg/l
Exposure time: 30 min

Comment: Mean of three assays. Methanol (<10%) was used to prepare the stock solutions. EC50 values were calculated from nominal concentrations.

Bibliographical reference: Kaiser, K.L.E., and Palabrica, V.S. (1991). *Water Poll. Res. J. Canada* **26**, 361-431.

ENDOSULFAN

CAS RN: 115-29-7
Synonym: (1,4,5,6,7,7-Hexachloro-8,9,10-trinorborn-5-en-2,3-ylene-bismethylene) sulfite

Temperature: 15°C
Test parameter: EC50
Effect: Reduction in light output
Concentration: Not toxic at 250 µg/l
Exposure time: 10 min
Comment: Chemical dissolved in 0.1 ml/l acetone. Test solutions were analyzed using gas chromatography.

Bibliographical reference: Bazin, C. (1985). DEA Ecologie Fondamentale et Appliquée des Eaux Continentales, University Lyon I, France.

EPICHLOROHYDRIN

CAS RN: 106-89-8
Synonym: 1-Chloro-2,3-epoxypropane

Temperature: 15°C
Test parameter: EC50
Effect: Reduction in light output

Concentration: 2310 mg/l
Exposure time: 5 min

Concentration: 1160 mg/l
Exposure time: 15 min

Concentration: 670 mg/l
Exposure time: 30 min

Comment: Chemical was diluted to give five sample concentrations in duplicate.

Bibliographical reference: Benson, W.H., and Stackhouse, R.A. (1986). *Drug Chem. Toxicol.* **9**, 275-283.

ESTERON 99

CAS RN: 3966-11-8
Synonym: Propylene glycol butylether ester of 2,4-D

Sample purity: Contained 2,4-D at a concentration of 485 mg/l
Test parameter: EC50
Effect: Reduction in light output

Concentration: 10.3 mg/l
Exposure times: 5 and 15 min

Concentration: 8.40 mg/l
Exposure time: 30 min

Bibliographical reference: Miller, W.E., Peterson, S.A., Greene, J.C., and Callahan, C.A. (1985). *J. Environ. Qual.* **14**, 569-574.

ETHANOL

CAS RN: 64-17-5
Synonym: Ethyl alcohol

Temperature: 15 ± 0.1°C
Test parameter: EC50
Effect: Reduction in light output
Concentration: 47000 mg/l
Exposure time: 5 min

Bibliographical reference: Chang, J.C., Taylor, P.B., and Leach, F.R. (1981). *Bull. Environ. Contam. Toxicol.* **26**, 150-156.

Temperature: 15°C
Test parameter: EC50
Effect: Reduction in light output
Concentration: 31000 mg/l
Exposure time: 5 min

Bibliographical reference: Bulich, A.A., Greene, M.W., and Isenberg, D.L. (1981). In: *Aquatic Toxicology and Hazard Assessment: Fourth Conference, ASTM STP 737*, D.R. Branson and K.L. Dickson (eds.), American Society for Testing and Materials, Philadelphia, p. 338-347.

Test parameter: EC50
Effect: Reduction in light output
Concentration: 44000 mg/l
Exposure time: 5 min
Comment: The EC50 value was calculated from nominal concentrations.

Bibliographical reference: Curtis, C., Lima, A., Lozano, S.J., and Veith, G.D. (1982). In: *Aquatic Toxicology and Hazard Assessment: Fifth Conference, ASTM STP 766*, J.G. Pearson, R.B. Foster, and W.E. Bishop (eds.), American Society for Testing and Materials, Philadelphia, p. 170-178.

Temperature: 15 ± 0.1°C
Test parameter: EC50
Effect: Reduction in light output
Concentration: 55575 mg/l

Exposure time: 5 min
Comment: Test was performed on *Photobacterium phosphoreum* NZ11D obtained from the Scripps Institute of Oceanography (La Jolla, CA).

Bibliographical reference: McFeters, G.A., Bond, P.J., Olson, S.B., and Tchan, Y.T. (1983). *Water Res.* **17**, 1757-1762.

Temperature: 15°C
Test parameter: EC50
Effect: Reduction in light output
Concentration: 29.11 µl/ml
Exposure time: 15 min
Comment: Four concentrations of chemical were tested in duplicate.

Bibliographical reference: Schiewe, M.H., Hawk, E.G., Actor, D.I., and Krahn, M.M. (1985). *Can. J. Fish. Aquat. Sci.* **42**, 1244-1248.

Temperature: 15°C
Test parameter: EC50
Effect: Reduction in light output

Concentration: 35470 mg/l
Exposure time: 5 min

Concentration: 34634 mg/l
Exposure time: 30 min

Bibliographical reference: Speece, R. (1987). Drexel University, Philadelphia, USA, private communication.

Temperature: 15°C
Test parameter: EC50
Effect: Reduction in light output
Concentration: 36000 mg/l
Exposure time: 5 min

Bibliographical reference: Kahru, A. (1993). *ATLA* **21**, 210-215.

ETHANOLAMINE

CAS RN: 141-43-5
Synonym: 2-Aminoethanol

Sample purity: 99%
Temperature: 15°C
Test parameter: EC50
Effect: Reduction in light output

Concentration: 14.3 mg/l
Exposure time: 5 min

Concentration: 12.2 mg/l
Exposure time: 15 min

Concentration: 13.7 mg/l
Exposure time: 30 min

Comment: Mean of three assays. Methanol (<10%) was used to prepare the stock solutions. EC50 values were calculated from nominal concentrations.

Bibliographical reference: Kaiser, K.L.E., and Palabrica, V.S. (1991). *Water Poll. Res. J. Canada* **26**, 361-431.

ETHIDIUM BROMIDE

CAS RN: 1239-45-8

Temperature: 15°C
Test parameter: EC50
Effect: Reduction in light output

Concentration: 5.46 ± 0.19 mg/l
Exposure time: 5 min

Concentration: 2.77 ± 0.15 mg/l
Exposure time: 10 min

Concentration: 1.97 ± 0.08 mg/l
Exposure time: 15 min

Concentration: 1.59 ± 0.05 mg/l

Exposure time: 30 min

Comment: Test was performed at least in triplicate. Chemical was solubilized in DMSO.

Bibliographical reference: Yates, I.E. (1985). *J. Microbiol. Meth.* **3**, 171-180.

Temperature: 15°C
Test parameter: EC50
Effect: Reduction in light output
Concentration: 2.93 ± 0.48 mg/l
Exposure time: 15 min

Bibliographical reference: Rychert, R., and Mortimer, M. (1991). *Environ. Toxicol. Water Qual.* **6**, 415-421.

4-ETHOXYBENZYL ALCOHOL

CAS RN: 6214-44-4

Sample purity: 98%
Temperature: 15°C
Test parameter: EC50
Effect: Reduction in light output

Concentration: 32.5 mg/l
Exposure time: 5 min

Concentration: 31.8 mg/l
Exposure time: 15 min

Concentration: 29.7 mg/l
Exposure time: 30 min

Comment: Mean of three assays. Methanol (<10%) was used to prepare the stock solutions. EC50 values were calculated from nominal concentrations.

Bibliographical reference: Kaiser, K.L.E., and Palabrica, V.S. (1991). *Water Poll. Res. J. Canada* **26**, 361-431.

2-ETHOXYETHANOL

CAS RN: 110-80-5

Sample purity: >99%
Temperature: 15°C
Test parameter: EC50
Effect: Reduction in light output

Concentration: 376 mg/l
Exposure time: 5 min

Concentration: 403 mg/l
Exposure time: 15 min

Concentration: 431 mg/l
Exposure time: 30 min

Comment: Mean of three assays. Methanol (<10%) was used to prepare the stock solutions. EC50 values were calculated from nominal concentrations.

Bibliographical reference: Kaiser, K.L.E., and Palabrica, V.S. (1991). *Water Poll. Res. J. Canada* **26**, 361-431.

2-(2-ETHOXYETHOXY)ETHANOL

CAS RN: 111-90-0
Synonym: Diethylene glycol monoethyl ether

Test parameter: EC50
Effect: Reduction in light output
Concentrations: 1000 mg/l
1290 mg/l
Exposure time: 5 min
Comment: The EC50 values were calculated from nominal concentrations.

Bibliographical reference: Curtis, C., Lima, A., Lozano, S.J., and Veith, G.D. (1982). In: *Aquatic Toxicology and Hazard Assessment: Fifth Conference, ASTM STP 766*, J.G. Pearson, R.B. Foster, and W.E. Bishop (eds.), American Society for Testing and Materials, Philadelphia, p. 170-178.

Temperature: 15°C
Test parameter: EC50
Effect: Reduction in light output
Concentration: 10954 mg/l (10592.8-11327.5)
Exposure time: 15 min
Comment: Single batches of Microtox® reagent were used for less than 2 h before being discarded. Four concentrations were tested. Concentrations were unmeasured.

Bibliographical reference: Nacci, D., Jackim, E., and Walsh, R. (1986). *Environ. Toxicol. Chem.* **5**, 521-525.

2-ETHOXYETHYL ACETATE

CAS RN: 115-15-9

Sample purity: 99%
Temperature: 15°C
Test parameter: EC50
Effect: Reduction in light output

Concentration: 1200 mg/l
Exposure time: 5 min

Concentration: 1400 mg/l
Exposure time: 15 min

Bibliographical reference: Cronin, M.T.D. (1993). Liverpool John Moores University, UK, private communication.

ETHYL ACETATE

CAS RN: 141-78-6

Sample purity: >98%
Temperature: 15 ± 0.1°C
Test parameter: EC50
Effect: Reduction in light output

Concentration: 5160 mg/l
Exposure time: 5 min

Concentration: 5870 mg/l
Exposure time: 15 min

Comment: Test was performed in duplicate. Toxicity values were based on nominal test concentrations.

Bibliographical reference: de Zwart, D., and Slooff, W. (1983). *Aquat. Toxicol.* **4**, 129-138.

Sample purity: >98%
Temperature: 15 ± 0.1°C
Test parameter: EC10
Effect: Reduction in light output

Concentration: 930 mg/l
Exposure time: 5 min

Concentration: 1650 mg/l
Exposure time: 15 min

Comment: Test was performed in duplicate. Toxicity values were based on nominal test concentrations.

Bibliographical reference: de Zwart, D., and Slooff, W. (1983). *Aquat. Toxicol.* **4**, 129-138.

ETHYL ACETOACETATE

CAS RN: 141-97-9

Sample purity: >99%
Temperature: 15°C
Test parameter: EC50
Effect: Reduction in light output

Concentration: 335 mg/l
Exposure time: 5 min

Concentration: 327 mg/l
Exposure time: 15 min

Concentration: 319 mg/l
Exposure time: 30 min

Comment: Mean of four assays. Methanol (<10%) was used to prepare the stock solutions. EC50 values were calculated from nominal concentrations.

Bibliographical reference: Kaiser, K.L.E., and Palabrica, V.S. (1991). *Water Poll. Res. J. Canada* **26**, 361-431.

4'-ETHYLACETOPHENONE

CAS RN: 937-30-4

Sample purity: 97%
Temperature: 15°C
Test parameter: EC50
Effect: Reduction in light output

Concentration: 0.62 mg/l
Exposure time: 5 min

Concentration: 0.65 mg/l
Exposure time: 15 min

Concentration: 0.63 mg/l
Exposure time: 30 min

Comment: Mean of three assays. Methanol (<10%) was used to prepare the stock solutions. EC50 values were calculated from nominal concentrations.

Bibliographical reference: Kaiser, K.L.E., and Palabrica, V.S. (1991). *Water Poll. Res. J. Canada* **26**, 361-431.

ETHYL ACRYLATE

CAS RN: 140-88-5

Sample purity: 99%
Temperature: 15°C
Test parameter: EC50
Effect: Reduction in light output

Concentration: 138 mg/l

Exposure time: 5 min

Concentration: 105 mg/l
Exposure time: 15 min

Concentration: 83 mg/l
Exposure time: 30 min

Comment: Mean of four assays. Methanol (<10%) was used to prepare the stock solutions. EC50 values were calculated from nominal concentrations.

Bibliographical reference: Kaiser, K.L.E., and Palabrica, V.S. (1992). National Water Research Institute, Burlington, Ontario, Canada, unpublished results.

N-ETHYLANILINE

CAS RN: 103-69-5

Sample purity: 97%
Temperature: 15°C
Test parameter: EC50
Effect: Reduction in light output

Concentration: 7.82 mg/l
Exposure time: 5 min

Concentration: 8.78 mg/l
Exposure time: 15 min

Concentration: 10.3 mg/l
Exposure time: 30 min

Comment: Mean of three assays. Methanol (<10%) was used to prepare the stock solutions. EC50 values were calculated from nominal concentrations.

Bibliographical reference: Kaiser, K.L.E., and Palabrica, V.S. (1991). *Water Poll. Res. J. Canada* **26**, 361-431.

4-ETHYLANILINE

CAS RN: 589-16-2

Sample purity: 99+%
Temperature: 15°C
Test parameter: EC50
Effect: Reduction in light output

Concentration: 0.20 mg/l
Exposure times: 5 and 15 min

Concentration: 0.21 mg/l*
Exposure time: 30 min

Comment: Mean of three assays. Methanol (<10%) was used to prepare the stock solutions. EC50 values were calculated from nominal concentrations.

Bibliographical references: Kaiser, K.L.E., and Palabrica, V.S. (1991). *Water Poll. Res. J. Canada* **26**, 361-431.
* Kaiser, K.L.E. (1987). In: *QSAR in Environmental Toxicology - II*, K.L.E. Kaiser (ed.), D. Reidel Publishing Company, Dordrecht, p. 169-188.

4-ETHYLBENZALDEHYDE

CAS RN: 4748-78-1

Sample purity: 98%
Temperature: 15°C
Test parameter: EC50
Effect: Reduction in light output

Concentration: 0.44 mg/l
Exposure time: 5 min

Concentration: 0.49 mg/l
Exposure time: 15 min

Concentration: 0.56 mg/l
Exposure time: 30 min

Comment: Mean of three assays. Methanol (<10%) was used to prepare the stock solutions. EC50 values were calculated from nominal concentrations.

Bibliographical reference: Kaiser, K.L.E., and Palabrica, V.S. (1991). *Water Poll. Res. J. Canada* **26**, 361-431.

ETHYLBENZENE

CAS RN: 100-41-4

Temperature: 15°C
Test parameter: EC50
Effect: Reduction in light output

Concentration: 5.5 ± 1.2 mg/l
Exposure time: 5 min

Concentration: 6.1 ± 1.7 mg/l
Exposure time: 10 min

Comment: Mean of four assays. Toxicity values were calculated from nominal concentrations.

Bibliographical reference: Ferard, J.F., Vasseur, P., Danoux, L., and Larbaigt, G. (1983). *Rev. Fr. Sci. Eau* **2**, 221-237.

Sample purity: 99+%
Temperature: 15°C
Test parameter: EC50
Effect: Reduction in light output

Concentration: 6.55 mg/l
Exposure time: 5 min

Concentration: 7.69 mg/l
Exposure time: 15 min

Concentration: 9.68 mg/l*
Exposure time: 30 min

Comment: Mean of three assays. Methanol (<10%) was used to prepare the stock solutions. EC50 values were calculated from nominal concentrations.

Bibliographical references: Kaiser, K.L.E., and Palabrica, V.S. (1991). *Water Poll. Res. J. Canada* **26**, 361-431.
* Kaiser, K.L.E., Palabrica, V.S., and Ribo, J.M. (1987). In: *QSAR in Environmental Toxicology - II*, K.L.E. Kaiser (ed.), D. Reidel Publishing Company, Dordrecht, p. 153-168.

S-ETHYL DIPROPYLTHIOCARBAMATE

CAS RN: 759-94-4
Synonym: EPTC

Sample purity: 98.8%
Temperature: 15°C
Test parameter: EC50
Effect: Reduction in light output

Concentration: 36.1 mg/l
Exposure time: 5 min

Concentration: 42.4 mg/l
Exposure time: 15 min

Concentration: 49.8 mg/l
Exposure time: 30 min

Comment: Mean of four assays. Methanol (<10%) was used to prepare the stock solutions. EC50 values were calculated from nominal concentrations.

Bibliographical reference: Kaiser, K.L.E., and Palabrica, V.S. (1991). *Water Poll. Res. J. Canada* **26**, 361-431.

ETHYLENEDIAMINE

CAS RN: 107-15-3

Test parameter: EC50
Effect: Reduction in light output

Concentration: 20.4 mg/l (18.1-23.0)
Exposure time: 15 min

Bibliographical reference: van Leeuwen, C.J., Maas-Diepeveen, J.L., Niebeek, G., Vergouw, W.H.A., Griffioen, P.S., and Luijken, M.W. (1985). *Aquat. Toxicol.* **7**, 145-164.

ETHYLENE GLYCOL

CAS RN: 107-21-1
Synonym: 1,2-Ethanediol

Sample purity: >99%
Temperature: 15°C
Test parameter: EC50
Effect: Reduction in light output

Concentration: 650 mg/l
Exposure times: 5 and 15 min

Concentration: 621 mg/l
Exposure time: 30 min

Comment: Mean of three assays. Methanol (<10%) was used to prepare the stock solutions. EC50 values were calculated from nominal concentrations.

Bibliographical reference: Kaiser, K.L.E., and Palabrica, V.S. (1991). *Water Poll. Res. J. Canada* **26**, 361-431.

ETHYLENETHIOUREA

CAS RN: 96-45-7
Synonym: 2-Imidazolidinethione

Sample purity: ≥99%
Test parameter: EC50
Effect: Reduction in light output
Concentration: 2100 mg/l (1500-2800)
Exposure time: 15 min

Bibliographical reference: van Leeuwen,. C.J., Maas-Diepeveen, J.L., Niebeek, G., Vergouw, W.H.A., Griffioen, P.S., and Luijken, M.W. (1985). *Aquat. Toxicol.* **7**, 145-164.

Sample purity: ≥97%
Test parameter: EC50
Effect: Reduction in light output
Concentration: 2447 mg/l
Exposure time: 15 min
Comment: EC50 value converted to mg/l from the original data expressed in log(1/µmol l^{-1}) and a rounded molecular weight of 102 given by the authors.

Bibliographical reference: Govers, H., Ruepert, C., Stevens, T., and van Leeuwen, C.J. (1986). *Chemosphere* **15**, 383-393.

ETHYLENEUREA

CAS RN: 120-93-4

Sample purity: ≥97%
Test parameter: EC50
Effect: Reduction in light output
Concentration: 3300 mg/l (2600-4200)
Exposure time: 15 min

Bibliographical reference: van Leeuwen, C.J., Maas-Diepeveen, J.L., Niebeek, G., Vergouw, W.H.A., Griffioen, P.S., and Luijken, M.W. (1985). *Aquat. Toxicol.* **7**, 145-164.

2-ETHYL-1,3-HEXANEDIOL

CAS RN: 94-96-2

Sample purity: 97%
Temperature: 15°C
Test parameter: EC50
Effect: Reduction in light output

Concentration: 25.4 mg/l
Exposure time: 5 min

Concentration: 28.5 mg/l
Exposure time: 15 min

Concentration: 33.5 mg/l
Exposure time: 30 min

Comment: Mean of three assays. Methanol (<10%) was used to prepare the stock solutions. EC50 values were calculated from nominal concentrations.

Bibliographical reference: Kaiser, K.L.E., and Palabrica, V.S. (1992). National Water Research Institute, Burlington, Ontario, Canada, unpublished results.

ETHYL HEXANOATE

CAS RN: 123-66-0
Synonym: Ethyl caproate

Sample purity: 99+%
Temperature: 15°C
Test parameter: EC50
Effect: Reduction in light output

Concentration: 42 mg/l
Exposure time: 5 min

Concentration: 60 mg/l
Exposure time: 15 min

Bibliographical reference: Cronin, M.T.D. (1993). Liverpool John Moores University, UK, private communication.

2-ETHYL-1-HEXANOL

CAS RN: 104-76-7

Sample purity: 99+%
Temperature: 15°C
Test parameter: EC50
Effect: Reduction in light output

Concentration: 4.95 mg/l
Exposure time: 5 min

Concentration: 5.68 mg/l
Exposure time: 15 min

Concentration: 6.38 mg/l
Exposure time: 30 min

Comment: Mean of four assays. Methanol (<10%) was used to prepare the stock solutions. EC50 values were calculated from nominal concentrations.

Bibliographical reference: Kaiser, K.L.E., and Palabrica, V.S. (1992). National Water Research Institute, Burlington, Ontario, Canada, unpublished results.

ETHYL 4-ISOCYANATOBENZOATE

CAS RN: 30806-83-8

Sample purity: 97%
Temperature: 15°C
Test parameter: EC50
Effect: Reduction in light output

Concentration: 17.4 mg/l
Exposure time: 5 min

Concentration: 17.0 mg/l
Exposure time: 15 min

Concentration: 22.0 mg/l
Exposure time: 30 min

Comment: Mean of three assays. Methanol (<10%) was used to prepare the stock solutions. EC50 values were calculated from nominal concentrations.

Bibliographical reference: Kaiser, K.L.E., and Palabrica, V.S. (1991). *Water Poll. Res. J. Canada* **26**, 361-431.

4-ETHYLMORPHOLINE

CAS RN: 100-74-3

Sample purity: 99+%
Temperature: 15°C
Test parameter: EC50
Effect: Reduction in light output

Concentration: 89.4 mg/l
Exposure time: 5 min

Concentration: 71.0 mg/l
Exposure time: 15 min

Concentration: 81.5 mg/l
Exposure time: 30 min

Comment: Mean of three assays. Methanol (<10%) was used to prepare the stock solutions. EC50 values were calculated from nominal concentrations.

Bibliographical reference: Kaiser, K.L.E., and Palabrica, V.S. (1991). *Water Poll. Res. J. Canada* **26**, 361-431.

1-ETHYLNAPHTHALENE

CAS RN: 1127-76-0

Sample purity: 99+%
Temperature: 15°C
Test parameter: EC50
Effect: Reduction in light output

Concentration: 0.15 mg/l
Exposure time: 5 min

Concentration: 0.19 mg/l
Exposure time: 15 min

Concentration: 0.24 mg/l
Exposure time: 30 min

Comment: Mean of three assays. Methanol (<10%) was used to prepare the stock solutions. EC50 values were calculated from nominal concentrations.

Bibliographical reference: Kaiser, K.L.E., and Palabrica, V.S. (1991). *Water Poll. Res. J. Canada* **26**, 361-431.

4-ETHYLNITROBENZENE

CAS RN: 100-12-9

Sample purity: 99%
Temperature: 15°C
Test parameter: EC50
Effect: Reduction in light output

Concentration: 1.29 mg/l
Exposure time: 5 min

Concentration: 1.35 mg/l
Exposure time: 15 min

Concentration: 1.44 mg/l*
Exposure time: 30 min

Comment: Mean of three assays. Methanol (<10%) was used to prepare the stock solutions. EC50 values were calculated from nominal concentrations.

Bibliographical references: Kaiser, K.L.E., and Palabrica, V.S. (1991). *Water Poll. Res. J. Canada* **26**, 361-431.
* Kaiser, K.L.E. (1987). In: *QSAR in Environmental Toxicology - II*, K.L.E. Kaiser (ed.), D. Reidel Publishing Company, Dordrecht, p. 169-188.

4-ETHYLPHENOL

CAS RN: 123-07-9

Sample purity: 99%
Temperature: 15°C
Test parameter: EC50

Effect: Reduction in light output

Concentration: 0.041 mg/l
Exposure time: 5 min

Concentration: 0.044 mg/l
Exposure time: 15 min

Concentration: 0.051 mg/l
Exposure time: 30 min

Comment: Mean of three assays. Methanol (<10%) was used to prepare the stock solutions. EC50 values were calculated from nominal concentrations.

Bibliographical reference: Kaiser, K.L.E., and Palabrica, V.S. (1991). *Water Poll. Res. J. Canada* **26**, 361-431.

4-ETHYLPHENYL ISOCYANATE

CAS RN: 23138-50-3

Temperature: 15°C
Test parameter: EC50
Effect: Reduction in light output

Concentration: 0.36 mg/l
Exposure time: 5 min

Concentration: 0.33 mg/l
Exposure time: 15 min

Concentration: 0.32 mg/l
Exposure time: 30 min

Comment: Mean of three assays. Methanol (<10%) was used to prepare the stock solutions. EC50 values were calculated from nominal concentrations.

Bibliographical reference: Kaiser, K.L.E., and Palabrica, V.S. (1991). *Water Poll. Res. J. Canada* **26**, 361-431.

ETHYL PROPIONATE

CAS RN: 105-37-3

Sample purity: >98%
Temperature: 15 ± 0.1°C
Test parameter: EC50
Effect: Reduction in light output

Concentration: 612 mg/l
Exposure time: 5 min

Concentration: 811 mg/l
Exposure time: 15 min

Comment: Test was performed in duplicate. Toxicity values were based on nominal test concentrations.

Bibliographical reference: de Zwart, D., and Slooff, W. (1983). *Aquat. Toxicol.* **4**, 129-138.

Sample purity: >98%
Temperature: 15 ± 0.1°C
Test parameter: EC10
Effect: Reduction in light output

Concentration: 80 mg/l
Exposure time: 5 min

Concentration: 125 mg/l
Exposure time: 15 min

Comment: Test was performed in duplicate. Toxicity values were based on nominal test concentrations.

Bibliographical reference: de Zwart, D., and Slooff, W. (1983). *Aquat. Toxicol.* **4**, 129-138.

4-ETHYLPYRIDINE

CAS RN: 536-75-4

Sample purity: 98%

Temperature: 15°C
Test parameter: EC50
Effect: Reduction in light output

Concentration: 4.79 mg/l
Exposure time: 5 min

Concentration: 4.90 mg/l
Exposure time: 15 min

Concentration: 5.50 mg/l
Exposure time: 30 min

Comment: Mean of three assays. Methanol (<10%) was used to prepare the stock solutions. EC50 values were calculated from nominal concentrations.

Bibliographical reference: Kaiser, K.L.E., and Palabrica, V.S. (1991). *Water Poll. Res. J. Canada* **26**, 361-431.

3-ETHYLQUINOLINE

CAS RN: 1873-54-7

Sample purity: >97%
Temperature: 15 ± 0.3°C
Test parameter: EC50
Effect: Reduction in light output

Concentration: 6.3 mg/l
Exposure time: 5 min

Concentration: 7.8 mg/l
Exposure time: 15 min

Comment: Test was performed in duplicate at pH = 5.1. Dichloromethane was used as carrier solvent. Concentrations of the chemical were determined by high performance liquid chromatography. Hexachloroethane was used for quality control/quality assurance (5-min EC50 = 0.31 mg/l; n = 16; sd = 0.08 mg/l).

Bibliographical reference: Birkholz, D.A., Coutts, R.T., Hrudey, S.E., Danell, R.W., and Lockhart, W.L. (1990). *Water Res.* **24**, 67-73.

N-ETHYLTHIOUREA

CAS RN: 625-53-6
Synonym: 1-Ethyl-2-thiourea

Sample purity: ≥97%
Test parameter: EC50
Effect: Reduction in light output
Concentration: 2173 mg/l
Exposure time: 15 min
Comment: EC50 value converted to mg/l from the original data expressed in log(1/μmol l^{-1}) and a rounded molecular weight of 104 given by the authors.

Bibliographical reference: Govers, H., Ruepert, C., Stevens, T., and van Leeuwen, C.J. (1986). *Chemosphere* **15**, 383-393.

4-ETHYLTOLUENE

CAS RN: 622-96-8

Sample purity: 98%
Temperature: 15°C
Test parameter: EC50
Effect: Reduction in light output

Concentration: 1.95 mg/l
Exposure time: 5 min

Concentration: 2.24 mg/l
Exposure time: 15 min

Concentration: 2.63 mg/l
Exposure time: 30 min

Comment: Mean of three assays. Methanol (<10%) was used to prepare the stock solutions. EC50 values were calculated from nominal concentrations.

Bibliographical reference: Kaiser, K.L.E., and Palabrica, V.S. (1991). *Water Poll. Res. J. Canada* **26**, 361-431.

ETHYL VINYL KETONE

CAS RN: 1629-58-9
Synonym: 1-Penten-3-one

Sample purity: 99%
Temperature: 15°C
Test parameter: EC50
Effect: Reduction in light output

Concentration: 2.21 mg/l
Exposure time: 5 min

Concentration: 1.08 mg/l
Exposure time: 15 min

Concentration: 0.75 mg/l
Exposure time: 30 min

Comment: Mean of three assays. Methanol (<10%) was used to prepare the stock solutions. EC50 values were calculated from nominal concentrations.

Bibliographical reference: Kaiser, K.L.E., and Palabrica, V.S. (1991). *Water Poll. Res. J. Canada* **26**, 361-431.

FENBUTATIN OXIDE

CAS RN: 13356-08-6

Temperature: 15°C
Test parameter: EC50
Effect: Reduction in light output
Concentration: >0.006 mg/l
Exposure time: 30 min

Bibliographical reference: Steinhäuser, K.G., Amann, W., Späth, A., and Polenz, A. (1985). *Vom Wasser* **65**, 203-214.

FERBAM

CAS RN: 14484-64-1

Synonym: Ferric dimethyldithiocarbamate

Sample purity: ≥95%
Test parameter: EC50
Effect: Reduction in light output
Concentration: 0.20 mg/l (0.14-0.28)
Exposure time: 15 min

Bibliographical reference: van Leeuwen, C.J., Maas-Diepeveen, J.L., Niebeek, G., Vergouw, W.H.A., Griffioen, P.S., and Luijken, M.W. (1985). *Aquat. Toxicol.* **7**, 145-164.

FERRIC CHLORIDE

CAS RN: 7705-08-0

Temperature: 15°C
Test parameter: EC50
Effect: Reduction in light output
Concentration: 1.77 ± 0.06 mg/l Fe^{+++}
Exposure time: 10 min
Comment: Mean of three assays. Toxicity values were calculated from nominal concentrations.

Bibliographical reference: Ferard, J.F., Vasseur, P., Danoux, L., and Larbaigt, G. (1983). *Rev. Fr. Sci. Eau* **2**, 221-237.

FLUAZIFOP-BUTYL

CAS RN: 69806-50-4
Synonym: Fusilade

Sample purity: 90%
Temperature: 15°C
Test parameter: EC50
Effect: Reduction in light output

Concentration: 2.65 mg/l
Exposure time: 5 min

Concentration: 5.67 mg/l
Exposure time: 15 min

Concentration: 10.6 mg/l
Exposure time: 30 min

Comment: Mean of three assays. Methanol (<10%) was used to prepare the stock solutions. EC50 values were calculated from nominal concentrations.

Bibliographical reference: Kaiser, K.L.E., and Palabrica, V.S. (1991). *Water Poll. Res. J. Canada* **26**, 361-431.

FLUORANTHENE

CAS RN: 206-44-0

Temperature: 15°C
Test parameter: EC50
Effect: Reduction in light output
Concentration: 2.02 mg/l (1.21-3.36)
Exposure time: 15 min
Comment: Chemical was added to 5 ml DMSO and placed in a bath sonicator until all visible material had dissolved.

Bibliographical reference: Jacobs, M.W., Coates, J.A., Delfino, J.J., Bitton, G., Davis, W.M., and Garcia, K.L. (1993). *Arch. Environ. Contam. Toxicol.* **24**, 461-468.

FLUORENE

CAS RN: 86-73-7

Temperature: 15°C
Test parameter: EC50
Effect: Reduction in light output
Concentration: 3.23 mg/l (2.24-4.66)
Exposure time: 15 min
Comment: Chemical was added to 5 ml DMSO and placed in a bath sonicator until all visible material had dissolved.

Bibliographical reference: Jacobs, M.W., Coates, J.A., Delfino, J.J., Bitton, G., Davis, W.M., and Garcia, K.L. (1993). *Arch. Environ. Contam. Toxicol.* **24**, 461-468.

FLUORENONE

CAS RN: 486-25-9

Sample purity: 98%
Temperature: 15°C
Test parameter: EC50
Effect: Reduction in light output

Concentration: 3.60 mg/l
Exposure time: 5 min

Concentration: 3.77 mg/l
Exposure time: 15 min

Concentration: 4.85 mg/l
Exposure time: 30 min

Comment: Mean of three assays. Methanol (<10%) was used to prepare the stock solutions. EC50 values were calculated from nominal concentrations.

Bibliographical reference: Kaiser, K.L.E., and Palabrica, V.S. (1991). *Water Poll. Res. J. Canada* **26**, 361-431.

4'-FLUOROACETOPHENONE

CAS RN: 403-42-9

Sample purity: 99%
Temperature: 15°C
Test parameter: EC50
Effect: Reduction in light output

Concentration: 45.7 mg/l
Exposure time: 5 min

Concentration: 49.0 mg/l
Exposure time: 15 min

Concentration: 52.5 mg/l*
Exposure time: 30 min

Comment: Mean of three assays. Methanol (<10%) was used to prepare the stock solutions. EC50 values were calculated from nominal concentrations.

Bibliographical references: Kaiser, K.L.E., and Palabrica, V.S. (1991). *Water Poll. Res. J. Canada* **26**, 361-431.
* Kaiser, K.L.E., and Gough, K.M. (1989). In: *Aquatic Toxicology and Environmental Fate: Eleventh Volume, ASTM STP 1007*, G.W. Suter and M.A. Lewis (eds.), American Society for Testing and Materials, Philadelphia, p. 424-441.

4-FLUOROANILINE

CAS RN: 371-40-4

Sample purity: 99%
Temperature: 15°C
Test parameter: EC50
Effect: Reduction in light output
Concentration: 82.4 mg/l
Exposure time: 5 min
Comment: Chemical was prepared in an initial solution of 2% methanol.

Bibliographical reference: Cronin, M.T.D., Dearden, J.C., and Dobbs, A.J. (1991). *Sci. Total Environ.* **109/110**, 431-439.

Temperature: 15°C
Test parameter: EC50
Effect: Reduction in light output

Concentration: 80.5 mg/l
Exposure time: 5 min

Concentration: 73.4 mg/l
Exposure time: 15 min

Concentration: 71.7 mg/l*
Exposure time: 30 min

Comment: Mean of three assays. Methanol (<10%) was used to prepare the stock solutions. EC50 values were calculated from nominal concentrations.

Bibliographical references: Kaiser, K.L.E., and Palabrica, V.S. (1991). *Water Poll. Res. J. Canada* **26**, 361-431.
* Kaiser, K.L.E. (1987). In: *QSAR in Environmental Toxicology - II*, K.L.E. Kaiser (ed.), D. Reidel Publishing Company, Dordrecht, p. 169-188.

4-FLUOROANISOLE

CAS RN: 459-60-9

Sample purity: 99%
Temperature: 15°C
Test parameter: EC50
Effect: Reduction in light output

Concentration: 18.2 mg/l
Exposure time: 5 min

Concentration: 20.0 mg/l
Exposure time: 15 min

Concentration: 21.9 mg/l*
Exposure time: 30 min

Comment: Mean of three assays. Methanol (<10%) was used to prepare the stock solutions. EC50 values were calculated from nominal concentrations.

Bibliographical references: Kaiser, K.L.E., and Palabrica, V.S. (1991). *Water Poll. Res. J. Canada* **26**, 361-431.
* Kaiser, K.L.E., and Gough, K.M. (1989). In: *Aquatic Toxicology and Environmental Fate: Eleventh Volume, ASTM STP 1007*, G.W. Suter and M.A. Lewis (eds.), American Society for Testing and Materials, Philadelphia, p. 424-441.

2-FLUOROBENZALDEHYDE

CAS RN: 446-52-6

Sample purity: 99%
Temperature: 15°C
Test parameter: EC50

Effect: Reduction in light output
Concentration: 20.0 mg/l
Exposure times: 5 and 15 min

Bibliographical reference: Cronin, M.T.D. (1993). Liverpool John Moores University, UK, private communication.

4-FLUOROBENZALDEHYDE

CAS RN: 459-57-4

Sample purity: 97%
Temperature: 15°C
Test parameter: EC50
Effect: Reduction in light output

Concentration: 62.2 mg/l
Exposure time: 5 min

Concentration: 68.2 mg/l
Exposure time: 15 min

Concentration: 71.4 mg/l
Exposure time: 30 min

Comment: Mean of three assays. Methanol (<10%) was used to prepare the stock solutions. EC50 values were calculated from nominal concentrations.

Bibliographical reference: Kaiser, K.L.E., and Palabrica, V.S. (1991). *Water Poll. Res. J. Canada* **26**, 361-431.

4-FLUOROBENZAMIDE

CAS RN: 824-75-9

Sample purity: 98%
Temperature: 15°C
Test parameter: EC50
Effect: Reduction in light output

Concentration: 226 mg/l

Exposure times: 5 and 15 min

Concentration: 231 mg/l
Exposure time: 30 min

Comment: Mean of three assays. Methanol (<10%) was used to prepare the stock solutions. EC50 values were calculated from nominal concentrations.

Bibliographical reference: Kaiser, K.L.E., and Palabrica, V.S. (1991). *Water Poll. Res. J. Canada* **26**, 361-431.

Sample purity: 98%
Temperature: 15°C
Test parameter: EC50
Effect: Reduction in light output

Concentration: 192 mg/l
Exposure time: 5 min

Concentration: 221 mg/l
Exposure time: 15 min

Concentration: 284 mg/l*
Exposure time: 30 min

Comment: Mean of three assays. Methanol (<10%) was used to prepare the stock solutions. EC50 values were calculated from nominal concentrations.

Bibliographical references: Kaiser, K.L.E., and Palabrica, V.S. (1991). *Water Poll. Res. J. Canada* **26**, 361-431.
* Kaiser, K.L.E., and Gough, K.M. (1989). In: *Aquatic Toxicology and Environmental Fate: Eleventh Volume, ASTM STP 1007*, G.W. Suter and M.A. Lewis (eds.), American Society for Testing and Materials, Philadelphia, p. 424-441.

FLUOROBENZENE

CAS RN: 462-06-6

Sample purity: 99%
Temperature: 15°C

Test parameter: EC50
Effect: Reduction in light output

Concentration: 192 mg/l
Exposure time: 5 min

Concentration: 163 mg/l
Exposure time: 15 min

Concentration: 183 mg/l*
Exposure time: 30 min

Comment: Mean of two assays. Methanol (<10%) was used to prepare the stock solutions. EC50 values were calculated from nominal concentrations.

Bibliographical references: Kaiser, K.L.E., and Palabrica, V.S. (1991). *Water Poll. Res. J. Canada* **26**, 361-431.
* Kaiser, K.L.E., Palabrica, V.S., and Ribo, J.M. (1987). In: *QSAR in Environmental Toxicology - II*, K.L.E. Kaiser (ed.), D. Reidel Publishing Company, Dordrecht, p. 153-168.

4-FLUOROBENZONITRILE

CAS RN: 1194-02-1

Sample purity: 99%
Temperature: 15°C
Test parameter: EC50
Effect: Reduction in light output

Concentration: 63.6 mg/l
Exposure time: 5 min

Concentration: 60.7 mg/l
Exposure time: 15 min

Concentration: 60.7 mg/l*
Exposure time: 30 min

Comment: Mean of three assays. Methanol (<10%) was used to prepare the stock solutions. EC50 values were calculated from nominal concentrations.

Bibliographical references: Kaiser, K.L.E., and Palabrica, V.S. (1991). *Water Poll. Res. J. Canada* **26**, 361-431.
* Kaiser, K.L.E., and Gough, K.M. (1989). In: *Aquatic Toxicology and Environmental Fate: Eleventh Volume, ASTM STP 1007*, G.W. Suter and M.A. Lewis (eds.), American Society for Testing and Materials, Philadelphia, p. 424-441.

4-FLUOROBENZOPHENONE

CAS RN: 345-83-5

Temperature: 15°C
Test parameter: EC50
Effect: Reduction in light output

Concentration: 4.48 mg/l
Exposure time: 5 min

Concentration: 4.91 mg/l
Exposure time: 15 min

Concentration: 5.39 mg/l*
Exposure time: 30 min

Comment: Mean of three assays. Methanol (<10%) was used to prepare the stock solutions. EC50 values were calculated from nominal concentrations.

Bibliographical references: Kaiser, K.L.E., and Palabrica, V.S. (1991). *Water Poll. Res. J. Canada* **26**, 361-431.
* Kaiser, K.L.E., and Gough, K.M. (1989). In: *Aquatic Toxicology and Environmental Fate: Eleventh Volume, ASTM STP 1007*, G.W. Suter and M.A. Lewis (eds.), American Society for Testing and Materials, Philadelphia, p. 424-441.

4-FLUOROBENZOTRIFLUORIDE

CAS RN: 402-44-8

Sample purity: 98+%
Temperature: 15°C
Test parameter: EC50

Effect: Reduction in light output

Concentration: 54.3 mg/l
Exposure time: 5 min

Concentration: 70.0 mg/l
Exposure time: 15 min

Concentration: 82.2 mg/l*
Exposure time: 30 min

Comment: Mean of three assays. Methanol (<10%) was used to prepare the stock solutions. EC50 values were calculated from nominal concentrations.

Bibliographical references: Kaiser, K.L.E., and Palabrica, V.S. (1991). *Water Poll. Res. J. Canada* **26**, 361-431.
* Kaiser, K.L.E., and Gough, K.M. (1989). In: *Aquatic Toxicology and Environmental Fate: Eleventh Volume, ASTM STP 1007*, G.W. Suter and M.A. Lewis (eds.), American Society for Testing and Materials, Philadelphia, p. 424-441.

4-FLUOROBENZOYL CHLORIDE

CAS RN: 403-43-0

Sample purity: 98%
Temperature: 15°C
Test parameter: EC50
Effect: Reduction in light output

Concentration: 10.7 mg/l
Exposure time: 5 min

Concentration: 13.2 mg/l
Exposure time: 15 min

Concentration: 15.5 mg/l
Exposure time: 30 min

Comment: Mean of three assays. Methanol (<10%) was used to prepare the stock solutions. EC50 values were calculated from nominal concentrations.

Bibliographical reference: Kaiser, K.L.E., and Palabrica, V.S. (1991). *Water Poll. Res. J. Canada* **26**, 361-431.

4-FLUOROBENZYL ALCOHOL

CAS RN: 459-56-3

Sample purity: 98%
Temperature: 15°C
Test parameter: EC50
Effect: Reduction in light output

Concentration: 162 mg/l
Exposure time: 5 min

Concentration: 152 mg/l
Exposure time: 15 min

Concentration: 138 mg/l
Exposure time: 30 min

Comment: Mean of three assays. Methanol (<10%) was used to prepare the stock solutions. EC50 values were calculated from nominal concentrations.

Bibliographical reference: Kaiser, K.L.E., and Palabrica, V.S. (1991). *Water Poll. Res. J. Canada* **26**, 361-431.

4-FLUOROBENZYLAMINE

CAS RN: 140-75-0

Sample purity: 97+%
Temperature: 15°C
Test parameter: EC50
Effect: Reduction in light output

Concentration: 33.7 mg/l
Exposure time: 5 min

Concentration: 29.3 mg/l
Exposure times: 15 and 30 min

Comment: Mean of three assays. Methanol (<10%) was used to prepare the stock solutions. EC50 values were calculated from nominal concentrations.

Bibliographical reference: Kaiser, K.L.E., and Palabrica, V.S. (1991). *Water Poll. Res. J. Canada* **26**, 361-431.

4-FLUOROBENZYL CHLORIDE

CAS RN: 352-11-4

Sample purity: 99%
Temperature: 15°C
Test parameter: EC50
Effect: Reduction in light output

Concentration: 4.17 mg/l
Exposure time: 5 min

Concentration: 4.90 mg/l
Exposure time: 15 min

Concentration: 5.37 mg/l
Exposure time: 30 min

Comment: Mean of three assays. Methanol (<10%) was used to prepare the stock solutions. EC50 values were calculated from nominal concentrations.

Bibliographical reference: Kaiser, K.L.E., and Palabrica, V.S. (1991). *Water Poll. Res. J. Canada* **26**, 361-431.

4-FLUOROCINNAMIC ACID

CAS RN: 459-32-5

Sample purity: 99%
Temperature: 15°C
Test parameter: EC50
Effect: Reduction in light output

Concentration: 44.7 mg/l

Exposure time: 5 min

Concentration: 51.3 mg/l
Exposure time: 15 min

Concentration: 56.3 mg/l
Exposure time: 30 min

Comment: Mean of four assays. Methanol (<10%) was used to prepare the stock solutions. EC50 values were calculated from nominal concentrations.

Bibliographical reference: Kaiser, K.L.E., and Palabrica, V.S. (1991). *Water Poll. Res. J. Canada* **26**, 361-431.

1-FLUORO-4-IODOBENZENE

CAS RN: 352-34-1

Sample purity: 99%
Temperature: 15°C
Test parameter: EC50
Effect: Reduction in light output

Concentration: 5.20 mg/l
Exposure time: 5 min

Concentration: 5.84 mg/l
Exposure time: 15 min

Concentration: 6.55 mg/l
Exposure time: 30 min

Comment: Mean of three assays. Methanol (<10%) was used to prepare the stock solutions. EC50 values were calculated from nominal concentrations.

Bibliographical reference: Kaiser, K.L.E., and Palabrica, V.S. (1991). *Water Poll. Res. J. Canada* **26**, 361-431.

4-FLUORO-*N*-METHYLANILINE

CAS RN: 459-59-6

Temperature: 15°C
Test parameter: EC50
Effect: Reduction in light output

Concentration: 7.20 mg/l
Exposure time: 5 min

Concentration: 7.37 mg/l
Exposure time: 15 min

Concentration: 7.54 mg/l
Exposure time: 30 min

Comment: Mean of three assays. Methanol (<10%) was used to prepare the stock solutions. EC50 values were calculated from nominal concentrations.

Bibliographical reference: Kaiser, K.L.E., and Palabrica, V.S. (1991). *Water Poll. Res. J. Canada* **26**, 361-431.

5-FLUORO-2-METHYL-BENZONITRILE

CAS RN: 77532-79-7

Sample purity: 98%
Temperature: 15°C
Test parameter: EC50
Effect: Reduction in light output

Concentration: 11.5 mg/l
Exposure time: 5 min

Concentration: 12.3 mg/l
Exposure time: 15 min

Concentration: 13.2 mg/l
Exposure time: 30 min

Comment: Mean of three assays. Methanol (<10%) was used to prepare the stock solutions. EC50 values were calculated from nominal concentrations.

Bibliographical reference: Kaiser, K.L.E., and Palabrica, V.S. (1991). *Water Poll. Res. J. Canada* **26**, 361-431.

2-FLUORO-4-NITROANISOLE

CAS RN: 455-93-6

Sample purity: 98%
Temperature: 15°C
Test parameter: EC50
Effect: Reduction in light output

Concentration: 39.2 mg/l
Exposure time: 5 min

Concentration: 43.0 mg/l
Exposure time: 15 min

Concentration: 49.4 mg/l
Exposure time: 30 min

Comment: Mean of three assays. Methanol (<10%) was used to prepare the stock solutions. EC50 values were calculated from nominal concentrations.

Bibliographical reference: Kaiser, K.L.E., and Palabrica, V.S. (1991). *Water Poll. Res. J. Canada* **26**, 361-431.

4-FLUORO-2-NITROANISOLE

CAS RN: 445-83-0

Sample purity: 98%
Temperature: 15°C
Test parameter: EC50
Effect: Reduction in light output

Concentration: 52.9 mg/l

Exposure time: 5 min

Concentration: 54.1 mg/l
Exposure time: 15 min

Concentration: 58.0 mg/l
Exposure time: 30 min

Comment: Mean of four assays. Methanol (<10%) was used to prepare the stock solutions. EC50 values were calculated from nominal concentrations.

Bibliographical reference: Kaiser, K.L.E., and Palabrica, V.S. (1991). *Water Poll. Res. J. Canada* **26**, 361-431.

1-FLUORO-2-NITROBENZENE

CAS RN: 1493-27-2

Sample purity: 99%
Temperature: 15°C
Test parameter: EC50
Effect: Reduction in light output

Concentration: 40.7 mg/l
Exposure time: 5 min

Concentration: 41.6 mg/l
Exposure time: 15 min

Concentration: 44.6 mg/l
Exposure time: 30 min

Comment: Mean of four assays. Methanol (<10%) was used to prepare the stock solutions. EC50 values were calculated from nominal concentrations.

Bibliographical reference: Kaiser, K.L.E., and Palabrica, V.S. (1991). *Water Poll. Res. J. Canada* **26**, 361-431.

1-FLUORO-3-NITROBENZENE

CAS RN: 402-67-5

Sample purity: 97%
Temperature: 15°C
Test parameter: EC50
Effect: Reduction in light output

Concentration: 72.4 mg/l
Exposure time: 5 min

Concentration: 74.1 mg/l
Exposure time: 15 min

Concentration: 79.3 mg/l
Exposure time: 30 min

Comment: Mean of three assays. Methanol (<10%) was used to prepare the stock solutions. EC50 values were calculated from nominal concentrations.

Bibliographical reference: Kaiser, K.L.E., and Palabrica, V.S. (1991). *Water Poll. Res. J. Canada* **26**, 361-431.

1-FLUORO-4-NITROBENZENE

CAS RN: 350-46-9

Sample purity: 99%
Temperature: 15°C
Test parameter: EC50
Effect: Reduction in light output

Concentration: 77.5 mg/l
Exposure time: 5 min

Concentration: 79.3 mg/l
Exposure time: 15 min

Concentration: 93.2 mg/l
Exposure time: 30 min

Comment: Mean of three assays. Methanol (<10%) was used to prepare the stock solutions. EC50 values were calculated from nominal concentrations.

Bibliographical reference: Kaiser, K.L.E., and Palabrica, V.S. (1991). *Water Poll. Res. J. Canada* **26**, 361-431.

2-FLUORO-5-NITROBENZOTRIFLUORIDE

CAS RN: 400-74-8
Synonym: α,α,α,2-Tetrafluoro-5-nitrotoluene

Sample purity: 98%
Temperature: 15°C
Test parameter: EC50
Effect: Reduction in light output

Concentration: 37.2 mg/l
Exposure time: 5 min

Concentration: 42.7 mg/l
Exposure time: 15 min

Concentration: 49.0 mg/l
Exposure time: 30 min

Comment: Mean of four assays. Methanol (<10%) was used to prepare the stock solutions. EC50 values were calculated from nominal concentrations.

Bibliographical reference: Kaiser, K.L.E., and Palabrica, V.S. (1991). *Water Poll. Res. J. Canada* **26**, 361-431.

2-FLUORO-4-NITROPHENOL

CAS RN: 403-19-0

Sample purity: 99%
Temperature: 15°C
Test parameter: EC50
Effect: Reduction in light output

Concentration: 9.69 mg/l
Exposure time: 5 min

Concentration: 9.47 mg/l
Exposure time: 15 min

Concentration: 9.91 mg/l
Exposure time: 30 min

Comment: Mean of four assays. Methanol (<10%) was used to prepare the stock solutions. EC50 values were calculated from nominal concentrations.

Bibliographical reference: Kaiser, K.L.E., and Palabrica, V.S. (1991). *Water Poll. Res. J. Canada* **26**, 361-431.

4-FLUORO-2-NITROPHENOL

CAS RN: 394-33-2

Sample purity: 99%
Temperature: 15°C
Test parameter: EC50
Effect: Reduction in light output

Concentration: 22.7 mg/l
Exposure time: 5 min

Concentration: 19.8 mg/l
Exposure time: 15 min

Concentration: 18.9 mg/l
Exposure time: 30 min

Comment: Mean of four assays. Methanol (<10%) was used to prepare the stock solutions. EC50 values were calculated from nominal concentrations.

Bibliographical reference: Kaiser, K.L.E., and Palabrica, V.S. (1991). *Water Poll. Res. J. Canada* **26**, 361-431.

4-FLUORO-3-NITROPHENYL ISOCYANATE

CAS RN: 65303-82-4

Sample purity: 98%
Temperature: 15°C
Test parameter: EC50
Effect: Reduction in light output

Concentration: 33.9 mg/l
Exposure time: 5 min

Concentration: 20.9 mg/l
Exposure time: 15 min

Concentration: 17.4 mg/l
Exposure time: 30 min

Comment: Mean of three assays. Methanol (<10%) was used to prepare the stock solutions. EC50 values were calculated from nominal concentrations.

Bibliographical reference: Kaiser, K.L.E., and Palabrica, V.S. (1991). *Water Poll. Res. J. Canada* **26**, 361-431.

4-FLUOROPHENETHYL ALCOHOL

CAS RN: 7589-27-7

Sample purity: 97%
Temperature: 15°C
Test parameter: EC50
Effect: Reduction in light output

Concentration: 9.26 mg/l
Exposure time: 5 min

Concentration: 10.4 mg/l
Exposure time: 15 min

Concentration: 11.4 mg/l
Exposure time: 30 min

Comment: Mean of three assays. Methanol (<10%) was used to prepare the stock solutions. EC50 values were calculated from nominal concentrations.

Bibliographical reference: Kaiser, K.L.E., and Palabrica, V.S. (1991). *Water Poll. Res. J. Canada* **26**, 361-431.

4-FLUOROPHENOL

CAS RN: 371-41-5

Temperature: 15°C
Test parameter: EC50
Effect: Reduction in light output
Concentration: 19.5 mg/l
Exposure time: 30 min
Comment: Mean of three assays. Methanol (<10%) was used to prepare the stock solutions. EC50 values were calculated from nominal concentrations.

Bibliographical reference: Kaiser, K.L.E. (1987). In: *QSAR in Environmental Toxicology - II*, K.L.E. Kaiser (ed.), D. Reidel Publishing Company, Dordrecht, p. 169-188.

4-FLUOROPHENOXYACETIC ACID

CAS RN: 405-79-8

Temperature: 15°C
Test parameter: EC50
Effect: Reduction in light output

Concentration: 118 mg/l
Exposure time: 5 min

Concentration: 120 mg/l
Exposure time: 15 min

Concentration: 138 mg/l
Exposure time: 30 min

Comment: Mean of three assays. Methanol (<10%) was used to prepare the stock solutions. EC50 values were calculated from nominal concentrations.

Bibliographical reference: Kaiser, K.L.E., and Palabrica, V.S. (1991). *Water Poll. Res. J. Canada* **26**, 361-431.

α-FLUOROPHENYLACETIC ACID

CAS RN: 1578-63-8

Sample purity: 97%
Temperature: 15°C
Test parameter: EC50
Effect: Reduction in light output

Concentration: 190 mg/l
Exposure time: 5 min

Concentration: 208 mg/l
Exposure time: 15 min

Concentration: 239 mg/l
Exposure time: 30 min

Comment: Mean of three assays. Methanol (<10%) was used to prepare the stock solutions. EC50 values were calculated from nominal concentrations.

Bibliographical reference: Kaiser, K.L.E., and Palabrica, V.S. (1991). *Water Poll. Res. J. Canada* **26**, 361-431.

4-FLUOROPHENYLACETIC ACID

CAS RN: 405-50-5

Sample purity: 98%
Temperature: 15°C
Test parameter: EC50
Effect: Reduction in light output

Concentration: 48.7 mg/l

Exposure time: 5 min

Concentration: 51.0 mg/l
Exposure time: 15 min

Concentration: 56.0 mg/l
Exposure time: 30 min

Comment: Mean of three assays. Methanol (<10%) was used to prepare the stock solutions. EC50 values were calculated from nominal concentrations.

Bibliographical reference: Kaiser, K.L.E., and Palabrica, V.S. (1991). *Water Poll. Res. J. Canada* **26**, 361-431.

(4-FLUOROPHENYL)ACETONE

CAS RN: 459-03-0

Sample purity: 98%
Temperature: 15°C
Test parameter: EC50
Effect: Reduction in light output

Concentration: 20.1 mg/l
Exposure time: 5 min

Concentration: 19.2 mg/l
Exposure time: 15 min

Concentration: 18.7 mg/l
Exposure time: 30 min

Comment: Mean of three assays. Methanol (<10%) was used to prepare the stock solutions. EC50 values were calculated from nominal concentrations.

Bibliographical reference: Kaiser, K.L.E., and Palabrica, V.S. (1991). *Water Poll. Res. J. Canada* **26**, 361-431.

4-FLUOROPHENYLACETONITRILE

CAS RN: 459-22-3

Sample purity: 97%
Temperature: 15°C
Test parameter: EC50
Effect: Reduction in light output

Concentrations: 2.30 mg/l
2.09 mg/l
Exposure time: 5 min

Concentrations: 2.35 mg/l
2.70 mg/l
Exposure time: 15 min

Concentrations: 2.64 mg/l
4.37 mg/l
Exposure time: 30 min

Comment: The two series of tests were performed in triplicate. Methanol (<10%) was used to prepare the stock solutions. EC50 values were calculated from nominal concentrations.

Bibliographical reference: Kaiser, K.L.E., and Palabrica, V.S. (1991). *Water Poll. Res. J. Canada* **26**, 361-431.

DL-4-FLUORO-α-PHENYLGLYCINE

CAS RN: 7292-73-1

Sample purity: 97%
Temperature: 15°C
Test parameter: EC50
Effect: Reduction in light output

Concentration: 67.3 mg/l
Exposure time: 5 min

Concentration: 70.5 mg/l
Exposure time: 15 min

Concentration: 72.2 mg/l

Exposure time: 30 min

Comment: Mean of three assays. Methanol (<10%) was used to prepare the stock solutions. EC50 values were calculated from nominal concentrations.

Bibliographical reference: Kaiser, K.L.E., and Palabrica, V.S. (1991). *Water Poll. Res. J. Canada* **26**, 361-431.

4-FLUOROPHENYL ISOCYANATE

CAS RN: 1195-45-5

Sample purity: 99%
Temperature: 15°C
Test parameter: EC50
Effect: Reduction in light output

Concentration: 32.1 mg/l
Exposure time: 5 min

Concentration: 26.1 mg/l
Exposure time: 15 min

Concentration: 20.3 mg/l*
Exposure time: 30 min

Comment: Mean of three assays. Methanol (<10%) was used to prepare the stock solutions. EC50 values were calculated from nominal concentrations.

Bibliographical references: Kaiser, K.L.E., and Palabrica, V.S. (1991). *Water Poll. Res. J. Canada* **26**, 361-431.
* Kaiser, K.L.E., and Gough, K.M. (1989). In: *Aquatic Toxicology and Environmental Fate: Eleventh Volume, ASTM STP 1007*, G.W. Suter and M.A. Lewis (eds.), American Society for Testing and Materials, Philadelphia, p. 424-441.

4-FLUOROPHENYL ISOTHIOCYANATE

CAS RN: 1544-68-9

Sample purity: 98%
Temperature: 15°C
Test parameter: EC50
Effect: Reduction in light output

Concentration: 1.01 mg/l
Exposure time: 5 min

Concentration: 0.88 mg/l
Exposure time: 15 min

Concentration: 0.80 mg/l
Exposure time: 30 min

Comment: Mean of three assays. Methanol (<10%) was used to prepare the stock solutions. EC50 values were calculated from nominal concentrations.

Bibliographical reference: Kaiser, K.L.E., and Palabrica, V.S. (1991). *Water Poll. Res. J. Canada* **26**, 361-431.

4'-FLUOROPROPIOPHENONE

CAS RN: 456-03-1

Sample purity: 99%
Temperature: 15°C
Test parameter: EC50
Effect: Reduction in light output

Concentration: 14.5 mg/l
Exposure times: 5 and 15 min

Concentration: 15.2 mg/l
Exposure time: 30 min

Comment: Mean of three assays. Methanol (<10%) was used to prepare the stock solutions. EC50 values were calculated from nominal concentrations.

Bibliographical reference: Kaiser, K.L.E., and Palabrica, V.S. (1991). *Water Poll. Res. J. Canada* **26**, 361-431.

3-FLUOROPYRIDINE

CAS RN: 372-47-4

Temperature: 15°C
Test parameter: EC50
Effect: Reduction in light output

Concentration: 321 mg/l
Exposure time: 5 min

Concentration: 307 mg/l
Exposure times: 15 and 30 min

Comment: Mean of four assays. Methanol (<10%) was used to prepare the stock solutions. EC50 values were calculated from nominal concentrations.

Bibliographical reference: Kaiser, K.L.E., and Palabrica, V.S. (1993). National Water Research Institute, Burlington, Ontario, Canada, unpublished results.

4-FLUOROSTYRENE

CAS RN: 405-99-2

Sample purity: 97%
Temperature: 15°C
Test parameter: EC50
Effect: Reduction in light output

Concentration: 4.86 mg/l
Exposure time: 5 min

Concentration: 6.26 mg/l
Exposure time: 15 min

Concentration: 9.05 mg/l
Exposure time: 30 min

Comment: Mean of three assays. Methanol (<10%) was used to prepare the stock solutions. EC50 values were calculated from nominal concentrations.

Bibliographical reference: Kaiser, K.L.E., and Palabrica, V.S. (1991). *Water Poll. Res. J. Canada* **26**, 361-431.

4-FLUOROTOLUENE

CAS RN: 352-32-9

Sample purity: 97%
Temperature: 15°C
Test parameter: EC50
Effect: Reduction in light output

Concentration: 23.5 mg/l
Exposure time: 5 min

Concentration: 28.3 mg/l
Exposure time: 15 min

Concentration: 38.2 mg/l*
Exposure time: 30 min

Comment: Mean of four assays. Methanol (<10%) was used to prepare the stock solutions. EC50 values were calculated from nominal concentrations.

Bibliographical references: Kaiser, K.L.E., and Palabrica, V.S. (1991). *Water Poll. Res. J. Canada* **26**, 361-431.
* Kaiser, K.L.E., and Gough, K.M. (1989). In: *Aquatic Toxicology and Environmental Fate: Eleventh Volume, ASTM STP 1007*, G.W. Suter and M.A. Lewis (eds.), American Society for Testing and Materials, Philadelphia, p. 424-441.

5-FLUOROURACIL

CAS RN: 51-21-8
Synonym: 5-Fluoro-2,4(1*H*,3*H*)-pyrimidinedione

Sample purity: 99%
Temperature: 15°C
Test parameter: EC50
Effect: Reduction in light output

Concentration: 2310 mg/l
Exposure time: 5 min

Concentration: 2060 mg/l
Exposure time: 15 min

Concentration: 1840 mg/l
Exposure time: 30 min

Comment: Mean of three assays. Methanol (<10%) was used to prepare the stock solutions. EC50 values were calculated from nominal concentrations.

Bibliographical reference: Kaiser, K.L.E., and Palabrica, V.S. (1992). National Water Research Institute, Burlington, Ontario, Canada, unpublished results.

FONOFOS

CAS RN: 944-22-9
Synonym: *O*-Ethyl *S*-phenyl ethylphosphonodithioate

Temperature: 15 ± 0.1°C
Test parameter: EC50
Effect: Reduction in light output
Concentration: 5.2 mg/l
Exposure time: 5 min

Bibliographical reference: Somasundaram, L., Coats, J.R., Racke, K.D., and Stahr, H.M. (1990). *Bull. Environ. Contam. Toxicol.* **44**, 254-259.

FORMALDEHYDE

CAS RN: 50-00-0

Temperature: 15°C
Test parameter: EC50
Effect: Reduction in light output
Concentration: 3.0 mg/l
Exposure time: 5 min

Bibliographical reference: Bulich, A.A., Greene, M.W., and Isenberg, D.L. (1981). In: *Aquatic Toxicology and Hazard Assessment: Fourth Conference, ASTM STP 737*, D.R. Branson and K.L. Dickson (eds.), American Society for Testing and Materials, Philadelphia, p. 338-347.

Temperature: 15 ± 0.1°C
Test parameter: EC50
Effect: Reduction in light output
Concentration: 8.7 mg/l
Exposure time: 5 min

Bibliographical reference: Chang, J.C., Taylor, P.B., and Leach, F.R. (1981). *Bull. Environ. Contam. Toxicol.* **26**, 150-156.

Temperature: 15 ± 0.1°C
Test parameter: EC50
Effect: Reduction in light output
Concentration: 904.17 mg/l
Exposure time: 5 min
Comment: Test was performed on *Photobacterium phosphoreum* NZ11D obtained from the Scripps Institute of Oceanography (La Jolla, CA).

Bibliographical reference: McFeters, G.A., Bond, P.J., Olson, S.B., and Tchan, Y.T. (1983). *Water Res.* **17**, 1757-1762.

Sample purity: Reagent grade
Test parameter: EC50
Effect: Reduction in light output

Concentration: 10 mg/l (7-15)
Exposure time: 5 min

Concentration: 8.5 mg/l (6.5-11)
Exposure times: 15 and 30 min

Bibliographical reference: Elnabarawy, M.T., Robideau, R.R., and Beach, S.A. (1988). *Tox. Assess.* **3**, 361-370.

Temperature: 15°C
Test parameter: EC50

Effect: Reduction in light output
Concentration: 10.1 mg/l
Exposure time: 5 min
Comment: The test was performed in duplicate. The pH of the chemical solutions was not adjusted.

Bibliographical reference: Awong, J., Bitton, G., Koopman, B., and Morel, J.L. (1989). *Bull. Environ. Contam. Toxicol.* **43**, 118-122.

Sample purity: 37% (w/w)
Temperature: 15°C
Test parameter: EC50
Effect: Reduction in light output

Concentration: 9.0 ± 0.69 mg/l
Exposure time: 5 min

Concentration: 7.26 ± 0.54 mg/l
Exposure time: 15 min

Concentration: 6.81 ± 0.42 mg/l
Exposure time: 25 min

Comment: Mean of five assays. The values were converted to mg/l from the original data expressed in µM and a rounded molecular weight of 30 given by the authors. Phenol solution was used for quality control/quality assurance. The 5-min EC50 value was 18.2 mg/l. The EC50 value at 15 min was 20.7 mg/l with a relative error of <5%.

Bibliographical reference: Chou, C.C., and Que Hee, S.S. (1992). *Ecotoxicol. Environ. Safety* **23**, 355-363.

FORMANILIDE

CAS RN: 103-70-8

Sample purity: 98%
Temperature: 15°C
Test parameter: EC50
Effect: Reduction in light output

Concentration: 4.30 mg/l
Exposure time: 5 min

Concentration: 4.40 mg/l
Exposure time: 15 min

Concentration: 4.71 mg/l*
Exposure time: 30 min

Comment: Mean of three assays. Methanol (<10%) was used to prepare the stock solutions. EC50 values were calculated from nominal concentrations.

Bibliographical references: Kaiser, K.L.E., and Palabrica, V.S. (1991). *Water Poll. Res. J. Canada* **26**, 361-431.
* Kaiser, K.L.E., Palabrica, V.S., and Ribo, J.M. (1987). In: *QSAR in Environmental Toxicology - II*, K.L.E. Kaiser (ed.), D. Reidel Publishing Company, Dordrecht, p. 153-168.

FORMIC ACID

CAS RN: 64-18-6

Sample purity: 96%
Temperature: 15°C
Test parameter: EC50
Effect: Reduction in light output

Concentration: 7.91 ± 0.22 mg/l
Exposure time: 5 min

Concentration: 7.96 ± 0.29 mg/l
Exposure time: 15 min

Concentration: 7.91 ± 0.35 mg/l
Exposure time: 25 min

Comment: Mean of six assays. The values were converted to mg/l from the original data expressed in μM and a rounded molecular weight of 46 given by the authors. Phenol solution was used for quality control/quality assurance. The 5-min EC50 value was 18.2 mg/l. The EC50 value at 15 min was 20.7 mg/l with a relative error of <5%.

Bibliographical reference: Chou, C.C., and Que Hee, S.S. (1992). *Ecotoxicol. Environ. Safety* **23**, 355-363.

FURAN

CAS RN: 110-00-9

Sample purity: >99%
Temperature: 15°C
Test parameter: EC50
Effect: Reduction in light output

Concentration: 171 mg/l
Exposure time: 5 min

Concentration: 142 mg/l
Exposure time: 15 min

Concentration: 103 mg/l
Exposure time: 30 min

Comment: Mean of three assays. Methanol (<10%) was used to prepare the stock solutions. EC50 values were calculated from nominal concentrations.

Bibliographical reference: Kaiser, K.L.E., and Palabrica, V.S. (1991). *Water Poll. Res. J. Canada* **26**, 361-431.

GLUCOSE

CAS RN: 492-62-6

Temperature: 15°C
Test parameter: EC50
Effect: Reduction in light output
Concentration: >100000 mg/l
Exposure time: 15 min

Bibliographical reference: Bulich, A.A., Tung, K.K., and Scheibner, G. (1990). *J. Biolumin. Chemilumin.* **5**, 71-77.

GLUTARALDEHYDE

CAS RN: 111-30-8
Synonym: Glutaric dialdehyde

Temperature: 15°C
Test parameter: EC50
Effect: Reduction in light output
Concentration: 75.7 mg/l
Exposure time: 5 min

Bibliographical reference: Microtox® Application Notes (1982). Beckman Instruments, Inc., Carlsbad, California.

Temperature: 15°C
Test parameter: EC50
Effect: Reduction in light output
Concentration: 3.2 mg/l
Exposure time: 15 min

Bibliographical reference: Bulich, A.A., Tung, K.K., and Scheibner, G. (1990). *J. Biolumin. Chemilumin.* **5**, 71-77.

GLYCEROL

CAS RN: 56-81-5

Temperature: 15°C
Test parameter: EC50
Effect: Reduction in light output
Concentration: 230000 mg/l
Exposure time: 15 min

Bibliographical reference: Bulich, A.A., Tung, K.K., and Scheibner, G. (1990). *J. Biolumin. Chemilumin.* **5**, 71-77.

GLYOXAL

CAS RN: 107-22-2

Sample purity: 40% in water
Temperature: 15°C
Test parameter: EC50
Effect: Reduction in light output

Concentration: 754 ± 55.1 mg/l

Exposure time: 5 min

Concentration: 554 ± 33.6 mg/l
Exposure time: 15 min

Concentration: 429 ± 25.5 mg/l
Exposure time: 25 min

Comment: Mean of five assays. The values were converted to mg/l from the original data expressed in μM and a rounded molecular weight of 58 given by the authors. Phenol solution was used for quality control/quality assurance. The 5-min EC50 value was 18.2 mg/l. The EC50 value at 15 min was 20.7 mg/l with a relative error of <5%.

Bibliographical reference: Chou, C.C., and Que Hee, S.S. (1992). *Ecotoxicol. Environ. Safety* **23**, 355-363.

GLYOXYLIC ACID

CAS RN: 298-12-4

Sample purity: 50% aqueous solution
Temperature: 15°C
Test parameter: EC50
Effect: Reduction in light output

Concentration: 11.2 ± 0.16 mg/l
Exposure time: 5 min

Concentration: 11.1 ± 0.25 mg/l
Exposure time: 15 min

Concentration: 11.2 ± 0.29 mg/l
Exposure time: 25 min

Comment: Mean of five assays. The values were converted to mg/l from the original data expressed in μM and a rounded molecular weight of 74 given by the authors. Phenol solution was used for quality control/quality assurance. The 5-min EC50 value was 18.2 mg/l. The EC50 value at 15 min was 20.7 mg/l with a relative error of <5%.

Bibliographical reference: Chou, C.C., and Que Hee, S.S. (1992). *Ecotoxicol. Environ. Safety* **23**, 355-363.

GLYPHOSATE

CAS RN: 1071-83-6
Synonym: *N*-(Phosphonomethyl)glycine

Temperature: 15 ± 0.1°C
Test parameter: EC50
Effect: Reduction in light output
Concentration: 7.7 mg/l
Exposure time: 5 min

Bibliographical reference: Chang, J.C., Taylor, P.B., and Leach, F.R. (1981). *Bull. Environ. Contam. Toxicol.* **26**, 150-156.

GROTAN

CAS RN: 4719-04-4

Sample purity: Hexahydro-1,3,5-tris(2-hydroxyethyl)-*s*-triazine 78%
Temperature: 15°C
Test parameter: EC50
Effect: Reduction in light output

Concentration: 43 mg/l
Exposure time: 5 min

Concentration: 29 mg/l
Exposure time: 15 min

Comment: Chemical was tested in duplicate at 6.25, 12.5, 25, and 50 mg/l. Verification that the controls were valid was accomplished by making sure that the difference between the duplicate blank ratios (5 minute light reading divided by the time 0 reading) was not greater than 0.02, otherwise the test was not continued. There was a rapid rise in the pH from 7.0 at 12.5 mg/l to 9.0 at 25 mg/l.

Bibliographical reference: Mallak, F.P., and Brunker, R.L. (1984). In: *Toxicity Screening Procedures Using Bacterial Systems*, D. Liu and B.J. Dutka (eds.), Marcel Dekker, New York, p. 65-76.

GROTAN HDII

CAS RN: 83245-20-9

Sample purity: 39% 2-Chloro-*N*-(hydroxymethyl) acetamide, 41% sodium tetraborate, and 0.39% potassium iodide
Temperature: 15°C
Test parameter: EC50
Effect: Reduction in light output

Concentration: 147 mg/l
Exposure time: 5 min

Concentration: 80 mg/l
Exposure time: 15 min

Comment: Chemical was tested in duplicate at 20, 80, 100, and 160 mg/l. Verification that the controls were valid was accomplished by making sure that the difference between the duplicate blank ratios (5 minute light reading divided by the time 0 reading) was not greater than 0.02, otherwise the test was not continued. There was a rapid rise in the pH from 7.1 at 20 mg/l to 8.5 at 160 mg/l.

Bibliographical reference: Mallak, F.P., and Brunker, R.L. (1984). In: *Toxicity Screening Procedures Using Bacterial Systems*, D. Liu and B.J. Dutka (eds.), Marcel Dekker, New York, p. 65-76.

HEPTALDEHYDE

CAS RN: 111-71-7
Synonym: Heptanal

Sample purity: 95%
Temperature: 15°C
Test parameter: EC50
Effect: Reduction in light output

Concentration: 22.1 ± 0.31 mg/l
Exposure time: 5 min

Concentration: 13.5 ± 1.11 mg/l
Exposure time: 15 min

Concentration: 11.5 ± 0.95 mg/l

Exposure time: 25 min

Comment: Mean of six assays. The values were converted to mg/l from the original data expressed in μM and a rounded molecular weight of 114 given by the authors. The EC50 values were determined in 0.45% (v/v) methanol in 2% (w/w) sodium chloride solution. The 5-min EC50 value for methanol was 43000 mg/l. Phenol solution was used for quality control/quality assurance. The 5-min EC50 value was 18.2 mg/l. The EC50 value at 15 min was 20.7 mg/l with a relative error of <5%.

Bibliographical reference: Chou, C.C., and Que Hee, S.S. (1992). *Ecotoxicol. Environ. Safety* **23**, 355-363.

HEPTANOIC ACID

CAS RN: 111-14-8

Sample purity: 99%
Temperature: 15°C
Test parameter: EC50
Effect: Reduction in light output

Concentration: 16.1 ± 0.42 mg/l
Exposure time: 5 min

Concentration: 17.0 ± 0.33 mg/l
Exposure time: 15 min

Concentration: 17.3 ± 0.36 mg/l
Exposure time: 25 min

Comment: Mean of five assays. The values were converted to mg/l from the original data expressed in μM and a rounded molecular weight of 130 given by the authors. The EC50 values were determined in 0.45% (v/v) methanol in 2% (w/w) sodium chloride solution. The 5-min EC50 value for methanol was 43000 mg/l. Phenol solution was used for quality control/quality assurance. The 5-min EC50 value was 18.2 mg/l. The EC50 value at 15 min was 20.7 mg/l with a relative error of <5%.

Bibliographical reference: Chou, C.C., and Que Hee, S.S. (1992). *Ecotoxicol. Environ. Safety* **23**, 355-363.

1-HEPTANOL

CAS RN: 111-70-6

Sample purity: >98%
Temperature: 15 ± 0.1°C
Test parameter: EC50
Effect: Reduction in light output

Concentrations: 14.5 mg/l
15.0 mg/l
Exposure time: 5 min

Concentrations: 19.9 mg/l
20.7 mg/l
Exposure time: 15 min

Comment: Test was performed in duplicate. Toxicity values were based on nominal test concentrations.

Bibliographical reference: de Zwart, D., and Slooff, W. (1983). *Aquat. Toxicol.* **4**, 129-138.

Sample purity: >98%
Temperature: 15 ± 0.1°C
Test parameter: EC10
Effect: Reduction in light output

Concentration: 2.25 mg/l
Exposure time: 5 min

Concentration: 3.8 mg/l
Exposure time: 15 min

Comment: Test was performed in duplicate. Toxicity values were based on nominal test concentrations.

Bibliographical reference: de Zwart, D., and Slooff, W. (1983). *Aquat. Toxicol.* **4**, 129-138.

Temperature: 15°C
Test parameter: EC50
Effect: Reduction in light output
Concentration: 9.89 mg/l

Exposure time: 15 min

Bibliographical reference: Hermens, J., Busser, F., Leeuwangh, P., and Musch, A. (1985). *Ecotoxicol. Environ. Safety* **9**, 17-25.

HEPTYLAMINE

CAS RN: 111-68-2

Sample purity: 99+%
Temperature: 15°C
Test parameter: EC50
Effect: Reduction in light output

Concentration: 27 mg/l
Exposure time: 5 min

Concentration: 24 mg/l*
Exposure time: 15 min

Bibliographical references: Cronin, M.T.D., Dearden, J.C., and Dobbs, A.J. (1991). *Sci. Total Environ.* **109/110**, 431-439.
* Cronin, M.T.D. (1990). Thesis, Liverpool Polytechnic, Liverpool, UK.

HEXACHLOROACETONE

CAS RN: 116-16-5

Temperature: 15°C
Test parameter: EC50
Effect: Reduction in light output
Concentration: 46 mg/l
Exposure time: 5 min

Bibliographical reference: Microtox® Application Notes (1982). Beckman Instruments, Inc., Carlsbad, California.

HEXACHLOROETHANE

CAS RN: 67-72-1
Synonym: Perchloroethane

Test parameter: EC50
Effect: Reduction in light output
Concentration: 0.14 mg/l
Exposure time: 5 min
Comment: Concentrations in the test were measured.

Bibliographical reference: Curtis, C., Lima, A., Lozano, S.J., and Veith, G.D. (1982). In: *Aquatic Toxicology and Hazard Assessment: Fifth Conference, ASTM STP 766*, J.G. Pearson, R.B. Foster, and W.E. Bishop (eds.), American Society for Testing and Materials, Philadelphia, p. 170-178.

Temperature: 15°C
Test parameter: EC50
Effect: Reduction in light output
Concentration: 8.3 mg/l (7.1-9.7)
Exposure time: 15 min
Comment: Single batches of Microtox® reagent were used for less than 2 h before being discarded. Four concentrations were tested. Hexachloroethane was first dissolved in ethanol (0.01%). Concentrations were unmeasured.

Bibliographical reference: Nacci, D., Jackim, E., and Walsh, R. (1986). *Environ. Toxicol. Chem.* **5**, 521-525.

Temperature: 15°C
Test parameter: EC50
Effect: Reduction in light output

Concentration: 0.45 mg/l
Exposure time: 5 min

Concentration: 0.21 mg/l
Exposure time: 30 min

Bibliographical reference: Speece, R. (1987). Drexel University, Philadelphia, USA, private communication.

Temperature: 15 ± 0.3°C
Test parameter: EC50
Effect: Reduction in light output
Concentration: 0.31 ± 0.08 mg/l
Exposure time: 5 min
Comment: Result derived from the average of 16 assays.

Bibliographical reference: Birkholz, D.A., Coutts, R.T., Hrudey, S.E., Danell, R.W., and Lockhart, W.L. (1990). *Water Res.* **24**, 67-73.

Sample purity: 98%
Temperature: 15°C
Test parameter: EC50
Effect: Reduction in light output

Concentration: 0.71 mg/l
Exposure time: 5 min

Concentration: 0.38 mg/l
Exposure time: 15 min

Concentration: 0.34 mg/l
Exposure time: 30 min

Comment: Mean of three assays. Methanol (<10%) was used to prepare the stock solutions. EC50 values were calculated from nominal concentrations.

Bibliographical reference: Kaiser, K.L.E., and Palabrica, V.S. (1991). *Water Poll. Res. J. Canada* **26**, 361-431.

HEXACHLOROPHENE

CAS RN: 70-30-4

Sample purity: 99%
Temperature: 15°C
Test parameter: EC50
Effect: Reduction in light output

Concentration: 1.15 mg/l
Exposure time: 5 min

Concentration: 0.42 mg/l
Exposure time: 15 min

Concentration: 0.35 mg/l
Exposure time: 30 min

Comment: Mean of three assays. Methanol (<10%) was used to prepare the stock solutions. EC50 values were calculated from nominal concentrations.

Bibliographical reference: Kaiser, K.L.E., and Palabrica, V.S. (1992). National Water Research Institute, Burlington, Ontario, Canada, unpublished results.

HEXANAL

CAS RN: 66-25-1

Sample purity: 99%
Temperature: 15°C
Test parameter: EC50
Effect: Reduction in light output

Concentration: 30.2 mg/l
Exposure time: 5 min

Concentration: 19.1 mg/l
Exposure time: 15 min

Concentration: 16.6 mg/l
Exposure time: 30 min

Comment: Mean of three assays. Methanol (<10%) was used to prepare the stock solutions. EC50 values were calculated from nominal concentrations.

Bibliographical reference: Kaiser, K.L.E., and Palabrica, V.S. (1992). National Water Research Institute, Burlington, Ontario, Canada, unpublished results.

Sample purity: 99%
Temperature: 15°C
Test parameter: EC50

Effect: Reduction in light output

Concentration: 32 mg/l
Exposure time: 5 min

Concentration: 22 mg/l
Exposure time: 15 min

Bibliographical reference: Cronin, M.T.D. (1993). Liverpool John Moores University, UK, private communication.

n-HEXANE

CAS RN: 110-54-3

Sample purity: 99+%
Temperature: 15°C
Test parameter: EC50
Effect: Reduction in light output
Concentration: 104 mg/l
Exposure times: 5, 15, and 30 min
Comment: Mean of four assays. Methanol (<10%) was used to prepare the stock solutions. EC50 values were calculated from nominal concentrations.

Bibliographical reference: Kaiser, K.L.E., and Palabrica, V.S. (1992). National Water Research Institute, Burlington, Ontario, Canada, unpublished results.

1,6-HEXANEDIOL

CAS RN: 629-11-8

Sample purity: >99%
Temperature: 15°C
Test parameter: EC50
Effect: Reduction in light output

Concentration: 167 mg/l
Exposure time: 5 min

Concentration: 163 mg/l

Exposure time: 15 min

Concentration: 205 mg/l
Exposure time: 30 min

Comment: Mean of three assays. Methanol (<10%) was used to prepare the stock solutions. EC50 values were calculated from nominal concentrations.

Bibliographical reference: Kaiser, K.L.E., and Palabrica, V.S. (1991). *Water Poll. Res. J. Canada* **26**, 361-431.

2,5-HEXANEDIOL

CAS RN: 2935-44-6

Sample purity: 99%
Temperature: 15°C
Test parameter: EC50
Effect: Reduction in light output

Concentration: 247 mg/l
Exposure time: 5 min

Concentration: 241 mg/l
Exposure time: 15 min

Concentration: 271 mg/l
Exposure time: 30 min

Comment: Mean of three assays. Methanol (<10%) was used to prepare the stock solutions. EC50 values were calculated from nominal concentrations.

Bibliographical reference: Kaiser, K.L.E., and Palabrica, V.S. (1991). *Water Poll. Res. J. Canada* **26**, 361-431.

1-HEXANOL

CAS RN: 111-27-3

Test parameter: EC50
Effect: Reduction in light output

Concentrations: 40.4 mg/l
42.7 mg/l
Exposure time: 5 min
Comment: The EC50 values were calculated from nominal concentrations.

Bibliographical reference: Curtis, C., Lima, A., Lozano, S.J., and Veith, G.D. (1982). In: *Aquatic Toxicology and Hazard Assessment: Fifth Conference, ASTM STP 766*, J.G. Pearson, R.B. Foster, and W.E. Bishop (eds.), American Society for Testing and Materials, Philadelphia, p. 170-178.

Temperature: 15°C
Test parameter: EC50
Effect: Reduction in light output
Concentration: 67.5 mg/l
Exposure time: 15 min

Bibliographical reference: Hermens, J., Busser, F., Leeuwangh, P., and Musch, A. (1985). *Ecotoxicol. Environ. Safety* **9**, 17-25.

Temperature: 15°C
Test parameter: EC50
Effect: Reduction in light output

Concentration: 21.5 mg/l
Exposure time: 5 min

Concentration: 27.6 mg/l
Exposure time: 30 min

Bibliographical reference: Speece, R. (1987). Drexel University, Philadelphia, USA, private communication.

HEXYL ACETATE

CAS RN: 142-92-7

Sample purity: 99%
Temperature: 15°C
Test parameter: EC50
Effect: Reduction in light output

Concentration: 9.31 mg/l
Exposure time: 5 min

Bibliographical reference: Cronin, M.T.D., Dearden, J.C., and Dobbs, A.J. (1991). *Sci. Total Environ.* **109/110**, 431-439.

n-HEXYLAMINE

CAS RN: 111-26-2

Temperature: 15°C
Test parameter: EC50
Effect: Reduction in light output

Concentration: 16.8 mg/l
Exposure time: 5 min

Concentration: 15.0 mg/l
Exposure time: 15 min

Concentration: 14.6 mg/l
Exposure time: 30 min

Comment: Mean of three assays. Methanol (<10%) was used to prepare the stock solutions. EC50 values were calculated from nominal concentrations.

Bibliographical reference: Kaiser, K.L.E., and Palabrica, V.S. (1991). *Water Poll. Res. J. Canada* **26**, 361-431.

1-HEXYNE

CAS RN: 693-02-7

Sample purity: 99%
Temperature: 15°C
Test parameter: EC50
Effect: Reduction in light output

Concentration: 164 mg/l
Exposure time: 5 min

Concentration: 172 mg/l
Exposure time: 15 min

Concentration: 206 mg/l
Exposure time: 30 min

Comment: Mean of three assays. Methanol (<10%) was used to prepare the stock solutions. EC50 values were calculated from nominal concentrations.

Bibliographical reference: Kaiser, K.L.E., and Palabrica, V.S. (1991). *Water Poll. Res. J. Canada* **26**, 361-431.

2-HEXYNE

CAS RN: 764-35-2

Sample purity: 99%
Temperature: 15°C
Test parameter: EC50
Effect: Reduction in light output

Concentration: 26.6 mg/l
Exposure time: 5 min

Concentration: 14.9 mg/l
Exposure time: 15 min

Concentration: 11.6 mg/l
Exposure time: 30 min

Comment: Mean of three assays. Methanol (<10%) was used to prepare the stock solutions. EC50 values were calculated from nominal concentrations.

Bibliographical reference: Kaiser, K.L.E., and Palabrica, V.S. (1991). *Water Poll. Res. J. Canada* **26**, 361-431.

HYDRAZINE

CAS RN: 302-01-2

Temperature: 15°C
Test parameter: EC50
Effect: Reduction in light output

Concentration: 0.02 mg/l
Exposure times: 5 and 10 min

Concentration: 0.01 mg/l
Exposure times: 15 and 20 min

Comment: Test was performed at least in triplicate. Chemical was solubilized in DMSO.

Bibliographical reference: Yates, I.E. (1985). *J. Microbiol. Meth.* **3**, 171-180.

4-HYDRAZINOBENZOIC ACID

CAS RN: 619-67-0

Sample purity: 97%
Temperature: 15°C
Test parameter: EC50
Effect: Reduction in light output

Concentration: 108 mg/l
Exposure time: 5 min

Concentration: 69.5 mg/l
Exposure time: 15 min

Concentration: 50.4 mg/l
Exposure time: 30 min

Comment: Mean of three assays. Methanol (<10%) was used to prepare the stock solutions. EC50 values were calculated from nominal concentrations.

Bibliographical reference: Kaiser, K.L.E., and Palabrica, V.S. (1991). *Water Poll. Res. J. Canada* **26**, 361-431.

HYDRAZOBENZENE

CAS RN: 122-66-7
Synonym: 1,2-Diphenylhydrazine

Sample purity: 97%
Temperature: 15°C
Test parameter: EC50
Effect: Reduction in light output

Concentration: 0.75 mg/l
Exposure time: 5 min

Concentration: 0.84 mg/l
Exposure time: 15 min

Concentration: 0.99 mg/l*
Exposure time: 30 min

Comment: Mean of three assays. Methanol (<10%) was used to prepare the stock solutions. EC50 values were calculated from nominal concentrations.

Bibliographical references: Kaiser, K.L.E., and Palabrica, V.S. (1991). *Water Poll. Res. J. Canada* **26**, 361-431.
* Kaiser, K.L.E., Palabrica, V.S., and Ribo, J.M. (1987). In: *QSAR in Environmental Toxicology - II*, K.L.E. Kaiser (ed.), D. Reidel Publishing Company, Dordrecht, p. 153-168.

HYDROGEN CYANIDE

CAS RN: 74-90-8

Temperature: 15°C
Test parameter: EC50
Effect: Reduction in light output
Concentration: 8.5 mg/l
Exposure time: 5 min

Bibliographical reference: Bulich, A.A., Greene, M.W., and Isenberg, D.L. (1981). In: *Aquatic Toxicology and Hazard Assessment: Fourth Conference, ASTM STP 737*, D.R. Branson and K.L. Dickson, (eds.), American Society for Testing and Materials, Philadelphia, p. 338-347.

HYDROGEN PEROXIDE

CAS RN: 7722-84-1

Temperature: 15°C
Test parameter: EC50
Effect: Reduction in light output
Concentration: 16 mg/l
Exposure time: 15 min

Bibliographical reference: Bulich, A.A., Tung, K.K., and Scheibner, G. (1990). *J. Biolumin. Chemilumin.* **5**, 71-77.

HYDROQUINONE

CAS RN: 123-31-9

Temperature: 15°C
Test parameter: EC50
Effect: Reduction in light output
Concentration: 0.079 mg/l
Exposure time: 5 min

Bibliographical reference: Lebsack, M.E., Anderson, A.D., DeGraeve, G.M., and Bergman, H.L. (1981). In: *Aquatic Toxicology and Hazard Assessment: Fourth Conference, ASTM STP 737*, D.R. Branson and K.L. Dickson, (eds.), American Society for Testing and Materials, Philadelphia, p. 348-356.

Temperature: 15°C
Test parameter: EC50
Effect: Reduction in light output

Concentration: 0.042 mg/l
Exposure time: 5 min

Concentration: 0.038 mg/l
Exposure times: 15 and 30 min

Comment: Mean of three assays. Methanol (<10%) was used to prepare the stock solutions. EC50 values were calculated from nominal concentrations.

Bibliographical reference: Ribo, J.M., and Kaiser, K.L.E. (1983). *Chemosphere* **12**, 1421-1442.

Temperature: 15°C

Test parameter: EC10
Effect: Reduction in light output
Concentration: 0.022 mg/l
Exposure time: 30 min

Test parameter: EC50
Effect: Reduction in light output
Concentration: 0.072 mg/l
Exposure time: 30 min

Test parameter: EC90
Effect: Reduction in light output
Concentration: 0.210 mg/l
Exposure time: 30 min

Comment: Toxicity values were calculated from nominal concentrations.

Bibliographical reference: Devillers, J., Boule, P., Vasseur, P., Prevot, P., Steiman, R., Seigle-Murandi, F., Benoit-Guyod, J.L., Nendza, M., Grioni, C., Dive, D., and Chambon, P. (1990). *Ecotoxicol. Environ. Safety* **19**, 327-354.

4'-HYDROXYACETOPHENONE

CAS RN: 99-93-4

Temperature: 15°C
Test parameter: EC50
Effect: Reduction in light output

Concentration: 4.51 mg/l
Exposure time: 5 min

Concentration: 4.94 mg/l
Exposure time: 15 min

Concentration: 4.41 mg/l

Exposure time: 30 min

Comment: Mean of three assays. Methanol (<10%) was used to prepare the stock solutions. EC50 values were calculated from nominal concentrations.

Bibliographical reference: Ribo, J.M., and Kaiser, K.L.E. (1983). *Chemosphere* **12**, 1421-1442.

4-HYDROXYBENZALDEHYDE

CAS RN: 123-08-0

Temperature: 15°C
Test parameter: EC50
Effect: Reduction in light output

Concentration: 5.98 mg/l
Exposure time: 5 min

Concentration: 7.71 mg/l
Exposure time: 15 min

Concentration: 8.85 mg/l
Exposure time: 30 min

Comment: Mean of three assays. Methanol (<10%) was used to prepare the stock solutions. EC50 values were calculated from nominal concentrations.

Bibliographical reference: Ribo, J.M., and Kaiser, K.L.E. (1983). *Chemosphere* **12**, 1421-1442.

4-HYDROXYBENZAMIDE

CAS RN: 619-57-8

Sample purity: 98%
Temperature: 15°C
Test parameter: EC50
Effect: Reduction in light output

Concentration: 64.1 mg/l
Exposure time: 5 min

Concentration: 68.7 mg/l
Exposure time: 15 min

Concentration: 78.9 mg/l*
Exposure time: 30 min

Comment: Mean of three assays. Methanol (<10%) was used to prepare the stock solutions. EC50 values were calculated from nominal concentrations.

Bibliographical references: Kaiser, K.L.E., and Palabrica, V.S. (1991). *Water Poll. Res. J. Canada* **26**, 361-431.
* Kaiser, K.L.E. (1987). In: *QSAR in Environmental Toxicology - II*, K.L.E. Kaiser (ed.), D. Reidel Publishing Company, Dordrecht, p. 169-188.

2-HYDROXYBENZIMIDAZOLE

CAS RN: 615-16-7

Sample purity: 97%
Temperature: 15°C
Test parameter: EC50
Effect: Reduction in light output

Concentration: 28.0 mg/l
Exposure time: 5 min

Concentration: 27.4 mg/l
Exposure time: 15 min

Concentration: 30.7 mg/l
Exposure time: 30 min

Comment: Mean of four assays. Methanol (<10%) was used to prepare the stock solutions. EC50 values were calculated from nominal concentrations.

Bibliographical reference: Kaiser, K.L.E., and Palabrica, V.S. (1991). *Water Poll. Res. J. Canada* **26**, 361-431.

4-HYDROXYBENZOIC ACID

CAS RN: 99-96-7

Temperature: 15°C
Test parameter: EC50
Effect: Reduction in light output

Concentration: 12.3 mg/l
Exposure time: 5 min

Concentration: 11.8 mg/l
Exposure time: 15 min

Concentration: 10.5 mg/l
Exposure time: 30 min

Comment: Mean of three assays. Methanol (<10%) was used to prepare the stock solutions. EC50 values were calculated from nominal concentrations.

Bibliographical reference: Ribo, J.M., and Kaiser, K.L.E. (1983). *Chemosphere* **12**, 1421-1442.

4-HYDROXYBENZONITRILE

CAS RN: 767-00-0

Temperature: 15°C
Test parameter: EC50
Effect: Reduction in light output

Concentration: 0.69 mg/l
Exposure time: 5 min

Concentration: 0.84 mg/l
Exposure time: 15 min

Concentration: 0.79 mg/l
Exposure time: 30 min

Comment: Mean of three assays. Methanol (<10%) was used to prepare the stock solutions. EC50 values were calculated from nominal concentrations.

Bibliographical reference: Ribo, J.M., and Kaiser, K.L.E. (1983). *Chemosphere* **12**, 1421-1442.

4-HYDROXYBENZOPHENONE

CAS RN: 1137-42-4

Temperature: 15°C
Test parameter: EC50
Effect: Reduction in light output

Concentration: 8.65 mg/l
Exposure time: 5 min

Concentration: 7.89 mg/l
Exposure times: 15 and 30 min

Comment: Mean of three assays. Methanol (<10%) was used to prepare the stock solutions. EC50 values were calculated from nominal concentrations.

Bibliographical reference: Ribo, J.M., and Kaiser, K.L.E. (1983). *Chemosphere* **12**, 1421-1442.

4-HYDROXYBENZOTRIFLUORIDE

CAS RN: 402-45-9

Temperature: 15°C
Test parameter: EC50
Effect: Reduction in light output

Concentration: 0.59 mg/l
Exposure time: 5 min

Concentration: 0.76 mg/l
Exposure time: 15 min

Concentration: 0.79 mg/l
Exposure time: 30 min

Comment: Mean of three assays. Methanol (<10%) was used to prepare the stock solutions. EC50 values were calculated from nominal concentrations.

Bibliographical reference: Ribo, J.M., and Kaiser, K.L.E. (1983). *Chemosphere* **12**, 1421-1442.

Temperature: 15°C
Test parameter: EC50
Effect: Reduction in light output
Concentration: 0.32 mg/l
Exposure time: 30 min
Comment: Mean of three assays. Methanol (<10%) was used to prepare the stock solutions. EC50 values were calculated from nominal concentrations.

Bibliographical reference: Kaiser, K.L.E. (1987). In: *QSAR in Environmental Toxicology - II*, K.L.E. Kaiser (ed.), D. Reidel Publishing Company, Dordrecht, p. 169-188.

4-HYDROXYBENZYL ALCOHOL

CAS RN: 623-05-2

Temperature: 15°C
Test parameter: EC50
Effect: Reduction in light output

Concentration: 5.55 mg/l
Exposure time: 5 min

Concentration: 5.18 mg/l
Exposure time: 15 min

Concentration: 5.06 mg/l
Exposure time: 30 min

Comment: Mean of three assays. Methanol (<10%) was used to prepare the stock solutions. EC50 values were calculated from nominal concentrations.

Bibliographical reference: Ribo, J.M., and Kaiser, K.L.E. (1983). *Chemosphere* **12**, 1421-1442.

Temperature: 15°C
Test parameter: EC50
Effect: Reduction in light output
Concentration: 6.22 mg/l
Exposure time: 30 min
Comment: Mean of three assays. Methanol (<10%) was used to prepare the stock solutions. EC50 values were calculated from nominal concentrations.

Bibliographical reference: Kaiser, K.L.E. (1987). In: *QSAR in Environmental Toxicology - II*, K.L.E. Kaiser (ed.), D. Reidel Publishing Company, Dordrecht, p. 169-188.

4-HYDROXYBIPHENYL

CAS RN: 92-69-3

Temperature: 15°C
Test parameter: EC50
Effect: Reduction in light output

Concentration: 1.74 mg/l
Exposure time: 5 min

Concentration: 2.30 mg/l
Exposure time: 15 min

Concentration: 2.35 mg/l
Exposure time: 30 min

Comment: Mean of three assays. Methanol (<10%) was used to prepare the stock solutions. EC50 values were calculated from nominal concentrations.

Bibliographical reference: Ribo, J.M., and Kaiser, K.L.E. (1983). *Chemosphere* **12**, 1421-1442.

2-HYDROXYDIBENZOFURAN

CAS RN: 86-77-1

Sample purity: 95%

Temperature: 15°C
Test parameter: EC50
Effect: Reduction in light output

Concentration: 0.65 mg/l
Exposure time: 5 min

Concentration: 0.68 mg/l
Exposure time: 15 min

Concentration: 0.77 mg/l
Exposure time: 30 min

Comment: Mean of four assays. Methanol (<10%) was used to prepare the stock solutions. EC50 values were calculated from nominal concentrations.

Bibliographical reference: Kaiser, K.L.E., and Palabrica, V.S. (1991). *Water Poll. Res. J. Canada* **26**, 361-431.

2-(2-HYDROXYETHYL)PYRIDINE

CAS RN: 103-74-2
Synonym: 2-Pyridineethanol

Sample purity: 98%
Temperature: 15°C
Test parameter: EC50
Effect: Reduction in light output

Concentration: 525 mg/l
Exposure time: 5 min

Concentration: 538 mg/l
Exposure times: 15 and 30 min

Comment: Mean of three assays. Methanol (<10%) was used to prepare the stock solutions. EC50 values were calculated from nominal concentrations.

Bibliographical reference: Kaiser, K.L.E., and Palabrica, V.S. (1991). *Water Poll. Res. J. Canada* **26**, 361-431.

2-HYDROXYFLUORENE

CAS RN: 2443-58-5
Synonym: 2-Fluorenol

Sample purity: 99%
Temperature: 15°C
Test parameter: EC50
Effect: Reduction in light output

Concentration: 0.42 mg/l
Exposure time: 5 min

Concentration: 0.50 mg/l
Exposure time: 15 min

Concentration: 0.68 mg/l
Exposure time: 30 min

Comment: Mean of three assays. Methanol (<10%) was used to prepare the stock solutions. EC50 values were calculated from nominal concentrations.

Bibliographical reference: Kaiser, K.L.E., and Palabrica, V.S. (1991). *Water Poll. Res. J. Canada* **26**, 361-431.

4-HYDROXY-3-METHOXYBENZONITRILE

CAS RN: 4421-08-3

Sample purity: 98%
Temperature: 15°C
Test parameter: EC50
Effect: Reduction in light output

Concentration: 52.9 mg/l
Exposure time: 5 min

Concentration: 55.4 mg/l
Exposure time: 15 min

Concentration: 54.2 mg/l
Exposure time: 30 min

Comment: Mean of four assays. Methanol (<10%) was used to prepare the stock solutions. EC50 values were calculated from nominal concentrations.

Bibliographical reference: Kaiser, K.L.E., and Palabrica, V.S. (1991). *Water Poll. Res. J. Canada* **26**, 361-431.

3-HYDROXY-4-METHOXYBENZYL ALCOHOL

CAS RN: 4383-06-6
Synonym: Isovanillyl alcohol

Sample purity: 98%
Temperature: 15°C
Test parameter: EC50
Effect: Reduction in light output

Concentration: 294 mg/l
Exposure time: 5 min

Concentration: 268 mg/l
Exposure time: 15 min

Concentration: 256 mg/l
Exposure time: 30 min

Comment: Mean of three assays. Methanol (<10%) was used to prepare the stock solutions. EC50 values were calculated from nominal concentrations.

Bibliographical reference: Kaiser, K.L.E., and Palabrica, V.S. (1991). *Water Poll. Res. J. Canada* **26**, 361-431.

4-HYDROXY-3-METHOXYBENZYL ALCOHOL

CAS RN: 498-00-0
Synonym: Vanillyl alcohol

Sample purity: 98%
Temperature: 15°C
Test parameter: EC50
Effect: Reduction in light output

Concentration: 154 mg/l
Exposure time: 5 min

Concentration: 144 mg/l
Exposure time: 15 min

Concentration: 134 mg/l
Exposure time: 30 min

Comment: Mean of three assays. Methanol (<10%) was used to prepare the stock solutions. EC50 values were calculated from nominal concentrations.

Bibliographical reference: Kaiser, K.L.E., and Palabrica, V.S. (1991). *Water Poll. Res. J. Canada* **26**, 361-431.

2-(HYDROXYMETHYL)ANTHRAQUINONE

CAS RN: 17241-59-7

Sample purity: 97%
Temperature: 15°C
Test parameter: EC50
Effect: Reduction in light output

Concentration: 30.7 mg/l
Exposure time: 5 min

Concentration: 21.7 mg/l
Exposure time: 15 min

Concentration: 22.2 mg/l
Exposure time: 30 min

Comment: Mean of three assays. Methanol (<10%) was used to prepare the stock solutions. EC50 values were calculated from nominal concentrations.

Bibliographical reference: Kaiser, K.L.E., and Palabrica, V.S. (1991). *Water Poll. Res. J. Canada* **26**, 361-431.

4-HYDROXY-3-NITROBENZENEARSONIC ACID

CAS RN: 121-19-7

Temperature: 15°C
Test parameter: EC50
Effect: Reduction in light output

Concentration: 891 mg/l
Exposure time: 5 min

Concentration: 955 mg/l
Exposure time: 15 min

Concentration: 832 mg/l
Exposure time: 30 min

Comment: Mean of three assays. Methanol (<10%) was used to prepare the stock solutions. EC50 values were calculated from nominal concentrations.

Bibliographical reference: Kaiser, K.L.E., and Palabrica, V.S. (1991). *Water Poll. Res. J. Canada* **26**, 361-431.

4-HYDROXY-3-NITROBENZONITRILE

CAS RN: 3272-08-0

Sample purity: 98%
Temperature: 15°C
Test parameter: EC50
Effect: Reduction in light output

Concentration: 13.0 mg/l
Exposure time: 5 min

Concentration: 12.7 mg/l
Exposure times: 15 and 30 min

Comment: Mean of five assays. Methanol (<10%) was used to prepare the stock solutions. EC50 values were calculated from nominal concentrations.

Bibliographical reference: Kaiser, K.L.E., and Palabrica, V.S. (1991). *Water Poll. Res. J. Canada* **26**, 361-431.

5-HYDROXY-2-NITROBENZYL ALCOHOL

CAS RN: 60463-12-9

Sample purity: 97%
Temperature: 15°C
Test parameter: EC50
Effect: Reduction in light output

Concentration: 35.3 mg/l
Exposure time: 5 min

Concentration: 33.7 mg/l
Exposure time: 15 min

Concentration: 33.0 mg/l
Exposure time: 30 min

Comment: Mean of four assays. Methanol (<10%) was used to prepare the stock solutions. EC50 values were calculated from nominal concentrations.

Bibliographical reference: Kaiser, K.L.E., and Palabrica, V.S. (1991). *Water Poll. Res. J. Canada* **26**, 361-431.

2-HYDROXY-5-NITROBENZYL BROMIDE

CAS RN: 772-33-8

Sample purity: 98%
Temperature: 15°C
Test parameter: EC50
Effect: Reduction in light output

Concentration: 22.2 mg/l
Exposure time: 5 min

Concentration: 19.7 mg/l
Exposure times: 15 and 30 min

Comment: Mean of three assays. Methanol (<10%) was used to prepare the stock solutions. EC50 values were calculated from nominal concentrations.

Bibliographical reference: Kaiser, K.L.E., and Palabrica, V.S. (1991). *Water Poll. Res. J. Canada* **26**, 361-431.

4-HYDROXYPHENETHYL ALCOHOL

CAS RN: 501-94-0

Sample purity: 98%
Temperature: 15°C
Test parameter: EC50
Effect: Reduction in light output

Concentration: 0.83 mg/l
Exposure times: 5 and 15 min

Concentration: 0.87 mg/l
Exposure time: 30 min

Comment: Mean of four assays. Methanol (<10%) was used to prepare the stock solutions. EC50 values were calculated from nominal concentrations.

Bibliographical reference: Kaiser, K.L.E., and Palabrica, V.S. (1991). *Water Poll. Res. J. Canada* **26**, 361-431.

4-HYDROXYPHENYLACETIC ACID

CAS RN: 156-38-7

Sample purity: 98%
Temperature: 15°C
Test parameter: EC50
Effect: Reduction in light output

Concentration: 5.52 mg/l
Exposure times: 5 and 15 min

Concentration: 7.63 mg/l

Exposure time: 30 min

Comment: Mean of three assays. Methanol (<10%) was used to prepare the stock solutions. EC50 values were calculated from nominal concentrations.

Bibliographical reference: Kaiser, K.L.E., and Palabrica, V.S. (1991). *Water Poll. Res. J. Canada* **26**, 361-431.

2-(4-HYDROXYPHENYLAZO)BENZOIC ACID

CAS RN: 1634-82-8

Sample purity: 98+%
Temperature: 15°C
Test parameter: EC50
Effect: Reduction in light output

Concentration: 55.5 mg/l
Exposure time: 5 min

Concentration: 34.2 mg/l
Exposure time: 15 min

Concentration: 25.4 mg/l
Exposure time: 30 min

Comment: Mean of three assays. Methanol (<10%) was used to prepare the stock solutions. EC50 values were calculated from nominal concentrations.

Bibliographical reference: Kaiser, K.L.E., and Palabrica, V.S. (1991). *Water Poll. Res. J. Canada* **26**, 361-431.

4-HYDROXY-5-PHENYL-4-CYCLOPENTENE-1,3-DIONE

CAS RN: 36394-22-6

Sample purity: 99%
Temperature: 15°C
Test parameter: EC50
Effect: Reduction in light output

Concentration: 42.1 mg/l
Exposure time: 5 min

Concentration: 43.1 mg/l
Exposure time: 15 min

Concentration: 48.4 mg/l
Exposure time: 30 min

Comment: Mean of three assays. Methanol (<10%) was used to prepare the stock solutions. EC50 values were calculated from nominal concentrations.

Bibliographical reference: Kaiser, K.L.E., and Palabrica, V.S. (1991). *Water Poll. Res. J. Canada* **26**, 361-431.

4-HYDROXY-3-PHENYL-2(5*H*)FURANONE

CAS RN: 23782-85-6

Sample purity: 98%
Temperature: 15°C
Test parameter: EC50
Effect: Reduction in light output

Concentration: 26.1 mg/l
Exposure time: 5 min

Concentration: 27.9 mg/l
Exposure time: 15 min

Concentration: 28.6 mg/l
Exposure time: 30 min

Comment: Mean of four assays. Methanol (<10%) was used to prepare the stock solutions. EC50 values were calculated from nominal concentrations.

Bibliographical reference: Kaiser, K.L.E., and Palabrica, V.S. (1992). National Water Research Institute, Burlington, Ontario, Canada, unpublished results.

D-4-HYDROXYPHENYLGLYCINE

CAS RN: 22818-40-2

Sample purity: 98+%
Temperature: 15°C
Test parameter: EC50
Effect: Reduction in light output

Concentration: 146 mg/l
Exposure times: 5 and 15 min

Concentration: 152 mg/l*
Exposure time: 30 min

Comment: Mean of three assays. Methanol (<10%) was used to prepare the stock solutions. EC50 values were calculated from nominal concentrations.

Bibliographical references: Kaiser, K.L.E., and Palabrica, V.S. (1991). *Water Poll. Res. J. Canada* **26**, 361-431.
* Kaiser, K.L.E. (1987). In: *QSAR in Environmental Toxicology - II*, K.L.E. Kaiser (ed.), D. Reidel Publishing Company, Dordrecht, p. 169-188.

3-HYDROXYPICOLINAMIDE

CAS RN: 933-90-4

Temperature: 15°C
Test parameter: EC50
Effect: Reduction in light output

Concentration: 93.4 mg/l
Exposure time: 5 min

Concentration: 105 mg/l
Exposure time: 15 min

Concentration: 120 mg/l
Exposure time: 30 min

Comment: Mean of three assays. Methanol (<10%) was used to prepare the stock solutions. EC50 values were calculated from nominal concentrations.

Bibliographical reference: Kaiser, K.L.E., and Palabrica, V.S. (1991). *Water Poll. Res. J. Canada* **26**, 361-431.

3-HYDROXYPICOLINIC ACID

CAS RN: 874-24-8

Sample purity: 98%
Temperature: 15°C
Test parameter: EC50
Effect: Reduction in light output

Concentration: 23.6 mg/l
Exposure time: 5 min

Concentration: 19.6 mg/l
Exposure time: 15 min

Concentration: 17.5 mg/l
Exposure time: 30 min

Comment: Mean of four assays. Methanol (<10%) was used to prepare the stock solutions. EC50 values were calculated from nominal concentrations.

Bibliographical reference: Kaiser, K.L.E., and Palabrica, V.S. (1991). *Water Poll. Res. J. Canada* **26**, 361-431.

4'-HYDROXYPROPIOPHENONE

CAS RN: 70-70-2

Sample purity: 97%
Temperature: 15°C
Test parameter: EC50
Effect: Reduction in light output

Concentration: 7.02 mg/l

Exposure time: 5 min

Concentration: 7.19 mg/l
Exposure time: 15 min

Concentration: 6.56 mg/l*
Exposure time: 30 min

Comment: Mean of three assays. Methanol (<10%) was used to prepare the stock solutions. EC50 values were calculated from nominal concentrations.

Bibliographical references: Kaiser, K.L.E., and Palabrica, V.S. (1991). *Water Poll. Res. J. Canada* **26**, 361-431.
* Kaiser, K.L.E. (1987). In: *QSAR in Environmental Toxicology - II*, K.L.E. Kaiser (ed.), D. Reidel Publishing Company, Dordrecht, p. 169-188.

2-HYDROXYPYRIDINE

CAS RN: 142-08-5

Sample purity: 97%
Temperature: 15°C
Test parameter: EC50
Effect: Reduction in light output

Concentration: 287 mg/l
Exposure time: 5 min

Concentration: 256 mg/l
Exposure time: 15 min

Concentration: 250 mg/l
Exposure time: 30 min

Comment: Mean of three assays. Methanol (<10%) was used to prepare the stock solutions. EC50 values were calculated from nominal concentrations.

Bibliographical reference: Kaiser, K.L.E., and Palabrica, V.S. (1991). *Water Poll. Res. J. Canada* **26**, 361-431.

3-HYDROXYPYRIDINE

CAS RN: 109-00-2

Sample purity: 98%
Temperature: 15°C
Test parameter: EC50
Effect: Reduction in light output

Concentration: 573 mg/l
Exposure time: 5 min

Concentration: 535 mg/l
Exposure time: 15 min

Concentration: 488 mg/l
Exposure time: 30 min

Comment: Mean of three assays. Methanol (<10%) was used to prepare the stock solutions. EC50 values were calculated from nominal concentrations.

Bibliographical reference: Kaiser, K.L.E., and Palabrica, V.S. (1991). *Water Poll. Res. J. Canada* **26**, 361-431.

4-HYDROXYPYRIMIDINE

CAS RN: 51953-17-4
Synonym: 4(3*H*)-Pyrimidone

Sample purity: >98%
Temperature: 15°C
Test parameter: EC50
Effect: Reduction in light output

Concentration: 680 mg/l
Exposure time: 5 min

Concentration: 606 mg/l
Exposure time: 15 min

Concentration: 650 mg/l
Exposure time: 30 min

Comment: Mean of four assays. Methanol (<10%) was used to prepare the stock solutions. EC50 values were calculated from nominal concentrations.

Bibliographical reference: Kaiser, K.L.E., and Palabrica, V.S. (1991). *Water Poll. Res. J. Canada* **26**, 361-431.

8-HYDROXYQUINALDINE

CAS RN: 826-81-3
Synonym: 2-Methyl-8-quinolinol

Sample purity: 98%
Temperature: 15°C
Test parameter: EC50
Effect: Reduction in light output

Concentration: 19.1 mg/l
Exposure time: 5 min

Concentration: 17.1 mg/l
Exposure times: 15 and 30 min

Comment: Mean of three assays. Methanol (<10%) was used to prepare the stock solutions. EC50 values were calculated from nominal concentrations.

Bibliographical reference: Kaiser, K.L.E., and Palabrica, V.S. (1991). *Water Poll. Res. J. Canada* **26**, 361-431.

2-HYDROXYQUINOLINE

CAS RN: 59-31-4
Synonyms: 2-Quinolinol; Carbostyril

Sample purity: >99%
Temperature: 15°C
Test parameter: EC50
Effect: Reduction in light output

Concentration: 0.90 mg/l
Exposure time: 5 min

Concentration: 0.96 mg/l
Exposure time: 15 min

Concentration: 1.00 mg/l
Exposure time: 30 min

Comment: Mean of four assays. Methanol (<10%) was used to prepare the stock solutions. EC50 values were calculated from nominal concentrations.

Bibliographical reference: Kaiser, K.L.E., and Palabrica, V.S. (1991). *Water Poll. Res. J. Canada* **26**, 361-431.

4-HYDROXYQUINOLINE

CAS RN: 611-36-9
Synonym: 4-Quinolinol

Sample purity: 98%
Temperature: 15°C
Test parameter: EC50
Effect: Reduction in light output

Concentration: 258 mg/l
Exposure time: 5 min

Concentration: 200 mg/l
Exposure time: 15 min

Concentration: 191 mg/l
Exposure time: 30 min

Comment: Mean of four assays. Methanol (<10%) was used to prepare the stock solutions. EC50 values were calculated from nominal concentrations.

Bibliographical reference: Kaiser, K.L.E., and Palabrica, V.S. (1991). *Water Poll. Res. J. Canada* **26**, 361-431.

8-HYDROXYQUINOLINE

CAS RN: 148-24-3

Synonym: 8-Quinolinol

Sample purity: >99%
Temperature: 15°C
Test parameter: EC50
Effect: Reduction in light output

Concentration: 12.1 mg/l
Exposure time: 5 min

Concentration: 7.80 mg/l
Exposure time: 15 min

Concentration: 2.30 mg/l
Exposure time: 30 min

Comment: Mean of three assays. Methanol (<10%) was used to prepare the stock solutions. EC50 values were calculated from nominal concentrations.

Bibliographical reference: Kaiser, K.L.E., and Palabrica, V.S. (1991). *Water Poll. Res. J. Canada* **26**, 361-431.

3-HYDROXYTETRAHYDROFURAN

CAS RN: 453-20-3

Sample purity: 97%
Temperature: 15°C
Test parameter: EC50
Effect: Reduction in light output

Concentration: 556 mg/l
Exposure time: 5 min

Concentration: 596 mg/l
Exposure time: 15 min

Concentration: 610 mg/l
Exposure time: 30 min

Comment: Mean of three assays. Methanol (<10%) was used to prepare the stock solutions. EC50 values were calculated from nominal concentrations.

Bibliographical reference: Kaiser, K.L.E., and Palabrica, V.S. (1991). *Water Poll. Res. J. Canada* **26**, 361-431.

4-HYDROXYTHIOPHENOL

CAS RN: 637-89-8
Synonyms: 4-Hydroxybenzenethiol; 4-Mercaptophenol

Sample purity: 90%
Temperature: 15°C
Test parameter: EC50
Effect: Reduction in light output

Concentration: 0.13 mg/l
Exposure time: 5 min

Concentration: 0.11 mg/l
Exposure time: 15 min

Concentration: 0.12 mg/l
Exposure time: 30 min

Comment: Mean of four assays. Methanol (<10%) was used to prepare the stock solutions. EC50 values were calculated from nominal concentrations.

Bibliographical reference: Kaiser, K.L.E., and Palabrica, V.S. (1991). *Water Poll. Res. J. Canada* **26**, 361-431.

HYDROXYUREA

CAS RN: 127-07-1

Sample purity: >98%
Temperature: 15°C
Test parameter: EC50
Effect: Reduction in light output

Concentration: 726 mg/l
Exposure time: 5 min

Concentration: 678 mg/l

Exposure time: 15 min

Concentration: 633 mg/l
Exposure time: 30 min

Comment: Mean of three assays. Methanol (<10%) was used to prepare the stock solutions. EC50 values were calculated from nominal concentrations.

Bibliographical reference: Kaiser, K.L.E., and Palabrica, V.S. (1991). *Water Poll. Res. J. Canada* **26**, 361-431.

IMIDAZOLE

CAS RN: 288-32-4

Sample purity: 99%
Temperature: 15°C
Test parameter: EC50
Effect: Reduction in light output

Concentration: 231 mg/l
Exposure time: 5 min

Concentration: 206 mg/l
Exposure time: 15 min

Concentration: 231 mg/l
Exposure time: 30 min

Comment: Mean of three assays. Methanol (<10%) was used to prepare the stock solutions. EC50 values were calculated from nominal concentrations.

Bibliographical reference: Kaiser, K.L.E., and Palabrica, V.S. (1991). *Water Poll. Res. J. Canada* **26**, 361-431.

4-(IMIDAZOL-1-YL)PHENOL

CAS RN: 10041-02-8

Sample purity: 97%

Temperature: 15°C
Test parameter: EC50
Effect: Reduction in light output

Concentration: 0.28 mg/l
Exposure time: 5 min

Concentration: 0.29 mg/l
Exposure time: 15 min

Concentration: 0.30 mg/l
Exposure time: 30 min

Comment: Mean of three assays. Methanol (<10%) was used to prepare the stock solutions. EC50 values were calculated from nominal concentrations.

Bibliographical reference: Kaiser, K.L.E., and Palabrica, V.S. (1991). *Water Poll. Res. J. Canada* **26**, 361-431.

INDOLE

CAS RN: 120-72-9

Sample purity: >99%
Temperature: 15°C
Test parameter: EC50
Effect: Reduction in light output

Concentration: 2.45 mg/l
Exposure time: 5 min

Concentration: 2.39 mg/l
Exposure times: 15 and 30 min

Comment: Mean of four assays. Methanol (<10%) was used to prepare the stock solutions. EC50 values were calculated from nominal concentrations.

Bibliographical reference: Kaiser, K.L.E., and Palabrica, V.S. (1991). *Water Poll. Res. J. Canada* **26**, 361-431.

INDOPHENOL SODIUM SALT

CAS RN: 5418-32-6

Temperature: 15°C
Test parameter: EC50
Effect: Reduction in light output

Concentration: 0.62 mg/l
Exposure time: 5 min

Concentration: 0.70 mg/l
Exposure time: 15 min

Concentration: 1.06 mg/l
Exposure time: 30 min

Comment: Mean of three assays. Methanol (<10%) was used to prepare the stock solutions. EC50 values were calculated from nominal concentrations.

Bibliographical reference: Kaiser, K.L.E., and Palabrica, V.S. (1991). *Water Poll. Res. J. Canada* **26**, 361-431.

4-IODOANILINE

CAS RN: 540-37-4

Sample purity: 99%
Temperature: 15°C
Test parameter: EC50
Effect: Reduction in light output

Concentration: 0.55 mg/l
Exposure time: 5 min

Concentration: 0.60 mg/l
Exposure time: 15 min

Concentration: 0.71 mg/l
Exposure time: 30 min

Comment: Mean of three assays. Methanol (<10%) was used to prepare the stock solutions. EC50 values were calculated from nominal concentrations.

Bibliographical reference: Kaiser, K.L.E., and Palabrica, V.S. (1991). *Water Poll. Res. J. Canada* **26**, 361-431.

4-IODOANISOLE

CAS RN: 696-62-8

Sample purity: 98%
Temperature: 15°C
Test parameter: EC50
Effect: Reduction in light output

Concentration: 1.17 mg/l
Exposure time: 5 min

Concentration: 1.29 mg/l
Exposure time: 15 min

Concentration: 1.26 mg/l*
Exposure time: 30 min

Comment: Mean of three assays. Methanol (<10%) was used to prepare the stock solutions. EC50 values were calculated from nominal concentrations.

Bibliographical references: Kaiser, K.L.E., and Palabrica, V.S. (1991). *Water Poll. Res. J. Canada* **26**, 361-431.
* Kaiser, K.L.E., and Gough, K.M. (1989). In: *Aquatic Toxicology and Environmental Fate: Eleventh Volume, ASTM STP 1007*, G.W. Suter and M.A. Lewis (eds.), American Society for Testing and Materials, Philadelphia, p. 424-441.

IODOBENZENE

CAS RN: 591-50-4

Temperature: 15°C
Test parameter: EC50

Effect: Reduction in light output
Concentration: 3.23 mg/l
Exposure time: 30 min
Comment: Mean of three assays. Methanol (<10%) was used to prepare the stock solutions. EC50 values were calculated from nominal concentrations.

Bibliographical reference: Kaiser, K.L.E., Palabrica, V.S., and Ribo, J.M. (1987). In: *QSAR in Environmental Toxicology - II*, K.L.E. Kaiser (ed.), D. Reidel Publishing Company, Dordrecht, p. 153-168.

4-IODOBENZOIC ACID

CAS RN: 619-58-9

Sample purity: 98%
Temperature: 15°C
Test parameter: EC50
Effect: Reduction in light output

Concentration: 86.0 mg/l
Exposure time: 5 min

Concentration: 90.1 mg/l
Exposure time: 15 min

Concentration: 88.0 mg/l*
Exposure time: 30 min

Comment: Mean of three assays. Methanol (<10%) was used to prepare the stock solutions. EC50 values were calculated from nominal concentrations.

Bibliographical references: Kaiser, K.L.E., and Palabrica, V.S. (1991). *Water Poll. Res. J. Canada* **26**, 361-431.
* Kaiser, K.L.E., and Gough, K.M. (1989). In: *Aquatic Toxicology and Environmental Fate: Eleventh Volume, ASTM STP 1007*, G.W. Suter and M.A. Lewis (eds.), American Society for Testing and Materials, Philadelphia, p. 424-441.

4-IODOBENZOYL CHLORIDE

CAS RN: 1711-02-0

Sample purity: 97%
Temperature: 15°C
Test parameter: EC50
Effect: Reduction in light output

Concentration: 3.94 mg/l
Exposure times: 5 and 15 min

Concentration: 3.68 mg/l
Exposure time: 30 min

Comment: Mean of three assays. Methanol (<10%) was used to prepare the stock solutions. EC50 values were calculated from nominal concentrations.

Bibliographical reference: Kaisèr, K.L.E., and Palabrica, V.S. (1991). *Water Poll. Res. J. Canada* **26**, 361-431.

1-IODO-4-NITROBENZENE

CAS RN: 636-98-6

Temperature: 15°C
Test parameter: EC50
Effect: Reduction in light output

Concentration: 5.84 mg/l
Exposure time: 5 min

Concentration: 5.57 mg/l
Exposure time: 15 min

Concentration: 5.84 mg/l*
Exposure time: 30 min

Comment: Mean of three assays. Methanol (<10%) was used to prepare the stock solutions. EC50 values were calculated from nominal concentrations.

Bibliographical references: Kaiser, K.L.E., and Palabrica, V.S. (1991). *Water Poll. Res. J. Canada* **26**, 361-431.
* Kaiser, K.L.E. (1987). In: *QSAR in Environmental Toxicology - II*, K.L.E. Kaiser (ed.), D. Reidel Publishing Company, Dordrecht, p. 169-188.

4-IODOPHENOL

CAS RN: 540-38-5

Temperature: 15°C
Test parameter: EC50
Effect: Reduction in light output
Concentration: 0.25 mg/l
Exposure time: 30 min
Comment: Mean of three assays. Methanol (<10%) was used to prepare the stock solutions. EC50 values were calculated from nominal concentrations.

Bibliographical reference: Kaiser, K.L.E. (1987). In: *QSAR in Environmental Toxicology - II*, K.L.E. Kaiser (ed.), D. Reidel Publishing Company, Dordrecht, p. 169-188.

ISOFENPHOS

CAS RN: 25311-71-1
Synonym: 1-Methylethyl 2-[[ethoxy[(1-methylethyl)amino]phosphinothioyl] oxy]benzoate

Temperature: 15 ± 0.1°C
Test parameter: EC50
Effect: Reduction in light output
Concentration: 97.8 mg/l
Exposure time: 5 min
Comment: Chemical was first dissolved in a solvent and then serially diluted with the diluent. The concentration of solvent did not exceed 8% (v/v) in the test samples.

Bibliographical reference: Somasundaram, L., Coats, J.R., Racke, K.D., and Stahr, H.M. (1990). *Bull. Environ. Contam. Toxicol.* **44**, 254-259.

ISONICOTINAMIDE

CAS RN: 1453-82-3

Sample purity: 99%
Temperature: 15°C
Test parameter: EC50
Effect: Reduction in light output

Concentration: 521 mg/l
Exposure time: 5 min

Concentration: 475 mg/l
Exposure time: 15 min

Concentration: 509 mg/l
Exposure time: 30 min

Comment: Mean of three assays. Methanol (<10%) was used to prepare the stock solutions. EC50 values were calculated from nominal concentrations.

Bibliographical reference: Kaiser, K.L.E., and Palabrica, V.S. (1991). *Water Poll. Res. J. Canada* **26**, 361-431.

ISONICOTINIC ACID

CAS RN: 55-22-1

Sample purity: >99%
Temperature: 15°C
Test parameter: EC50
Effect: Reduction in light output

Concentration: 93.4 mg/l
Exposure time: 5 min

Concentration: 97.8 mg/l
Exposure time: 15 min

Concentration: 120 mg/l
Exposure time: 30 min

Comment: Mean of three assays. Methanol (<10%) was used to prepare the stock solutions. EC50 values were calculated from nominal concentrations.

Bibliographical reference: Kaiser, K.L.E., and Palabrica, V.S. (1991). *Water Poll. Res. J. Canada* **26**, 361-431.

ISOPHTHALALDEHYDE

CAS RN: 626-19-7

Sample purity: 97%
Temperature: 15°C
Test parameter: EC50
Effect: Reduction in light output

Concentration: 21.3 mg/l
Exposure time: 5 min

Concentration: 16.5 mg/l
Exposure time: 15 min

Concentration: 12.5 mg/l
Exposure time: 30 min

Comment: Mean of three assays. Methanol (<10%) was used to prepare the stock solutions. EC50 values were calculated from nominal concentrations.

Bibliographical reference: Kaiser, K.L.E., and Palabrica, V.S. (1991). *Water Poll. Res. J. Canada* **26**, 361-431.

ISOPHTHALONITRILE

CAS RN: 626-17-5
Synonym: 1,3-Dicyanobenzene

Sample purity: 99+%
Temperature: 15°C
Test parameter: EC50
Effect: Reduction in light output

Concentration: 131 mg/l
Exposure time: 5 min

Concentration: 120 mg/l
Exposure time: 15 min

Concentration: 107 mg/l
Exposure time: 30 min

Comment: Mean of three assays. Methanol (<10%) was used to prepare the stock solutions. EC50 values were calculated from nominal concentrations.

Bibliographical reference: Kaiser, K.L.E., and Palabrica, V.S. (1991). *Water Poll. Res. J. Canada* **26**, 361-431.

ISOPROPYL ETHER

CAS RN: 108-20-3

Temperature: 15°C
Test parameter: EC50
Effect: Reduction in light output
Concentration: 505 mg/l
Exposure time: 5 min

Bibliographical reference: Microtox® Application Notes (1982). Beckman Instruments, Inc., Carlsbad, California.

3-ISOPROPYLQUINOLINE

CAS RN: 72359-42-3

Sample purity: >97%
Temperature: 15 ± 0.3°C
Test parameter: EC50
Effect: Reduction in light output

Concentration: 10 mg/l
Exposure time: 5 min

Concentration: 9.5 mg/l

Exposure time: 15 min

Comment: Test was performed in duplicate at pH = 5.1. Dichloromethane was used as carrier solvent. Concentrations of the chemical were determined by high performance liquid chromatography. Hexachloroethane was used for quality control/quality assurance (5-min EC50 = 0.31 mg/l; n = 16; sd = 0.08 mg/l).

Bibliographical reference: Birkholz, D.A., Coutts, R.T., Hrudey, S.E., Danell, R.W., and Lockhart, W.L. (1990). *Water Res.* **24**, 67-73.

ISOPROPYL SALICYLATE

CAS RN: 607-85-2

Temperature: 15 ± 0.1°C
Test parameter: EC50
Effect: Reduction in light output
Concentration: 5.6 mg/l
Exposure time: 5 min

Bibliographical reference: Somasundaram, L., Coats, J.R., Racke, K.D., and Stahr, H.M. (1990). *Bull. Environ. Contam. Toxicol.* **44**, 254-259.

ISOQUINOLINE

CAS RN: 119-65-3

Sample purity: >97%
Temperature: 15 ± 0.3°C
Test parameter: EC50
Effect: Reduction in light output

Concentration: 1.7 mg/l
Exposure time: 5 min

Concentration: 2.2 mg/l
Exposure time: 15 min

Comment: Test was performed in duplicate at pH = 5.1. Dichloromethane was used as carrier solvent. Concentrations of the chemical were determined by high performance liquid chromatography. Hexachloroethane was used for quality control/quality assurance (5-min EC50 = 0.31 mg/l; n = 16; sd = 0.08 mg/l).

Bibliographical reference: Birkholz, D.A., Coutts, R.T., Hrudey, S.E., Danell, R.W., and Lockhart, W.L. (1990). *Water Res.* **24**, 67-73.

ISOTHIOCYANATOBENZENE

CAS RN: 103-72-0

Temperature: 15°C
Test parameter: EC50
Effect: Reduction in light output
Concentration: 2.09 mg/l
Exposure time: 30 min
Comment: Mean of three assays. Methanol (<10%) was used to prepare the stock solutions. EC50 values were calculated from nominal concentrations.

Bibliographical reference: Kaiser, K.L.E., Palabrica, V.S., and Ribo, J.M. (1987). In: *QSAR in Environmental Toxicology - II*, K.L.E. Kaiser (ed.), D. Reidel Publishing Company, Dordrecht, p. 153-168.

KATHON 886-MW

CAS RN: 55965-84-9

Sample purity: 2.6% 2-Methyl-4-isothiazolin-3-one and 8.6% 5-chloro-2-methyl-4-isothiazolin-3-one
Temperature: 15°C
Test parameter: EC50
Effect: Reduction in light output

Concentration: 0.09 mg/l
Exposure time: 5 min

Concentration: 0.05 mg/l
Exposure time: 15 min

Comment: Chemical was tested in duplicate at 0.0125, 0.025, 0.1, 0.2, 0.5, and 1 mg/l. Verification that the controls were valid was accomplished by making sure that the difference between the duplicate blank ratios (5 minute light reading divided by the time 0 reading) was not greater than 0.02, otherwise the test was not continued.

Bibliographical reference: Mallak, F.P., and Brunker, R.L. (1984). In: *Toxicity Screening Procedures Using Bacterial Systems*, D. Liu and B.J. Dutka (eds.), Marcel Dekker, New York, p. 65-76.

KELTHANE

CAS RN: 115-32-2
Synonyms: Dicofol; 2,2,2-Trichloro-1,1-di-(4-chlorophenyl)ethanol

Test parameter: EC50
Effect: Reduction in light output
Concentration: 0.45 mg/l
Exposure time: 5 min
Comment: Concentrations in the test were measured.

Bibliographical reference: Curtis, C., Lima, A., Lozano, S.J., and Veith, G.D. (1982). In: *Aquatic Toxicology and Hazard Assessment: Fifth Conference, ASTM STP 766*, J.G. Pearson, R.B. Foster, and W.E. Bishop (eds.), American Society for Testing and Materials, Philadelphia, p. 170-178.

LAURIC ACID

CAS RN: 143-07-7

Sample purity: 99.5%
Temperature: 15°C
Test parameter: EC50
Effect: Reduction in light output

Concentration: 4.28 ± 0.26 mg/l
Exposure time: 5 min

Concentration: 3.40 ± 0.18 mg/l
Exposure time: 15 min

Concentration: 3.26 ± 0.22 mg/l
Exposure time: 25 min

Comment: Mean of five assays. The values were converted to mg/l from the original data expressed in µM and a rounded molecular weight of 200 given by the authors. The EC50 values were determined in 0.45% (v/v) methanol in 2% (w/w) sodium chloride solution. The 5-min EC50 value for methanol was 43000 mg/l. Phenol solution was used for quality control/quality assurance. The 5-min EC50 value was 18.2 mg/l. The EC50 value at 15 min was 20.7 mg/l with a relative error of <5%.

Bibliographical reference: Chou, C.C., and Que Hee, S.S. (1992). *Ecotoxicol. Environ. Safety* **23**, 355-363.

LEAD(II) ACETATE TRIHYDRATE

CAS RN: 6080-56-4

Sample purity: Analytical grade
Temperature: 15°C
Test parameter: EC50
Effect: Reduction in light output

Concentration: 2.56 mg/l Pb^{++}
Exposure time: 5 min

Concentration: 0.78 mg/l Pb^{++}
Exposure time: 10 min

Concentration: 0.46 mg/l Pb^{++}
Exposure time: 15 min

Concentration: 0.39 mg/l Pb^{++}
Exposure time: 20 min

Concentration: 0.31 mg/l Pb^{++}
Exposure time: 30 min

Comment: Results derived from the average of two replicates.

Bibliographical reference: Qureshi, A.A., Coleman, R.N., and Paran, J.H. (1984). In: *Toxicity Screening Procedures Using Bacterial Systems*, D. Liu and B.J. Dutka (eds.), Marcel Dekker, New York, p. 1-22.

LEAD NITRATE

CAS RN: 10099-74-8

Temperature: 20°C
Test parameter: EC50
Effect: Reduction in light output
Concentration: 0.08 ± 0.03 mg/l Pb^{++}
Exposure time: 10 min
Comment: Mean of three assays. Toxicity values were calculated from nominal concentrations.

Bibliographical reference: Ferard, J.F., Vasseur, P., Danoux, L., and Larbaigt, G. (1983). *Rev. Fr. Sci. Eau* **2**, 221-237.

LINDANE

CAS RN: 58-89-9
Synonym: 1,2,3,4,5,6-Hexachlorocyclohexane (gamma isomer)

Temperature: 15°C
Test parameter: EC50
Effect: Reduction in light output
Concentration: 11 mg/l
Exposure time: 15 min
Comment: Extrapolated EC50. No concentration with more than 50% inhibition could be tested.

Bibliographical reference: Hermens, J., Busser, F., Leeuwangh, P., and Musch, A. (1985). *Ecotoxicol. Environ. Safety* **9**, 17-25.

Temperature: 15°C
Test parameter: EC50
Effect: Reduction in light output
Concentration: 5.7 mg/l
Exposure time: 10 min
Comment: Mean of three assays. Chemical was dissolved in 0.1 ml/l acetone. Test solutions were analyzed using gas chromatography.

Bibliographical reference: Bazin, C., Chambon, P., Bonnefille, M., and Larbaigt, G. (1987). *Sci. Eau* **6**, 403-413.

2,4-LUTIDINE

CAS RN: 108-47-4
Synonym: 2,4-Dimethylpyridine

Sample purity: 96+%
Temperature: 15°C
Test parameter: EC50
Effect: Reduction in light output

Concentration: 17.8 mg/l
Exposure time: 5 min

Concentration: 18.2 mg/l
Exposure time: 15 min

Concentration: 18.6 mg/l
Exposure time: 30 min

Comment: Mean of four assays. Methanol (<10%) was used to prepare the stock solutions. EC50 values were calculated from nominal concentrations.

Bibliographical reference: Kaiser, K.L.E., and Palabrica, V.S. (1991). *Water Poll. Res. J. Canada* **26**, 361-431.

2,6-LUTIDINE

CAS RN: 108-48-5
Synonym: 2,6-Dimethylpyridine

Sample purity: 97%
Temperature: 15°C
Test parameter: EC50
Effect: Reduction in light output
Concentration: 117 mg/l
Exposure times: 5, 15, and 30 min
Comment: Mean of four assays. Methanol (<10%) was used to prepare the stock solutions. EC50 values were calculated from nominal concentrations.

Bibliographical reference: Kaiser, K.L.E., and Palabrica, V.S. (1991). *Water Poll. Res. J. Canada* **26**, 361-431.

3,4-LUTIDINE

CAS RN: 583-58-4
Synonym: 3,4-Dimethylpyridine

Sample purity: 98%
Temperature: 15°C
Test parameter: EC50
Effect: Reduction in light output

Concentration: 7.59 mg/l
Exposure time: 5 min

Concentration: 8.51 mg/l
Exposure time: 15 min

Concentration: 10.0 mg/l
Exposure time: 30 min

Comment: Mean of three assays. Methanol (<10%) was used to prepare the stock solutions. EC50 values were calculated from nominal concentrations.

Bibliographical reference: Kaiser, K.L.E., and Palabrica, V.S. (1991). *Water Poll. Res. J. Canada* **26**, 361-431.

MALATHION

CAS RN: 121-75-5
Synonym: Diethyl (dimethoxyphosphinothioyl)thiobutanedioate

Temperature: 15°C
Test parameter: EC50
Effect: Reduction in light output
Concentration: 3.0 mg/l
Exposure time: 5 min

Bibliographical reference: Bulich, A.A., Greene, M.W., and Isenberg, D.L. (1981). In: *Aquatic Toxicology and Hazard Assessment: Fourth Conference*, *ASTM STP 737*, D.R. Branson and K.L. Dickson (eds.), American Society for Testing and Materials, Philadelphia, p. 338-347.

Temperature: 15 ± 0.1°C
Test parameter: EC50
Effect: Reduction in light output
Concentration: 10 mg/l
Exposure time: 5 min

Bibliographical reference: Chang, J.C., Taylor, P.B., and Leach, F.R. (1981). *Bull. Environ. Contam. Toxicol.* **26**, 150-156.

Sample purity: Agrichemical grade
Temperature: 15 ± 0.1°C
Test parameter: EC50
Effect: Reduction in light output
Concentration: 59.7 mg/l
Exposure time: 5 min
Comment: Test was performed on *Photobacterium phosphoreum* NZ11D obtained from the Scripps Institute of Oceanography (La Jolla, CA).

Bibliographical reference: McFeters, G.A., Bond, P.J., Olson, S.B., and Tchan, Y.T. (1983). *Water Res.* **17**, 1757-1762.

Temperature: 15°C
Test parameter: EC50
Effect: Reduction in light output
Concentration: 84 mg/l
Exposure time: 5 min
Comment: Chemical was dissolved in acetone (0.01%). EC50 value was calculated from nominal concentrations.

Bibliographical reference: Reteuna, C. (1988). Thesis, University of Metz, Metz, France.

MALEIC ANHYDRIDE

CAS RN: 108-31-6

Sample purity: 99%
Temperature: 15°C
Test parameter: EC50
Effect: Reduction in light output

Concentration: 38.1 mg/l
Exposure time: 5 min

Concentration: 41.8 mg/l
Exposure time: 15 min

Concentration: 43.8 mg/l
Exposure time: 30 min

Comment: Mean of three assays. Methanol (<10%) was used to prepare the stock solutions. EC50 values were calculated from nominal concentrations.

Bibliographical reference: Kaiser, K.L.E., and Palabrica, V.S. (1991). *Water Poll. Res. J. Canada* **26**, 361-431.

L-MALIC ACID

CAS RN: 97-67-6

Sample purity: 99%
Temperature: 15°C
Test parameter: EC50
Effect: Reduction in light output

Concentration: 53.4 mg/l
Exposure time: 5 min

Concentration: 55.9 mg/l
Exposure time: 15 min

Concentration: 58.5 mg/l
Exposure time: 30 min

Comment: Mean of three assays. Methanol (<10%) was used to prepare the stock solutions. EC50 values were calculated from nominal concentrations.

Bibliographical reference: Kaiser, K.L.E., and Palabrica, V.S. (1991). *Water Poll. Res. J. Canada* **26**, 361-431.

MALONAMIDE

CAS RN: 108-13-4
Synonym: Malonodiamide

Sample purity: 97%
Temperature: 15°C
Test parameter: EC50
Effect: Reduction in light output

Concentration: 644 mg/l
Exposure time: 5 min

Concentration: 587 mg/l
Exposure time: 15 min

Concentration: 601 mg/l
Exposure time: 30 min

Comment: Mean of three assays. Methanol (<10%) was used to prepare the stock solutions. EC50 values were calculated from nominal concentrations.

Bibliographical reference: Kaiser, K.L.E., and Palabrica, V.S. (1991). *Water Poll. Res. J. Canada* **26**, 361-431.

MALONONITRILE

CAS RN: 109-77-3

Sample purity: 99%
Temperature: 15°C
Test parameter: EC50
Effect: Reduction in light output

Concentration: 390 mg/l
Exposure time: 5 min

Concentration: 150 mg/l*
Exposure time: 15 min

Comment: Chemical was prepared in an initial solution of 3% methanol.

Bibliographical references: Cronin, M.T.D., Dearden, J.C., and Dobbs, A.J. (1991). *Sci. Total Environ.* **109/110**, 431-439.
* Cronin, M.T.D. (1993). Liverpool John Moores University, UK, private communication.

Sample purity: 99%
Temperature: 15°C
Test parameter: EC50
Effect: Reduction in light output

Concentration: 240 mg/l
Exposure time: 5 min

Concentration: 145 mg/l
Exposure time: 15 min

Concentration: 102 mg/l
Exposure time: 30 min

Comment: Mean of three assays. Methanol (<10%) was used to prepare the stock solutions. EC50 values were calculated from nominal concentrations.

Bibliographical reference: Kaiser, K.L.E., and Palabrica, V.S. (1992). National Water Research Institute, Burlington, Ontario, Canada, unpublished results.

MANCOZEB

CAS RN: 8018-01-7

Sample purity: Lamers & Indemans (36% CS_2)
Test parameter: EC50
Effect: Reduction in light output
Concentration: 0.08 mg/l (0.07-0.08)
Exposure time: 15 min

Bibliographical reference: van Leeuwen, C.J., Maas-Diepeveen, J.L., Niebeek, G., Vergouw, W.H.A., Griffioen, P.S., and Luijken, M.W. (1985). *Aquat. Toxicol.* **7**, 145-164.

MANEB

CAS RN: 12427-38-2
Synonym: Manganese 1,2-ethanediylbis(carbamodithioate) complex

Sample purity: ≥90%
Test parameter: EC50
Effect: Reduction in light output
Concentration: 1.2 mg/l (1.2-1.3)
Exposure time: 15 min

Bibliographical reference: van Leeuwen, C.J., Maas-Diepeveen, J.L., Niebeek, G., Vergouw, W.H.A., Griffioen, P.S., and Luijken, M.W. (1985). *Aquat. Toxicol.* **7**, 145-164.

Sample purity: 95%
Temperature: 15°C
Test parameter: EC50
Effect: Reduction in light output
Concentration: 0.13 ± 0.03 mg/l
Exposure time: 30 min
Comment: The assay was run in triplicate (pH = 6.2-6.6) using different bacterial reagents. The toxicity values were calculated from nominal concentrations. Synergistic effects were observed with $CuSO_4.5H_2O$ (20 µg/l Cu^{++} and 9 µg/l maneb).

Bibliographical reference: Vasseur, P., Dive, D., Sokar, Z., and Bonnemain, H. (1988). *Chemosphere* **17**, 767-782.

2-MERCAPTOBENZIMIDAZOLE

CAS RN: 583-39-1
Synonym: 2-Benzimidazolethiol

Sample purity: 98%
Temperature: 15°C
Test parameter: EC50
Effect: Reduction in light output

Concentration: 70.3 mg/l
Exposure time: 5 min

Concentration: 77.0 mg/l

Exposure time: 15 min

Concentration: 86.4 mg/l
Exposure time: 30 min

Comment: Mean of three assays. Methanol (<10%) was used to prepare the stock solutions. EC50 values were calculated from nominal concentrations.

Bibliographical reference: Kaiser, K.L.E., and Palabrica, V.S. (1991). *Water Poll. Res. J. Canada* **26**, 361-431.

2-MERCAPTOBENZOTHIAZOLE

CAS RN: 149-30-4
Synonym: 2-Benzothiazolethiol

Sample purity: 98%
Temperature: 15°C
Test parameter: EC50
Effect: Reduction in light output

Concentration: 1.18 mg/l
Exposure time: 5 min

Concentration: 0.78 mg/l
Exposure time: 15 min

Concentration: 0.68 mg/l
Exposure time: 30 min

Comment: Mean of three assays. Methanol (<10%) was used to prepare the stock solutions. EC50 values were calculated from nominal concentrations.

Bibliographical reference: Kaiser, K.L.E., and Palabrica, V.S. (1991). *Water Poll. Res. J. Canada* **26**, 361-431.

2-MERCAPTOBENZOXAZOLE

CAS RN: 2382-96-9
Synonym: 2-Benzoxazolethiol

Sample purity: 95%
Temperature: 15°C
Test parameter: EC50
Effect: Reduction in light output

Concentrations: 3.09 mg/l
12.3 mg/l
Exposure time: 5 min

Concentrations: 2.95 mg/l
11.5 mg/l
Exposure time: 15 min

Concentrations: 2.88 mg/l
11.5 mg/l
Exposure time: 30 min

Comment: The two series of tests were performed four times. Methanol (<10%) was used to prepare the stock solutions. EC50 values were calculated from nominal concentrations.

Bibliographical reference: Kaiser, K.L.E., and Palabrica, V.S. (1991). *Water Poll. Res. J. Canada* **26**, 361-431.

2-MERCAPTOETHANOL

CAS RN: 60-24-2

Sample purity: 98%
Temperature: 15°C
Test parameter: EC50
Effect: Reduction in light output

Concentration: 541 mg/l
Exposure times: 5 and 15 min

Concentration: 579 mg/l
Exposure time: 30 min

Comment: Mean of three assays. Methanol (<10%) was used to prepare the stock solutions. EC50 values were calculated from nominal concentrations.

Bibliographical reference: Kaiser, K.L.E., and Palabrica, V.S. (1992). National Water Research Institute, Burlington, Ontario, Canada, unpublished results.

2-MERCAPTOIMIDAZOLE

CAS RN: 872-35-5
Synonym: 2-Imidazolethiol

Sample purity: 98%
Temperature: 15°C
Test parameter: EC50
Effect: Reduction in light output

Concentration: 252 mg/l
Exposure times: 5 and 15 min

Concentration: 296 mg/l
Exposure time: 30 min

Comment: Mean of three assays. Methanol (<10%) was used to prepare the stock solutions. EC50 values were calculated from nominal concentrations.

Bibliographical reference: Kaiser, K.L.E., and Palabrica, V.S. (1991). *Water Poll. Res. J. Canada* **26**, 361-431.

2-MERCAPTO-1-METHYLIMIDAZOLE

CAS RN: 60-56-0
Synonym: Methimazole

Sample purity: 98%
Temperature: 15°C
Test parameter: EC50
Effect: Reduction in light output

Concentration: 522 mg/l
Exposure times: 5 and 15 min

Concentration: 546 mg/l
Exposure time: 30 min

Comment: Mean of four assays. Methanol (<10%) was used to prepare the stock solutions. EC50 values were calculated from nominal concentrations.

Bibliographical reference: Kaiser, K.L.E., and Palabrica, V.S. (1991). *Water Poll. Res. J. Canada* **26**, 361-431.

Sample purity: >99%
Temperature: 15°C
Test parameter: EC50
Effect: Reduction in light output

Concentration: 337 mg/l
Exposure time: 5 min

Concentration: 293 mg/l
Exposure time: 15 min

Concentration: 244 mg/l
Exposure time: 30 min

Comment: Mean of four assays. Methanol (<10%) was used to prepare the stock solutions. EC50 values were calculated from nominal concentrations.

Bibliographical reference: Kaiser, K.L.E., and Palabrica, V.S. (1991). *Water Poll. Res. J. Canada* **26**, 361-431.

2-MERCAPTOPYRIDINE

CAS RN: 2637-34-5
Synonym: 2-Pyridinethiol

Sample purity: 99%
Temperature: 15°C
Test parameter: EC50
Effect: Reduction in light output

Concentration: 2.43 mg/l
Exposure time: 5 min

Concentration: 2.32 mg/l
Exposure time: 15 min

Concentration: 3.21 mg/l
Exposure time: 30 min

Comment: Mean of four assays. Methanol (<10%) was used to prepare the stock solutions. EC50 values were calculated from nominal concentrations.

Bibliographical reference: Kaiser, K.L.E., and Palabrica, V.S. (1991). *Water Poll. Res. J. Canada* **26**, 361-431.

MERCAPTOTRIAZOLE

CAS RN: 3179-31-5

Temperature: 15°C
Test parameter: EC50
Effect: Reduction in light output
Concentrations: 100 mg/l
148 mg/l
Exposure time: 5 min

Bibliographical reference: King, E.F., and Painter, H.A. (1981). In: *Les Tests de Toxicité Aiguë en Milieu Aquatique. Acute Aquatic Ecotoxicological Tests*, H. Leclerc and D. Dive (eds.), INSERM 106, Paris, p. 143-153.

MERCURY(II) CHLORIDE

CAS RN: 7487-94-7

Temperature: 15°C
Test parameter: EC50
Effect: Reduction in light output

Concentration: 0.064 mg/l Hg^{++}
Exposure time: 5 min

Concentration: 0.049 mg/l Hg^{++}
Exposure time: 10 min

Concentration: 0.046 mg/l Hg^{++}
Exposure time: 15 min

Comment: Chemical was tested at pH 6.7.

Bibliographical references: Dutka, B.J., and Kwan, K.K. (1982). *Environ. Pollut. Ser. A* **29**, 125-134.
Dutka, B.J., Nyholm, N., and Petersen, J. (1983). *Water Res.* **17**, 1363-1368.

Effect: Percentage light inhibition
Result:

Chemical	incubation time		
ppm	5-min	10-min	15-min
0.025 Hg^{++}	2%	7%	8%
0.05 Hg^{++}	32%	38%	36%
0.075 Hg^{++}	64%	93%	96%
5.0 Pb^{++}($PbCl_2$)	19%	17%	17%
5 Ni^{++}($NiCl_2.6H_2O$)	18%	18%	26%
0.05 Hg^{++} + 5 Pb^{++}	19%	25%	20%
0.075 Hg^{++} + 5 Pb^{++}	54%	84%	94%
0.05 Hg^{++} + 5 Ni^{++} + 5 Pb^{++}	27%	30%	30%
0.075 Hg^{++} + 5 Ni^{++} + 5 Pb^{++}	58%	86%	95%

Comment: Mean of duplicated tests. Chemicals were tested at pH 6.7.

Bibliographical reference: Dutka, B.J., and Kwan, K.K. (1982). *Environ. Pollut. Ser. A* **29**, 125-134.

Sample purity: Reagent grade
Test parameter: EC50
Effect: Reduction in light output

Concentration: 0.044 mg/l Hg^{++} (0.040-0.049)
Exposure time: 5 min

Concentration: 0.029 mg/l Hg^{++} (0.024-0.036)
Exposure times: 15 and 30 min

Bibliographical reference: Elnabarawy, M.T., Robideau, R.R., and Beach, S.A. (1988). *Tox. Assess.* **3**, 361-370.

METHANOL

CAS RN: 67-56-1

Test parameter: EC50
Effect: Reduction in light output
Concentration: 125000 mg/l
Exposure time: 5 min
Comment: The EC50 value was calculated from nominal concentrations.

Bibliographical reference: Curtis, C., Lima, A., Lozano, S.J., and Veith, G.D. (1982). In: *Aquatic Toxicology and Hazard Assessment: Fifth Conference, ASTM STP 766*, J.G. Pearson, R.B. Foster, and W.E. Bishop (eds.), American Society for Testing and Materials, Philadelphia, p. 170-178.

Temperature: 15 ± 0.1°C
Test parameter: EC50
Effect: Reduction in light output
Concentration: 56715 mg/l
Exposure time: 5 min
Comment: Test was performed on *Photobacterium phosphoreum* NZ11D obtained from the Scripps Institute of Oceanography (La Jolla, CA).

Bibliographical reference: McFeters, G.A., Bond, P.J., Olson, S.B., and Tchan, Y.T. (1983). *Water Res.* **17**, 1757-1762.

Test parameter: EC50
Effect: Reduction in light output

Concentration: 158000 mg/l
Exposure time: 5 min

Concentration: 284400 mg/l
Exposure time: 15 min

Concentration: 323900 mg/l
Exposure time: 30 min

Comment: Two replicates of each concentration were tested.

Bibliographical reference: Greene, J.C., Miller, W.E., Debacon, M.K., Long, M.A., and Bartels, C.L. (1985). *Arch. Environ. Contam. Toxicol.* **14**, 659-667.

Temperature: 15°C
Test parameter: EC50
Effect: Reduction in light output
Concentration: 42237 mg/l
Exposure time: 15 min

Bibliographical reference: Hermens, J., Busser, F., Leeuwangh, P., and Musch, A. (1985). *Ecotoxicol. Environ. Safety* **9**, 17-25.

Temperature: 15°C
Test parameter: EC50
Effect: Reduction in light output
Concentration: 18.62 μl/ml
Exposure time: 15 min
Comment: Four concentrations of chemical were tested in duplicate.

Bibliographical reference: Schiewe, M.H., Hawk, E.G., Actor, D.I., and Krahn, M.M. (1985). *Can. J. Fish. Aquat. Sci.* **42**, 1244-1248.

Temperature: 15°C
Test parameter: EC50
Effect: Reduction in light output

Concentration: 54294 mg/l
Exposure time: 5 min

Concentration: 50420 mg/l
Exposure time: 30 min

Bibliographical reference: Speece, R. (1987). Drexel University, Philadelphia, USA, private communication.

Temperature: 15 ± 0.5°C
Test parameter: EC20
Effect: Reduction in light output

Concentration: 3600 mg/l (3500-3700)
Exposure time: 15 min
Comment: Osmotic adjustment was achieved using NaCl (2.0%).

Concentration: 1700 mg/l (1500-1900)
Exposure time: 15 min
Comment: Osmotic adjustment was achieved using sucrose (20.4%).

Bibliographical reference: Ankley, G.T., Peterson, G.S., Amato, J.R., and Jenson, J.J. (1990). *Environ. Toxicol. Chem.* **9**, 1305-1310.

Sample purity: Optima grade, Fisher
Temperature: 15°C
Test parameter: EC50
Effect: Reduction in light output

Concentration: 43000 mg/l
Exposure time: 5 min

Concentration: 40000 mg/l
Exposure time: 15 min

Concentration: 39000 mg/l
Exposure time: 25 min

Bibliographical reference: Chou, C.C., and Que Hee, S.S. (1992). *Ecotoxicol. Environ. Safety* **23**, 355-363.

Temperature: 15°C
Test parameter: EC50
Effect: Reduction in light output
Concentration: 105200 mg/l
Exposure time: 5 min

Bibliographical reference: Kahru, A. (1993). *ATLA* **21**, 210-215.

3'-METHOXYACETOPHENONE

CAS RN: 586-37-8

Sample purity: 99%
Temperature: 15°C
Test parameter: EC50
Effect: Reduction in light output

Concentration: 3.52 mg/l
Exposure time: 5 min

Concentration: 3.69 mg/l
Exposure time: 15 min

Concentration: 3.44 mg/l
Exposure time: 30 min

Comment: Mean of three assays. Methanol (<10%) was used to prepare the stock solutions. EC50 values were calculated from nominal concentrations.

Bibliographical reference: Kaiser, K.L.E., and Palabrica, V.S. (1991). *Water Poll. Res. J. Canada* **26**, 361-431.

4'-METHOXYACETOPHENONE

CAS RN: 100-06-1

Sample purity: 98%
Temperature: 15°C
Test parameter: EC50
Effect: Reduction in light output

Concentration: 8.45 mg/l
Exposure time: 5 min

Concentration: 8.25 mg/l
Exposure time: 15 min

Concentration: 7.53 mg/l
Exposure time: 30 min

Comment: Mean of three assays. Methanol (<10%) was used to prepare the stock solutions. EC50 values were calculated from nominal concentrations.

Bibliographical reference: Kaiser, K.L.E., and Palabrica, V.S. (1991). *Water Poll. Res. J. Canada* **26**, 361-431.

4-METHOXYAZOBENZENE

CAS RN: 2396-60-3

Sample purity: 99%
Temperature: 15°C
Test parameter: EC50

Effect: Reduction in light output

Concentration: 0.10 mg/l
Exposure time: 5 min

Concentration: 0.11 mg/l
Exposure time: 15 min

Concentration: 0.11 mg/l*
Exposure time: 30 min

Comment: Mean of three assays. Methanol (<10%) was used to prepare the stock solutions. EC50 values were calculated from nominal concentrations.

Bibliographical references: Kaiser, K.L.E., and Palabrica, V.S. (1991). *Water Poll. Res. J. Canada* **26**, 361-431.
* Kaiser, K.L.E., Palabrica, V.S., and Ribo, J.M. (1987). In: *QSAR in Environmental Toxicology - II*, K.L.E. Kaiser (ed.), D. Reidel Publishing Company, Dordrecht, p. 153-168.

4-METHOXYBENZAMIDINE

CAS RN: 22265-37-8

Temperature: 15°C
Test parameter: EC50
Effect: Reduction in light output

Concentration: 7.53 mg/l
Exposure time: 5 min

Concentration: 7.19 mg/l
Exposure time: 15 min

Concentration: 8.64 mg/l
Exposure time: 30 min

Comment: Mean of three assays. Methanol (<10%) was used to prepare the stock solutions. EC50 values were calculated from nominal concentrations.

Bibliographical reference: Kaiser, K.L.E., and Palabrica, V.S. (1991). *Water Poll. Res. J. Canada* **26**, 361-431.

4-METHOXYBENZENE SULFONAMIDE

CAS RN: 1129-26-6

Temperature: 15°C
Test parameter: EC50
Effect: Reduction in light output

Concentration: 89.6 mg/l
Exposure time: 5 min

Concentration: 93.8 mg/l
Exposure time: 15 min

Concentration: 113 mg/l
Exposure time: 30 min

Comment: Mean of three assays. Methanol (<10%) was used to prepare the stock solutions. EC50 values were calculated from nominal concentrations.

Bibliographical reference: Kaiser, K.L.E., and Palabrica, V.S. (1991). *Water Poll. Res. J. Canada* **26**, 361-431.

4-METHOXYBENZONITRILE

CAS RN: 874-90-8

Sample purity: 99%
Temperature: 15°C
Test parameter: EC50
Effect: Reduction in light output

Concentration: 7.32 mg/l
Exposure time: 5 min

Concentration: 6.99 mg/l
Exposure time: 15 min

Concentration: 5.95 mg/l*
Exposure time: 30 min

Comment: Mean of three assays. Methanol (<10%) was used to prepare the stock solutions. EC50 values were calculated from nominal concentrations.

Bibliographical references: Kaiser, K.L.E., and Palabrica, V.S. (1991). *Water Poll. Res. J. Canada* **26**, 361-431.
* Kaiser, K.L.E., and Gough, K.M. (1989). In: *Aquatic Toxicology and Environmental Fate: Eleventh Volume, ASTM STP 1007*, G.W. Suter and M.A. Lewis (eds.), American Society for Testing and Materials, Philadelphia, p. 424-441.

4-METHOXYBENZYL ALCOHOL

CAS RN: 105-13-5
Synonym: *p*-Anisyl alcohol

Sample purity: 98%
Temperature: 15°C
Test parameter: EC50
Effect: Reduction in light output

Concentrations: 1.07 mg/l
1.70 mg/l
Exposure time: 5 min

Concentrations: 1.05 mg/l
1.78 mg/l
Exposure time: 15 min

Concentrations: 1.02 mg/l
1.78 mg/l
Exposure time: 30 min

Comment: The two series of tests were performed in triplicate. Methanol (<10%) was used to prepare the stock solutions. EC50 values were calculated from nominal concentrations.

Bibliographical reference: Kaiser, K.L.E., and Palabrica, V.S. (1991). *Water Poll. Res. J. Canada* **26**, 361-431.

4-METHOXYBENZYLAMINE

CAS RN: 2393-23-9

Sample purity: 98%
Temperature: 15°C
Test parameter: EC50
Effect: Reduction in light output

Concentration: 25.0 mg/l
Exposure times: 5 and 15 min

Concentration: 25.5 mg/l
Exposure time: 30 min

Comment: Mean of three assays. Methanol (<10%) was used to prepare the stock solutions. EC50 values were calculated from nominal concentrations.

Bibliographical reference: Kaiser, K.L.E., and Palabrica, V.S. (1991). *Water Poll. Res. J. Canada* **26**, 361-431.

4-METHOXYBENZYL CHLORIDE

CAS RN: 824-94-2
Synonym: 4-(Chloromethyl)anisole

Sample purity: 98%
Temperature: 15°C
Test parameter: EC50
Effect: Reduction in light output

Concentration: 6.23 mg/l
Exposure time: 5 min

Concentration: 6.38 mg/l
Exposure time: 15 min

Concentration: 6.84 mg/l
Exposure time: 30 min

Comment: Mean of four assays. Methanol (<10%) was used to prepare the stock solutions. EC50 values were calculated from nominal concentrations.

Bibliographical reference: Kaiser, K.L.E., and Palabrica, V.S. (1991). *Water Poll. Res. J. Canada* **26**, 361-431.

4-METHOXYCINNAMIC ACID

CAS RN: 830-09-1

Sample purity: 97%
Temperature: 15°C
Test parameter: EC50
Effect: Reduction in light output

Concentration: 13.5 mg/l
Exposure times: 5 and 15 min

Concentration: 13.2 mg/l
Exposure time: 30 min

Comment: Mean of four assays. Methanol (<10%) was used to prepare the stock solutions. EC50 values were calculated from nominal concentrations.

Bibliographical reference: Kaiser, K.L.E., and Palabrica, V.S. (1991). *Water Poll. Res. J. Canada* **26**, 361-431.

2-METHOXYETHYLAMINE

CAS RN: 109-85-3

Sample purity: 99+%
Temperature: 15°C
Test parameter: EC50
Effect: Reduction in light output
Concentration: 24.9 mg/l
Exposure time: 5 min

Bibliographical reference: Cronin, M.T.D., Dearden, J.C., and Dobbs, A.J. (1991). *Sci. Total Environ.* **109/110**, 431-439.

4-METHOXY-3-NITROANILINE

CAS RN: 577-72-0

Sample purity: 95%
Temperature: 15°C
Test parameter: EC50
Effect: Reduction in light output

Concentration: 37.6 mg/l
Exposure time: 5 min

Concentration: 31.3 mg/l
Exposure time: 15 min

Concentration: 28.6 mg/l
Exposure time: 30 min

Comment: Mean of three assays. Methanol (<10%) was used to prepare the stock solutions. EC50 values were calculated from nominal concentrations.

Bibliographical reference: Kaiser, K.L.E., and Palabrica, V.S. (1991). *Water Poll. Res. J. Canada* **26**, 361-431.

4-METHOXYPHENETHYL ALCOHOL

CAS RN: 702-23-8

Sample purity: 98%
Temperature: 15°C
Test parameter: EC50
Effect: Reduction in light output

Concentration: 5.04 mg/l
Exposure time: 5 min

Concentration: 5.16 mg/l
Exposure time: 15 min

Concentration: 5.04 mg/l
Exposure time: 30 min

Comment: Mean of three assays. Methanol (<10%) was used to prepare the stock solutions. EC50 values were calculated from nominal concentrations.

Bibliographical reference: Kaiser, K.L.E., and Palabrica, V.S. (1991). *Water Poll. Res. J. Canada* **26**, 361-431.

3-METHOXYPHENOL

CAS RN: 150-19-6

Temperature: 15°C
Test parameter: EC50
Effect: Reduction in light output
Concentration: 55 mg/l
Exposure time: 5 min

Bibliographical reference: Microtox® Application Notes (1982). Beckman Instruments, Inc., Carlsbad, California.

4-METHOXYPHENOL

CAS RN: 150-76-5

Temperature: 15°C
Test parameter: EC50
Effect: Reduction in light output

Concentration: 3.66 mg/l
Exposure time: 5 min

Concentration: 4.30 mg/l
Exposure time: 15 min

Concentration: 4.61 mg/l
Exposure time: 30 min

Comment: Mean of three assays. Methanol (<10%) was used to prepare the stock solutions. EC50 values were calculated from nominal concentrations.

Bibliographical reference: Ribo, J.M., and Kaiser, K.L.E. (1983). *Chemosphere* **12**, 1421-1442.

2-(4-METHOXYPHENOXY)ACETIC ACID

CAS RN: 1877-75-4

Temperature: 15°C
Test parameter: EC50
Effect: Reduction in light output

Concentration: 148 mg/l
Exposure time: 5 min

Concentration: 155 mg/l
Exposure time: 15 min

Concentration: 159 mg/l
Exposure time: 30 min

Comment: Mean of four assays. Methanol (<10%) was used to prepare the stock solutions. EC50 values were calculated from nominal concentrations.

Bibliographical reference: Kaiser, K.L.E., and Palabrica, V.S. (1991). *Water Poll. Res. J. Canada* **26**, 361-431.

2-(4-METHOXYPHENOXY)ETHANOL

CAS RN: 5394-57-0

Temperature: 15°C
Test parameter: EC50
Effect: Reduction in light output

Concentration: 24.3 mg/l
Exposure time: 5 min

Concentration: 23.8 mg/l
Exposure time: 15 min

Concentration: 22.7 mg/l

Exposure time: 30 min

Comment: Mean of three assays. Methanol (<10%) was used to prepare the stock solutions. EC50 values were calculated from nominal concentrations.

Bibliographical reference: Kaiser, K.L.E., and Palabrica, V.S. (1991). *Water Poll. Res. J. Canada* **26**, 361-431.

4-METHOXYPHENYLACETIC ACID

CAS RN: 104-01-8

Sample purity: 99%
Temperature: 15°C
Test parameter: EC50
Effect: Reduction in light output

Concentration: 57.6 mg/l
Exposure time: 5 min

Concentration: 56.3 mg/l
Exposure times: 15 and 30 min

Comment: Mean of three assays. Methanol (<10%) was used to prepare the stock solutions. EC50 values were calculated from nominal concentrations.

Bibliographical reference: Kaiser, K.L.E., and Palabrica, V.S. (1991). *Water Poll. Res. J. Canada* **26**, 361-431.

4-METHOXYPHENYLACETONE

CAS RN: 122-84-9

Sample purity: 95%
Temperature: 15°C
Test parameter: EC50
Effect: Reduction in light output

Concentration: 7.17 mg/l
Exposure time: 5 min

Concentration: 7.33 mg/l
Exposure time: 15 min

Concentration: 5.96 mg/l
Exposure time: 30 min

Comment: Mean of three assays. Methanol (<10%) was used to prepare the stock solutions. EC50 values were calculated from nominal concentrations.

Bibliographical reference: Kaiser, K.L.E., and Palabrica, V.S. (1991). *Water Poll. Res. J. Canada* **26**, 361-431.

(4-METHOXYPHENYL)ACETONITRILE

CAS RN: 104-47-2

Sample purity: 97%
Temperature: 15°C
Test parameter: EC50
Effect: Reduction in light output

Concentration: 0.18 mg/l
Exposure times: 5 and 15 min

Concentration: 0.17 mg/l
Exposure time: 30 min

Comment: Mean of three assays. Methanol (<10%) was used to prepare the stock solutions. EC50 values were calculated from nominal concentrations.

Bibliographical reference: Kaiser, K.L.E., and Palabrica, V.S. (1991). *Water Poll. Res. J. Canada* **26**, 361-431.

4-METHOXYPHENYL ISOCYANATE

CAS RN: 5416-93-3

Sample purity: 99%
Temperature: 15°C
Test parameter: EC50

Effect: Reduction in light output

Concentration: 7.83 mg/l
Exposure time: 5 min

Concentration: 7.14 mg/l
Exposure time: 15 min

Concentration: 5.94 mg/l
Exposure time: 30 min

Comment: Mean of three assays. Methanol (<10%) was used to prepare the stock solutions. EC50 values were calculated from nominal concentrations.

Bibliographical reference: Kaiser, K.L.E., and Palabrica, V.S. (1991). *Water Poll. Res. J. Canada* **26**, 361-431.

4'-METHOXYPROPIOPHENONE

CAS RN: 121-97-1

Sample purity: 99+%
Temperature: 15°C
Test parameter: EC50
Effect: Reduction in light output

Concentration: 6.10 mg/l
Exposure time: 5 min

Concentration: 6.69 mg/l
Exposure time: 15 min

Concentration: 7.17 mg/l
Exposure time: 30 min

Comment: Mean of three assays. Methanol (<10%) was used to prepare the stock solutions. EC50 values were calculated from nominal concentrations.

Bibliographical reference: Kaiser, K.L.E., and Palabrica, V.S. (1991). *Water Poll. Res. J. Canada* **26**, 361-431.

2-METHOXYPYRAZINE

CAS RN: 3149-28-8

Sample purity: 95%
Temperature: 15°C
Test parameter: EC50
Effect: Reduction in light output

Concentration: 235 mg/l
Exposure time: 5 min

Concentration: 241 mg/l
Exposure time: 15 min

Concentration: 252 mg/l
Exposure time: 30 min

Comment: Mean of three assays. Methanol (<10%) was used to prepare the stock solutions. EC50 values were calculated from nominal concentrations.

Bibliographical reference: Kaiser, K.L.E., and Palabrica, V.S. (1991). *Water Poll. Res. J. Canada* **26**, 361-431.

2-METHOXYPYRIDINE

CAS RN: 1628-89-3

Sample purity: 98%
Temperature: 15°C
Test parameter: EC50
Effect: Reduction in light output

Concentration: 274 mg/l
Exposure time: 5 min

Concentration: 262 mg/l
Exposure times: 15 and 30 min

Comment: Mean of three assays. Methanol (<10%) was used to prepare the stock solutions. EC50 values were calculated from nominal concentrations.

Bibliographical reference: Kaiser, K.L.E., and Palabrica, V.S. (1991). *Water Poll. Res. J. Canada* **26**, 361-431.

6-METHOXYQUINOLINE

CAS RN: 5263-87-6

Sample purity: 98%
Temperature: 15°C
Test parameter: EC50
Effect: Reduction in light output

Concentration: 0.31 mg/l
Exposure time: 5 min

Concentration: 0.32 mg/l
Exposure time: 15 min

Concentration: 0.34 mg/l
Exposure time: 30 min

Comment: Mean of three assays. Methanol (<10%) was used to prepare the stock solutions. EC50 values were calculated from nominal concentrations.

Bibliographical reference: Kaiser, K.L.E., and Palabrica, V.S. (1991). *Water Poll. Res. J. Canada* **26**, 361-431.

4-METHOXYTHIOPHENOL

CAS RN: 696-63-9

Sample purity: 95%
Temperature: 15°C
Test parameter: EC50
Effect: Reduction in light output

Concentration: 0.42 mg/l
Exposure time: 5 min

Concentration: 0.34 mg/l
Exposure time: 15 min

Concentration: 0.33 mg/l
Exposure time: 30 min

Comment: Mean of three assays. Methanol (<10%) was used to prepare the stock solutions. EC50 values were calculated from nominal concentrations.

Bibliographical reference: Kaiser, K.L.E., and Palabrica, V.S. (1991). *Water Poll. Res. J. Canada* **26**, 361-431.

4-METHOXY-α-TOLUENETHIOL

CAS RN: 6258-60-2

Temperature: 15°C
Test parameter: EC50
Effect: Reduction in light output

Concentration: 0.24 mg/l
Exposure time: 5 min

Concentration: 0.23 mg/l
Exposure time: 15 min

Concentration: 0.38 mg/l
Exposure time: 30 min

Comment: Mean of three assays. Methanol (<10%) was used to prepare the stock solutions. EC50 values were calculated from nominal concentrations.

Bibliographical reference: Kaiser, K.L.E., and Palabrica, V.S. (1991). *Water Poll. Res. J. Canada* **26**, 361-431.

4'-METHOXY-2,2,2-TRIFLUOROACETOPHENONE

CAS RN: 711-38-6

Temperature: 15°C
Test parameter: EC50
Effect: Reduction in light output

Concentration: 22.9 mg/l
Exposure time: 5 min

Concentration: 24.5 mg/l
Exposure time: 15 min

Concentration: 27.5 mg/l
Exposure time: 30 min

Comment: Mean of three assays. Methanol (<10%) was used to prepare the stock solutions. EC50 values were calculated from nominal concentrations.

Bibliographical reference: Kaiser, K.L.E., and Palabrica, V.S. (1991). *Water Poll. Res. J. Canada* **26**, 361-431.

METHYL ACETATE

CAS RN: 79-20-9

Sample purity: 99%
Temperature: 15°C
Test parameter: EC50
Effect: Reduction in light output
Concentration: 12000 mg/l
Exposure time: 5 min

Bibliographical reference: Cronin, M.T.D., Dearden, J.C., and Dobbs, A.J. (1991). *Sci. Total Environ.* **109/110**, 431-439.

4'-METHYLACETOPHENONE

CAS RN: 122-00-9

Sample purity: 90%
Temperature: 15°C
Test parameter: EC50
Effect: Reduction in light output

Concentration: 3.29 mg/l
Exposure times: 5 and 15 min

Concentration: 3.15 mg/l
Exposure time: 30 min

Comment: Mean of three assays. Methanol (<10%) was used to prepare the stock solutions. EC50 values were calculated from nominal concentrations.

Bibliographical reference: Kaiser, K.L.E., and Palabrica, V.S. (1991). *Water Poll. Res. J. Canada* **26**, 361-431.

METHYL 4-ACETYLBENZOATE

CAS RN: 3609-53-8

Temperature: 15°C
Test parameter: EC50
Effect: Reduction in light output

Concentration: 10.3 mg/l
Exposure time: 5 min

Concentration: 11.2 mg/l
Exposure time: 15 min

Concentration: 14.2 mg/l
Exposure time: 30 min

Comment: Mean of three assays. Methanol (<10%) was used to prepare the stock solutions. EC50 values were calculated from nominal concentrations.

Bibliographical reference: Kaiser, K.L.E., and Palabrica, V.S. (1991). *Water Poll. Res. J. Canada* **26**, 361-431.

METHYLAMINE

CAS RN: 74-89-5

Temperature: 15 ± 0.1°C
Test parameter: EC50
Effect: Reduction in light output
Concentration: 34.6 mg/l

Exposure time: 5 min

Bibliographical reference: Somasundaram, L., Coats, J.R., Racke, K.D., and Stahr, H.M. (1990). *Bull. Environ. Contam. Toxicol.* **44**, 254-259.

METHYL 4-AMINOBENZOATE

CAS RN: 619-45-4

Sample purity: 98%
Temperature: 15°C
Test parameter: EC50
Effect: Reduction in light output

Concentration: 12.6 mg/l
Exposure time: 5 min

Concentration: 12.3 mg/l
Exposure time: 15 min

Concentration: 11.0 mg/l*
Exposure time: 30 min

Comment: Mean of three assays. Methanol (<10%) was used to prepare the stock solutions. EC50 values were calculated from nominal concentrations.

Bibliographical references: Kaiser, K.L.E., and Palabrica, V.S. (1991). *Water Poll. Res. J. Canada* **26**, 361-431.
* Kaiser, K.L.E. (1987). In: *QSAR in Environmental Toxicology - II*, K.L.E. Kaiser (ed.), D. Reidel Publishing Company, Dordrecht, p. 169-188.

4-(METHYLAMINO)BENZOIC ACID

CAS RN: 10541-83-0

Sample purity: 97%
Temperature: 15°C
Test parameter: EC50
Effect: Reduction in light output

Concentration: 85.0 mg/l
Exposure times: 5 and 15 min

Concentration: 79.3 mg/l
Exposure time: 30 min

Comment: Mean of three assays. Methanol (<10%) was used to prepare the stock solutions. EC50 values were calculated from nominal concentrations.

Bibliographical reference: Kaiser, K.L.E., and Palabrica, V.S. (1991). *Water Poll. Res. J. Canada* **26**, 361-431.

N-METHYLANILINE

CAS RN: 100-61-8

Sample purity: 99%
Temperature: 15°C
Test parameter: EC50
Effect: Reduction in light output

Concentration: 11.7 mg/l
Exposure time: 5 min

Concentration: 12.3 mg/l
Exposure time: 15 min

Concentration: 13.8 mg/l*
Exposure time: 30 min

Comment: Mean of four assays. Methanol (<10%) was used to prepare the stock solutions. EC50 values were calculated from nominal concentrations.

Bibliographical references: Kaiser, K.L.E., and Palabrica, V.S. (1991). *Water Poll. Res. J. Canada* **26**, 361-431.
* Kaiser, K.L.E., Palabrica, V.S., and Ribo, J.M. (1987). In: *QSAR in Environmental Toxicology - II*, K.L.E. Kaiser (ed.), D. Reidel Publishing Company, Dordrecht, p. 153-168.

METHYL 4-ANISATE

CAS RN: 121-98-2
Synonym: Methyl 4-methoxybenzoate

Sample purity: >98%
Temperature: 15°C
Test parameter: EC50
Effect: Reduction in light output

Concentration: 1.55 mg/l
Exposure time: 5 min

Concentration: 1.66 mg/l
Exposure time: 15 min

Concentration: 2.29 mg/l
Exposure time: 30 min

Comment: Mean of three assays. Methanol (<10%) was used to prepare the stock solutions. EC50 values were calculated from nominal concentrations.

Bibliographical reference: Kaiser, K.L.E., and Palabrica, V.S. (1991). *Water Poll. Res. J. Canada* **26**, 361-431.

4-METHYLANISOLE

CAS RN: 104-93-8

Sample purity: 99%
Temperature: 15°C
Test parameter: EC50
Effect: Reduction in light output

Concentration: 2.80 mg/l
Exposure time: 5 min

Concentration: 3.00 mg/l
Exposure time: 15 min

Concentration: 3.52 mg/l
Exposure time: 30 min

Comment: Mean of three assays. Methanol (<10%) was used to prepare the stock solutions. EC50 values were calculated from nominal concentrations.

Bibliographical reference: Kaiser, K.L.E., and Palabrica, V.S. (1991). *Water Poll. Res. J. Canada* **26**, 361-431.

4-METHYLBENZALDEHYDE

CAS RN: 104-87-0

Sample purity: 98%
Temperature: 15°C
Test parameter: EC50
Effect: Reduction in light output

Concentration: 4.57 mg/l
Exposure time: 5 min

Concentration: 5.01 mg/l
Exposure time: 15 min

Concentration: 6.45 mg/l
Exposure time: 30 min

Comment: Mean of three assays. Methanol (<10%) was used to prepare the stock solutions. EC50 values were calculated from nominal concentrations.

Bibliographical reference: Kaiser, K.L.E., and Palabrica, V.S. (1991). *Water Poll. Res. J. Canada* **26**, 361-431.

4-METHYLBENZENE ACETONITRILE

CAS RN: 2947-61-7

Sample purity: 98%
Temperature: 15°C
Test parameter: EC50
Effect: Reduction in light output

Concentration: 0.11 mg/l

Exposure time: 5 min

Concentration: 0.12 mg/l
Exposure times: 15 and 30 min

Comment: Mean of three assays. Methanol (<10%) was used to prepare the stock solutions. EC50 values were calculated from nominal concentrations.

Bibliographical reference: Kaiser, K.L.E., and Palabrica, V.S. (1991). *Water Poll. Res. J. Canada* **26**, 361-431.

METHYL BENZOATE

CAS RN: 93-58-3

Sample purity: 99%
Temperature: 15°C
Test parameter: EC50
Effect: Reduction in light output

Concentration: 4.41 mg/l
Exposure time: 5 min

Concentration: 4.21 mg/l
Exposure time: 15 min

Concentration: 4.61 mg/l*
Exposure time: 30 min

Comment: Mean of three assays. Methanol (<10%) was used to prepare the stock solutions. EC50 values were calculated from nominal concentrations.

Bibliographical references: Kaiser, K.L.E., and Palabrica, V.S. (1991). *Water Poll. Res. J. Canada* **26**, 361-431.
* Kaiser, K.L.E., Palabrica, V.S., and Ribo, J.M. (1987). In: *QSAR in Environmental Toxicology - II*, K.L.E. Kaiser (ed.), D. Reidel Publishing Company, Dordrecht, p. 153-168.

4-METHYLBENZYL ALCOHOL

CAS RN: 589-18-4

Sample purity: 98%
Temperature: 15°C
Test parameter: EC50
Effect: Reduction in light output

Concentration: 14.0 mg/l
Exposure time: 5 min

Concentration: 14.4 mg/l
Exposure time: 15 min

Concentration: 12.8 mg/l
Exposure time: 30 min

Comment: Mean of three assays. Methanol (<10%) was used to prepare the stock solutions. EC50 values were calculated from nominal concentrations.

Bibliographical reference: Kaiser, K.L.E., and Palabrica, V.S. (1991). *Water Poll. Res. J. Canada* **26**, 361-431.

4-METHYLBENZYLAMINE

CAS RN: 104-84-7

Sample purity: 98%
Temperature: 15°C
Test parameter: EC50
Effect: Reduction in light output

Concentration: 20.6 mg/l
Exposure time: 5 min

Concentration: 22.1 mg/l
Exposure time: 15 min

Concentration: 25.9 mg/l
Exposure time: 30 min

Comment: Mean of three assays. Methanol (<10%) was used to prepare the stock solutions. EC50 values were calculated from nominal concentrations.

Bibliographical reference: Kaiser, K.L.E., and Palabrica, V.S. (1991). *Water Poll. Res. J. Canada* **26**, 361-431.

METHYL 4-BROMOBENZOATE

CAS RN: 619-42-1

Temperature: 15°C
Test parameter: EC50
Effect: Reduction in light output

Concentration: 2.73 mg/l
Exposure time: 5 min

Concentration: 2.93 mg/l
Exposure time: 15 min

Concentration: 3.52 mg/l
Exposure time: 30 min

Comment: Mean of three assays. Methanol (<10%) was used to prepare the stock solutions. EC50 values were calculated from nominal concentrations.

Bibliographical reference: Kaiser, K.L.E., and Palabrica, V.S. (1991). *Water Poll. Res. J. Canada* **26**, 361-431.

3-METHYL-2-BUTANONE

CAS RN: 563-80-4
Synonym: Methyl isopropyl ketone

Sample purity: 99%
Temperature: 15°C
Test parameter: EC50
Effect: Reduction in light output

Concentration: 64 mg/l

Exposure time: 5 min

Concentration: 85 mg/l
Exposure time: 15 min

Bibliographical reference: Cronin, M.T.D. (1993). Liverpool John Moores University, UK, private communication.

9-METHYLCARBAZOLE

CAS RN: 1484-12-4

Sample purity: 99%
Temperature: 15°C
Test parameter: EC50
Effect: Reduction in light output

Concentration: 1.28 mg/l
Exposure time: 5 min

Concentration: 1.65 mg/l
Exposure time: 15 min

Concentration: 1.81 mg/l
Exposure time: 30 min

Comment: Mean of three assays. Methanol (<10%) was used to prepare the stock solutions. EC50 values were calculated from nominal concentrations.

Bibliographical reference: Kaiser, K.L.E., and Palabrica, V.S. (1991). *Water Poll. Res. J. Canada* **26**, 361-431.

4-METHYLCATECHOL

CAS RN: 452-86-8

Sample purity: 98+%
Temperature: 15°C
Test parameter: EC50
Effect: Reduction in light output

Concentration: 0.54 mg/l
Exposure time: 5 min

Concentration: 0.46 mg/l
Exposure time: 15 min

Concentration: 0.43 mg/l
Exposure time: 30 min

Comment: Mean of three assays. Methanol (<10%) was used to prepare the stock solutions. EC50 values were calculated from nominal concentrations.

Bibliographical reference: Kaiser, K.L.E., and Palabrica, V.S. (1991). *Water Poll. Res. J. Canada* **26**, 361-431.

METHYL 4-CHLOROBENZOATE

CAS RN: 1126-46-1

Sample purity: 99%
Temperature: 15°C
Test parameter: EC50
Effect: Reduction in light output

Concentration: 3.25 mg/l
Exposure time: 5 min

Concentration: 3.40 mg/l
Exposure time: 15 min

Concentration: 3.65 mg/l*
Exposure time: 30 min

Comment: Mean of three assays. Methanol (<10%) was used to prepare the stock solutions. EC50 values were calculated from nominal concentrations.

Bibliographical references: Kaiser, K.L.E., and Palabrica, V.S. (1991). *Water Poll. Res. J. Canada* **26**, 361-431.
* Kaiser, K.L.E. (1987). In: *QSAR in Environmental Toxicology - II*, K.L.E. Kaiser (ed.), D. Reidel Publishing Company, Dordrecht, p. 169-188.

METHYL 4-CYANOBENZOATE

CAS RN: 1129-35-7

Sample purity: 99%
Temperature: 15°C
Test parameter: EC50
Effect: Reduction in light output

Concentration: 20.3 mg/l
Exposure time: 5 min

Concentration: 17.3 mg/l
Exposure times: 15 and 30 min

Comment: Mean of three assays. Methanol (<10%) was used to prepare the stock solutions. EC50 values were calculated from nominal concentrations.

Bibliographical reference: Kaiser, K.L.E., and Palabrica, V.S. (1991). *Water Poll. Res. J. Canada* **26**, 361-431.

METHYLENE DITHIOCYANATE

CAS RN: 6317-18-6

Temperature: 15°C
Test parameter: EC50
Effect: Reduction in light output
Concentrations: 0.10 mg/l
0.24 mg/l
Exposure time: 5 min

Bibliographical reference: King, E.F., and Painter, H.A. (1981). In: *Les Tests de Toxicité Aiguë en Milieu Aquatique. Acute Aquatic Ecotoxicological Tests*, H. Leclerc and D. Dive (eds.), INSERM 106, Paris, p. 143-153.

Sample purity: 99%
Temperature: 15°C
Test parameter: EC50
Effect: Reduction in light output
Concentration: 0.023 mg/l

Exposure time: 15 min
Comment: Different stock solutions of 0.03 g/l were prepared by shaking suspensions of chemical and distilled water overnight.

Bibliographical reference: King, E.F. (1984). In: *Toxicity Screening Procedures Using Bacterial Systems*, D. Liu and B.J. Dutka (eds.), Marcel Dekker, New York, p. 175-194.

Sample purity: 95%
Test parameter: EC50
Effect: Reduction in light output
Concentration: 0.054 mg/l (0.049-0.059)
Exposure time: 15 min
Comment: Stock solution was made in acetone.

Bibliographical reference: Maas-Diepeveen, J.L., and van Leeuwen, C.J. (1988). *Bull. Environ. Contam. Toxicol.* **40**, 517-524.

2-METHYL-8-ETHYLQUINOLINE

CAS RN: 72804-93-4

Sample purity: >97%
Temperature: 15 ± 0.3°C
Test parameter: EC50
Effect: Reduction in light output

Concentration: 14 mg/l
Exposure time: 5 min

Concentration: 13 mg/l
Exposure time: 15 min

Comment: Test was performed in duplicate at pH = 5.1. Dichloromethane was used as carrier solvent. Concentrations of the chemical were determined by high performance liquid chromatography. Hexachloroethane was used for quality control/quality assurance (5-min EC50 = 0.31 mg/l; n = 16; sd = 0.08 mg/l).

Bibliographical reference: Birkholz, D.A., Coutts, R.T., Hrudey, S.E., Danell, R.W., and Lockhart, W.L. (1990). *Water Res.* **24**, 67-73.

METHYL 4-FLUOROBENZOATE

CAS RN: 403-33-8

Temperature: 15°C
Test parameter: EC50
Effect: Reduction in light output

Concentration: 10.4 mg/l
Exposure time: 5 min

Concentration: 11.4 mg/l
Exposure time: 15 min

Concentration: 12.2 mg/l
Exposure time: 30 min

Comment: Mean of three assays. Methanol (<10%) was used to prepare the stock solutions. EC50 values were calculated from nominal concentrations.

Bibliographical reference: Kaiser, K.L.E., and Palabrica, V.S. (1991). *Water Poll. Res. J. Canada* **26**, 361-431.

6-METHYL-5-HEPTEN-2-ONE

CAS RN: 110-93-0

Test parameter: EC50
Effect: Reduction in light output
Concentration: 17.5 mg/l
Exposure time: 5 min
Comment: The EC50 value was calculated from nominal concentrations.

Bibliographical reference: Curtis, C., Lima, A., Lozano, S.J., and Veith, G.D. (1982). In: *Aquatic Toxicology and Hazard Assessment: Fifth Conference, ASTM STP 766*, J.G. Pearson, R.B. Foster, and W.E. Bishop (eds.), American Society for Testing and Materials, Philadelphia, p. 170-178.

METHYLHYDRAZINE

CAS RN: 60-34-4

Sample purity: 98%
Temperature: 15°C
Test parameter: EC50
Effect: Reduction in light output

Concentration: 27.1 mg/l
Exposure time: 5 min

Concentration: 16.3 mg/l
Exposure time: 15 min

Concentration: 15.3 mg/l
Exposure time: 30 min

Comment: Mean of three assays. Methanol (<10%) was used to prepare the stock solutions. EC50 values were calculated from nominal concentrations.

Bibliographical reference: Kaiser, K.L.E., and Palabrica, V.S. (1991). *Water Poll. Res. J. Canada* **26**, 361-431.

METHYLHYDROQUINONE

CAS RN: 95-71-6

Sample purity: 99+%
Temperature: 15°C
Test parameter: EC50
Effect: Reduction in light output

Concentration: 0.40 mg/l
Exposure time: 5 min

Concentration: 0.39 mg/l
Exposure time: 15 min

Concentration: 0.43 mg/l
Exposure time: 30 min

Comment: Mean of three assays. Methanol (<10%) was used to prepare the stock solutions. EC50 values were calculated from nominal concentrations.

Bibliographical reference: Kaiser, K.L.E., and Palabrica, V.S. (1991). *Water Poll. Res. J. Canada* **26**, 361-431.

METHYL 4-HYDROXYBENZOATE

CAS RN: 99-76-3

Sample purity: 99%
Temperature: 15°C
Test parameter: EC50
Effect: Reduction in light output

Concentration: 6.06 mg/l
Exposure time: 5 min

Concentration: 6.20 mg/l
Exposure time: 15 min

Concentration: 6.34 mg/l*
Exposure time: 30 min

Comment: Mean of three assays. Methanol (<10%) was used to prepare the stock solutions. EC50 values were calculated from nominal concentrations.

Bibliographical references: Kaiser, K.L.E., and Palabrica, V.S. (1991). *Water Poll. Res. J. Canada* **26**, 361-431.
* Kaiser, K.L.E. (1987). In: *QSAR in Environmental Toxicology - II*, K.L.E. Kaiser (ed.), D. Reidel Publishing Company, Dordrecht, p. 169-188.

METHYL ISONICOTINATE

CAS RN: 2459-09-8

Sample purity: 98%
Temperature: 15°C
Test parameter: EC50

Effect: Reduction in light output

Concentration: 198 mg/l
Exposure time: 5 min

Concentration: 208 mg/l
Exposure time: 15 min

Concentration: 217 mg/l
Exposure time: 30 min

Comment: Mean of three assays. Methanol (<10%) was used to prepare the stock solutions. EC50 values were calculated from nominal concentrations.

Bibliographical reference: Kaiser, K.L.E., and Palabrica, V.S. (1991). *Water Poll. Res. J. Canada* **26**, 361-431.

METHYL METHACRYLATE

CAS RN: 80-62-6

Sample purity: 99%
Temperature: 15°C
Test parameter: EC50
Effect: Reduction in light output

Concentration: 229 mg/l
Exposure time: 5 min

Concentration: 263 mg/l
Exposure time: 15 min

Concentration: 309 mg/l
Exposure time: 30 min

Comment: Mean of three assays. Methanol (<10%) was used to prepare the stock solutions. EC50 values were calculated from nominal concentrations.

Bibliographical reference: Kaiser, K.L.E., and Palabrica, V.S. (1992). National Water Research Institute, Burlington, Ontario, Canada, unpublished results.

METHYL 4-METHYLBENZOATE

CAS RN: 99-75-2

Sample purity: 99+%
Temperature: 15°C
Test parameter: EC50
Effect: Reduction in light output

Concentration: 1.22 mg/l
Exposure time: 5 min

Concentration: 1.47 mg/l
Exposure time: 15 min

Concentration: 1.89 mg/l
Exposure time: 30 min

Comment: Mean of three assays. Methanol (<10%) was used to prepare the stock solutions. EC50 values were calculated from nominal concentrations.

Bibliographical reference: Kaiser, K.L.E., and Palabrica, V.S. (1991). *Water Poll. Res. J. Canada* **26**, 361-431.

2-METHYL-1,4-NAPHTHOQUINONE

CAS RN: 58-27-5

Sample purity: 98%
Temperature: 15°C
Test parameter: EC50
Effect: Reduction in light output

Concentration: 0.49 mg/l
Exposure time: 5 min

Concentration: 0.27 mg/l
Exposure time: 15 min

Concentration: 0.25 mg/l
Exposure time: 30 min

Comment: Mean of three assays. Methanol (<10%) was used to prepare the stock solutions. EC50 values were calculated from nominal concentrations.

Bibliographical reference: Kaiser, K.L.E., and Palabrica, V.S. (1991). *Water Poll. Res. J. Canada* **26**, 361-431.

N-METHYL-4-NITROANILINE

CAS RN: 100-15-2

Sample purity: 97%
Temperature: 15°C
Test parameter: EC50
Effect: Reduction in light output

Concentration: 1.33 mg/l
Exposure time: 5 min

Concentration: 1.27 mg/l
Exposure time: 15 min

Concentration: 1.21 mg/l*
Exposure time: 30 min

Comment: Mean of three assays. Methanol (<10%) was used to prepare the stock solutions. EC50 values were calculated from nominal concentrations.

Bibliographical references: Kaiser, K.L.E., and Palabrica, V.S. (1991). *Water Poll. Res. J. Canada* **26**, 361-431.
* Kaiser, K.L.E. (1987). In: *QSAR in Environmental Toxicology - II*, K.L.E. Kaiser (ed.), D. Reidel Publishing Company, Dordrecht, p. 169-188.

2-METHYL-3-NITROANILINE

CAS RN: 603-83-8
Synonym: 3-Nitro-*o*-toluidine

Sample purity: 97%
Temperature: 15°C

Test parameter: EC50
Effect: Reduction in light output

Concentration: 2.90 mg/l
Exposure time: 5 min

Concentration: 2.97 mg/l
Exposure time: 15 min

Concentration: 3.18 mg/l
Exposure time: 30 min

Comment: Mean of four assays. Methanol (<10%) was used to prepare the stock solutions. EC50 values were calculated from nominal concentrations.

Bibliographical reference: Kaiser, K.L.E., and Palabrica, V.S. (1991). *Water Poll. Res. J. Canada* **26**, 361-431.

2-METHYL-4-NITROANILINE

CAS RN: 99-52-5
Synonym: 4-Nitro-*o*-toluidine

Sample purity: 97%
Temperature: 15°C
Test parameter: EC50
Effect: Reduction in light output

Concentration: 3.11 mg/l
Exposure time: 5 min

Concentration: 3.18 mg/l
Exposure time: 15 min

Concentration: 3.57 mg/l
Exposure time: 30 min

Comment: Mean of three assays. Methanol (<10%) was used to prepare the stock solutions. EC50 values were calculated from nominal concentrations.

Bibliographical reference: Kaiser, K.L.E., and Palabrica, V.S. (1991). *Water Poll. Res. J. Canada* **26**, 361-431.

2-METHYL-5-NITROANILINE

CAS RN: 99-55-8
Synonym: 5-Nitro-*o*-toluidine

Sample purity: 99%
Temperature: 15°C
Test parameter: EC50
Effect: Reduction in light output

Concentration: 14.5 mg/l
Exposure time: 5 min

Concentration: 13.9 mg/l
Exposure time: 15 min

Concentration: 15.2 mg/l
Exposure time: 30 min

Comment: Mean of three assays. Methanol (<10%) was used to prepare the stock solutions. EC50 values were calculated from nominal concentrations.

Bibliographical reference: Kaiser, K.L.E., and Palabrica, V.S. (1991). *Water Poll. Res. J. Canada* **26**, 361-431.

2-METHYL-6-NITROANILINE

CAS RN: 570-24-1
Synonym: 6-Nitro-*o*-toluidine

Sample purity: 99%
Temperature: 15°C
Test parameter: EC50
Effect: Reduction in light output

Concentration: 1.21 mg/l
Exposure time: 5 min

Concentration: 1.33 mg/l
Exposure time: 15 min

Concentration: 1.59 mg/l
Exposure time: 30 min

Comment: Mean of three assays. Methanol (<10%) was used to prepare the stock solutions. EC50 values were calculated from nominal concentrations.

Bibliographical reference: Kaiser, K.L.E., and Palabrica, V.S. (1992). National Water Research Institute, Burlington, Ontario, Canada, unpublished results.

4-METHYL-2-NITROANILINE

CAS RN: 89-62-3
Synonym: 2-Nitro-*p*-toluidine

Temperature: 15°C
Test parameter: EC50
Effect: Reduction in light output

Concentration: 5.40 mg/l
Exposure times: 5 and 15 min

Concentration: 5.92 mg/l
Exposure time: 30 min

Comment: Mean of three assays. Methanol (<10%) was used to prepare the stock solutions. EC50 values were calculated from nominal concentrations.

Bibliographical reference: Kaiser, K.L.E., and Palabrica, V.S. (1991). *Water Poll. Res. J. Canada* **26**, 361-431.

4-METHYL-3-NITROANILINE

CAS RN: 119-32-4
Synonym: 3-Nitro-*p*-toluidine

Sample purity: 97%
Temperature: 15°C
Test parameter: EC50
Effect: Reduction in light output

Concentration: 5.78 mg/l
Exposure time: 5 min

Concentration: 5.04 mg/l
Exposure time: 15 min

Concentration: 5.40 mg/l
Exposure time: 30 min

Comment: Mean of four assays. Methanol (<10%) was used to prepare the stock solutions. EC50 values were calculated from nominal concentrations.

Bibliographical reference: Kaiser, K.L.E., and Palabrica, V.S. (1991). *Water Poll. Res. J. Canada* **26**, 361-431.

5-METHYL-2-NITROANILINE

CAS RN: 578-46-1
Synonym: 6-Nitro-*m*-toluidine

Sample purity: 97%
Temperature: 15°C
Test parameter: EC50
Effect: Reduction in light output

Concentration: 0.82 mg/l
Exposure time: 5 min

Concentration: 0.90 mg/l
Exposure time: 15 min

Concentration: 1.08 mg/l
Exposure time: 30 min

Comment: Mean of three assays. Methanol (<10%) was used to prepare the stock solutions. EC50 values were calculated from nominal concentrations.

Bibliographical reference: Kaiser, K.L.E., and Palabrica, V.S. (1991). *Water Poll. Res. J. Canada* **26**, 361-431.

4-METHYL-2-NITROANISOLE

CAS RN: 119-10-8

Synonym: 1-Methoxy-4-methyl-2-nitrobenzene

Sample purity: 99%
Temperature: 15°C
Test parameter: EC50
Effect: Reduction in light output

Concentration: 9.84 mg/l
Exposure time: 5 min

Concentration: 11.0 mg/l
Exposure time: 15 min

Concentration: 13.0 mg/l
Exposure time: 30 min

Comment: Mean of three assays. Methanol (<10%) was used to prepare the stock solutions. EC50 values were calculated from nominal concentrations.

Bibliographical reference: Kaiser, K.L.E., and Palabrica, V.S. (1991). *Water Poll. Res. J. Canada* **26**, 361-431.

METHYL 3-NITROBENZOATE

CAS RN: 618-95-1

Sample purity: 99%
Temperature: 15°C
Test parameter: EC50
Effect: Reduction in light output

Concentration: 10.4 mg/l
Exposure time: 5 min

Concentration: 10.9 mg/l
Exposure time: 15 min

Concentration: 12.0 mg/l
Exposure time: 30 min

Comment: Mean of three assays. Methanol (<10%) was used to prepare the stock solutions. EC50 values were calculated from nominal concentrations.

Bibliographical reference: Kaiser, K.L.E., and Palabrica, V.S. (1991). *Water Poll. Res. J. Canada* **26**, 361-431.

METHYL 4-NITROBENZOATE

CAS RN: 619-50-1

Temperature: 15°C
Test parameter: EC50
Effect: Reduction in light output
Concentration: 14.1 mg/l
Exposure time: 30 min
Comment: Mean of three assays. Methanol (<10%) was used to prepare the stock solutions. EC50 values were calculated from nominal concentrations.

Bibliographical reference: Kaiser, K.L.E. (1987). In: *QSAR in Environmental Toxicology - II*, K.L.E. Kaiser (ed.), D. Reidel Publishing Company, Dordrecht, p. 169-188.

3-METHYL-3-NITRO-2-BUTANOL

CAS RN: 20575-38-6

Temperature: 15°C
Test parameter: EC50
Effect: Reduction in light output

Concentration: 2.66 mg/l
Exposure time: 5 min

Concentration: 2.85 mg/l
Exposure time: 15 min

Concentration: 3.19 mg/l
Exposure time: 30 min

Comment: Mean of four assays. Methanol (<10%) was used to prepare the stock solutions. EC50 values were calculated from nominal concentrations.

Bibliographical reference: Kaiser, K.L.E., and Palabrica, V.S. (1992). National Water Research Institute, Burlington, Ontario, Canada, unpublished results.

1-METHYL-3-NITRO-1-NITROSOGUANIDINE

CAS RN: 70-25-7

Temperature: 15°C
Test parameter: EC50
Effect: Reduction in light output

Concentration: 25.6 ± 1.1 mg/l
Exposure time: 5 min

Concentration: 12.8 ± 0.6 mg/l
Exposure time: 10 min

Concentration: 8.78 ± 0.2 mg/l
Exposure time: 15 min

Concentration: 6.53 ± 0.1 mg/l
Exposure time: 20 min

Comment: Test was performed at least in triplicate. Chemical was solubilized in DMSO.

Bibliographical reference: Yates, I.E. (1985). *J. Microbiol. Meth.* **3**, 171-180.

3-METHYL-4-NITROPHENOL

CAS RN: 2581-34-2

Sample purity: 98%
Temperature: 15°C
Test parameter: EC50
Effect: Reduction in light output

Concentration: 1.19 mg/l
Exposure time: 5 min

Concentration: 1.22 mg/l
Exposure time: 15 min

Concentration: 1.30 mg/l
Exposure time: 30 min

Comment: Mean of four assays. Methanol (<10%) was used to prepare the stock solutions. EC50 values were calculated from nominal concentrations.

Bibliographical reference: Kaiser, K.L.E., and Palabrica, V.S. (1991). *Water Poll. Res. J. Canada* **26**, 361-431.

4-METHYL-2-NITROPHENOL

CAS RN: 119-33-5

Temperature: 15°C
Test parameter: EC50
Effect: Reduction in light output

Concentration: 7.68 mg/l
Exposure time: 5 min

Concentration: 8.81 mg/l
Exposure time: 15 min

Concentration: 9.66 mg/l
Exposure time: 30 min

Comment: Mean of three assays. Methanol (<10%) was used to prepare the stock solutions. EC50 values were calculated from nominal concentrations.

Bibliographical reference: Kaiser, K.L.E., and Palabrica, V.S. (1992). National Water Research Institute, Burlington, Ontario, Canada, unpublished results.

METHYL ORANGE

CAS RN: 547-58-0

Sample purity: 96%
Temperature: 15°C
Test parameter: EC50
Effect: Reduction in light output

Concentration: 9.23 mg/l
Exposure time: 5 min

Concentration: 8.61 mg/l
Exposure time: 15 min

Concentration: 10.4 mg/l
Exposure time: 30 min

Comment: Mean of three assays. Methanol (<10%) was used to prepare the stock solutions. EC50 values were calculated from nominal concentrations.

Bibliographical reference: Kaiser, K.L.E., and Palabrica, V.S. (1992). National Water Research Institute, Burlington, Ontario, Canada, unpublished results.

2-METHYL-2,4-PENTANEDIOL

CAS RN: 107-41-5

Test parameter: EC50
Effect: Reduction in light output
Concentrations: 3300 mg/l
3200 mg/l
2710 mg/l
Exposure time: 5 min
Comment: Toxicity values were calculated from nominal concentrations.

Bibliographical reference: Curtis, C., Lima, A., Lozano, S.J., and Veith, G.D. (1982). In: *Aquatic Toxicology and Hazard Assessment: Fifth Conference, ASTM STP 766*, J.G. Pearson, R.B. Foster, and W.E. Bishop (eds.), American Society for Testing and Materials, Philadelphia, p. 170-178.

Temperature: 15°C
Test parameter: EC50

Effect: Reduction in light output
Concentration: 1447.5 mg/l (1386.5-1511.2)
Exposure time: 15 min
Comment: Single batches of Microtox® reagent were used for less than 2 h before being discarded. Four concentrations were tested. Concentrations were unmeasured.

Bibliographical reference: Nacci, D., Jackim, E., and Walsh, R. (1986). *Environ. Toxicol. Chem.* **5**, 521-525.

4-METHYL-2-PENTANONE

CAS RN: 108-10-1
Synonym: Methyl isobutyl ketone

Test parameter: EC50
Effect: Reduction in light output
Concentration: 80 mg/l
Exposure time: 5 min
Comment: The EC50 value was calculated from nominal concentrations.

Bibliographical reference: Curtis, C., Lima, A., Lozano, S.J., and Veith, G.D. (1982). In: *Aquatic Toxicology and Hazard Assessment: Fifth Conference, ASTM STP 766*, J.G. Pearson, R.B. Foster, and W.E. Bishop (eds.), American Society for Testing and Materials, Philadelphia, p. 170-178.

METHYL PHENYL SULFONE

CAS RN: 3112-85-4

Temperature: 15 ± 0.1°C
Test parameter: EC50
Effect: Reduction in light output
Concentration: 3.2 mg/l
Exposure time: 5 min

Bibliographical reference: Somasundaram, L., Coats, J.R., Racke, K.D., and Stahr, H.M. (1990). *Bull. Environ. Contam. Toxicol.* **44**, 254-259.

2-METHYL-1-PROPANOL

CAS RN: 78-83-1
Synonym: Isobutyl alcohol

Test parameter: EC50
Effect: Reduction in light output
Concentration: 1670 mg/l
Exposure time: 5 min
Comment: The EC50 value was calculated from nominal concentrations.

Bibliographical reference: Curtis, C., Lima, A., Lozano, S.J., and Veith, G.D. (1982). In: *Aquatic Toxicology and Hazard Assessment: Fifth Conference, ASTM STP 766*, J.G. Pearson, R.B. Foster, and W.E. Bishop (eds.), American Society for Testing and Materials, Philadelphia, p. 170-178.

Temperature: 15°C
Test parameter: EC50
Effect: Reduction in light output
Concentration: 1224.6 mg/l (1148.1-1308.3)
Exposure time: 15 min
Comment: Single batches of Microtox® reagent were used for less than 2 h before being discarded. Four concentrations were tested. Concentrations were unmeasured.

Bibliographical reference: Nacci, D., Jackim, E., and Walsh, R. (1986). *Environ. Toxicol. Chem.* **5**, 521-525.

2-METHYLPYRAZINE

CAS RN: 109-08-0

Sample purity: >99%
Temperature: 15°C
Test parameter: EC50
Effect: Reduction in light output

Concentration: 440 mg/l
Exposure time: 5 min

Concentration: 420 mg/l

Exposure time: 15 min

Concentration: 430 mg/l
Exposure time: 30 min

Comment: Mean of three assays. Methanol (<10%) was used to prepare the stock solutions. EC50 values were calculated from nominal concentrations.

Bibliographical reference: Kaiser, K.L.E., and Palabrica, V.S. (1991). *Water Poll. Res. J. Canada* **26**, 361-431.

4-METHYLPYRIMIDINE

CAS RN: 3438-46-8

Sample purity: 99%
Temperature: 15°C
Test parameter: EC50
Effect: Reduction in light output

Concentration: 236 mg/l
Exposure time: 5 min

Concentration: 259 mg/l
Exposure time: 15 min

Concentration: 265 mg/l
Exposure time: 30 min

Comment: Mean of three assays. Methanol (<10%) was used to prepare the stock solutions. EC50 values were calculated from nominal concentrations.

Bibliographical reference: Kaiser, K.L.E., and Palabrica, V.S. (1991). *Water Poll. Res. J. Canada* **26**, 361-431.

1-METHYLPYRROLIDINE

CAS RN: 120-94-5

Temperature: 15°C

Test parameter: EC50
Effect: Reduction in light output

Concentration: 8041 mg/l
Exposure time: 5 min

Concentration: 8058 mg/l
Exposure time: 30 min

Bibliographical reference: Speece, R. (1987). Drexel University, Philadelphia, USA, private communication.

Sample purity: 97%
Temperature: 15°C
Test parameter: EC50
Effect: Reduction in light output

Concentration: 41.7 mg/l
Exposure time: 5 min

Concentration: 33.1 mg/l
Exposure time: 15 min

Concentration: 33.9 mg/l
Exposure time: 30 min

Comment: Mean of three assays. Methanol (<10%) was used to prepare the stock solutions. EC50 values were calculated from nominal concentrations.

Bibliographical reference: Kaiser, K.L.E., and Palabrica, V.S. (1992). National Water Research Institute, Burlington, Ontario, Canada, unpublished results.

4-METHYLQUINOLINE

CAS RN: 491-35-0
Synonym: Lepidine

Sample purity: 99%
Temperature: 15°C
Test parameter: EC50
Effect: Reduction in light output

Concentration: 3.85 mg/l
Exposure time: 5 min

Concentration: 4.23 mg/l
Exposure time: 15 min

Concentration: 5.08 mg/l
Exposure time: 30 min

Comment: Mean of three assays. Methanol (<10%) was used to prepare the stock solutions. EC50 values were calculated from nominal concentrations.

Bibliographical reference: Kaiser, K.L.E., and Palabrica, V.S. (1991). *Water Poll. Res. J. Canada* **26**, 361-431.

5-METHYLQUINOLINE

CAS RN: 7661-55-4

Sample purity: >97%
Temperature: 15 ± 0.3°C
Test parameter: EC50
Effect: Reduction in light output

Concentration: 0.95 mg/l
Exposure time: 5 min

Concentration: 0.98 mg/l
Exposure time: 15 min

Comment: Test was performed in duplicate at pH = 5.1. Dichloromethane was used as carrier solvent. Concentrations of the chemical were determined by high performance liquid chromatography. Hexachloroethane was used for quality control/quality assurance (5-min EC50 = 0.31 mg/l; n = 16; sd = 0.08 mg/l).

Bibliographical reference: Birkholz, D.A., Coutts, R.T., Hrudey, S.E., Danell, R.W., and Lockhart, W.L. (1990). *Water Res.* **24**, 67-73.

6-METHYLQUINOLINE

CAS RN: 91-62-3

Sample purity: >97%
Temperature: 15 ± 0.3°C
Test parameter: EC50
Effect: Reduction in light output

Concentration: 2.2 mg/l
Exposure time: 5 min

Concentration: 2.8 mg/l
Exposure time: 15 min

Comment: Test was performed in duplicate at pH = 5.1. Dichloromethane was used as carrier solvent. Concentrations of the chemical were determined by high performance liquid chromatography. Hexachloroethane was used for quality control/quality assurance (5-min EC50 = 0.31 mg/l; n = 16; sd = 0.08 mg/l).

Bibliographical reference: Birkholz, D.A., Coutts, R.T., Hrudey, S.E., Danell, R.W., and Lockhart, W.L. (1990). *Water Res.* **24**, 67-73.

8-METHYLQUINOLINE

CAS RN: 611-32-5

Sample purity: >97%
Temperature: 15 ± 0.3°C
Test parameter: EC50
Effect: Reduction in light output

Concentration: 8.8 mg/l
Exposure time: 5 min

Concentration: 9.1 mg/l
Exposure time: 15 min

Comment: Test was performed in duplicate at pH = 5.1. Dichloromethane was used as carrier solvent. Concentrations of the chemical were determined by high performance liquid chromatography. Hexachloroethane was used for quality control/quality assurance (5-min EC50 = 0.31 mg/l; n = 16; sd = 0.08 mg/l).

Bibliographical reference: Birkholz, D.A., Coutts, R.T., Hrudey, S.E., Danell, R.W., and Lockhart, W.L. (1990). *Water Res.* **24**, 67-73.

2-METHYLQUINOXALINE

CAS RN: 7251-61-8

Sample purity: 97%
Temperature: 15°C
Test parameter: EC50
Effect: Reduction in light output

Concentration: 56.1 mg/l
Exposure time: 5 min

Concentration: 57.4 mg/l
Exposure time: 15 min

Concentration: 61.5 mg/l
Exposure time: 30 min

Comment: Mean of three assays. Methanol (<10%) was used to prepare the stock solutions. EC50 values were calculated from nominal concentrations.

Bibliographical reference: Kaiser, K.L.E., and Palabrica, V.S. (1991). *Water Poll. Res. J. Canada* **26**, 361-431.

N-METHYLTHIOUREA

CAS RN: 598-52-7

Sample purity: ≥97%
Test parameter: EC50
Effect: Reduction in light output
Concentration: 1796 mg/l
Exposure time: 15 min
Comment: EC50 value converted to mg/l from the original data expressed in log(1/μmol l^{-1}) and a rounded molecular weight of 90 given by the authors.

Bibliographical reference: Govers, H., Ruepert, C., Stevens, T., and van Leeuwen, C.J. (1986). *Chemosphere* **15**, 383-393.

METHYLTINTRIS(THIOGLYCOLIC ACID ISOOCTYL ESTER)

CAS RN: 57583-34-3

Temperature: 15°C
Test parameter: EC50
Effect: Reduction in light output
Concentration: 2.30 mg/l
Exposure time: 30 min

Bibliographical reference: Steinhäuser, K.G., Amann, W., Späth, A., and Polenz, A. (1985). *Vom Wasser* **65**, 203-214.

METHYL 4-(TRIFLUOROMETHYL)BENZOATE

CAS RN: 2967-66-0

Sample purity: 99%
Temperature: 15°C
Test parameter: EC50
Effect: Reduction in light output

Concentration: 0.81 mg/l
Exposure time: 5 min

Concentration: 0.87 mg/l
Exposure time: 15 min

Concentration: 1.00 mg/l
Exposure time: 30 min

Comment: Mean of three assays. Methanol (<10%) was used to prepare the stock solutions. EC50 values were calculated from nominal concentrations.

Bibliographical reference: Kaiser, K.L.E., and Palabrica, V.S. (1991). *Water Poll. Res. J. Canada* **26**, 361-431.

METHYL VIOLET B BASE

CAS RN: 52080-58-7
Synonyms: C.I. 42535B; Solvent violet 8

Temperature: 15°C
Test parameter: EC50
Effect: Reduction in light output

Concentration: 0.27 mg/l
Exposure time: 5 min

Concentration: 0.14 mg/l
Exposure time: 15 min

Concentration: 0.12 mg/l
Exposure time: 30 min

Comment: Mean of three assays. Methanol (<10%) was used to prepare the stock solutions. EC50 values were calculated from nominal concentrations.

Bibliographical reference: Kaiser, K.L.E., and Palabrica, V.S. (1991). *Water Poll. Res. J. Canada* **26**, 361-431.

METIRAM

CAS RN: 9006-42-2

Sample purity: Lamers & Indemans (38.5% CS_2)
Test parameter: EC50
Effect: Reduction in light output
Concentration: 0.37 mg/l (0.35-0.39)
Exposure time: 15 min

Bibliographical reference: van Leeuwen, C.J., Maas-Diepeveen, J.L., Niebeek, G., Vergouw, W.H.A., Griffioen, P.S., and Luijken, M.W. (1985). *Aquat. Toxicol.* **7**, 145-164.

MITOMYCIN C

CAS RN: 50-07-7

Temperature: 15°C
Test parameter: EC50
Effect: Reduction in light output

Concentration: <16.0 mg/l
Exposure time: 5 min

Concentration: <16.1 mg/l
Exposure time: 10 min

Concentration: <15.2 mg/l
Exposure time: 15 min

Concentration: <13.7 mg/l
Exposure time: 20 min

Comment: Chemical was solubilized in DMSO.

Bibliographical reference: Yates, I.E. (1985). *J. Microbiol. Meth.* **3**, 171-180.

MONURON

CAS RN: 150-68-5
Synonym: 3-(4-Chlorophenyl)-1,1-dimethylurea

Sample purity: Agrichemical grade
Temperature: 15 ± 0.1°C
Test parameter: EC50
Effect: Reduction in light output
Concentration: 228.67 mg/l
Exposure time: 5 min
Comment: Test was performed on *Photobacterium phosphoreum* NZ11D obtained from the Scripps Institute of Oceanography (La Jolla, CA).

Bibliographical reference: McFeters, G.A., Bond, P.J., Olson, S.B., and Tchan, Y.T. (1983). *Water Res.* **17**, 1757-1762.

MORPHOLINE

CAS RN: 110-91-8

Sample purity: >99%
Temperature: 15°C
Test parameter: EC50
Effect: Reduction in light output

Concentration: 60.3 mg/l
Exposure time: 5 min

Concentration: 51.3 mg/l
Exposure time: 15 min

Concentration: 57.6 mg/l
Exposure time: 30 min

Comment: Mean of three assays. Methanol (<10%) was used to prepare the stock solutions. EC50 values were calculated from nominal concentrations.

Bibliographical reference: Kaiser, K.L.E., and Palabrica, V.S. (1991). *Water Poll. Res. J. Canada* **26**, 361-431.

4-MORPHOLINEPROPIONITRILE

CAS RN: 4542-47-6

Sample purity: 98%
Temperature: 15°C
Test parameter: EC50
Effect: Reduction in light output

Concentration: 584 mg/l
Exposure time: 5 min

Concentration: 598 mg/l
Exposure time: 15 min

Concentration: 671 mg/l
Exposure time: 30 min

Comment: Mean of three assays. Methanol (<10%) was used to prepare the stock solutions. EC50 values were calculated from nominal concentrations.

Bibliographical reference: Kaiser, K.L.E., and Palabrica, V.S. (1991). *Water Poll. Res. J. Canada* **26**, 361-431.

NABAM

CAS RN: 142-59-6
Synonym: Ethylene (bisdithiocarbamate) disodium salt

Sample purity: ≥99%
Test parameter: EC50
Effect: Reduction in light output
Concentration: 102 mg/l (96-110)
Exposure time: 15 min

Bibliographical reference: van Leeuwen, C.J., Maas-Diepeveen, J.L., Niebeek, G., Vergouw, W.H.A., Griffioen, P.S., and Luijken, M.W. (1985). *Aquat. Toxicol.* **7**, 145-164.

NAPHTHALENE

CAS RN: 91-20-3

Temperature: 15°C
Test parameter: EC50
Effect: Reduction in light output
Concentration: 2 mg/l
Exposure time: 5 min
Comment: Value derived by extrapolation.

Bibliographical reference: Samak, Q.M., and Noiseux, R. (1981). *Can. Tech. Rep. Fish. Aquat. Sci.* **990**, 288-308.

Sample purity: 99%
Temperature: 15°C
Test parameter: EC50
Effect: Reduction in light output

Concentration: 0.81 mg/l
Exposure time: 5 min

Concentration: 0.91 mg/l
Exposure time: 15 min

Concentration: 0.93 mg/l
Exposure time: 30 min

Comment: Mean of three assays. Methanol (<10%) was used to prepare the stock solutions. EC50 values were calculated from nominal concentrations.

Bibliographical reference: Kaiser, K.L.E., and Palabrica, V.S. (1991). *Water Poll. Res. J. Canada* **26**, 361-431.

Temperature: 15°C
Test parameter: EC50
Effect: Reduction in light output
Concentration: 0.68 mg/l (0.53-0.87)
Exposure time: 15 min
Comment: Chemical was added to 5 ml DMSO and placed in a bath sonicator until all visible material had dissolved.

Bibliographical reference: Jacobs, M.W., Coates, J.A., Delfino, J.J., Bitton, G., Davis, W.M., and Garcia, K.L. (1993). *Arch. Environ. Contam. Toxicol.* **24**, 461-468.

1-NAPHTHOL

CAS RN: 90-15-3

Test parameter: EC50
Effect: Reduction in light output
Concentration: 3.8 mg/l
Exposure time: 15 min
Comment: Test was performed in duplicate at pH = 6.7.

Bibliographical references: Dutka, B.J., and Kwan, K.K. (1981). *Bull. Environ. Contam. Toxicol.* **27**, 753-757.
Dutka, B.J., and Kwan, K.K. (1984). In: *Toxicity Screening Procedures Using Bacterial Systems*, D. Liu and B.J. Dutka (eds.), Marcel Dekker, New York, p. 125-138.

Test parameter: EC50
Effect: Reduction in light output
Concentration: 5.66 mg/l
Exposure time: 5 min
Comment: Concentrations in the test were measured.

Bibliographical reference: Curtis, C., Lima, A., Lozano, S.J., and Veith, G.D. (1982). In: *Aquatic Toxicology and Hazard Assessment: Fifth Conference, ASTM STP 766*, J.G. Pearson, R.B. Foster, and W.E. Bishop (eds.), American Society for Testing and Materials, Philadelphia, p. 170-178.

Temperature: 15 ± 0.1°C
Test parameter: EC50
Effect: Reduction in light output
Concentration: 3.7 mg/l
Exposure time: 5 min

Bibliographical reference: Somasundaram, L., Coats, J.R., Racke, K.D., and Stahr, H.M. (1990). *Bull. Environ. Contam. Toxicol.* **44**, 254-259.

Sample purity: 99+%
Temperature: 15°C
Test parameter: EC50
Effect: Reduction in light output

Concentration: 2.23 mg/l
Exposure time: 5 min

Concentration: 2.34 mg/l
Exposure time: 15 min

Concentration: 2.81 mg/l
Exposure time: 30 min

Comment: Mean of three assays. Methanol (<10%) was used to prepare the stock solutions. EC50 values were calculated from nominal concentrations.

Bibliographical reference: Kaiser, K.L.E., and Palabrica, V.S. (1991). *Water Poll. Res. J. Canada* **26**, 361-431.

2-NAPHTHOL

CAS RN: 135-19-3

Sample purity: 99%
Temperature: 15°C
Test parameter: EC50
Effect: Reduction in light output

Concentration: 0.22 mg/l
Exposure time: 5 min

Concentration: 0.24 mg/l
Exposure time: 15 min

Concentration: 0.27 mg/l
Exposure time: 30 min

Comment: Mean of three assays. Methanol (<10%) was used to prepare the stock solutions. EC50 values were calculated from nominal concentrations.

Bibliographical reference: Kaiser, K.L.E., and Palabrica, V.S. (1991). *Water Poll. Res. J. Canada* **26**, 361-431.

1-NAPHTHYL ISOCYANATE

CAS RN: 86-84-0

Sample purity: 98%
Temperature: 15°C
Test parameter: EC50
Effect: Reduction in light output

Concentration: 0.61 mg/l
Exposure time: 5 min

Concentration: 1.58 mg/l
Exposure time: 15 min

Concentration: 1.77 mg/l
Exposure time: 30 min

Comment: Mean of three assays. Methanol (<10%) was used to prepare the stock solutions. EC50 values were calculated from nominal concentrations.

Bibliographical reference: Kaiser, K.L.E., and Palabrica, V.S. (1991). *Water Poll. Res. J. Canada* **26**, 361-431.

NICKEL CHLORIDE HEXAHYDRATE

CAS RN: 7791-20-0

Temperature: 15 ± 0.1°C
Test parameter: EC50
Effect: Reduction in light output
Concentration: 22900 mg/l
Exposure time: 5 min
Comment: Test was performed on *Photobacterium phosphoreum* NZ11D obtained from the Scripps Institute of Oceanography (La Jolla, CA).

Bibliographical reference: McFeters, G.A., Bond, P.J., Olson, S.B., and Tchan, Y.T. (1983). *Water Res.* **17**, 1757-1762.

Sample purity: Analytical grade
Temperature: 15°C
Test parameter: EC50
Effect: Reduction in light output

Concentrations: 917 mg/l Ni^{++} (time after reconstitution = 0.5 h)
1196 mg/l Ni^{++} (time after reconstitution = 4 h)
Exposure time: 5 min

Concentrations: 246 mg/l Ni^{++} (time after reconstitution = 0.5 h)
256 mg/l Ni^{++} (time after reconstitution = 4 h)
Exposure time: 15 min

Bibliographical reference: Qureshi, A.A., Coleman, R.N., and Paran, J.H. (1984). In: *Toxicity Screening Procedures Using Bacterial Systems*, D. Liu and B.J. Dutka (eds.), Marcel Dekker, New York, p. 1-22.

Sample purity: Analytical grade

Temperature: 15°C
Test parameter: EC50
Effect: Reduction in light output

Concentration: 1056 mg/l Ni^{++}
Exposure time: 5 min

Concentration: 1073 mg/l Ni^{++}
Exposure time: 10 min

Concentration: 251 mg/l Ni^{++}
Exposure time: 15 min

Concentration: 136 mg/l Ni^{++}
Exposure time: 20 min

Concentration: 42.2 mg/l Ni^{++}
Exposure time: 30 min

Comment: Results derived from the average of two replicates.

Bibliographical reference: Qureshi, A.A., Coleman, R.N., and Paran, J.H. (1984). In: *Toxicity Screening Procedures Using Bacterial Systems*, D. Liu and B.J. Dutka (eds.), Marcel Dekker, New York, p. 1-22.

NICOTINE

CAS RN: 54-11-5

Sample purity: 98% (Eastman Kodak Company, Rochester, NY)
Test parameter: EC50
Effect: Reduction in light output
Result:

Chemical	toxicity (10^{-4} mol/l)		
	5 min	15 min	25 min
Nicotine	7.77 ± 1.54	7.31 ± 1.83	7.39 ± 1.46
Cotinine*	10.6 ± 1.05	11.6 ± 1.25	11.8 ± 1.35
Nicotine/Cotinine = 0.4	15.9 ± 3.72	14.9 ± 2.37	15.3 ± 2.50
Nicotine/Cotinine = 1.0	14.1 ± 3.26	13.4 ± 2.15	13.5 ± 1.77
Nicotine/Cotinine = 1.6	14.1 ± 2.94	12.1 ± 2.44	12.0 ± 2.24

* purity 98% (Aldrich Chemical Company, Milwaukee, WI)

Bibliographical reference: Chou, C.C., and Que Hee, S.S. (1993). *J. Biolumin. Chemilumin.* **8**, 39-48.

NICOTINIC ACID

CAS RN: 59-67-6
Synonym: Niacin

Sample purity: 98%
Temperature: 15°C
Test parameter: EC50
Effect: Reduction in light output

Concentration: 162 mg/l
Exposure time: 5 min

Concentration: 186 mg/l
Exposure time: 15 min

Concentration: 214 mg/l
Exposure time: 30 min

Comment: Mean of three assays. Methanol (<10%) was used to prepare the stock solutions. EC50 values were calculated from nominal concentrations.

Bibliographical reference: Kaiser, K.L.E., and Palabrica, V.S. (1991). *Water Poll. Res. J. Canada* **26**, 361-431.

NILE BLUE A

CAS RN: 2381-85-3

Temperature: 15°C
Test parameter: EC50
Effect: Reduction in light output

Concentration: 1.22 mg/l
Exposure time: 5 min

Concentration: 0.86 mg/l
Exposure time: 15 min

Concentration: 0.75 mg/l
Exposure time: 30 min

Comment: Mean of three assays. Methanol (<10%) was used to prepare the stock solutions. EC50 values were calculated from nominal concentrations.

Bibliographical reference: Kaiser, K.L.E., and Palabrica, V.S. (1991). *Water Poll. Res. J. Canada* **26**, 361-431.

NITRILOTRIACETIC ACID

CAS RN: 139-13-9

Test parameter: EC50
Effect: Reduction in light output
Concentrations: >1000 mg/l
1000 mg/l*
Exposure time: 15 min
Comment: Test was performed in duplicate at pH = 6.7.

Bibliographical references: Dutka, B.J., and Kwan, K.K. (1981). *Bull. Environ. Contam. Toxicol.* **27**, 753-757.
* Dutka, B.J., and Kwan, K.K. (1984). In: *Toxicity Screening Procedures Using Bacterial Systems*, D. Liu and B.J. Dutka (eds.), Marcel Dekker, New York, p. 125-138.

4'-NITROACETOPHENONE

CAS RN: 100-19-6

Temperature: 15°C
Test parameter: EC50
Effect: Reduction in light output
Concentration: 27.4 mg/l
Exposure time: 30 min
Comment: Mean of three assays. Methanol (<10%) was used to prepare the stock solutions. EC50 values were calculated from nominal concentrations.

Bibliographical reference: Kaiser, K.L.E. (1987). In: *QSAR in Environmental Toxicology - II*, K.L.E. Kaiser (ed.), D. Reidel Publishing Company, Dordrecht, p. 169-188.

4-NITRO-2-AMINOPHENOL

CAS RN: 99-57-0

Temperature: 15°C
Test parameter: EC50
Effect: Reduction in light output

Concentration: 1.41 mg/l
Exposure time: 5 min

Concentration: 1.90 mg/l
Exposure time: 15 min

Concentration: 2.68 mg/l
Exposure time: 30 min

Comment: Mean of three assays. Methanol (<10%) was used to prepare the stock solutions. EC50 values were calculated from nominal concentrations.

Bibliographical reference: Kaiser, K.L.E., and Palabrica, V.S. (1992). National Water Research Institute, Burlington, Ontario, Canada, unpublished results.

2-NITROANILINE

CAS RN: 88-74-4

Sample purity: 98%
Temperature: 15°C
Test parameter: EC50
Effect: Reduction in light output

Concentration: 6.61 mg/l
Exposure time: 5 min

Concentration: 6.92 mg/l

Exposure time: 15 min

Concentration: 7.25 mg/l
Exposure time: 30 min

Comment: Mean of three assays. Methanol (<10%) was used to prepare the stock solutions. EC50 values were calculated from nominal concentrations.

Bibliographical reference: Kaiser, K.L.E., and Palabrica, V.S. (1991). *Water Poll. Res. J. Canada* **26**, 361-431.

3-NITROANILINE

CAS RN: 99-09-2

Sample purity: 98%
Temperature: 15°C
Test parameter: EC50
Effect: Reduction in light output

Concentration: 24.0 mg/l
Exposure time: 5 min

Concentration: 25.7 mg/l
Exposure time: 15 min

Concentration: 30.2 mg/l
Exposure time: 30 min

Comment: Mean of three assays. Methanol (<10%) was used to prepare the stock solutions. EC50 values were calculated from nominal concentrations.

Bibliographical reference: Kaiser, K.L.E., and Palabrica, V.S. (1991). *Water Poll. Res. J. Canada* **26**, 361-431.

4-NITROANILINE

CAS RN: 100-01-6

Sample purity: 99%

Temperature: 15°C
Test parameter: EC50
Effect: Reduction in light output

Concentration: 0.81 mg/l
Exposure time: 5 min

Concentration: 0.87 mg/l
Exposure time: 15 min

Concentrations: 1.00 mg/l
1.02 mg/l*
Exposure time: 30 min

Comment: Mean of three assays. Methanol (<10%) was used to prepare the stock solutions. EC50 values were calculated from nominal concentrations.

Bibliographical references: Kaiser, K.L.E., and Palabrica, V.S. (1991). *Water Poll. Res. J. Canada* **26**, 361-431.
* Kaiser, K.L.E. (1987). In: *QSAR in Environmental Toxicology - II*, K.L.E. Kaiser (ed.), D. Reidel Publishing Company, Dordrecht, p. 169-188.

4-NITROANISOLE

CAS RN: 100-17-4

Temperature: 15°C
Test parameter: EC50
Effect: Reduction in light output
Concentration: 16.4 mg/l
Exposure time: 30 min
Comment: Mean of three assays. Methanol (<10%) was used to prepare the stock solutions. EC50 values were calculated from nominal concentrations.

Bibliographical reference: Kaiser, K.L.E. (1987). In: *QSAR in Environmental Toxicology - II*, K.L.E. Kaiser (ed.), D. Reidel Publishing Company, Dordrecht, p. 169-188.

4-NITROBENZALDEHYDE

CAS RN: 555-16-8

Temperature: 15°C
Test parameter: EC50
Effect: Reduction in light output
Concentration: 7.07 mg/l
Exposure time: 30 min
Comment: Mean of three assays. Methanol (<10%) was used to prepare the stock solutions. EC50 values were calculated from nominal concentrations.

Bibliographical reference: Kaiser, K.L.E., and Zaruk, B.M. (1984). National Water Research Institute, Burlington, Ontario, Canada, unpublished results.

Sample purity: 99%
Temperature: 15°C
Test parameter: EC50
Effect: Reduction in light output

Concentration: 11.7 mg/l
Exposure time: 5 min

Concentration: 8.30 mg/l
Exposure time: 15 min

Concentration: 6.75 mg/l*
Exposure time: 30 min

Comment: Mean of three assays. Methanol (<10%) was used to prepare the stock solutions. EC50 values were calculated from nominal concentrations.

Bibliographical references: Kaiser, K.L.E., and Palabrica, V.S. (1991). *Water Poll. Res. J. Canada* **26**, 361-431.
* Kaiser, K.L.E. (1987). In: *QSAR in Environmental Toxicology - II*, K.L.E. Kaiser (ed.), D. Reidel Publishing Company, Dordrecht, p. 169-188.

4-NITROBENZALDOXIME

CAS RN: 1129-37-9

Sample purity: 99%
Temperature: 15°C
Test parameter: EC50
Effect: Reduction in light output

Concentration: 1.12 mg/l
Exposure time: 5 min

Concentration: 1.23 mg/l
Exposure time: 15 min

Concentration: 1.38 mg/l
Exposure time: 30 min

Comment: Mean of three assays. Methanol (<10%) was used to prepare the stock solutions. EC50 values were calculated from nominal concentrations.

Bibliographical reference: Kaiser, K.L.E., and Palabrica, V.S. (1992). National Water Research Institute, Burlington, Ontario, Canada, unpublished results.

4-NITROBENZAMIDE

CAS RN: 619-80-7

Temperature: 15°C
Test parameter: EC50
Effect: Reduction in light output
Concentration: 75.5 mg/l
Exposure time: 30 min
Comment: Mean of three assays. Methanol (<10%) was used to prepare the stock solutions. EC50 values were calculated from nominal concentrations.

Bibliographical reference: Kaiser, K.L.E. (1987). In: *QSAR in Environmental Toxicology - II*, K.L.E. Kaiser (ed.), D. Reidel Publishing Company, Dordrecht, p. 169-188.

NITROBENZENE

CAS RN: 98-95-3

Test parameter: EC50
Effect: Reduction in light output
Concentration: 17.8 mg/l
Exposure time: 15 min
Comment: The concentrations of the compound increased geometrically with a factor of 3.2. The concentrations were calculated on the basis of added amounts of material.

Bibliographical reference: Deneer, J.W., van Leeuwen, C.J., Seinen, W., Maas-Diepeveen, J.L., and Hermens, J.L.M. (1989). *Aquat. Toxicol.* **15**, 83-98.

Temperature: 15°C
Test parameter: EC50
Effect: Reduction in light output

Concentration: 28.2 mg/l
Exposure time: 5 min

Concentration: 29.5 mg/l
Exposure time: 15 min

Concentration: 34.7 mg/l*
Exposure time: 30 min

Comment: Mean of three assays. Methanol (<10%) was used to prepare the stock solutions. EC50 values were calculated from nominal concentrations.

Bibliographical references: Kaiser, K.L.E., and Palabrica, V.S. (1991). *Water Poll. Res. J. Canada* **26**, 361-431.
* Kaiser, K.L.E., and Ribo, J.M. (1985). In: *QSAR in Toxicology and Xenobiochemistry*, M. Tichy (ed.), Elsevier, Amsterdam, p. 27-38.

Temperature: 15°C
Test parameter: EC50
Effect: Reduction in light output
Concentration: 54 mg/l
Exposure time: 5 min

Bibliographical reference: Kahru, A. (1993). *ATLA* **21**, 210-215.

2-NITROBENZENESULFENYL CHLORIDE

CAS RN: 7669-54-7

Sample purity: 99%
Temperature: 15°C
Test parameter: EC50
Effect: Reduction in light output

Concentration: 3.96 mg/l
Exposure time: 5 min

Concentration: 3.37 mg/l
Exposure times: 15 and 30 min

Comment: Mean of three assays. Methanol (<10%) was used to prepare the stock solutions. EC50 values were calculated from nominal concentrations.

Bibliographical reference: Kaiser, K.L.E., and Palabrica, V.S. (1991). *Water Poll. Res. J. Canada* **26**, 361-431.

4-NITROBENZENESULFONAMIDE

CAS RN: 6325-93-5

Temperature: 15°C
Test parameter: EC50
Effect: Reduction in light output
Concentration: 0.61 mg/l
Exposure time: 30 min
Comment: Mean of three assays. Methanol (<10%) was used to prepare the stock solutions. EC50 values were calculated from nominal concentrations.

Bibliographical reference: Kaiser, K.L.E. (1987). In: *QSAR in Environmental Toxicology - II*, K.L.E. Kaiser (ed.), D. Reidel Publishing Company, Dordrecht, p. 169-188.

5-NITROBENZIMIDAZOLE

CAS RN: 94-52-0

Sample purity: 98%
Temperature: 15°C
Test parameter: EC50
Effect: Reduction in light output

Concentration: 10.5 mg/l
Exposure time: 5 min

Concentration: 10.8 mg/l
Exposure time: 15 min

Concentration: 11.3 mg/l
Exposure time: 30 min

Comment: Mean of three assays. Methanol (<10%) was used to prepare the stock solutions. EC50 values were calculated from nominal concentrations.

Bibliographical reference: Kaiser, K.L.E., and Palabrica, V.S. (1991). *Water Poll. Res. J. Canada* **26**, 361-431.

4-NITROBENZOIC ACID

CAS RN: 62-23-7

Temperature: 15°C
Test parameter: EC50
Effect: Reduction in light output
Concentration: 16.0 mg/l
Exposure time: 30 min
Comment: Mean of three assays. Methanol (<10%) was used to prepare the stock solutions. EC50 values were calculated from nominal concentrations.

Bibliographical reference: Kaiser, K.L.E. (1987). In: *QSAR in Environmental Toxicology - II*, K.L.E. Kaiser (ed.), D. Reidel Publishing Company, Dordrecht, p. 169-188.

4-NITROBENZOIC HYDRAZIDE

CAS RN: 636-97-5

Temperature: 15°C
Test parameter: EC50
Effect: Reduction in light output

Concentration: 38.7 mg/l
Exposure times: 5 and 15 min

Concentration: 39.6 mg/l
Exposure time: 30 min

Comment: Mean of four assays. Methanol (<10%) was used to prepare the stock solutions. EC50 values were calculated from nominal concentrations.

Bibliographical reference: Kaiser, K.L.E., and Palabrica, V.S. (1992). National Water Research Institute, Burlington, Ontario, Canada, unpublished results.

4-NITROBENZOPHENONE

CAS RN: 1144-74-7

Temperature: 15°C
Test parameter: EC50
Effect: Reduction in light output
Concentration: 7.02 mg/l
Exposure time: 30 min
Comment: Mean of three assays. Methanol (<10%) was used to prepare the stock solutions. EC50 values were calculated from nominal concentrations.

Bibliographical reference: Kaiser, K.L.E. (1987). In: *QSAR in Environmental Toxicology - II*, K.L.E. Kaiser (ed.), D. Reidel Publishing Company, Dordrecht, p. 169-188.

5-NITROBENZOTRIAZOLE

CAS RN: 2338-12-7

Sample purity: 98%
Temperature: 15°C
Test parameter: EC50
Effect: Reduction in light output

Concentration: 15.7 mg/l
Exposure time: 5 min

Concentration: 14.3 mg/l
Exposure time: 15 min

Concentration: 12.7 mg/l
Exposure time: 30 min

Comment: Mean of three assays. Methanol (<10%) was used to prepare the stock solutions. EC50 values were calculated from nominal concentrations.

Bibliographical reference: Kaiser, K.L.E., and Palabrica, V.S. (1991). *Water Poll. Res. J. Canada* **26**, 361-431.

4-NITROBENZOYL CHLORIDE

CAS RN: 122-04-3

Temperature: 15°C
Test parameter: EC50
Effect: Reduction in light output
Concentration: 19.0 mg/l
Exposure time: 30 min
Comment: Mean of three assays. Methanol (<10%) was used to prepare the stock solutions. EC50 values were calculated from nominal concentrations.

Bibliographical reference: Kaiser, K.L.E. (1987). In: *QSAR in Environmental Toxicology - II*, K.L.E. Kaiser (ed.), D. Reidel Publishing Company, Dordrecht, p. 169-188.

4-NITROBENZYL ACETATE

CAS RN: 619-90-9

Temperature: 15°C
Test parameter: EC50
Effect: Reduction in light output

Concentration: 9.13 mg/l
Exposure time: 5 min

Concentration: 9.34 mg/l
Exposure times: 15 and 30 min

Comment: Mean of three assays. Methanol (<10%) was used to prepare the stock solutions. EC50 values were calculated from nominal concentrations.

Bibliographical reference: Kaiser, K.L.E., and Palabrica, V.S. (1991). *Water Poll. Res. J. Canada* **26**, 361-431.

4-NITROBENZYL ALCOHOL

CAS RN: 619-73-8

Sample purity: 97%
Temperature: 15°C
Test parameter: EC50
Effect: Reduction in light output

Concentration: 32.0 mg/l
Exposure time: 5 min

Concentration: 33.5 mg/l
Exposure time: 15 min

Concentration: 35.9 mg/l*
Exposure time: 30 min

Comment: Mean of three assays. Methanol (<10%) was used to prepare the stock solutions. EC50 values were calculated from nominal concentrations.

Bibliographical references: Kaiser, K.L.E., and Palabrica, V.S. (1991). *Water Poll. Res. J. Canada* **26**, 361-431.
* Kaiser, K.L.E. (1987). In: *QSAR in Environmental Toxicology - II*, K.L.E. Kaiser (ed.), D. Reidel Publishing Company, Dordrecht, p. 169-188.

4-NITROBENZYL BROMIDE

CAS RN: 100-11-8

Sample purity: 99%
Temperature: 15°C
Test parameter: EC50
Effect: Reduction in light output

Concentration: 0.57 mg/l
Exposure time: 5 min

Concentration: 0.34 mg/l
Exposure time: 15 min

Concentration: 0.20 mg/l
Exposure time: 30 min

Comment: Mean of three assays. Methanol (<10%) was used to prepare the stock solutions. EC50 values were calculated from nominal concentrations.

Bibliographical reference: Kaiser, K.L.E., and Palabrica, V.S. (1992). National Water Research Institute, Burlington, Ontario, Canada, unpublished results.

4-NITROBENZYL CHLORIDE

CAS RN: 100-14-1

Sample purity: 99%
Temperature: 15°C
Test parameter: EC50
Effect: Reduction in light output

Concentration: 2.31 mg/l
Exposure time: 5 min

Concentration: 2.48 mg/l
Exposure time: 15 min

Concentration: 2.60 mg/l*
Exposure time: 30 min

Comment: Mean of two assays. Methanol (<10%) was used to prepare the stock solutions. EC50 values were calculated from nominal concentrations.

Bibliographical references: Kaiser, K.L.E., and Palabrica, V.S. (1991). *Water Poll. Res. J. Canada* **26**, 361-431.
* Kaiser, K.L.E. (1987). In: *QSAR in Environmental Toxicology - II*, K.L.E. Kaiser (ed.), D. Reidel Publishing Company, Dordrecht, p. 169-188.

4-NITROBENZYL CHLOROFORMATE

CAS RN: 4457-32-3

Sample purity: 97%
Temperature: 15°C
Test parameter: EC50
Effect: Reduction in light output

Concentration: 7.83 mg/l
Exposure time: 5 min

Concentration: 8.58 mg/l
Exposure time: 15 min

Concentration: 8.99 mg/l
Exposure time: 30 min

Comment: Mean of two assays. Methanol (<10%) was used to prepare the stock solutions. EC50 values were calculated from nominal concentrations.

Bibliographical reference: Kaiser, K.L.E., and Palabrica, V.S. (1991). *Water Poll. Res. J. Canada* **26**, 361-431.

4-(4-NITROBENZYL)PYRIDINE

CAS RN: 1083-48-3

Sample purity: 98%
Temperature: 15°C
Test parameter: EC50

Effect: Reduction in light output

Concentration: 2.58 mg/l
Exposure times: 5 and 15 min

Concentration: 2.70 mg/l
Exposure time: 30 min

Comment: Mean of four assays. Methanol (<10%) was used to prepare the stock solutions. EC50 values were calculated from nominal concentrations.

Bibliographical reference: Kaiser, K.L.E., and Palabrica, V.S. (1991). *Water Poll. Res. J. Canada* **26**, 361-431.

3-NITROBIPHENYL

CAS RN: 2113-58-8

Sample purity: 98+%
Temperature: 15°C
Test parameter: EC50
Effect: Reduction in light output

Concentration: 0.58 mg/l
Exposure time: 5 min

Concentration: 0.72 mg/l
Exposure time: 15 min

Concentration: 0.91 mg/l
Exposure time: 30 min

Comment: Mean of three assays. Methanol (<10%) was used to prepare the stock solutions. EC50 values were calculated from nominal concentrations.

Bibliographical reference: Kaiser, K.L.E., and Palabrica, V.S. (1992). National Water Research Institute, Burlington, Ontario, Canada, unpublished results.

1-NITROBUTANE

CAS RN: 627-05-4

Sample purity: 98%
Temperature: 15°C
Test parameter: EC50
Effect: Reduction in light output

Concentration: 54.1 mg/l
Exposure times: 5 and 15 min

Concentration: 58.0 mg/l
Exposure time: 30 min

Comment: Mean of three assays. Methanol (<10%) was used to prepare the stock solutions. EC50 values were calculated from nominal concentrations.

Bibliographical reference: Kaiser, K.L.E., and Palabrica, V.S. (1991). *Water Poll. Res. J. Canada* **26**, 361-431.

Sample purity: Aldrich, Steinheim (Germany)
Temperature: 15°C
Test parameter: EC50
Effect: Reduction in light output

Concentrations: 54.1 mg/l
58.0 mg/l
49.9 mg/l
Exposure time: 5 min

Concentrations: 46.9 mg/l
55.6 mg/l
32.7 mg/l
Exposure time: 15 min

Concentrations: 58.6 mg/l
59.4 mg/l
36.1 mg/l
Exposure time: 30 min

Comment: The photoluminometer "Lumistox" and the photobacteria were purchased from Dr. Lange, GmbH, Germany. The tests were performed in open cuvettes. The pH of the solutions was 7.0 ± 0.2.

Bibliographical reference: Thumm, W., Brüggemann, R., Freitag, D., and Kettrup, A. (1992). *Chemosphere* **24**, 1835-1843.

2-NITROBUTANE

CAS RN: 600-24-8

Sample purity: K&K Laboratories, Cleveland (USA)
Temperature: 15°C
Test parameter: EC50
Effect: Reduction in light output

Concentrations: 24.3 mg/l
34.2 mg/l
28.2 mg/l
Exposure time: 5 min

Concentrations: 33.5 mg/l
31.8 mg/l
27.5 mg/l
Exposure time: 15 min

Concentrations: 40.3 mg/l
37.3 mg/l
29.6 mg/l
Exposure time: 30 min

Comment: The photoluminometer "Lumistox" and the photobacteria were purchased from Dr. Lange, GmbH, Germany. The tests were performed in open cuvettes. The pH of the solutions was 7.0 ± 0.2.

Bibliographical reference: Thumm, W., Brüggemann, R., Freitag, D., and Kettrup, A. (1992). *Chemosphere* **24**, 1835-1843.

tert-NITROBUTANE

CAS RN: 594-70-7
Synonym: 2-Methyl-2-nitropropane

Sample purity: Aldrich, Steinheim (Germany)
Temperature: 15°C
Test parameter: EC50

Effect: Reduction in light output

Concentrations: 0.64 mg/l
0.61 mg/l
0.48 mg/l
Exposure time: 5 min

Concentrations: 0.79 mg/l
0.61 mg/l
0.56 mg/l
Exposure time: 15 min

Concentrations: 0.89 mg/l
0.78 mg/l
0.67 mg/l
Exposure time: 30 min

Comment: The photoluminometer "Lumistox" and the photobacteria were purchased from Dr. Lange, GmbH, Germany. The tests were performed in open cuvettes. The pH of the solutions was 7.0 ± 0.2.

Bibliographical reference: Thumm, W., Brüggemann, R., Freitag, D., and Kettrup, A. (1992). *Chemosphere* **24**, 1835-1843.

4-NITROCATECHOL

CAS RN: 3316-09-4

Sample purity: 98%
Temperature: 15°C
Test parameter: EC50
Effect: Reduction in light output

Concentration: 9.13 mg/l
Exposure time: 5 min

Concentration: 7.95 mg/l
Exposure time: 15 min

Concentration: 7.77 mg/l
Exposure time: 30 min

Comment: Mean of three assays. Methanol (<10%) was used to prepare the stock solutions. EC50 values were calculated from nominal concentrations.

Bibliographical reference: Kaiser, K.L.E., and Palabrica, V.S. (1991). *Water Poll. Res. J. Canada* **26**, 361-431.

3-NITROCHALCONE

CAS RN: 614-48-2

Sample purity: 98+%
Temperature: 15°C
Test parameter: EC50
Effect: Reduction in light output

Concentration: 5.80 mg/l
Exposure time: 5 min

Concentration: 5.41 mg/l
Exposure time: 15 min

Concentration: 5.54 mg/l
Exposure time: 30 min

Comment: Mean of three assays. Methanol (<10%) was used to prepare the stock solutions. EC50 values were calculated from nominal concentrations.

Bibliographical reference: Kaiser, K.L.E., and Palabrica, V.S. (1992). National Water Research Institute, Burlington, Ontario, Canada, unpublished results.

4-NITROCHALCONE

CAS RN: 1222-98-6

Temperature: 15°C
Test parameter: EC50
Effect: Reduction in light output

Concentration: 1.36 mg/l

Exposure time: 5 min

Concentration: 1.18 mg/l
Exposure time: 15 min

Concentration: 1.21 mg/l
Exposure time: 30 min

Comment: Mean of three assays. Methanol (<10%) was used to prepare the stock solutions. EC50 values were calculated from nominal concentrations.

Bibliographical reference: Kaiser, K.L.E., and Palabrica, V.S. (1992). National Water Research Institute, Burlington, Ontario, Canada, unpublished results.

trans-4-NITROCINNAMIC ACID

CAS RN: 619-89-6

Sample purity: 98%
Temperature: 15°C
Test parameter: EC50
Effect: Reduction in light output

Concentration: 94.6 mg/l
Exposure time: 5 min

Concentration: 92.5 mg/l
Exposure time: 15 min

Concentration: 90.3 mg/l*
Exposure time: 30 min

Comment: Mean of three assays. Methanol (<10%) was used to prepare the stock solutions. EC50 values were calculated from nominal concentrations.

Bibliographical references: Kaiser, K.L.E., and Palabrica, V.S. (1991). *Water Poll. Res. J. Canada* **26**, 361-431.
* Kaiser, K.L.E. (1987). In: *QSAR in Environmental Toxicology - II*, K.L.E. Kaiser (ed.), D. Reidel Publishing Company, Dordrecht, p. 169-188.

4-NITROCINNAMYL ALCOHOL

CAS RN: 1504-63-8
Synonym: 3-(*p*-Nitrophenyl)-2-propen-1-ol

Sample purity: 98%
Temperature: 15°C
Test parameter: EC50
Effect: Reduction in light output

Concentration: 17.1 mg/l
Exposure time: 5 min

Concentration: 17.5 mg/l
Exposure time: 15 min

Concentration: 19.2 mg/l
Exposure time: 30 min

Comment: Mean of three assays. Methanol (<10%) was used to prepare the stock solutions. EC50 values were calculated from nominal concentrations.

Bibliographical reference: Kaiser, K.L.E., and Palabrica, V.S. (1991). *Water Poll. Res. J. Canada* **26**, 361-431.

NITROCYCLOPENTANE

CAS RN: 2562-38-1

Sample purity: 99%
Temperature: 15°C
Test parameter: EC50
Effect: Reduction in light output

Concentration: 3.99 mg/l
Exposure time: 5 min

Concentration: 4.38 mg/l
Exposure time: 15 min

Concentration: 5.03 mg/l
Exposure time: 30 min

Comment: Mean of three assays. Methanol (<10%) was used to prepare the stock solutions. EC50 values were calculated from nominal concentrations.

Bibliographical reference: Kaiser, K.L.E., and Palabrica, V.S. (1991). *Water Poll. Res. J. Canada* **26**, 361-431.

2-NITRODIPHENYLAMINE

CAS RN: 119-75-5

Temperature: 15°C
Test parameter: EC50
Effect: Reduction in light output

Concentration: 2.00 mg/l
Exposure time: 5 min

Concentration: 2.64 mg/l
Exposure time: 15 min

Concentration: 3.56 mg/l
Exposure time: 30 min

Comment: Mean of four assays. Methanol (<10%) was used to prepare the stock solutions. EC50 values were calculated from nominal concentrations.

Bibliographical reference: Kaiser, K.L.E., and Palabrica, V.S. (1992). National Water Research Institute, Burlington, Ontario, Canada, unpublished results.

NITROETHANE

CAS RN: 79-24-3

Sample purity: Angus Chemie, Ibbenbüren (Germany)
Temperature: 15°C
Test parameter: EC50
Effect: Reduction in light output

Concentrations: 1021.6 mg/l

2145.0 mg/l
1996.2 mg/l
Exposure time: 5 min

Concentrations: 1493.8 mg/l
1104.3 mg/l
1350.5 mg/l
1404.7 mg/l
Exposure time: 15 min

Concentrations: 791.8 mg/l
924.5 mg/l
1205 mg/l
Exposure time: 30 min

Comment: The photoluminometer "Lumistox" and the photobacteria were purchased from Dr. Lange, GmbH, Germany. The tests were performed in open cuvettes. The pH of the solutions was 7.0 ± 0.2.

Bibliographical reference: Thumm, W., Brüggemann, R., Freitag, D., and Kettrup, A. (1992). *Chemosphere* **24**, 1835-1843.

5-NITRO-2-FURALDEHYDE DIACETATE

CAS RN: 92-55-7

Sample purity: 98%
Temperature: 15°C
Test parameter: EC50
Effect: Reduction in light output

Concentration: 1.16 mg/l
Exposure time: 5 min

Concentration: 0.34 mg/l
Exposure time: 15 min

Concentration: 0.21 mg/l
Exposure time: 30 min

Comment: Mean of three assays. Methanol (<10%) was used to prepare the stock solutions. EC50 values were calculated from nominal concentrations.

Bibliographical reference: Kaiser, K.L.E., and Palabrica, V.S. (1991). *Water Poll. Res. J. Canada* **26**, 361-431.

1-NITROHEXANE

CAS RN: 646-14-0

Sample purity: Aldrich, Steinheim (Germany)
Temperature: 15°C
Test parameter: EC50
Effect: Reduction in light output

Concentrations: 2.3 mg/l
2.3 mg/l
2.2 mg/l
Exposure time: 5 min

Concentrations: 2.3 mg/l
2.9 mg/l
2.9 mg/l
Exposure time: 15 min

Concentrations: 2.9 mg/l
3.5 mg/l
3.5 mg/l
Exposure time: 30 min

Comment: The photoluminometer "Lumistox" and the photobacteria were purchased from Dr. Lange, GmbH, Germany. The tests were performed in open cuvettes. The pH of the solutions was 7.0 ± 0.2.

Bibliographical reference: Thumm, W., Brüggemann, R., Freitag, D., and Kettrup, A. (1992). *Chemosphere* **24**, 1835-1843.

4-NITROIMIDAZOLE

CAS RN: 3034-38-6

Temperature: 15°C
Test parameter: EC50
Effect: Reduction in light output

Concentration: 68.1 mg/l
Exposure time: 5 min

Concentration: 59.3 mg/l
Exposure time: 15 min

Concentration: 54.1 mg/l
Exposure time: 30 min

Comment: Mean of four assays. Methanol (<10%) was used to prepare the stock solutions. EC50 values were calculated from nominal concentrations.

Bibliographical reference: Kaiser, K.L.E., and Palabrica, V.S. (1992). National Water Research Institute, Burlington, Ontario, Canada, unpublished results.

5-NITROINDAZOLE

CAS RN: 5401-94-5

Sample purity: 97%
Temperature: 15°C
Test parameter: EC50
Effect: Reduction in light output

Concentration: 1.59 mg/l
Exposure time: 5 min

Concentration: 1.71 mg/l
Exposure time: 15 min

Concentration: 2.01 mg/l
Exposure time: 30 min

Comment: Mean of three assays. Methanol (<10%) was used to prepare the stock solutions. EC50 values were calculated from nominal concentrations.

Bibliographical reference: Kaiser, K.L.E., and Palabrica, V.S. (1991). *Water Poll. Res. J. Canada* **26**, 361-431.

4-NITROINDOLE

CAS RN: 4769-97-5

Sample purity: 97%
Temperature: 15°C
Test parameter: EC50
Effect: Reduction in light output

Concentration: 0.55 mg/l
Exposure time: 5 min

Concentration: 0.59 mg/l
Exposure time: 15 min

Concentration: 0.63 mg/l
Exposure time: 30 min

Comment: Mean of three assays. Methanol (<10%) was used to prepare the stock solutions. EC50 values were calculated from nominal concentrations.

Bibliographical reference: Kaiser, K.L.E., and Palabrica, V.S. (1991). *Water Poll. Res. J. Canada* **26**, 361-431.

5-NITROINDOLE

CAS RN: 6146-52-7

Sample purity: 98%
Temperature: 15°C
Test parameter: EC50
Effect: Reduction in light output

Concentration: 0.085 mg/l
Exposure time: 5 min

Concentration: 0.089 mg/l
Exposure time: 15 min

Concentration: 0.093 mg/l
Exposure time: 30 min

Comment: Mean of three assays. Methanol (<10%) was used to prepare the stock solutions. EC50 values were calculated from nominal concentrations.

Bibliographical reference: Kaiser, K.L.E., and Palabrica, V.S. (1991). *Water Poll. Res. J. Canada* **26**, 361-431.

5-NITROISOQUINOLINE

CAS RN: 607-32-9

Sample purity: 98%
Temperature: 15°C
Test parameter: EC50
Effect: Reduction in light output

Concentration: 3.17 mg/l
Exposure times: 5 and 15 min

Concentration: 3.47 mg/l
Exposure time: 30 min

Comment: Mean of four assays. Methanol (<10%) was used to prepare the stock solutions. EC50 values were calculated from nominal concentrations.

Bibliographical reference: Kaiser, K.L.E., and Palabrica, V.S. (1991). *Water Poll. Res. J. Canada* **26**, 361-431.

2-NITROMESITYLENE

CAS RN: 603-71-4
Synonym: 1,3,5-Trimethyl-2-nitrobenzene

Temperature: 15°C
Test parameter: EC50
Effect: Reduction in light output

Concentration: 0.13 mg/l
Exposure time: 5 min

Concentration: 0.14 mg/l

Exposure time: 15 min

Concentration: 0.16 mg/l
Exposure time: 30 min

Comment: Mean of three assays. Methanol (<10%) was used to prepare the stock solutions. EC50 values were calculated from nominal concentrations.

Bibliographical reference: Kaiser, K.L.E., and Palabrica, V.S. (1992). National Water Research Institute, Burlington, Ontario, Canada, unpublished results.

NITROMETHANE

CAS RN: 75-52-5

Sample purity: Angus Chemie, Ibbenbüren (Germany)
Temperature: 15°C
Test parameter: EC50
Effect: Reduction in light output

Concentrations: 4829.4 mg/l
6516.1 mg/l
4843.0 mg/l
Exposure time: 5 min

Concentrations: 6927.1 mg/l
4454.0 mg/l
5742.2 mg/l
Exposure time: 15 min

Concentrations: 6353.8 mg/l
4462.5 mg/l
6047.4 mg/l
Exposure time: 30 min

Comment: The photoluminometer "Lumistox" and the photobacteria were purchased from Dr. Lange, GmbH, Germany. The tests were performed in open cuvettes. The pH of the solutions was 7.0 ± 0.2.

Bibliographical reference: Thumm, W., Brüggemann, R., Freitag, D., and Kettrup, A. (1992). *Chemosphere* **24**, 1835-1843.

1-NITRO-2-METHYLPROPANE

CAS RN: 625-74-1
Synonym: Isonitrobutane

Sample purity: Synthesized in the Institute of Ecological Chemistry (Germany)
Temperature: 15°C
Test parameter: EC50
Effect: Reduction in light output

Concentrations: 63.0 mg/l
65.9 mg/l
79.4 mg/l
79.4 mg/l
58.5 mg/l
65.8 mg/l
Exposure time: 5 min

Concentrations: 175.9 mg/l
131.2 mg/l
327.5 mg/l
52.5 mg/l
77.4 mg/l
101.4 mg/l
81.1 mg/l
54.9 mg/l
60.4 mg/l
Exposure time: 15 min

Concentrations: 232.5 mg/l
152.3 mg/l
383.0 mg/l
67.0 mg/l
95.5 mg/l
99.2 mg/l
88.5 mg/l
57.1 mg/l
62.7 mg/l
Exposure time: 30 min

Comment: The photoluminometer "Lumistox" and the photobacteria were purchased from Dr. Lange, GmbH, Germany. The tests were performed in open cuvettes. The pH of the solutions was 7.0 ± 0.2. The first three values for EC15 and EC30 are higher than the other values.

According to the authors, the difference could be due to the fact that these values were measured with other preparations of the substance.

Bibliographical reference: Thumm, W., Brüggemann, R., Freitag, D., and Kettrup, A. (1992). *Chemosphere* **24**, 1835-1843.

1-NITRONAPHTHALENE

CAS RN: 86-57-7

Sample purity: 99%
Temperature: 15°C
Test parameter: EC50
Effect: Reduction in light output

Concentration: 0.20 mg/l
Exposure times: 5 and 15 min

Concentration: 0.21 mg/l
Exposure time: 30 min

Comment: Mean of four assays. Methanol (<10%) was used to prepare the stock solutions. EC50 values were calculated from nominal concentrations.

Bibliographical reference: Kaiser, K.L.E., and Palabrica, V.S. (1991). *Water Poll. Res. J. Canada* **26**, 361-431.

Sample purity: 99%
Temperature: 15°C
Test parameter: EC50
Effect: Reduction in light output

Concentration: 0.25 mg/l
Exposure time: 5 min

Concentration: 0.27 mg/l
Exposure times: 15 and 30 min

Comment: Mean of four assays. Methanol (<10%) was used to prepare the stock solutions. EC50 values were calculated from nominal concentrations.

Bibliographical reference: Kaiser, K.L.E., and Palabrica, V.S. (1991). *Water Poll. Res. J. Canada* **26**, 361-431.

1-NITROPENTANE

CAS RN: 628-05-7

Sample purity: Aldrich, Steinheim (Germany)
Temperature: 15°C
Test parameter: EC50
Effect: Reduction in light output

Concentrations: 16.3 mg/l
14.8 mg/l
22.4 mg/l
Exposure time: 5 min

Concentrations: 24.6 mg/l
16.0 mg/l
13.5 mg/l
19.4 mg/l
Exposure time: 15 min

Concentrations: 30.4 mg/l
14.9 mg/l
22.3 mg/l
Exposure time: 30 min

Comment: The photoluminometer "Lumistox" and the photobacteria were purchased from Dr. Lange, GmbH, Germany. The tests were performed in open cuvettes. The pH of the solutions was 7.0 ± 0.2.

Bibliographical reference: Thumm, W., Brüggemann, R., Freitag, D., and Kettrup, A. (1992). *Chemosphere* **24**, 1835-1843.

2-NITROPHENOL

CAS RN: 88-75-5

Temperature: 15°C
Test parameter: EC50
Effect: Reduction in light output

Concentration: 33.6 mg/l
Exposure time: 5 min

Concentration: 35.1 mg/l
Exposure time: 30 min

Bibliographical reference: Speece, R. (1987). Drexel University, Philadelphia, USA, private communication.

4-NITROPHENOL

CAS RN: 100-02-7

Test parameter: EC50
Effect: Reduction in light output
Concentration: 13.0 mg/l
Exposure time: 5 min
Comment: Concentrations in the test were measured.

Bibliographical reference: Curtis, C., Lima, A., Lozano, S.J., and Veith, G.D. (1982). In: *Aquatic Toxicology and Hazard Assessment: Fifth Conference, ASTM STP 766*, J.G. Pearson, R.B. Foster, and W.E. Bishop (eds.), American Society for Testing and Materials, Philadelphia, p. 170-178.

Temperature: 15°C
Test parameter: EC50
Effect: Reduction in light output

Concentration: 10.9 mg/l
Exposure time: 5 min

Concentration: 13.4 mg/l
Exposure time: 30 min

Bibliographical reference: Speece, R. (1987). Drexel University, Philadelphia, USA, private communication.

Temperature: 15°C
Test parameter: EC50
Effect: Reduction in light output
Concentration: 9.40 mg/l

Exposure time: 30 min
Comment: Mean of three assays. Methanol (<10%) was used to prepare the stock solutions. EC50 values were calculated from nominal concentrations.

Bibliographical reference: Kaiser, K.L.E. (1987). In: *QSAR in Environmental Toxicology - II*, K.L.E. Kaiser (ed.), D. Reidel Publishing Company, Dordrecht, p. 169-188.

Temperature: 15 ± 0.1°C
Test parameter: EC50
Effect: Reduction in light output
Concentration: 13.7 mg/l
Exposure time: 5 min

Bibliographical reference: Somasundaram, L., Coats, J.R., Racke, K.D., and Stahr, H.M. (1990). *Bull. Environ. Contam. Toxicol.* **44**, 254-259.

4-NITROPHENOXYACETONITRILE

CAS RN: 33901-46-1

Temperature: 15°C
Test parameter: EC50
Effect: Reduction in light output

Concentration: 15.5 mg/l
Exposure time: 5 min

Concentration: 17.9 mg/l
Exposure time: 15 min

Concentration: 21.5 mg/l
Exposure time: 30 min

Comment: Mean of three assays. Methanol (<10%) was used to prepare the stock solutions. EC50 values were calculated from nominal concentrations.

Bibliographical reference: Kaiser, K.L.E., and Palabrica, V.S. (1992). National Water Research Institute, Burlington, Ontario, Canada, unpublished results.

2-(4-NITROPHENOXY)ETHANOL

CAS RN: 16365-27-8

Temperature: 15°C
Test parameter: EC50
Effect: Reduction in light output

Concentration: 127 mg/l
Exposure time: 5 min

Concentration: 118 mg/l
Exposure time: 15 min

Concentration: 108 mg/l*
Exposure time: 30 min

Comment: Recrystallized from hexane. Mean of three assays. Methanol (<10%) was used to prepare the stock solutions. EC50 values were calculated from nominal concentrations.

Bibliographical references: Kaiser, K.L.E., and Palabrica, V.S. (1991). *Water Poll. Res. J. Canada* **26**, 361-431.
* Kaiser, K.L.E. (1987). In: *QSAR in Environmental Toxicology - II*, K.L.E. Kaiser (ed.), D. Reidel Publishing Company, Dordrecht, p. 169-188.

4-NITROPHENYL ACETATE

CAS RN: 830-03-5

Temperature: 15°C
Test parameter: EC50
Effect: Reduction in light output
Concentration: 10.2 mg/l
Exposure time: 30 min
Comment: Mean of three assays. Methanol (<10%) was used to prepare the stock solutions. EC50 values were calculated from nominal concentrations.

Bibliographical reference: Kaiser, K.L.E. (1987). In: *QSAR in Environmental Toxicology - II*, K.L.E. Kaiser (ed.), D. Reidel Publishing Company, Dordrecht, p. 169-188.

4-NITROPHENYLACETIC ACID

CAS RN: 104-03-0

Temperature: 15°C
Test parameter: EC50
Effect: Reduction in light output
Concentration: 194 mg/l
Exposure time: 30 min
Comment: Mean of three assays. Methanol (<10%) was used to prepare the stock solutions. EC50 values were calculated from nominal concentrations.

Bibliographical reference: Kaiser, K.L.E. (1987). In: *QSAR in Environmental Toxicology - II*, K.L.E. Kaiser (ed.), D. Reidel Publishing Company, Dordrecht, p. 169-188.

1-(4-NITROPHENYL)ACETONE

CAS RN: 5332-96-7

Temperature: 15°C
Test parameter: EC50
Effect: Reduction in light output

Concentration: 33.4 mg/l
Exposure time: 5 min

Concentration: 32.6 mg/l
Exposure times: 15 and 30 min

Comment: Mean of three assays. Methanol (<10%) was used to prepare the stock solutions. EC50 values were calculated from nominal concentrations.

Bibliographical reference: Kaiser, K.L.E., and Palabrica, V.S. (1991). *Water Poll. Res. J. Canada* **26**, 361-431.

4-NITROPHENYLACETONITRILE

CAS RN: 555-21-5

Temperature: 15°C
Test parameter: EC50
Effect: Reduction in light output

Concentration: 6.16 mg/l
Exposure time: 5 min

Concentration: 6.46 mg/l
Exposure time: 15 min

Concentration: 7.08 mg/l*
Exposure time: 30 min

Comment: Recrystallized from toluene. Mean of three assays. Methanol (<10%) was used to prepare the stock solutions. EC50 values were calculated from nominal concentrations.

Bibliographical references: Kaiser, K.L.E., and Palabrica, V.S. (1991). *Water Poll. Res. J. Canada* **26**, 361-431.
* Kaiser, K.L.E. (1987). In: *QSAR in Environmental Toxicology - II*, K.L.E. Kaiser (ed.), D. Reidel Publishing Company, Dordrecht, p. 169-188.

DL-4-NITROPHENYLALANINE

CAS RN: 2922-40-9

Temperature: 15°C
Test parameter: EC50
Effect: Reduction in light output

Concentration: 8.37 mg/l
Exposure time: 5 min

Concentration: 5.04 mg/l
Exposure time: 15 min

Concentration: 3.82 mg/l*
Exposure time: 30 min

Comment: Mean of three assays. Methanol (<10%) was used to prepare the stock solutions. EC50 values were calculated from nominal concentrations.

Bibliographical references: Kaiser, K.L.E., and Palabrica, V.S. (1991). *Water Poll. Res. J. Canada* **26**, 361-431.
* Kaiser, K.L.E. (1987). In: *QSAR in Environmental Toxicology - II*, K.L.E. Kaiser (ed.), D. Reidel Publishing Company, Dordrecht, p. 169-188.

4-(4-NITROPHENYLAZO)RESORCINOL

CAS RN: 74-39-5

Temperature: 15°C
Test parameter: EC50
Effect: Reduction in light output

Concentration: 4.40 mg/l
Exposure time: 5 min

Concentration: 4.30 mg/l
Exposure time: 15 min

Concentration: 4.40 mg/l
Exposure time: 30 min

Comment: Mean of four assays. Methanol (<10%) was used to prepare the stock solutions. EC50 values were calculated from nominal concentrations.

Bibliographical reference: Kaiser, K.L.E., and Palabrica, V.S. (1992). National Water Research Institute, Burlington, Ontario, Canada, unpublished results.

3-NITROPHENYLBORONIC ACID

CAS RN: 13331-27-6

Temperature: 15°C
Test parameter: EC50
Effect: Reduction in light output

Concentration: 3.11 mg/l
Exposure time: 5 min

Concentration: 2.97 mg/l
Exposure times: 15 and 30 min

Comment: Mean of three assays. Methanol (<10%) was used to prepare the stock solutions. EC50 values were calculated from nominal concentrations.

Bibliographical reference: Kaiser, K.L.E., and Palabrica, V.S. (1991). *Water Poll. Res. J. Canada* **26**, 361-431.

4-NITROPHENYL CHLOROFORMATE

CAS RN: 7693-46-1

Sample purity: 97%
Temperature: 15°C
Test parameter: EC50
Effect: Reduction in light output

Concentration: 1.46 mg/l
Exposure time: 5 min

Concentration: 0.59 mg/l
Exposure time: 15 min

Concentration: 0.30 mg/l
Exposure time: 30 min

Comment: Mean of four assays. Methanol (<10%) was used to prepare the stock solutions. EC50 values were calculated from nominal concentrations.

Bibliographical reference: Kaiser, K.L.E., and Palabrica, V.S. (1991). *Water Poll. Res. J. Canada* **26**, 361-431.

4-NITRO-1,2-PHENYLENEDIAMINE

CAS RN: 99-56-9
Synonym: 1,2-Diamino-4-nitrobenzene

Sample purity: 97%
Temperature: 15°C

Test parameter: EC50
Effect: Reduction in light output

Concentration: 27.9 mg/l
Exposure time: 5 min

Concentration: 22.1 mg/l
Exposure time: 15 min

Concentration: 21.6 mg/l
Exposure time: 30 min

Comment: Mean of three assays. Methanol (<10%) was used to prepare the stock solutions. EC50 values were calculated from nominal concentrations.

Bibliographical reference: Kaiser, K.L.E., and Palabrica, V.S. (1991). *Water Poll. Res. J. Canada* **26**, 361-431.

4-NITRO-1,3-PHENYLENEDIAMINE

CAS RN: 5131-58-8
Synonym: 1,3-Diamino-4-nitrobenzene

Temperature: 15°C
Test parameter: EC50
Effect: Reduction in light output

Concentration: 0.19 mg/l
Exposure time: 5 min

Concentration: 0.20 mg/l
Exposure time: 15 min

Concentration: 0.23 mg/l
Exposure time: 30 min

Comment: Mean of three assays. Methanol (<10%) was used to prepare the stock solutions. EC50 values were calculated from nominal concentrations.

Bibliographical reference: Kaiser, K.L.E., and Palabrica, V.S. (1992). National Water Research Institute, Burlington, Ontario, Canada, unpublished results.

2-(4-NITROPHENYL)ETHANOL

CAS RN: 100-27-6

Sample purity: 98%
Temperature: 15°C
Test parameter: EC50
Effect: Reduction in light output

Concentration: 33.4 mg/l
Exposure time: 5 min

Concentration: 34.9 mg/l
Exposure time: 15 min

Concentration: 38.3 mg/l*
Exposure time: 30 min

Comment: Mean of three assays. Methanol (<10%) was used to prepare the stock solutions. EC50 values were calculated from nominal concentrations.

Bibliographical references: Kaiser, K.L.E., and Palabrica, V.S. (1991). *Water Poll. Res. J. Canada* **26**, 361-431.
* Kaiser, K.L.E. (1987). In: *QSAR in Environmental Toxicology - II*, K.L.E. Kaiser (ed.), D. Reidel Publishing Company, Dordrecht, p. 169-188.

Sample purity: 98%
Temperature: 15°C
Test parameter: EC50
Effect: Reduction in light output

Concentration: 51.7 mg/l
Exposure time: 5 min

Concentration: 48.2 mg/l
Exposure time: 15 min

Concentration: 45.0 mg/l
Exposure time: 30 min

Comment: Mean of three assays. Methanol (<10%) was used to prepare the stock solutions. EC50 values were calculated from nominal concentrations.

Bibliographical reference: Kaiser, K.L.E., and Palabrica, V.S. (1991). *Water Poll. Res. J. Canada* **26**, 361-431.

4-NITROPHENYL FORMATE

CAS RN: 1865-01-6

Sample purity: >99%
Temperature: 15°C
Test parameter: EC50
Effect: Reduction in light output

Concentration: 1.21 mg/l
Exposure time: 5 min

Concentration: 1.33 mg/l
Exposure time: 15 min

Concentration: 2.20 mg/l
Exposure time: 30 min

Comment: Mean of three assays. Methanol (<10%) was used to prepare the stock solutions. EC50 values were calculated from nominal concentrations.

Bibliographical reference: Kaiser, K.L.E., and Palabrica, V.S. (1991). *Water Poll. Res. J. Canada* **26**, 361-431.

4-NITROPHENYLHYDRAZINE

CAS RN: 100-16-3

Temperature: 15°C
Test parameter: EC50
Effect: Reduction in light output
Concentration: 1.22 mg/l
Exposure time: 30 min
Comment: Mean of three assays. Methanol (<10%) was used to prepare the stock solutions. EC50 values were calculated from nominal concentrations.

Bibliographical reference: Kaiser, K.L.E. (1987). In: *QSAR in Environmental Toxicology - II*, K.L.E. Kaiser (ed.), D. Reidel Publishing Company, Dordrecht, p. 169-188.

4-NITROPHENYLPHENYLAMINE

CAS RN: 836-30-6

Temperature: 15°C
Test parameter: EC50
Effect: Reduction in light output
Concentration: 0.73 mg/l
Exposure time: 30 min
Comment: Mean of three assays. Methanol (<10%) was used to prepare the stock solutions. EC50 values were calculated from nominal concentrations.

Bibliographical reference: Kaiser, K.L.E. (1987). In: *QSAR in Environmental Toxicology - II*, K.L.E. Kaiser (ed.), D. Reidel Publishing Company, Dordrecht, p. 169-188.

4-NITROPHENYL PHENYL ETHER

CAS RN: 620-88-2

Temperature: 15°C
Test parameter: EC50
Effect: Reduction in light output
Concentration: 1.10 mg/l
Exposure time: 30 min
Comment: Mean of three assays. Methanol (<10%) was used to prepare the stock solutions. EC50 values were calculated from nominal concentrations.

Bibliographical reference: Kaiser, K.L.E. (1987). In: *QSAR in Environmental Toxicology - II*, K.L.E. Kaiser (ed.), D. Reidel Publishing Company, Dordrecht, p. 169-188.

4-NITROPHENYL PROPIONATE

CAS RN: 1956-06-5

Sample purity: 98%
Temperature: 15°C
Test parameter: EC50
Effect: Reduction in light output

Concentration: 2.24 mg/l
Exposure time: 5 min

Concentration: 2.14 mg/l
Exposure time: 15 min

Concentration: 2.00 mg/l
Exposure time: 30 min

Comment: Mean of three assays. Methanol (<10%) was used to prepare the stock solutions. EC50 values were calculated from nominal concentrations.

Bibliographical reference: Kaiser, K.L.E., and Palabrica, V.S. (1991). *Water Poll. Res. J. Canada* **26**, 361-431.

4-NITROPHENYL SULFATE POTASSIUM SALT

CAS RN: 6217-68-1

Sample purity: 98%
Temperature: 15°C
Test parameter: EC50
Effect: Reduction in light output

Concentration: 46.8 mg/l
Exposure time: 5 min

Concentration: 38.9 mg/l
Exposure time: 15 min

Concentration: 37.2 mg/l
Exposure time: 30 min

Comment: Mean of three assays. Methanol (<10%) was used to prepare the stock solutions. EC50 values were calculated from nominal concentrations.

Bibliographical reference: Kaiser, K.L.E., and Palabrica, V.S. (1991). *Water Poll. Res. J. Canada* **26**, 361-431.

4-(4-NITROPHENYLSULFONYL)ANILINE

CAS RN: 1948-92-1

Temperature: 15°C
Test parameter: EC50
Effect: Reduction in light output
Concentration: 6.83 mg/l
Exposure time: 30 min
Comment: Mean of three assays. Methanol (<10%) was used to prepare the stock solutions. EC50 values were calculated from nominal concentrations.

Bibliographical reference: Kaiser, K.L.E., and Palabrica, V.S. (1991). *Water Poll. Res. J. Canada* **26**, 361-431.

Sample purity: 96%
Temperature: 15°C
Test parameter: EC50
Effect: Reduction in light output

Concentration: 8.03 mg/l
Exposure time: 5 min

Concentration: 5.95 mg/l
Exposure time: 15 min

Concentration: 5.43 mg/l
Exposure time: 30 min

Comment: Mean of three assays. Methanol (<10%) was used to prepare the stock solutions. EC50 values were calculated from nominal concentrations.

Bibliographical reference: Kaiser, K.L.E., and Palabrica, V.S. (1991). *Water Poll. Res. J. Canada* **26**, 361-431.

2-(3-NITROPHENYLSULFONYL)ETHANOL

CAS RN: 41687-30-3

Sample purity: 98%
Temperature: 15°C
Test parameter: EC50
Effect: Reduction in light output

Concentration: 149 mg/l
Exposure time: 5 min

Concentration: 63.7 mg/l
Exposure time: 15 min

Concentration: 48.3 mg/l
Exposure time: 30 min

Comment: Mean of three assays. Methanol (<10%) was used to prepare the stock solutions. EC50 values were calculated from nominal concentrations.

Bibliographical reference: Kaiser, K.L.E., and Palabrica, V.S. (1991). *Water Poll. Res. J. Canada* **26**, 361-431.

4-NITROPHENYL TRIFLUOROACETATE

CAS RN: 658-78-6

Sample purity: 98%
Temperature: 15°C
Test parameter: EC50
Effect: Reduction in light output

Concentration: 13.2 mg/l
Exposure times: 5 and 15 min

Concentration: 13.8 mg/l
Exposure time: 30 min

Comment: Mean of three assays. Methanol (<10%) was used to prepare the stock solutions. EC50 values were calculated from nominal concentrations.

Bibliographical reference: Kaiser, K.L.E., and Palabrica, V.S. (1991). *Water Poll. Res. J. Canada* **26**, 361-431.

1-NITROPROPANE

CAS RN: 108-03-2

Sample purity: Merck, Darmstadt (Germany)
Temperature: 15°C
Test parameter: EC50
Effect: Reduction in light output

Concentrations: 42.8 mg/l
44.2 mg/l
82.5 mg/l
71.6 mg/l
Exposure time: 5 min

Concentrations: 65.2 mg/l
57.9 mg/l
45.4 mg/l
Exposure time: 15 min

Concentrations: 89.6 mg/l
69.6 mg/l
50.8 mg/l
Exposure time: 30 min

Comment: The photoluminometer "Lumistox" and the photobacteria were purchased from Dr. Lange, GmbH, Germany. The tests were performed in open cuvettes. The pH of the solutions was 7.0 ± 0.2. Chemical was redistilled before use.

Bibliographical reference: Thumm, W., Brüggemann, R., Freitag, D., and Kettrup, A. (1992). *Chemosphere* **24**, 1835-1843.

2-NITROPROPANE

CAS RN: 79-46-9

Sample purity: Merck, Darmstadt (Germany)
Temperature: 15°C

Test parameter: EC50
Effect: Reduction in light output

Concentrations: 68.3 mg/l
69.3 mg/l
64.5 mg/l
Exposure time: 5 min

Concentrations: 27.4 mg/l
53.3 mg/l
42.1 mg/l
44.6 mg/l
Exposure time: 15 min

Concentrations: 35.2 mg/l
47.6 mg/l
56.4 mg/l
Exposure time: 30 min

Comment: The photoluminometer "Lumistox" and the photobacteria were purchased from Dr. Lange, GmbH, Germany. The tests were performed in open cuvettes. Mass balance performed by HPLC showed that after one hour of leaving the cuvettes open at least 95% of compound was still present. The pH of the solutions was 7.0 ± 0.2. Chemical was redistilled before use.

Bibliographical reference: Thumm, W., Brüggemann, R., Freitag, D., and Kettrup, A. (1992). *Chemosphere* **24**, 1835-1843.

4-NITROPYRIDINE *N*-OXIDE

CAS RN: 1124-33-0

Temperature: 15°C
Test parameter: EC50
Effect: Reduction in light output

Concentration: 3.60 mg/l
Exposure time: 5 min

Concentration: 1.31 mg/l
Exposure time: 15 min

Concentration: 0.93 mg/l

Exposure time: 30 min

Comment: Mean of three assays. Methanol (<10%) was used to prepare the stock solutions. EC50 values were calculated from nominal concentrations.

Bibliographical reference: Kaiser, K.L.E., and Palabrica, V.S. (1991). *Water Poll. Res. J. Canada* **26**, 361-431.

8-NITROQUINALDINE

CAS RN: 881-07-2
Synonym: 2-Methyl-8-nitroquinoline

Sample purity: 98%
Temperature: 15°C
Test parameter: EC50
Effect: Reduction in light output

Concentration: 54.3 mg/l
Exposure time: 5 min

Concentration: 56.8 mg/l
Exposure time: 15 min

Concentration: 59.5 mg/l
Exposure time: 30 min

Comment: Mean of three assays. Methanol (<10%) was used to prepare the stock solutions. EC50 values were calculated from nominal concentrations.

Bibliographical reference: Kaiser, K.L.E., and Palabrica, V.S. (1991). *Water Poll. Res. J. Canada* **26**, 361-431.

6-NITROQUINOLINE

CAS RN: 613-50-3

Sample purity: 98%
Temperature: 15°C
Test parameter: EC50

Effect: Reduction in light output

Concentration: 39.0 mg/l
Exposure times: 5 and 15 min

Concentration: 42.8 mg/l
Exposure time: 30 min

Comment: Mean of three assays. Methanol (<10%) was used to prepare the stock solutions. EC50 values were calculated from nominal concentrations.

Bibliographical reference: Kaiser, K.L.E., and Palabrica, V.S. (1991). *Water Poll. Res. J. Canada* **26**, 361-431.

4-NITROQUINOLINE *N*-OXIDE

CAS RN: 56-57-5

Temperature: 15°C
Test parameter: EC50
Effect: Reduction in light output

Concentration: 1.09 ± 0.05 mg/l
Exposure time: 5 min

Concentration: 0.79 ± 0.03 mg/l
Exposure time: 10 min

Concentration: 0.67 ± 0.01 mg/l
Exposure time: 15 min

Concentration: 0.61 ± 0.01 mg/l
Exposure time: 20 min

Comment: Test was performed at least in triplicate. Chemical was solubilized in DMSO.

Bibliographical reference: Yates, I.E. (1985). *J. Microbiol. Meth.* **3**, 171-180.

N-NITROSODIETHYLAMINE

CAS RN: 55-18-5
Synonym: *N,N*-Diethylnitrosamine

Test parameter: EC50
Effect: Reduction in light output
Concentration: 140 mg/l
Exposure time: 15 min
Comment: Test was performed in duplicate at pH = 6.7.

Bibliographical reference: Dutka, B.J., and Kwan, K.K. (1981). *Bull. Environ. Contam. Toxicol.* **27**, 753-757.

4-NITROSO-*N,N*-DIMETHYLANILINE

CAS RN: 138-89-6
Synonym: *N,N*-Dimethyl-4-nitrosoaniline

Sample purity: 99%
Temperature: 15°C
Test parameter: EC50
Effect: Reduction in light output

Concentration: 0.13 mg/l
Exposure time: 5 min

Concentration: 0.041 mg/l
Exposure time: 15 min

Concentration: 0.017 mg/l
Exposure time: 30 min

Comment: Mean of three assays. Methanol (<10%) was used to prepare the stock solutions. EC50 values were calculated from nominal concentrations.

Bibliographical reference: Kaiser, K.L.E., and Palabrica, V.S. (1991). *Water Poll. Res. J. Canada* **26**, 361-431.

4-NITROTHIOPHENOL

CAS RN: 1849-36-1

Sample purity: 80% (adjusted to 100%)
Temperature: 15°C
Test parameter: EC50
Effect: Reduction in light output

Concentration: 0.42 mg/l
Exposure time: 5 min

Concentration: 0.34 mg/l
Exposure times: 15 and 30 min

Comment: Mean of three assays. Methanol (<10%) was used to prepare the stock solutions. EC50 values were calculated from nominal concentrations.

Bibliographical reference: Kaiser, K.L.E., and Palabrica, V.S. (1991). *Water Poll. Res. J. Canada* **26**, 361-431.

2-NITROTOLUENE

CAS RN: 88-72-2

Test parameter: EC50
Effect: Reduction in light output
Concentration: 1.85 mg/l
Exposure time: 15 min
Comment: The concentrations of the compound increased geometrically with a factor of 3.2. The concentrations were calculated on the basis of added amounts of material.

Bibliographical reference: Deneer, J.W., van Leeuwen, C.J., Seinen, W., Maas-Diepeveen, J.L., and Hermens, J.L.M. (1989). *Aquat. Toxicol.* **15**, 83-98.

3-NITROTOLUENE

CAS RN: 99-08-1

Test parameter: EC50
Effect: Reduction in light output
Concentration: 3.96 mg/l
Exposure time: 15 min
Comment: The concentrations of the compound increased geometrically with a factor of 3.2. The concentrations were calculated on the basis of added amounts of material.

Bibliographical reference: Deneer, J.W., van Leeuwen, C.J., Seinen, W., Maas-Diepeveen, J.L., and Hermens, J.L.M. (1989). *Aquat. Toxicol.* **15**, 83-98.

4-NITROTOLUENE

CAS RN: 99-99-0

Temperature: 15°C
Test parameter: EC50
Effect: Reduction in light output
Concentration: 12.5 mg/l
Exposure time: 30 min
Comment: Mean of three assays. Methanol (<10%) was used to prepare the stock solutions. EC50 values were calculated from nominal concentrations.

Bibliographical reference: Kaiser, K.L.E. (1987). In: *QSAR in Environmental Toxicology - II*, K.L.E. Kaiser (ed.), D. Reidel Publishing Company, Dordrecht, p. 169-188.

Test parameter: EC50
Effect: Reduction in light output
Concentration: 10.9 mg/l
Exposure time: 15 min
Comment: The concentrations of the compound increased geometrically with a factor of 3.2. The concentrations were calculated on the basis of added amounts of material.

Bibliographical reference: Deneer, J.W., van Leeuwen, C.J., Seinen, W., Maas-Diepeveen, J.L., and Hermens, J.L.M. (1989). *Aquat. Toxicol.* **15**, 83-98.

2-NITRO-4-(TRIFLUOROMETHYL)BENZONITRILE

CAS RN: 778-94-9

Sample purity: 99%
Temperature: 15°C
Test parameter: EC50
Effect: Reduction in light output

Concentration: 0.32 mg/l
Exposure time: 5 min

Concentration: 0.090 mg/l
Exposure time: 15 min

Concentration: 0.060 mg/l
Exposure time: 30 min

Comment: Mean of four assays. Methanol (<10%) was used to prepare the stock solutions. EC50 values were calculated from nominal concentrations.

Bibliographical reference: Kaiser, K.L.E., and Palabrica, V.S. (1991). *Water Poll. Res. J. Canada* **26**, 361-431.

4-NITRO-α,α,α-TRIFLUOROTOLUENE

CAS RN: 402-54-0

Sample purity: 96%
Temperature: 15°C
Test parameter: EC50
Effect: Reduction in light output

Concentration: 16.3 mg/l
Exposure time: 5 min

Concentration: 14.8 mg/l
Exposure time: 15 min

Concentration: 15.2 mg/l*
Exposure time: 30 min

Comment: Mean of three assays. Methanol (<10%) was used to prepare the stock solutions. EC50 values were calculated from nominal concentrations.

Bibliographical references: Kaiser, K.L.E., and Palabrica, V.S. (1991). *Water Poll. Res. J. Canada* **26**, 361-431.
* Kaiser, K.L.E. (1987). In: *QSAR in Environmental Toxicology - II*, K.L.E. Kaiser (ed.), D. Reidel Publishing Company, Dordrecht, p. 169-188.

5-NONANONE

CAS RN: 502-56-7
Synonym: Dibutyl ketone

Sample purity: 98%
Temperature: 15°C
Test parameter: EC50
Effect: Reduction in light output

Concentration: 25 mg/l
Exposure time: 5 min

Concentration: 35 mg/l*
Exposure time: 15 min

Bibliographical references: Cronin, M.T.D., Dearden, J.C., and Dobbs, A.J. (1991). *Sci. Total Environ.* **109/110**, 431-439.
* Cronin, M.T.D. (1990). Thesis, Liverpool Polytechnic, Liverpool, UK.

NOVOBIOCIN

CAS RN: 303-81-1

Temperature: 15°C
Test parameter: EC50
Effect: Reduction in light output

Concentration: 28.8 ± 0.5 mg/l
Exposure time: 5 min

Concentration: 26.0 ± 0.2 mg/l
Exposure time: 10 min

Concentration: 25.5 ± 0.3 mg/l
Exposure time: 15 min

Concentration: 21.4 ± 0.5 mg/l
Exposure time: 20 min

Comment: Test was performed at least in triplicate. Chemical was solubilized in DMSO.

Bibliographical reference: Yates, I.E. (1985). *J. Microbiol. Meth.* **3**, 171-180.

OCHRATOXIN A

CAS RN: 303-47-9

Temperature: 15°C
Test parameter: EC50
Effect: Reduction in light output

Concentration: 18.49 mg/l
Exposure time: 5 min

Concentration: 16.60 mg/l
Exposure time: 10 min

Concentration: 16.17 mg/l
Exposure time: 15 min

Concentration: 16.27 mg/l
Exposure time: 20 min

Comment: Freshly reconstituted bacterial suspensions. Chemical was dissolved in methanol.

Bibliographical reference: Yates, I.E., and Porter, J.K. (1982). *Appl. Environ. Microbiol.* **44**, 1072-1075.

Temperature: 15°C
Test parameter: EC20

Effect: Reduction in light output

Concentration: 12.61 mg/l
Exposure time: 5 min

Concentration: 12.56 mg/l
Exposure time: 20 min

Comment: Freshly reconstituted bacterial suspensions. Chemical was dissolved in methanol.

Bibliographical reference: Yates, I.E., and Porter, J.K. (1982). *Appl. Environ. Microbiol.* **44**, 1072-1075.

Temperature: 15°C
Test parameter: EC50
Effect: Reduction in light output

Concentration: 18.53 mg/l
Exposure time: 5 min

Concentration: 16.40 mg/l
Exposure time: 10 min

Concentration: 16.39 mg/l
Exposure time: 15 min

Concentration: 16.68 mg/l
Exposure time: 20 min

Comment: Performed on bacterial suspensions maintained at 3°C for 5 h after reconstitution. Chemical was dissolved in methanol.

Bibliographical reference: Yates, I.E., and Porter, J.K. (1982). *Appl. Environ. Microbiol.* **44**, 1072-1075.

OCTADECYL SULFATE SODIUM SALT

CAS RN: 1120-04-3

Sample purity: 93%
Temperature: 15°C
Test parameter: EC50

Effect: Reduction in light output

Concentration: 34.8 mg/l
Exposure time: 5 min

Concentration: 38.1 mg/l
Exposure time: 15 min

Concentration: 53.8 mg/l
Exposure time: 30 min

Comment: Mean of three assays. Methanol (<10%) was used to prepare the stock solutions. EC50 values were calculated from nominal concentrations.

Bibliographical reference: Kaiser, K.L.E., and Palabrica, V.S. (1991). *Water Poll. Res. J. Canada* **26**, 361-431.

1-OCTANOL

CAS RN: 111-87-5

Test parameter: EC50
Effect: Reduction in light output
Concentration: 6.3 mg/l
Exposure time: 5 min
Comment: Concentrations in the test were measured.

Bibliographical reference: Curtis, C., Lima, A., Lozano, S.J., and Veith, G.D. (1982). In: *Aquatic Toxicology and Hazard Assessment: Fifth Conference, ASTM STP 766*, J.G. Pearson, R.B. Foster, and W.E. Bishop (eds.), American Society for Testing and Materials, Philadelphia, p. 170-178.

Temperature: 15°C
Test parameter: EC50
Effect: Reduction in light output
Concentration: 4.73 mg/l
Exposure time: 15 min

Bibliographical reference: Hermens, J., Busser, F., Leeuwangh, P., and Musch, A. (1985). *Ecotoxicol. Environ. Safety* **9**, 17-25.

Sample purity: >98%
Temperature: 15 ± 0.1°C
Test parameter: EC50
Effect: Reduction in light output

Concentration: 5.93 mg/l
Exposure time: 5 min

Concentration: 7.93 mg/l
Exposure time: 15 min

Concentration: 7.26 mg/l
Exposure time: 30 min

Comment: The pH was not adjusted.

Bibliographical reference: Tarkpea, M., Hansson, M., and Samuelsson, B. (1986). *Ecotoxicol. Environ. Safety* **11**, 127-143.

Temperature: 15°C
Test parameter: EC50
Effect: Reduction in light output

Concentration: 3.40 mg/l
Exposure time: 5 min

Concentration: 3.71 mg/l
Exposure time: 30 min

Bibliographical reference: Speece, R. (1987). Drexel University, Philadelphia, USA, private communication.

2-OCTANONE

CAS RN: 111-13-7

Test parameter: EC50
Effect: Reduction in light output
Concentrations: 15.0 mg/l
20.5 mg/l
Exposure time: 5 min
Comment: Concentrations in the tests were measured.

Bibliographical reference: Curtis, C., Lima, A., Lozano, S.J., and Veith, G.D. (1982). In: *Aquatic Toxicology and Hazard Assessment: Fifth Conference, ASTM STP 766*, J.G. Pearson, R.B. Foster, and W.E. Bishop (eds.), American Society for Testing and Materials, Philadelphia, p. 170-178.

OCTYLAMINE

CAS RN: 111-86-4

Sample purity: 97%
Temperature: 15°C
Test parameter: EC50
Effect: Reduction in light output

Concentration: 29 mg/l
Exposure time: 5 min

Concentration: 27 mg/l*
Exposure time: 15 min

Bibliographical references: Cronin, M.T.D., Dearden, J.C., and Dobbs, A.J. (1991). *Sci. Total Environ.* **109/110**, 431-439.
* Cronin, M.T.D. (1990). Thesis, Liverpool Polytechnic, Liverpool, UK.

OCTYL CYANIDE

CAS RN: 2243-27-8
Synonym: Pelargononitrile

Sample purity: 98%
Temperature: 15°C
Test parameter: EC50
Effect: Reduction in light output

Concentration: 1.16 mg/l
Exposure time: 5 min

Concentration: 0.88 mg/l*
Exposure time: 15 min

Comment: Chemical was prepared in an initial solution of 3% methanol.

Bibliographical references: Cronin, M.T.D., Dearden, J.C., and Dobbs, A.J. (1991). *Sci. Total Environ.* **109/110**, 431-439.
* Cronin, M.T.D. (1990). Thesis, Liverpool Polytechnic, Liverpool, UK.

OXALIC ACID

CAS RN: 144-62-7

Sample purity: 99+%
Temperature: 15°C
Test parameter: EC50
Effect: Reduction in light output

Concentration: 11.3 ± 0.22 mg/l
Exposure time: 5 min

Concentration: 11.3 ± 0.25 mg/l
Exposure time: 15 min

Concentration: 11.4 ± 0.26 mg/l
Exposure time: 25 min

Comment: Mean of six assays. The values were converted to mg/l from the original data expressed in μM and a rounded molecular weight of 90 given by the authors. Phenol solution was used for quality control/quality assurance. The 5-min EC50 value was 18.2 mg/l. The EC50 value at 15 min was 20.7 mg/l with a relative error of <5%.

Bibliographical reference: Chou, C.C., and Que Hee, S.S. (1992). *Ecotoxicol. Environ. Safety* **23**, 355-363.

4,4'-OXYDIBENZENESULFONYL HYDRAZIDE

CAS RN: 80-51-3

Sample purity: 90%
Temperature: 15°C
Test parameter: EC50

Effect: Reduction in light output

Concentration: 8.60 mg/l
Exposure time: 5 min

Concentration: 9.00 mg/l
Exposure time: 15 min

Concentration: 10.3 mg/l
Exposure time: 30 min

Comment: Mean of three assays. Methanol (<10%) was used to prepare the stock solutions. EC50 values were calculated from nominal concentrations.

Bibliographical reference: Kaiser, K.L.E., and Palabrica, V.S. (1992). National Water Research Institute, Burlington, Ontario, Canada, unpublished results.

PARAQUAT

CAS RN: 1910-42-5

Temperature: 15 ± 0.1°C
Test parameter: EC50
Effect: Reduction in light output
Concentration: 780 mg/l
Exposure time: 5 min

Bibliographical reference: Chang, J.C., Taylor, P.B., and Leach, F.R. (1981). *Bull. Environ. Contam. Toxicol.* **26**, 150-156.

PARATHION

CAS RN: 56-38-2
Synonym: *O*,*O*-Diethyl *O*-4-nitrophenyl phosphorothioate

Temperature: 15 ± 0.1°C
Test parameter: EC50
Effect: Reduction in light output
Concentration: 8.5 mg/l
Exposure time: 5 min

Bibliographical reference: Somasundaram, L., Coats, J.R., Racke, K.D., and Stahr, H.M. (1990). *Bull. Environ. Contam. Toxicol.* **44**, 254-259.

PATULIN

CAS RN: 149-29-1

Temperature: 15°C
Test parameter: EC50
Effect: Reduction in light output

Concentration: 7.53 mg/l
Exposure time: 5 min

Concentration: 3.87 mg/l
Exposure time: 10 min

Concentration: 2.67 mg/l
Exposure time: 15 min

Concentration: 2.16 mg/l
Exposure time: 20 min

Comment: Freshly reconstituted bacterial suspensions. Chemical was dissolved in methanol.

Bibliographical reference: Yates, I.E., and Porter, J.K. (1982). *Appl. Environ. Microbiol.* **44**, 1072-1075.

Temperature: 15°C
Test parameter: EC20
Effect: Reduction in light output

Concentration: 2.56 mg/l
Exposure time: 5 min

Concentration: 0.89 mg/l
Exposure time: 20 min

Comment: Freshly reconstituted bacterial suspensions. Chemical was dissolved in methanol.

Bibliographical reference: Yates, I.E., and Porter, J.K. (1982). *Appl. Environ. Microbiol.* **44**, 1072-1075.

Temperature: 15°C
Test parameter: EC50
Effect: Reduction in light output

Concentration: 6.17 mg/l
Exposure time: 5 min

Concentration: 3.45 mg/l
Exposure time: 10 min

Concentration: 2.36 mg/l
Exposure time: 15 min

Concentration: 1.82 mg/l
Exposure time: 20 min

Comment: Performed on bacterial suspensions maintained at 3°C for 5 h after reconstitution. Chemical was dissolved in methanol.

Bibliographical reference: Yates, I.E., and Porter, J.K. (1982). *Appl. Environ. Microbiol.* **44**, 1072-1075.

Temperature: 30°C
Test parameter: EC50
Effect: Reduction in light output

Concentration: 3.54 mg/l
Exposure time: 5 min

Concentration: 1.82 mg/l
Exposure time: 10 min

Concentration: 1.32 mg/l
Exposure time: 15 min

Concentration: 1.10 mg/l
Exposure time: 20 min

Comment: Chemical was dissolved in methanol. Test was performed at pH = 8.0 units.

Bibliographical reference: Yates, I.E., and Porter, J.K. (1984). In: *Toxicity Screening Procedures Using Bacterial Systems*, D. Liu and B.J. Dutka (eds.), Marcel Dekker, New York, p. 77-88.

PENICILLIC ACID

CAS RN: 90-65-3

Temperature: 15°C
Test parameter: EC50
Effect: Reduction in light output

Concentration: 15.95 mg/l
Exposure time: 5 min

Concentration: 10.65 mg/l
Exposure time: 10 min

Concentration: 8.72 mg/l
Exposure time: 15 min

Concentration: 7.44 mg/l
Exposure time: 20 min

Comment: Freshly reconstituted bacterial suspensions. Chemical was dissolved in methanol.

Bibliographical reference: Yates, I.E., and Porter, J.K. (1982). *Appl. Environ. Microbiol.* **44**, 1072-1075.

Temperature: 15°C
Test parameter: EC20
Effect: Reduction in light output

Concentration: 5.34 mg/l
Exposure time: 5 min

Concentration: 3.14 mg/l
Exposure time: 20 min

Comment: Freshly reconstituted bacterial suspensions. Chemical was dissolved in methanol.

Bibliographical reference: Yates, I.E., and Porter, J.K. (1982). *Appl. Environ. Microbiol.* **44**, 1072-1075.

Temperature: 15°C
Test parameter: EC50
Effect: Reduction in light output

Concentration: 14.67 mg/l
Exposure time: 5 min

Concentration: 9.79 mg/l
Exposure time: 10 min

Concentration: 7.60 mg/l
Exposure time: 15 min

Concentration: 5.91 mg/l
Exposure time: 20 min

Comment: Performed on bacterial suspensions maintained at 3°C for 5 h after reconstitution. Chemical was dissolved in methanol.

Bibliographical reference: Yates, I.E., and Porter, J.K. (1982). *Appl. Environ. Microbiol.* **44**, 1072-1075.

Temperature: 25°C
Test parameter: EC50
Effect: Reduction in light output

Concentration: 4.70 mg/l
Exposure time: 5 min

Concentration: 2.03 mg/l
Exposure time: 10 min

Concentration: 1.45 mg/l
Exposure time: 15 min

Concentration: 1.25 mg/l
Exposure time: 20 min

Comment: Chemical was dissolved in methanol. Test was performed at pH = 6.5 units.

Bibliographical reference: Yates, I.E., and Porter, J.K. (1984). In: *Toxicity Screening Procedures Using Bacterial Systems*, D. Liu and B.J. Dutka (eds.), Marcel Dekker, New York, p. 77-88.

PENTABROMOPHENOL

CAS RN: 608-71-9

Temperature: 15°C
Test parameter: EC50
Effect: Reduction in light output

Concentration: 0.89 mg/l
Exposure time: 5 min

Concentration: 0.45 mg/l*
Exposure time: 30 min

* EC50 value from linear regression fit to data.

Bibliographical reference: Speece, R. (1987). Drexel University, Philadelphia, USA, private communication.

2,2',4,4',5-PENTACHLORODIPHENYL ETHER

CAS RN: 60123-64-0

Temperature: 15°C
Test parameter: EC50
Effect: Reduction in light output

Concentration: 14.0 mg/l
Exposure time: 5 min

Concentration: 12.7 mg/l
Exposure time: 15 min

Concentration: 13.3 mg/l
Exposure time: 30 min

Comment: Mean of three assays. Methanol (<10%) was used to prepare the stock solutions. EC50 values were calculated from nominal concentrations.

Bibliographical reference: Kaiser, K.L.E., and Palabrica, V.S. (1991). *Water Poll. Res. J. Canada* **26**, 361-431.

PENTACHLOROETHANE

CAS RN: 76-01-7

Test parameter: EC50
Effect: Reduction in light output
Concentration: 0.75 mg/l
Exposure time: 5 min
Comment: Concentrations in the test were measured.

Bibliographical reference: Curtis, C., Lima, A., Lozano, S.J., and Veith, G.D. (1982). In: *Aquatic Toxicology and Hazard Assessment: Fifth Conference, ASTM STP 766*, J.G. Pearson, R.B. Foster, and W.E. Bishop (eds.), American Society for Testing and Materials, Philadelphia, p. 170-178.

Temperature: 15°C
Test parameter: EC50
Effect: Reduction in light output

Concentration: 0.63 mg/l
Exposure time: 5 min

Concentration: 0.90 mg/l
Exposure time: 30 min

Bibliographical reference: Speece, R. (1987). Drexel University, Philadelphia, USA, private communication.

PENTACHLORONITROBENZENE

CAS RN: 82-68-8

Temperature: 15°C

Test parameter: EC50
Effect: Reduction in light output
Concentration: 3.80 mg/l
Exposure times: 5, 15, and 30 min
Comment: Mean of three assays. Methanol (<10%) was used to prepare the stock solutions. EC50 values were calculated from nominal concentrations.

Bibliographical reference: Kaiser, K.L.E., and Ribo, J.M. (1985). In: *QSAR in Toxicology and Xenobiochemistry*, M. Tichy (ed.), Elsevier, Amsterdam, p. 27-38.

PENTACHLOROPHENOL

CAS RN: 87-86-5
Synonym: PCP

Temperature: 15°C
Test parameter: EC50
Effect: Reduction in light output
Concentrations: 0.70 mg/l
2.2 mg/l
Exposure time: 5 min

Bibliographical reference: King, E.F., and Painter, H.A. (1981). In: *Les Tests de Toxicité Aiguë en Milieu Aquatique. Acute Aquatic Ecotoxicological Tests*, H. Leclerc and D. Dive (eds.), INSERM 106, Paris, p. 143-153.

Test parameter: EC50
Effect: Reduction in light output
Concentration: 0.08 mg/l
Exposure time: 5 min
Comment: The EC50 value was calculated from nominal concentrations.

Bibliographical reference: Curtis, C., Lima, A., Lozano, S.J., and Veith, G.D. (1982). In: *Aquatic Toxicology and Hazard Assessment: Fifth Conference, ASTM STP 766*, J.G. Pearson, R.B. Foster, and W.E. Bishop (eds.), American Society for Testing and Materials, Philadelphia, p. 170-178.

Temperature: 15°C
Test parameter: EC50
Effect: Reduction in light output

Concentration: 0.92 mg/l
Exposure time: 5 min

Concentration: 0.61 mg/l
Exposure time: 15 min

Concentration: 0.52 mg/l
Exposure time: 30 min

Comment: Mean of three assays. Methanol (<10%) was used to prepare the stock solutions. EC50 values were calculated from nominal concentrations.

Bibliographical reference: Ribo, J.M., and Kaiser, K.L.E. (1983). *Chemosphere* **12**, 1421-1442.

Temperature: 20°C
Test parameter: EC50
Effect: Reduction in light output
Concentration: 0.99 ± 0.15 mg/l
Exposure time: 10 min
Comment: Solutions were prepared 12 hours before the beginning of the tests, to insure complete solubility. Toxicity values were calculated from nominal concentrations.

Bibliographical reference: Vasseur, P., Bois, F., Ferard, J.F., Rast, C., and Larbaigt, G. (1986). *Tox. Assess.* **1**, 283-300.

Temperature: 15 ± 0.1°C
Test parameter: EC50
Effect: Reduction in light output

Concentration: 1.03 mg/l
Exposure time: 5 min

Concentration: 0.61 mg/l
Exposure time: 15 min

Concentration: 0.48 mg/l
Exposure time: 30 min

Comment: The pH was not adjusted.

Bibliographical reference: Tarkpea, M., Hansson, M., and Samuelsson, B. (1986). *Ecotoxicol. Environ. Safety* **11**, 127-143.

Temperature: 15°C
Test parameter: EC50
Effect: Reduction in light output
Concentration: 1.0 mg/l (0.9-1.1)
Exposure time: 15 min
Comment: Single batches of Microtox® reagent were used for less than 2 h before being discarded. Four concentrations were tested. Pentachlorophenol was first dissolved in ethanol (0.01%). Concentrations were unmeasured.

Bibliographical reference: Nacci, D., Jackim, E., and Walsh, R. (1986). *Environ. Toxicol. Chem.* **5**, 521-525.

Temperature: 20°C
Test parameter: Responses to zinc ($ZnSO_4.7H_2O$) and pentachlorophenol (PCP) alone or in combination, expressed by $\Delta\% = 100\ (I_{0(x)}.BR - I_{t(x)})/(I_{0(x)}.BR)$ where BR is the blank ratio
Result:

Zn	PCP		Δ%	
mg/l	mg/l	n° 1	n° 2	n° 3
0	0	-4.9	-0.1	-10.8
0	0.25	15.0	18.9	13.7
0	0.81	51.9	48.1	52.7
0	2.08	83.5	83.5	81.7
0.2	0	22.9	26.5	26.5
0.2	0.25	27.8	33.9	32.4
0.2	0.81	66.3	62.0	61.2
0.2	2.08	87.2	86.1	85.4
0.6	0	58.8	60.0	62.9
0.6	0.25	60.3	62.5	62.8
0.6	0.81	82.5	80.7	76.1
0.6	2.08	91.2	92.1	91.8
1.4	0	90.6	87.1	87.0
1.4	0.25	92.6	84.1	83.4
1.4	0.8	94.6	90.0	89.8
1.4	2.08	96.7	97.3	95.0

The model computed was:

$\hat{\Delta}\% = 83 + 46\ X_1 + 50\ X_2 + 19\ X_2^2 - 84\ X_2^3 - 88\ X_1\ X_2$

where: $X_1 = \log((Zn^{++}) + 0.2)$ and $X_2 = \log((PCP) + 0.2)$
Exposure time: 30 min

Bibliographical reference: Bois, F., Vaillant, M., and Vasseur, P. (1986). *Bull. Environ. Contam. Toxicol.* **36**, 707-714.

Temperature: 20°C
Test parameter: Responses of to zinc ($ZnSO_4.7H_2O$) and pentachlorophenol (PCP) alone or in combination, expressed by log β = log ($I_{t(x)}/(I_{0(x)}.BR)$) where BR is the blank ratio
Result:

Zn	PCP		- log β	
mg/l	mg/l	n° 1	n° 2	n° 3
0	0	-0.021	-0.004	-0.045
0	0.25	0.071	0.091	0.064
0	0.81	0.318	0.285	0.325
0	2.08	0.784	0.780	0.740
0.2	0	0.113	0.130	0.130
0.2	0.25	0.142	0.181	0.170
0.2	0.81	0.472	0.420	0.410
0.2	2.08	0.892	0.856	0.839
0.6	0	0.385	0.410	0.430
0.6	0.25	0.400	0.429	0.430
0.6	0.81	0.756	0.710	0.620
0.6	2.08	1.06	1.10	1.10
1.4	0	1.03	0.890	0.870
1.4	0.25	1.13	0.800	0.780
1.4	0.8	1.30	1.00	0.990
1.4	2.08	1.49	1.57	1.30

The model computed was:

$-\log \hat{\beta} = 0.8 + 1.19\ X_1 + 0.62\ X_1^2 + 0.79\ X_2 + 0.645\ X_2^2$

where: $X_1 = \log((Zn^{++}) + 0.2)$ and $X_2 = \log((PCP) + 0.2)$
Exposure time: 30 min

Bibliographical reference: Bois, F., Vaillant, M., and Vasseur, P. (1986). *Bull. Environ. Contam. Toxicol.* **36**, 707-714.

Temperature: 15°C
Test parameter: EC50

Effect: Reduction in light output
Concentration: 1.8 mg/l
Exposure time: 10 min
Comment: Mean of three assays. Prior testing, chemical was dissolved at pH = 10. Test solutions were analyzed using gas chromatography.

Bibliographical reference: Bazin, C., Chambon, P., Bonnefille, M., and Larbaigt, G. (1987). *Sci. Eau* **6**, 403-413.

Sample purity: Reagent grade
Test parameter: EC50
Effect: Reduction in light output

Concentration: 1.3 mg/l (1.2-1.4)
Exposure time: 5 min

Concentration: 1.1 mg/l (1.0-1.1)
Exposure time: 15 min

Concentration: 1.0 mg/l (0.9-1.1)
Exposure time: 30 min

Bibliographical reference: Elnabarawy, M.T., Robideau, R.R., and Beach, S.A. (1988). *Tox. Assess.* **3**, 361-370.

Test parameter: EC50
Effect: Reduction in light output
Concentration: 0.55 mg/l
Exposure time: 5 min

Bibliographical reference: Middaugh, D.P., Resnick, S.M., Lantz, S.E., Heard, C.S., and Mueller, J.G. (1993). *Arch. Environ. Contam. Toxicol.* **24**, 165-172.

2,4-PENTANEDIONE

CAS RN: 123-54-6
Synonym: Acetylacetone

Test parameter: EC50
Effect: Reduction in light output
Concentration: 1050 mg/l

Exposure time: 5 min
Comment: The EC50 value was calculated from nominal concentrations.

Bibliographical reference: Curtis, C., Lima, A., Lozano, S.J., and Veith, G.D. (1982). In: *Aquatic Toxicology and Hazard Assessment: Fifth Conference, ASTM STP 766*, J.G. Pearson, R.B. Foster, and W.E. Bishop (eds.), American Society for Testing and Materials, Philadelphia, p. 170-178.

Temperature: 15°C
Test parameter: EC50
Effect: Reduction in light output
Concentration: 373 mg/l (346.6-401.5)
Exposure time: 15 min
Comment: Single batches of Microtox® reagent were used for less than 2 h before being discarded. Four concentrations were tested. Concentrations were unmeasured.

Bibliographical reference: Nacci, D., Jackim, E., and Walsh, R. (1986). *Environ. Toxicol. Chem.* **5**, 521-525.

1-PENTANOL

CAS RN: 71-41-0

Temperature: 15°C
Test parameter: EC50
Effect: Reduction in light output

Concentration: 298 mg/l
Exposure time: 5 min

Concentration: 393 mg/l
Exposure time: 30 min

Bibliographical reference: Speece, R. (1987). Drexel University, Philadelphia, USA, private communication.

3-PENTANOL

CAS RN: 584-02-1

Temperature: 15°C
Test parameter: EC50
Effect: Reduction in light output
Concentration: 1500 mg/l
Exposure time: 15 min

Bibliographical reference: Hermens, J., Busser, F., Leeuwangh, P., and Musch, A. (1985). *Ecotoxicol. Environ. Safety* **9**, 17-25.

3-PENTANONE

CAS RN: 96-22-0

Temperature: 15°C
Test parameter: EC50
Effect: Reduction in light output
Concentration: 849 mg/l
Exposure time: 5 min

Bibliographical reference: Microtox® Application Notes (1982). Beckman Instruments, Inc., Carlsbad, California.

Sample purity: 99%
Temperature: 15°C
Test parameter: EC50
Effect: Reduction in light output

Concentration: 1870 mg/l
Exposure time: 5 min

Concentration: 2120 mg/l
Exposure time: 15 min

Bibliographical reference: Cronin, M.T.D. (1993). Liverpool John Moores University, UK, private communication.

PERINAPHTHENONE

CAS RN: 548-39-0

Temperature: 15°C
Test parameter: EC50
Effect: Reduction in light output

Concentration: 0.75 mg/l
Exposure times: 5 and 15 min

Concentration: 0.84 mg/l
Exposure time: 30 min

Comment: Mean of four assays. Methanol (<10%) was used to prepare the stock solutions. EC50 values were calculated from nominal concentrations.

Bibliographical reference: Kaiser, K.L.E., and Palabrica, V.S. (1991). *Water Poll. Res. J. Canada* **26**, 361-431.

PERMETHRIN

CAS RN: 52645-53-1

Test parameter: EC50
Effect: Reduction in light output
Concentration: 0.56 mg/l
Exposure time: 5 min
Comment: Concentrations in the test were measured.

Bibliographical reference: Curtis, C., Lima, A., Lozano, S.J., and Veith, G.D. (1982). In: *Aquatic Toxicology and Hazard Assessment: Fifth Conference, ASTM STP 766*, J.G. Pearson, R.B. Foster, and W.E. Bishop (eds.), American Society for Testing and Materials, Philadelphia, p. 170-178.

Sample purity: 91%
Temperature: 15°C
Test parameter: EC50
Effect: Reduction in light output

Concentration: 23.0 mg/l

Exposure time: 5 min

Concentration: 27.7 mg/l
Exposure time: 15 min

Concentration: 31.8 mg/l
Exposure time: 30 min

Comment: Mean of three assays. Methanol (<10%) was used to prepare the stock solutions. EC50 values were calculated from nominal concentrations.

Bibliographical reference: Kaiser, K.L.E., and Palabrica, V.S. (1991). *Water Poll. Res. J. Canada* **26**, 361-431.

PHENACETIN

CAS RN: 62-44-2

Sample purity: 97%
Temperature: 15°C
Test parameter: EC50
Effect: Reduction in light output

Concentration: 103 mg/l
Exposure time: 5 min

Concentration: 111 mg/l
Exposure times: 15 and 30 min

Comment: Mean of three assays. Methanol (<10%) was used to prepare the stock solutions. EC50 values were calculated from nominal concentrations.

Bibliographical reference: Kaiser, K.L.E., and Palabrica, V.S. (1991). *Water Poll. Res. J. Canada* **26**, 361-431.

PHENANTHRENE

CAS RN: 85-01-8

Temperature: 15°C

Test parameter: EC50
Effect: Reduction in light output

Concentration: 0.042 mg/l
Exposure time: 5 min

Concentration: 0.049 mg/l
Exposure time: 15 min

Concentration: 0.073 mg/l
Exposure time: 30 min

Comment: Mean of three assays. Methanol (<10%) was used to prepare the stock solutions. EC50 values were calculated from nominal concentrations.

Bibliographical reference: Kaiser, K.L.E., and Palabrica, V.S. (1991). *Water Poll. Res. J. Canada* **26**, 361-431.

Temperature: 15°C
Test parameter: EC50
Effect: Reduction in light output
Concentration: 0.53 mg/l (0.35-0.79)
Exposure time: 15 min
Comment: Chemical was added to 5 ml DMSO and placed in a bath sonicator until all visible material had dissolved.

Bibliographical reference: Jacobs, M.W., Coates, J.A., Delfino, J.J., Bitton, G., Davis, W.M., and Garcia, K.L. (1993). *Arch. Environ. Contam. Toxicol.* **24**, 461-468.

PHENANTHRENEQUINONE

CAS RN: 84-11-7

Sample purity: >99.9%
Temperature: 15°C
Test parameter: EC50
Effect: Reduction in light output

Concentration: 0.087 mg/l
Exposure time: 5 min

Concentration: 0.067 mg/l
Exposure time: 15 min

Concentration: 0.066 mg/l
Exposure time: 30 min

Comment: Mean of three assays. Methanol (<10%) was used to prepare the stock solutions. EC50 values were calculated from nominal concentrations.

Bibliographical reference: Kaiser, K.L.E., and Palabrica, V.S. (1991). *Water Poll. Res. J. Canada* **26**, 361-431.

PHENAZINE

CAS RN: 92-82-0

Sample purity: 98%
Temperature: 15°C
Test parameter: EC50
Effect: Reduction in light output

Concentration: 26.7 mg/l
Exposure time: 5 min

Concentration: 23.2 mg/l
Exposure time: 15 min

Concentration: 21.7 mg/l
Exposure time: 30 min

Comment: Mean of three assays. Methanol (<10%) was used to prepare the stock solutions. EC50 values were calculated from nominal concentrations.

Bibliographical reference: Kaiser, K.L.E., and Palabrica, V.S. (1991). *Water Poll. Res. J. Canada* **26**, 361-431.

PHENETHYL ALCOHOL

CAS RN: 60-12-8

Sample purity: 99%
Temperature: 15°C
Test parameter: EC50
Effect: Reduction in light output

Concentration: 5.46 mg/l
Exposure time: 5 min

Concentration: 4.98 mg/l
Exposure time: 15 min

Concentration: 5.33 mg/l*
Exposure time: 30 min

Comment: Mean of five assays. Methanol (<10%) was used to prepare the stock solutions. EC50 values were calculated from nominal concentrations.

Bibliographical references: Kaiser, K.L.E., and Palabrica, V.S. (1991). *Water Poll. Res. J. Canada* **26**, 361-431.
* Kaiser, K.L.E., Palabrica, V.S., and Ribo, J.M. (1987). In: *QSAR in Environmental Toxicology - II*, K.L.E. Kaiser (ed.), D. Reidel Publishing Company, Dordrecht, p. 153-168.

PHENETHYLAMINE

CAS RN: 64-04-0

Sample purity: 99%
Temperature: 15°C
Test parameter: EC50
Effect: Reduction in light output

Concentration: 14.9 mg/l
Exposure time: 5 min

Concentration: 13.0 mg/l
Exposure time: 15 min

Concentration: 12.4 mg/l*
Exposure time: 30 min

Comment: Mean of two assays. Methanol (<10%) was used to prepare the stock solutions. EC50 values were calculated from nominal concentrations.

Bibliographical references: Kaiser, K.L.E., and Palabrica, V.S. (1991). *Water Poll. Res. J. Canada* **26**, 361-431.
* Kaiser, K.L.E., Palabrica, V.S., and Ribo, J.M. (1987). In: *QSAR in Environmental Toxicology - II*, K.L.E. Kaiser (ed.), D. Reidel Publishing Company, Dordrecht, p. 153-168.

PHENETOLE

CAS RN: 103-73-1
Synonyms: Ethoxybenzene; Ethyl phenyl ether

Sample purity: 99%
Temperature: 15°C
Test parameter: EC50
Effect: Reduction in light output

Concentration: 5.33 mg/l
Exposure time: 5 min

Concentration: 6.27 mg/l
Exposure time: 15 min

Concentration: 7.53 mg/l
Exposure time: 30 min

Comment: Mean of three assays. Methanol (<10%) was used to prepare the stock solutions. EC50 values were calculated from nominal concentrations.

Bibliographical reference: Kaiser, K.L.E., and Palabrica, V.S. (1991). *Water Poll. Res. J. Canada* **26**, 361-431.

PHENOL

CAS RN: 108-95-2

Temperature: 15°C
Test parameter: EC50

Effect: Reduction in light output
Concentration: 42 mg/l
Exposure time: 5 min

Bibliographical reference: Samak, Q.M., and Noiseux, R. (1981). *Can. Tech. Rep. Fish. Aquat. Sci.* **990**, 288-308.

Temperature: 15°C
Test parameter: EC50
Effect: Reduction in light output
Concentration: 25 mg/l
Exposure time: 5 min

Bibliographical reference: Bulich, A.A., Greene, M.W., and Isenberg, D.L. (1981). In: *Aquatic Toxicology and Hazard Assessment: Fourth Conference, ASTM STP 737*, D.R. Branson and K.L. Dickson (eds.), American Society for Testing and Materials, Philadelphia, p. 338-347.

Temperature: 15°C
Test parameter: EC50
Effect: Reduction in light output
Concentration: 25 mg/l
Exposure time: 5 min

Bibliographical reference: Lebsack, M.E., Anderson, A.D., DeGraeve, G.M., and Bergman, H.L. (1981). In: *Aquatic Toxicology and Hazard Assessment: Fourth Conference, ASTM STP 737*, D.R. Branson and K.L. Dickson, (eds.), American Society for Testing and Materials, Philadelphia, p. 348-356.

Temperature: 15 ± 0.1°C
Test parameter: EC50
Effect: Reduction in light output
Concentration: 26 mg/l
Exposure time: 5 min

Bibliographical reference: Chang, J.C., Taylor, P.B., and Leach, F.R. (1981). *Bull. Environ. Contam. Toxicol.* **26**, 150-156.

Temperature: 15 ± 0.3°C
Test parameter: EC50

Effect: Reduction in light output
Concentrations: 25.8 mg/l
18.3 mg/l
Exposure time: 5 min
Comment: Test solutions were analyzed.

Bibliographical reference: Qureshi, A.A., Flood, K.W., Thompson, S.R., Janhurst, S.M., Inniss, C.S., and Rokosh, D.A. (1982). In: *Aquatic Toxicology and Hazard Assessment: Fifth Conference, ASTM STP 766*, J.G. Pearson, R.B. Foster, and W.E. Bishop (eds.), American Society for Testing and Materials, Philadelphia, p. 179-195.

Effect: Percentage light inhibition
Result:

Chemical	incubation time		
ppm	5-min	10-min	15-min
2.5 phenol	4%	2%	1%
20 phenol	41%	35%	35%
0.5 sodium lauryl sulfate (SLS)	21%	22%	22%
2.5 phenol + 0.5 SLS	46%	57%	64%
20 phenol + 0.5 SLS	59%	63%	66%

Comment: Mean of duplicated tests. Chemicals were tested at pH 6.7.

Bibliographical reference: Dutka, B.J., and Kwan, K.K. (1982). *Environ. Pollut. Ser. A* **29**, 125-134.

Temperature: 15°C
Test parameter: EC50
Effect: Reduction in light output

Concentration: 28 mg/l
Exposure time: 5 min

Concentration: 31.9 mg/l
Exposure time: 10 min

Concentration: 34.3 mg/l
Exposure time: 15 min

Comment: Chemical was tested in duplicate at pH 6.7.

Bibliographical references: Dutka, B.J., and Kwan, K.K. (1982). *Environ. Pollut. Ser. A* **29**, 125-134.

Dutka, B.J., Nyholm, N., and Petersen, J. (1983). *Water Res.* **17**, 1363-1368.

Test parameter: EC50
Effect: Reduction in light output
Concentrations: 39.8 mg/l
40.7 mg/l
Exposure time: 5 min
Comment: The EC50 values were calculated from nominal concentrations.

Bibliographical reference: Curtis, C., Lima, A., Lozano, S.J., and Veith, G.D. (1982). In: *Aquatic Toxicology and Hazard Assessment: Fifth Conference, ASTM STP 766*, J.G. Pearson, R.B. Foster, and W.E. Bishop (eds.), American Society for Testing and Materials, Philadelphia, p. 170-178.

Temperature: 15 ± 0.1°C
Test parameter: EC50
Effect: Reduction in light output
Concentration: 39.5 mg/l
Exposure time: 5 min
Comment: Test was performed on *Photobacterium phosphoreum* NZ11D obtained from the Scripps Institute of Oceanography (La Jolla, CA).

Bibliographical reference: McFeters, G.A., Bond, P.J., Olson, S.B., and Tchan, Y.T. (1983). *Water Res.* **17**, 1757-1762.

Temperature: 15°C
Test parameter: EC50
Effect: Reduction in light output

Concentration: 29.8 mg/l
Exposure time: 5 min

Concentration: 34.2 mg/l
Exposure time: 15 min

Concentration: 35.8 mg/l
Exposure time: 30 min

Comment: Mean of three assays. Methanol (<10%) was used to prepare the stock solutions. EC50 values were calculated from nominal concentrations.

Bibliographical reference: Ribo, J.M., and Kaiser, K.L.E. (1983). *Chemosphere* **12**, 1421-1442.

Temperature: 15°C
Test parameter: EC50
Effect: Reduction in light output
Concentration: 34 ± 8 mg/l
Exposure time: 5 min
Comment: Mean of three assays. Toxicity values were calculated from nominal concentrations.

Bibliographical reference: Férard, J.F., Vasseur, P., Danoux, L., and Larbaigt, G. (1983). *Rev. Fr. Sci. Eau* **2**, 221-237.

Temperature: 15°C
Test parameter: EC50
Effect: Reduction in light output

Concentration: 42 mg/l
Exposure time: 5 min

Concentration: 62 mg/l
Exposure time: 15 min

Bibliographical reference: Mallak, F.P., and Brunker, R.L. (1984). In: *Toxicity Screening Procedures Using Bacterial Systems*, D. Liu and B.J. Dutka (eds.), Marcel Dekker, New York, p. 65-76.

Temperature: 15°C
Test parameter: EC50
Effect: Reduction in light output

Concentrations: 23.6 mg/l (time after reconstitution = 0.5 h)
25.4 mg/l (time after reconstitution = 4 h)
Exposure time: 5 min

Concentrations: 26.0 mg/l (time after reconstitution = 0.5 h)
26.6 mg/l (time after reconstitution = 4 h)
Exposure time: 15 min

Bibliographical reference: Qureshi, A.A., Coleman, R.N., and Paran, J.H. (1984). In: *Toxicity Screening Procedures Using Bacterial Systems*, D. Liu and B.J. Dutka (eds.), Marcel Dekker, New York, p. 1-22.

Temperature: 15°C
Test parameter: EC50
Effect: Reduction in light output

Concentration: 24.5 mg/l
Exposure time: 5 min

Concentration: 28.1 mg/l
Exposure time: 10 min

Concentration: 26.3 mg/l
Exposure time: 15 min

Concentration: 26.4 mg/l
Exposure time: 20 min

Concentration: 26.0 mg/l
Exposure time: 30 min

Comment: Results derived from the average of two replicates.

Bibliographical reference: Qureshi, A.A., Coleman, R.N., and Paran, J.H. (1984). In: *Toxicity Screening Procedures Using Bacterial Systems*, D. Liu and B.J. Dutka (eds.), Marcel Dekker, New York, p. 1-22.

Temperature: 15°C
Test parameter: EC50
Effect: Reduction in light output
Concentration: 35.5 ± 3.5 mg/l
Exposure time: 10 min

Bibliographical reference: Bazin, C. (1985). DEA Ecologie Fondamentale et Appliquée des Eaux Continentales. University Lyon I, Lyon, France.

Sample purity: Analytical grade
Temperature: 15°C

Test parameter: EC50
Effect: Reduction in light output
Concentration: 30.1 mg/l
Exposure time: 5 min

Bibliographical reference: Beaubien, A., Lapierre, L., Bouchard, A., and Jolicoeur, C. (1986). *Tox. Assess.* **1**, 187-200.

Temperature: 15°C
Test parameter: EC50
Effect: Reduction in light output

Concentration: 18.2 mg/l
Exposure time: 5 min

Concentration: 26.1 mg/l
Exposure time: 30 min

Bibliographical reference: Speece, R. (1987). Drexel University, Philadelphia, USA, private communication.

Temperature: 15°C
Test parameter: EC50
Effect: Reduction in light output
Concentration: 21.1 mg/l
Exposure time: 30 min
Comment: Mean of three assays. Methanol (<10%) was used to prepare the stock solutions. EC50 values were calculated from nominal concentrations.

Bibliographical reference: Kaiser, K.L.E., Palabrica, V.S., and Ribo, J.M. (1987). In: *QSAR in Environmental Toxicology - II*, K.L.E. Kaiser (ed.), D. Reidel Publishing Company, Dordrecht, p. 153-168.

Sample purity: Reagent grade
Test parameter: EC50
Effect: Reduction in light output

Concentration: 20 mg/l (19-21)
Exposure time: 5 min

Concentration: 21 mg/l (20-22)
Exposure times: 15 and 30 min

Bibliographical reference: Elnabarawy, M.T., Robideau, R.R., and Beach, S.A. (1988). *Tox. Assess.* **3**, 361-370.

Test parameter: EC50
Effect: Reduction in light output

Concentration: 28.8 mg/l
Exposure time: 5 min

Concentration: 31.6 mg/l
Exposure time: 15 min

Bibliographical reference: Ribo, J.M., and Rogers, F. (1990). *Tox. Assess.* **5**, 135-152.

Temperature: 15 ± 0.5°C
Test parameter: EC20
Effect: Reduction in light output

Concentration: 10.40 mg/l (8.50-12.60)
Exposure time: 15 min
Comment: Osmotic adjustment was achieved using NaCl (2.0%).

Concentration: 11.60 mg/l (10.30-13.10)
Exposure time: 15 min
Comment: Osmotic adjustment was achieved using sucrose (20.4%).

Bibliographical reference: Ankley, G.T., Peterson, G.S., Amato, J.R., and Jenson, J.J. (1990). *Environ. Toxicol. Chem.* **9**, 1305-1310.

Temperature: 15°C
Test parameter: EC50
Effect: Reduction in light output
Concentrations: 32.3 ± 1.8 mg/l (Microbics strain)
20.5 ± 2.6 mg/l (Dr. Lange strain)
Exposure time: 15 min
Comment: EC50 values were calculated from nominal concentrations.

Bibliographical reference: Vasseur, P. (1992). Centre des Sciences de l'Environnement, Metz, France, private communication.

Temperature: 15°C

Test parameter: EC50
Effect: Reduction in light output
Concentration: 27 mg/l
Exposure time: 5 min

Bibliographical reference: Kahru, A. (1993). *ATLA* **21**, 210-215.

Test parameter: EC50
Effect: Reduction in light output
Concentration: 16.35 ± 1.60 mg/l
Exposure time: 5 min

Bibliographical reference: Middaugh, D.P., Resnick, S.M., Lantz, S.E., Heard, C.S., and Mueller, J.G. (1993). *Arch. Environ. Contam. Toxicol.* **24**, 165-172.

PHENOL RED

CAS RN: 143-74-8

Temperature: 15°C
Test parameter: EC50
Effect: Reduction in light output

Concentration: 19.9 mg/l
Exposure time: 5 min

Concentration: 23.4 mg/l
Exposure time: 15 min

Concentration: 28.1 mg/l
Exposure time: 30 min

Comment: Mean of five assays. Methanol (<10%) was used to prepare the stock solutions. EC50 values were calculated from nominal concentrations.

Bibliographical reference: Kaiser, K.L.E., and Palabrica, V.S. (1991). *Water Poll. Res. J. Canada* **26**, 361-431.

PHENOL-4-SULFONIC ACID

CAS RN: 98-67-9

Sample purity: 65%
Temperature: 15°C
Test parameter: EC50
Effect: Reduction in light output

Concentration: 230 mg/l
Exposure time: 5 min

Concentration: 282 mg/l
Exposure time: 15 min

Concentration: 381 mg/l
Exposure time: 30 min

Comment: Mean of three assays. Methanol (<10%) was used to prepare the stock solutions. EC50 values were calculated from nominal concentrations.

Bibliographical reference: Kaiser, K.L.E., and Palabrica, V.S. (1991). *Water Poll. Res. J. Canada* **26**, 361-431.

PHENOTHIAZINE

CAS RN: 92-84-2

Sample purity: >98%
Temperature: 15°C
Test parameter: EC50
Effect: Reduction in light output

Concentration: 14.1 mg/l
Exposure time: 5 min

Concentration: 15.5 mg/l
Exposure time: 15 min

Concentration: 19.9 mg/l
Exposure time: 30 min

Comment: Mean of three assays. Methanol (<10%) was used to prepare the stock solutions. EC50 values were calculated from nominal concentrations.

Bibliographical reference: Kaiser, K.L.E., and Palabrica, V.S. (1991). *Water Poll. Res. J. Canada* **26**, 361-431.

PHENOXYACETIC ACID

CAS RN: 122-59-8

Sample purity: 98%
Temperature: 15°C
Test parameter: EC50
Effect: Reduction in light output

Concentration: 66.4 mg/l
Exposure time: 5 min

Concentration: 69.5 mg/l
Exposure time: 15 min

Concentration: 74.5 mg/l*
Exposure time: 30 min

Comment: Mean of two assays. Methanol (<10%) was used to prepare the stock solutions. EC50 values were calculated from nominal concentrations.

Bibliographical references: Kaiser, K.L.E., and Palabrica, V.S. (1991). *Water Poll. Res. J. Canada* **26**, 361-431.
* Kaiser, K.L.E., Palabrica, V.S., and Ribo, J.M. (1987). In: *QSAR in Environmental Toxicology - II*, K.L.E. Kaiser (ed.), D. Reidel Publishing Company, Dordrecht, p. 153-168.

4-PHENOXYANILINE

CAS RN: 139-59-3

Temperature: 15°C
Test parameter: EC50
Effect: Reduction in light output

Concentration: 0.56 mg/l
Exposure time: 30 min
Comment: Mean of three assays. Methanol (<10%) was used to prepare the stock solutions. EC50 values were calculated from nominal concentrations.

Bibliographical reference: Kaiser, K.L.E. (1987). In: *QSAR in Environmental Toxicology - II*, K.L.E. Kaiser (ed.), D. Reidel Publishing Company, Dordrecht, p. 169-188.

4-PHENOXYBENZOIC ACID

CAS RN: 2215-77-2

Sample purity: 98%
Temperature: 15°C
Test parameter: EC50
Effect: Reduction in light output

Concentration: 19.5 mg/l
Exposure time: 5 min

Concentration: 19.1 mg/l
Exposure time: 15 min

Concentration: 18.2 mg/l
Exposure time: 30 min

Comment: Mean of four assays. Methanol (<10%) was used to prepare the stock solutions. EC50 values were calculated from nominal concentrations.

Bibliographical reference: Kaiser, K.L.E., and Palabrica, V.S. (1991). *Water Poll. Res. J. Canada* **26**, 361-431.

4-PHENOXYBUTYRIC ACID

CAS RN: 6303-58-8

Sample purity: 98%
Temperature: 15°C
Test parameter: EC50

Effect: Reduction in light output

Concentration: 485 mg/l
Exposure time: 5 min

Concentration: 520 mg/l
Exposure time: 15 min

Concentration: 544 mg/l*
Exposure time: 30 min

Comment: Mean of three assays. Methanol (<10%) was used to prepare the stock solutions. EC50 values were calculated from nominal concentrations.

Bibliographical references: Kaiser, K.L.E., and Palabrica, V.S. (1991). *Water Poll. Res. J. Canada* **26**, 361-431.
* Kaiser, K.L.E., Palabrica, V.S., and Ribo, J.M. (1987). In: *QSAR in Environmental Toxicology - II*, K.L.E. Kaiser (ed.), D. Reidel Publishing Company, Dordrecht, p. 153-168.

2-PHENOXYETHANOL

CAS RN: 122-99-6

Test parameter: EC50
Effect: Reduction in light output
Concentration: 32.7 mg/l
Exposure time: 5 min
Comment: The EC50 value was calculated from nominal concentrations.

Bibliographical reference: Curtis, C., Lima, A., Lozano, S.J., and Veith, G.D. (1982). In: *Aquatic Toxicology and Hazard Assessment: Fifth Conference, ASTM STP 766*, J.G. Pearson, R.B. Foster, and W.E. Bishop (eds.), American Society for Testing and Materials, Philadelphia, p. 170-178.

4-PHENOXYPHENOL

CAS RN: 831-82-3

Temperature: 15°C
Test parameter: EC50
Effect: Reduction in light output

Concentration: 2.09 mg/l
Exposure time: 5 min

Concentration: 3.63 mg/l
Exposure time: 15 min

Concentration: 6.17 mg/l
Exposure time: 30 min

Comment: Mean of three assays. Methanol (<10%) was used to prepare the stock solutions. EC50 values were calculated from nominal concentrations.

Bibliographical reference: Ribo, J.M., and Kaiser, K.L.E. (1983). *Chemosphere* **12**, 1421-1442.

5-PHENOXY-1-*H*-TETRAZOLE

CAS RN: 6489-09-4

Temperature: 15°C
Test parameter: EC50
Effect: Reduction in light output

Concentration: 15.5 mg/l
Exposure time: 5 min

Concentration: 15.8 mg/l
Exposure time: 15 min

Concentration: 17.4 mg/l
Exposure time: 30 min

Comment: Mean of three assays. Methanol (<10%) was used to prepare the stock solutions. EC50 values were calculated from nominal concentrations.

Bibliographical reference: Kaiser, K.L.E., and Palabrica, V.S. (1991). *Water Poll. Res. J. Canada* **26**, 361-431.

N-PHENYLACETAMIDE

CAS RN: 103-84-4

Sample purity: 97%
Temperature: 15°C
Test parameter: EC50
Effect: Reduction in light output

Concentration: 332 mg/l
Exposure time: 5 min

Concentration: 270 mg/l
Exposure time: 15 min

Concentration: 282 mg/l*
Exposure time: 30 min

Comment: Mean of four assays. Methanol (<10%) was used to prepare the stock solutions. EC50 values were calculated from nominal concentrations.

Bibliographical references: Kaiser, K.L.E., and Palabrica, V.S. (1991). *Water Poll. Res. J. Canada* **26**, 361-431.
* Kaiser, K.L.E., Palabrica, V.S., and Ribo, J.M. (1987). In: *QSAR in Environmental Toxicology - II*, K.L.E. Kaiser (ed.), D. Reidel Publishing Company, Dordrecht, p. 153-168.

PHENYL ACETATE

CAS RN: 122-79-2

Sample purity: 97%
Temperature: 15°C
Test parameter: EC50
Effect: Reduction in light output

Concentration: 11.6 mg/l
Exposure time: 5 min

Concentration: 10.6 mg/l
Exposure time: 15 min

Concentration: 11.1 mg/l*

Exposure time: 30 min

Comment: Mean of four assays. Methanol (<10%) was used to prepare the stock solutions. EC50 values were calculated from nominal concentrations.

Bibliographical references: Kaiser, K.L.E., and Palabrica, V.S. (1991). *Water Poll. Res. J. Canada* **26**, 361-431.
* Kaiser, K.L.E., Palabrica, V.S., and Ribo, J.M. (1987). In: *QSAR in Environmental Toxicology - II*, K.L.E. Kaiser (ed.), D. Reidel Publishing Company, Dordrecht, p. 153-168.

PHENYLACETIC ACID

CAS RN: 103-82-2

Sample purity: 99%
Temperature: 15°C
Test parameter: EC50
Effect: Reduction in light output

Concentration: 461 mg/l
Exposure time: 5 min

Concentration: 431 mg/l
Exposure time: 15 min

Concentration: 542 mg/l*
Exposure time: 30 min

Comment: Mean of two assays. Methanol (<10%) was used to prepare the stock solutions. EC50 values were calculated from nominal concentrations.

Bibliographical references: Kaiser, K.L.E., and Palabrica, V.S. (1991). *Water Poll. Res. J. Canada* **26**, 361-431.
* Kaiser, K.L.E., Palabrica, V.S., and Ribo, J.M. (1987). In: *QSAR in Environmental Toxicology - II*, K.L.E. Kaiser (ed.), D. Reidel Publishing Company, Dordrecht, p. 153-168.

PHENYLACETYL CHLORIDE

CAS RN: 103-80-0

Sample purity: 98%
Temperature: 15°C
Test parameter: EC50
Effect: Reduction in light output

Concentration: 4.56 mg/l
Exposure time: 5 min

Concentration: 3.62 mg/l
Exposure time: 15 min

Concentration: 3.38 mg/l*
Exposure time: 30 min

Comment: Mean of three assays. Methanol (<10%) was used to prepare the stock solutions. EC50 values were calculated from nominal concentrations.

Bibliographical references: Kaiser, K.L.E., and Palabrica, V.S. (1991). *Water Poll. Res. J. Canada* **26**, 361-431.
* Kaiser, K.L.E., Palabrica, V.S., and Ribo, J.M. (1987). In: *QSAR in Environmental Toxicology - II*, K.L.E. Kaiser (ed.), D. Reidel Publishing Company, Dordrecht, p. 153-168.

PHENYLACETYLENE

CAS RN: 536-74-3
Synonym: Ethynylbenzene

Sample purity: 98%
Temperature: 15°C
Test parameter: EC50
Effect: Reduction in light output

Concentration: 4.16 mg/l
Exposure time: 5 min

Concentration: 4.56 mg/l
Exposure time: 15 min

Concentration: 5.74 mg/l
Exposure time: 30 min

Comment: Mean of four assays. Methanol (<10%) was used to prepare the stock solutions. EC50 values were calculated from nominal concentrations.

Bibliographical reference: Kaiser, K.L.E., and Palabrica, V.S. (1991). *Water Poll. Res. J. Canada* **26**, 361-431.

DL-PHENYLALANINE

CAS RN: 150-30-1

Sample purity: 99%
Temperature: 15°C
Test parameter: EC50
Effect: Reduction in light output

Concentration: 445 mg/l
Exposure time: 5 min

Concentration: 434 mg/l
Exposure time: 15 min

Concentration: 405 mg/l*
Exposure time: 30 min

Comment: Mean of three assays. Methanol (<10%) was used to prepare the stock solutions. EC50 values were calculated from nominal concentrations.

Bibliographical references: Kaiser, K.L.E., and Palabrica, V.S. (1991). *Water Poll. Res. J. Canada* **26**, 361-431.
* Kaiser, K.L.E., Palabrica, V.S., and Ribo, J.M. (1987). In: *QSAR in Environmental Toxicology - II*, K.L.E. Kaiser (ed.), D. Reidel Publishing Company, Dordrecht, p. 153-168.

PHENYLARSONIC ACID

CAS RN: 98-05-5
Synonym: Benzenearsonic acid

Temperature: 15°C
Test parameter: EC50
Effect: Reduction in light output
Concentration: 583 mg/l
Exposure time: 30 min
Comment: Mean of four assays. Methanol (<10%) was used to prepare the stock solutions. EC50 values were calculated from nominal concentrations.

Bibliographical reference: Kaiser, K.L.E., Palabrica, V.S., and Ribo, J.M. (1987). In: *QSAR in Environmental Toxicology - II*, K.L.E. Kaiser (ed.), D. Reidel Publishing Company, Dordrecht, p. 153-168.

4-PHENYLAZOPHENOL

CAS RN: 1689-82-3
Synonym: 4-Hydroxyazobenzene

Temperature: 15°C
Test parameter: EC50
Effect: Reduction in light output
Concentration: 0.93 mg/l
Exposure time: 30 min
Comment: Mean of three assays. Methanol (<10%) was used to prepare the stock solutions. EC50 values were calculated from nominal concentrations.

Bibliographical reference: Kaiser, K.L.E. (1987). In: *QSAR in Environmental Toxicology - II*, K.L.E. Kaiser (ed.), D. Reidel Publishing Company, Dordrecht, p. 169-188.

PHENYL BENZOATE

CAS RN: 93-99-2

Sample purity: 99%
Temperature: 15°C
Test parameter: EC50
Effect: Reduction in light output
Concentration: 1.44 mg/l
Exposure time: 30 min

Comment: Mean of three assays. Methanol (<10%) was used to prepare the stock solutions. EC50 values were calculated from nominal concentrations.

Bibliographical reference: Kaiser, K.L.E., Palabrica, V.S., and Ribo, J.M. (1987). In: *QSAR in Environmental Toxicology - II*, K.L.E. Kaiser (ed.), D. Reidel Publishing Company, Dordrecht, p. 153-168.

N-PHENYLBENZYLAMINE

CAS RN: 103-32-2

Sample purity: 99+%
Temperature: 15°C
Test parameter: EC50
Effect: Reduction in light output

Concentration: 1.83 mg/l
Exposure time: 5 min

Concentration: 2.20 mg/l
Exposure time: 15 min

Concentration: 2.47 mg/l
Exposure time: 30 min

Comment: Mean of three assays. Methanol (<10%) was used to prepare the stock solutions. EC50 values were calculated from nominal concentrations.

Bibliographical reference: Kaiser, K.L.E., and Palabrica, V.S. (1991). *Water Poll. Res. J. Canada* **26**, 361-431.

PHENYL CHLOROFORMATE

CAS RN: 1885-14-9

Sample purity: 97%
Temperature: 15°C
Test parameter: EC50
Effect: Reduction in light output

Concentration: 5.43 mg/l
Exposure time: 5 min

Concentration: 5.18 mg/l
Exposure time: 15 min

Concentration: 5.68 mg/l*
Exposure time: 30 min

Comment: Mean of four assays. Methanol (<10%) was used to prepare the stock solutions. EC50 values were calculated from nominal concentrations.

Bibliographical references: Kaiser, K.L.E., and Palabrica, V.S. (1991). *Water Poll. Res. J. Canada* **26**, 361-431.
* Kaiser, K.L.E., Palabrica, V.S., and Ribo, J.M. (1987). In: *QSAR in Environmental Toxicology - II*, K.L.E. Kaiser (ed.), D. Reidel Publishing Company, Dordrecht, p. 153-168.

1,4-PHENYLENEDIACETONITRILE

CAS RN: 622-75-3

Sample purity: 99%
Temperature: 15°C
Test parameter: EC50
Effect: Reduction in light output
Concentration: 0.10 mg/l
Exposure times: 5, 15, and 30 min
Comment: Mean of four assays. Methanol (<10%) was used to prepare the stock solutions. EC50 values were calculated from nominal concentrations.

Bibliographical reference: Kaiser, K.L.E., and Palabrica, V.S. (1991). *Water Poll. Res. J. Canada* **26**, 361-431.

1,4-PHENYLENEDIAMINE

CAS RN: 106-50-3

Temperature: 15°C
Test parameter: EC50

Effect: Reduction in light output
Concentration: 37.5 mg/l
Exposure time: 30 min
Comment: Mean of three assays. Methanol (<10%) was used to prepare the stock solutions. EC50 values were calculated from nominal concentrations.

Bibliographical reference: Kaiser, K.L.E. (1987). In: *QSAR in Environmental Toxicology - II*, K.L.E. Kaiser (ed.), D. Reidel Publishing Company, Dordrecht, p. 169-188.

1,4-PHENYLENE DIISOCYANIDE

CAS RN: 935-16-0
Synonym: 1,4-Diisocyanobenzene

Sample purity: 99%
Temperature: 15°C
Test parameter: EC50
Effect: Reduction in light output

Concentration: 0.039 mg/l
Exposure time: 5 min

Concentration: 0.019 mg/l
Exposure time: 15 min

Concentration: 0.015 mg/l
Exposure time: 30 min

Comment: Mean of three assays. Methanol (<10%) was used to prepare the stock solutions. EC50 values were calculated from nominal concentrations.

Bibliographical reference: Kaiser, K.L.E., and Palabrica, V.S. (1991). *Water Poll. Res. J. Canada* **26**, 361-431.

1,4-PHENYLENE DIISOTHIOCYANATE

CAS RN: 4044-65-9

Sample purity: 98%

Temperature: 15°C
Test parameter: EC50
Effect: Reduction in light output
Concentration: 0.018 mg/l
Exposure time: 30 min
Comment: Mean of three assays. Methanol (<10%) was used to prepare the stock solutions. EC50 values were calculated from nominal concentrations.

Bibliographical reference: Ribo, J.M., and Kaiser, K.L.E. National Water Research Institute, Burlington, Ontario, Canada, unpublished results (value later published by Kaiser, K.L.E., and Palabrica, V.S. (1991). *Water Poll. Res. J. Canada* **26**, 361-431).

DL-α-PHENYLGLYCINE

CAS RN: 2835-06-5

Sample purity: 98%
Temperature: 15°C
Test parameter: EC50
Effect: Reduction in light output

Concentration: 6.02 mg/l
Exposure time: 5 min

Concentration: 5.49 mg/l
Exposure time: 15 min

Concentration: 5.49 mg/l*
Exposure time: 30 min

Comment: Mean of three assays. Methanol (<10%) was used to prepare the stock solutions. EC50 values were calculated from nominal concentrations.

Bibliographical references: Kaiser, K.L.E., and Palabrica, V.S. (1991). *Water Poll. Res. J. Canada* **26**, 361-431.
* Kaiser, K.L.E., Palabrica, V.S., and Ribo, J.M. (1987). In: *QSAR in Environmental Toxicology - II*, K.L.E. Kaiser (ed.), D. Reidel Publishing Company, Dordrecht, p. 153-168.

PHENYLHYDRAZINE

CAS RN: 100-63-0

Sample purity: 99%
Temperature: 15°C
Test parameter: EC50
Effect: Reduction in light output

Concentration: 78.3 mg/l
Exposure time: 5 min

Concentration: 73.1 mg/l
Exposure time: 15 min

Concentration: 66.7 mg/l*
Exposure time: 30 min

Comment: Mean of three assays. Methanol (<10%) was used to prepare the stock solutions. EC50 values were calculated from nominal concentrations.

Bibliographical references: Kaiser, K.L.E., and Palabrica, V.S. (1991). *Water Poll. Res. J. Canada* **26**, 361-431.
* Kaiser, K.L.E., Palabrica, V.S., and Ribo, J.M. (1987). In: *QSAR in Environmental Toxicology - II*, K.L.E. Kaiser (ed.), D. Reidel Publishing Company, Dordrecht, p. 153-168.

PHENYL ISOCYANATE

CAS RN: 103-71-9

Sample purity: 98%
Temperature: 15°C
Test parameter: EC50
Effect: Reduction in light output

Concentration: 19.8 mg/l
Exposure time: 5 min

Concentration: 17.2 mg/l
Exposure time: 15 min

Concentration: 20.2 mg/l*

Exposure time: 30 min

Comment: Mean of three assays. Methanol (<10%) was used to prepare the stock solutions. EC50 values were calculated from nominal concentrations.

Bibliographical references: Kaiser, K.L.E., and Palabrica, V.S. (1991). *Water Poll. Res. J. Canada* **26**, 361-431.
* Kaiser, K.L.E., Palabrica, V.S., and Ribo, J.M. (1987). In: *QSAR in Environmental Toxicology - II*, K.L.E. Kaiser (ed.), D. Reidel Publishing Company, Dordrecht, p. 153-168.

4-PHENYLMORPHOLINE

CAS RN: 92-53-5

Sample purity: 98%
Temperature: 15°C
Test parameter: EC50
Effect: Reduction in light output

Concentration: 36.5 mg/l
Exposure time: 5 min

Concentration: 39.2 mg/l
Exposure time: 15 min

Concentration: 46.0 mg/l
Exposure time: 30 min

Comment: Mean of three assays. Methanol (<10%) was used to prepare the stock solutions. EC50 values were calculated from nominal concentrations.

Bibliographical reference: Kaiser, K.L.E., and Palabrica, V.S. (1991). *Water Poll. Res. J. Canada* **26**, 361-431.

2-PHENYLPHENOL

CAS RN: 90-43-7

Test parameter: EC50

Effect: Reduction in light output
Concentration: 2.05 mg/l
Exposure time: 5 min
Comment: Concentrations in the test were measured.

Bibliographical reference: Curtis, C., Lima, A., Lozano, S.J., and Veith, G.D. (1982). In: *Aquatic Toxicology and Hazard Assessment: Fifth Conference, ASTM STP 766*, J.G. Pearson, R.B. Foster, and W.E. Bishop (eds.), American Society for Testing and Materials, Philadelphia, p. 170-178.

1-PHENYL-1-PROPYNE

CAS RN: 673-32-5

Sample purity: 99%
Temperature: 15°C
Test parameter: EC50
Effect: Reduction in light output

Concentration: 1.27 mg/l
Exposure time: 5 min

Concentration: 1.50 mg/l
Exposure time: 15 min

Concentration: 1.84 mg/l
Exposure time: 30 min

Comment: Mean of three assays. Methanol (<10%) was used to prepare the stock solutions. EC50 values were calculated from nominal concentrations.

Bibliographical reference: Kaiser, K.L.E., and Palabrica, V.S. (1991). *Water Poll. Res. J. Canada* **26**, 361-431.

3-PHENYLPYRIDINE

CAS RN: 1008-88-4

Sample purity: 97%
Temperature: 15°C

Test parameter: EC50
Effect: Reduction in light output

Concentration: 0.94 mg/l
Exposure time: 5 min

Concentration: 1.03 mg/l
Exposure time: 15 min

Concentration: 1.10 mg/l
Exposure time: 30 min

Comment: Mean of three assays. Methanol (<10%) was used to prepare the stock solutions. EC50 values were calculated from nominal concentrations.

Bibliographical reference: Kaiser, K.L.E., and Palabrica, V.S. (1991). *Water Poll. Res. J. Canada* **26**, 361-431.

4-PHENYLPYRIDINE

CAS RN: 939-23-1

Sample purity: 99%
Temperature: 15°C
Test parameter: EC50
Effect: Reduction in light output

Concentration: 1.91 mg/l
Exposure time: 5 min

Concentration: 2.05 mg/l
Exposure time: 15 min

Concentration: 2.24 mg/l*
Exposure time: 30 min

Comment: Mean of three assays. Methanol (<10%) was used to prepare the stock solutions. EC50 values were calculated from nominal concentrations.

Bibliographical references: Kaiser, K.L.E., and Palabrica, V.S. (1991). *Water Poll. Res. J. Canada* **26**, 361-431.

* Kaiser, K.L.E., Palabrica, V.S., and Ribo, J.M. (1987). In: *QSAR in Environmental Toxicology - II*, K.L.E. Kaiser (ed.), D. Reidel Publishing Company, Dordrecht, p. 153-168.

PHENYLSELENYL CHLORIDE

CAS RN: 5707-04-0

Sample purity: 97%
Temperature: 15°C
Test parameter: EC50
Effect: Reduction in light output

Concentration: 1.67 mg/l
Exposure time: 5 min

Concentration: 1.13 mg/l
Exposure time: 15 min

Concentration: 1.45 mg/l*
Exposure time: 30 min

Comment: Mean of four assays. Methanol (<10%) was used to prepare the stock solutions. EC50 values were calculated from nominal concentrations.

Bibliographical references: Kaiser, K.L.E., and Palabrica, V.S. (1991). *Water Poll. Res. J. Canada* **26**, 361-431.
* Kaiser, K.L.E., Palabrica, V.S., and Ribo, J.M. (1987). In: *QSAR in Environmental Toxicology - II*, K.L.E. Kaiser (ed.), D. Reidel Publishing Company, Dordrecht, p. 153-168.

1-PHENYLSEMICARBAZIDE

CAS RN: 103-03-7

Sample purity: 99%
Temperature: 15°C
Test parameter: EC50
Effect: Reduction in light output

Concentration: 8.90 mg/l

Exposure time: 5 min

Concentration: 9.32 mg/l
Exposure time: 15 min

Concentration: 10.7 mg/l*
Exposure time: 30 min

Comment: Mean of three assays. Methanol (<10%) was used to prepare the stock solutions. EC50 values were calculated from nominal concentrations.

Bibliographical references: Kaiser, K.L.E., and Palabrica, V.S. (1991). *Water Poll. Res. J. Canada* **26**, 361-431.
* Kaiser, K.L.E., Palabrica, V.S., and Ribo, J.M. (1987). In: *QSAR in Environmental Toxicology - II*, K.L.E. Kaiser (ed.), D. Reidel Publishing Company, Dordrecht, p. 153-168.

4-PHENYLSEMICARBAZIDE

CAS RN: 537-47-3

Sample purity: 97%
Temperature: 15°C
Test parameter: EC50
Effect: Reduction in light output

Concentration: 85.0 mg/l
Exposure time: 5 min

Concentration: 69.1 mg/l
Exposure time: 15 min

Concentration: 57.5 mg/l*
Exposure time: 30 min

Comment: Mean of four assays. Methanol (<10%) was used to prepare the stock solutions. EC50 values were calculated from nominal concentrations.

Bibliographical references: Kaiser, K.L.E., and Palabrica, V.S. (1991). *Water Poll. Res. J. Canada* **26**, 361-431.

* Kaiser, K.L.E., Palabrica, V.S., and Ribo, J.M. (1987). In: *QSAR in Environmental Toxicology - II*, K.L.E. Kaiser (ed.), D. Reidel Publishing Company, Dordrecht, p. 153-168.

PHENYL SULFONE

CAS RN: 127-63-9

Sample purity: 97%
Temperature: 15°C
Test parameter: EC50
Effect: Reduction in light output

Concentration: 17.3 mg/l
Exposure time: 5 min

Concentration: 15.8 mg/l
Exposure time: 15 min

Concentration: 15.8 mg/l*
Exposure time: 30 min

Comment: Mean of three assays. Methanol (<10%) was used to prepare the stock solutions. EC50 values were calculated from nominal concentrations.

Bibliographical references: Kaiser, K.L.E., and Palabrica, V.S. (1991). *Water Poll. Res. J. Canada* **26**, 361-431.
* Kaiser, K.L.E., Palabrica, V.S., and Ribo, J.M. (1987). In: *QSAR in Environmental Toxicology - II*, K.L.E. Kaiser (ed.), D. Reidel Publishing Company, Dordrecht, p. 153-168.

(PHENYLSULFONYL)ACETONITRILE

CAS RN: 7605-28-9

Sample purity: 98%
Temperature: 15°C
Test parameter: EC50
Effect: Reduction in light output

Concentration: 33.7 mg/l

Exposure time: 5 min

Concentration: 35.3 mg/l
Exposure time: 15 min

Concentration: 42.5 mg/l
Exposure time: 30 min

Comment: Mean of four assays. Methanol (<10%) was used to prepare the stock solutions. EC50 values were calculated from nominal concentrations.

Bibliographical reference: Kaiser, K.L.E., and Palabrica, V.S. (1991). *Water Poll. Res. J. Canada* **26**, 361-431.

4-PHENYL-3-THIOSEMICARBAZIDE

CAS RN: 5351-69-9

Sample purity: 98%
Temperature: 15°C
Test parameter: EC50
Effect: Reduction in light output

Concentration: 8.00 mg/l
Exposure time: 5 min

Concentration: 4.50 mg/l
Exposure time: 15 min

Concentration: 3.26 mg/l*
Exposure time: 30 min

Comment: Mean of three assays. Methanol (<10%) was used to prepare the stock solutions. EC50 values were calculated from nominal concentrations.

Bibliographical references: Kaiser, K.L.E., and Palabrica, V.S. (1991). *Water Poll. Res. J. Canada* **26**, 361-431.
* Kaiser, K.L.E., Palabrica, V.S., and Ribo, J.M. (1987). In: *QSAR in Environmental Toxicology - II*, K.L.E. Kaiser (ed.), D. Reidel Publishing Company, Dordrecht, p. 153-168.

1-PHENYL-2-THIOUREA

CAS RN: 103-85-5

Sample purity: ≥97%
Test parameter: EC50
Effect: Reduction in light output
Concentration: 13.5 mg/l
Exposure time: 15 min
Comment: EC50 value converted to mg/l from the original data expressed in log(1/μmol l^{-1}) and a rounded molecular weight of 152 given by the authors.

Bibliographical reference: Govers, H., Ruepert, C., Stevens, T., and van Leeuwen, C.J. (1986). *Chemosphere* **15**, 383-393.

Sample purity: 97%
Temperature: 15°C
Test parameter: EC50
Effect: Reduction in light output

Concentration: 5.92 mg/l
Exposure time: 5 min

Concentration: 4.19 mg/l
Exposure time: 15 min

Concentration: 3.49 mg/l*
Exposure time: 30 min

Comment: Mean of three assays. Methanol (<10%) was used to prepare the stock solutions. EC50 values were calculated from nominal concentrations.

Bibliographical references: Kaiser, K.L.E., and Palabrica, V.S. (1991). *Water Poll. Res. J. Canada* **26**, 361-431.
* Kaiser, K.L.E., Palabrica, V.S., and Ribo, J.M. (1987). In: *QSAR in Environmental Toxicology - II*, K.L.E. Kaiser (ed.), D. Reidel Publishing Company, Dordrecht, p. 153-168.

3-PHENYLTOLUENE

CAS RN: 643-93-6

Synonym: 3-Methylbiphenyl

Sample purity: 95%
Temperature: 15°C
Test parameter: EC50
Effect: Reduction in light output

Concentration: 0.90 mg/l
Exposure time: 5 min

Concentration: 1.37 mg/l
Exposure time: 15 min

Concentration: 1.61 mg/l
Exposure time: 30 min

Comment: Mean of three assays. Methanol (<10%) was used to prepare the stock solutions. EC50 values were calculated from nominal concentrations.

Bibliographical reference: Kaiser, K.L.E., and Palabrica, V.S. (1991). *Water Poll. Res. J. Canada* **26**, 361-431.

4-PHENYLTOLUENE

CAS RN: 644-08-6
Synonym: 4-Methylbiphenyl

Sample purity: 98%
Temperature: 15°C
Test parameter: EC50
Effect: Reduction in light output

Concentration: 1.01 mg/l
Exposure time: 5 min

Concentration: 1.57 mg/l
Exposure time: 15 min

Concentration: 2.22 mg/l*
Exposure time: 30 min

Comment: Mean of three assays. Methanol (<10%) was used to prepare the stock solutions. EC50 values were calculated from nominal concentrations.

Bibliographical references: Kaiser, K.L.E., and Palabrica, V.S. (1991). *Water Poll. Res. J. Canada* **26**, 361-431.
* Kaiser, K.L.E., Palabrica, V.S., and Ribo, J.M. (1987). In: *QSAR in Environmental Toxicology - II*, K.L.E. Kaiser (ed.), D. Reidel Publishing Company, Dordrecht, p. 153-168.

Sample purity: 98%
Temperature: 15°C
Test parameter: EC50
Effect: Reduction in light output

Concentration: 1.22 mg/l
Exposure time: 5 min

Concentration: 1.80 mg/l
Exposure time: 15 min

Concentration: 2.38 mg/l
Exposure time: 30 min

Comment: Mean of three assays. Methanol (<10%) was used to prepare the stock solutions. EC50 values were calculated from nominal concentrations.

Bibliographical reference: Kaiser, K.L.E., and Palabrica, V.S. (1991). *Water Poll. Res. J. Canada* **26**, 361-431.

N-PHENYLTRIFLUOROMETHANESULFONIMIDE

CAS RN: 37595-74-7
Synonym: 1,1,1-Trifluoro-*N*-phenyl-*N*-[(trifluoromethyl)sulfonyl]-methanesulfonamide

Sample purity: 99%
Temperature: 15°C
Test parameter: EC50
Effect: Reduction in light output

Concentration: 47.1 mg/l

Exposure time: 5 min

Concentration: 45.0 mg/l
Exposure time: 15 min

Concentration: 42.0 mg/l
Exposure time: 30 min

Comment: Mean of four assays. Methanol (<10%) was used to prepare the stock solutions. EC50 values were calculated from nominal concentrations.

Bibliographical reference: Kaiser, K.L.E., and Palabrica, V.S. (1991). *Water Poll. Res. J. Canada* **26**, 361-431.

PHTHALAZINE

CAS RN: 253-52-1

Sample purity: 98%
Temperature: 15°C
Test parameter: EC50
Effect: Reduction in light output

Concentration: 4.41 mg/l
Exposure time: 5 min

Concentration: 5.43 mg/l
Exposure time: 15 min

Concentration: 6.67 mg/l
Exposure time: 30 min

Comment: Mean of three assays. Methanol (<10%) was used to prepare the stock solutions. EC50 values were calculated from nominal concentrations.

Bibliographical reference: Kaiser, K.L.E., and Palabrica, V.S. (1991). *Water Poll. Res. J. Canada* **26**, 361-431.

PHTHALONITRILE

CAS RN: 91-15-6
Synonym: 1,2-Dicyanobenzene

Sample purity: 98%
Temperature: 15°C
Test parameter: EC50
Effect: Reduction in light output

Concentration: 90.7 mg/l
Exposure time: 5 min

Concentration: 95.0 mg/l
Exposure time: 15 min

Concentration: 97.2 mg/l
Exposure time: 30 min

Comment: Mean of four assays. Methanol (<10%) was used to prepare the stock solutions. EC50 values were calculated from nominal concentrations.

Bibliographical reference: Kaiser, K.L.E., and Palabrica, V.S. (1991). *Water Poll. Res. J. Canada* **26**, 361-431.

2-PICOLINE

CAS RN: 109-06-8

Sample purity: 98+%
Temperature: 15°C
Test parameter: EC50
Effect: Reduction in light output

Concentration: 132 mg/l
Exposure time: 5 min

Concentration: 126 mg/l
Exposure time: 15 min

Concentration: 109 mg/l
Exposure time: 30 min

Comment: Mean of three assays. Methanol (<10%) was used to prepare the stock solutions. EC50 values were calculated from nominal concentrations.

Bibliographical reference: Kaiser, K.L.E., and Palabrica, V.S. (1991). *Water Poll. Res. J. Canada* **26**, 361-431.

3-PICOLINE

CAS RN: 108-99-6

Sample purity: 97%
Temperature: 15°C
Test parameter: EC50
Effect: Reduction in light output

Concentration: 70.6 mg/l
Exposure time: 5 min

Concentration: 74.0 mg/l
Exposure times: 15 and 30 min

Comment: Mean of three assays. Methanol (<10%) was used to prepare the stock solutions. EC50 values were calculated from nominal concentrations.

Bibliographical reference: Kaiser, K.L.E., and Palabrica, V.S. (1991). *Water Poll. Res. J. Canada* **26**, 361-431.

4-PICOLINE

CAS RN: 108-89-4

Sample purity: 98+%
Temperature: 15°C
Test parameter: EC50
Effect: Reduction in light output

Concentration: 28.8 mg/l
Exposure time: 5 min

Concentration: 28.1 mg/l

Exposure time: 15 min

Concentration: 26.9 mg/l
Exposure time: 30 min

Comment: Mean of three assays. Methanol (<10%) was used to prepare the stock solutions. EC50 values were calculated from nominal concentrations.

Bibliographical reference: Kaiser, K.L.E., and Palabrica, V.S. (1991). *Water Poll. Res. J. Canada* **26**, 361-431.

PICOLINIC ACID

CAS RN: 98-98-6

Sample purity: 99%
Temperature: 15°C
Test parameter: EC50
Effect: Reduction in light output

Concentration: 40.8 mg/l
Exposure time: 5 min

Concentration: 45.7 mg/l
Exposure time: 15 min

Concentration: 41.7 mg/l
Exposure time: 30 min

Comment: Mean of three assays. Methanol (<10%) was used to prepare the stock solutions. EC50 values were calculated from nominal concentrations.

Bibliographical reference: Kaiser, K.L.E., and Palabrica, V.S. (1991). *Water Poll. Res. J. Canada* **26**, 361-431.

PICRIC ACID

CAS RN: 88-89-1
Synonym: 2,4,6-Trinitrophenol

Temperature: 15 ± 0.2°C
Test parameter: EC50
Effect: Reduction in light output
Concentration: 535 mg/l
Exposure times: 5 and 30 min
Comment: The test was run in duplicate with a minimum of five serial concentration dilutions. Samples were pH adjusted to 7 ± 0.5 pH units.

Bibliographical reference: Indorato, A.M., Snyder, K.B., and Usinowicz, P.J. (1984). In: *Toxicity Screening Procedures Using Bacterial Systems*, D. Liu and B.J. Dutka (eds.), Marcel Dekker, New York, p. 37-53.

Temperature: 15°C
Test parameter: EC50
Effect: Reduction in light output
Concentration: 121 mg/l
Exposure time: 5 min

Bibliographical reference: Kahru, A. (1993). *ATLA* **21**, 210-215.

POLYETHYLENE GLYCOL

CAS RN: 25322-68-3

Temperature: 15°C
Test parameter: EC50
Effect: Reduction in light output
Concentration: >100000 mg/l
Exposure time: 15 min

Bibliographical reference: Bulich, A.A., Tung, K.K., and Scheibner, G. (1990). *J. Biolumin. Chemilumin.* **5**, 71-77.

POLYMERIC ETHYLENETHIURAM DISULFIDE

Sample purity: 3% Ethylenethiourea and 0.1% 5,6-dihydro-3*H*-imidazo(2,1-c)-1,2,4-dithiazole-3-thione
Test parameter: EC50
Effect: Reduction in light output
Concentration: 0.06 mg/l (0.05-0.06)

Exposure time: 15 min

Bibliographical reference: van Leeuwen, C.J., Maas-Diepeveen, J.L., Niebeek, G., Vergouw, W.H.A., Griffioen, P.S., and Luijken, M.W. (1985). *Aquat. Toxicol.* **7**, 145-164.

POTASSIUM CYANIDE

CAS RN: 151-50-8

Sample purity: Reagent grade
Test parameter: EC50
Effect: Reduction in light output

Concentration: 4.77 mg/l CN^-
Exposure time: 5 min

Concentration: 1.40 mg/l CN^-
Exposure time: 15 min

Concentration: 0.61 mg/l CN^-
Exposure time: 30 min

Comment: Two replicates of each concentration were tested.

Bibliographical reference: Greene, J.C., Miller, W.E., Debacon, M.K., Long, M.A., and Bartels, C.L. (1985). *Arch. Environ. Contam. Toxicol.* **14**, 659-667.

Sample purity: Reagent grade
Test parameter: EC50
Effect: Reduction in light output

Concentration: 10 mg/l CN^- (8-12)
Exposure time: 5 min

Concentration: 4.3 mg/l CN^- (3.6-5.2)
Exposure time: 15 min

Concentration: 3.2 mg/l CN^- (2.9-3.6)
Exposure time: 30 min

Bibliographical reference: Elnabarawy, M.T., Robideau, R.R., and Beach, S.A. (1988). *Tox. Assess.* **3**, 361-370.

POTASSIUM DICHROMATE

CAS RN: 7778-50-9
Synonym: Chromic acid dipotassium salt

Temperature: 15°C
Test parameter: EC50
Effect: Reduction in light output
Concentration: 28.5 ± 7.4 mg/l Cr^{6+}
Exposure time: 10 min
Comment: Mean of six assays. Toxicity values were calculated from nominal concentrations.

Bibliographical reference: Ferard, J.F., Vasseur, P., Danoux, L., and Larbaigt, G. (1983). *Rev. Fr. Sci. Eau* **2**, 221-237.

Temperature: 20°C
Test parameter: EC50
Effect: Reduction in light output
Concentration: 33 ± 13 mg/l Cr^{6+}
Exposure time: 10 min
Comment: Mean of four assays. Toxicity values were calculated from nominal concentrations.

Bibliographical reference: Ferard, J.F., Vasseur, P., Danoux, L., and Larbaigt, G. (1983). *Rev. Fr. Sci. Eau* **2**, 221-237.

Sample purity: Analytical grade
Temperature: 15°C
Test parameter: EC50
Effect: Reduction in light output

Concentration: 38.6 mg/l Cr^{6+}
Exposure time: 5 min

Concentration: 39.5 mg/l Cr^{6+}
Exposure time: 10 min

Concentration: 26.9 mg/l Cr^{6+}

Exposure time: 15 min

Concentration: 24.0 mg/l Cr^{6+}
Exposure time: 20 min

Concentration: 16.8 mg/l Cr^{6+}
Exposure time: 30 min

Comment: Results derived from the average of two replicates.

Bibliographical reference: Qureshi, A.A., Coleman, R.N., and Paran, J.H. (1984). In: *Toxicity Screening Procedures Using Bacterial Systems*, D. Liu and B.J. Dutka (eds.), Marcel Dekker, New York, p. 1-22.

Temperature: 15°C

Test parameter: EC20
Effect: Reduction in light output
Concentration: 3.4 mg/l Cr^{6+}
Exposure time: 30 min

Test parameter: EC50
Effect: Reduction in light output
Concentration: 13 mg/l Cr^{6+}
Exposure time: 30 min

Comment: Toxicity values were calculated from nominal concentrations.

Bibliographical reference: Reteuna, C. (1988). Thesis, University of Metz, Metz, France.

Sample purity: Reagent grade
Test parameter: EC50
Effect: Reduction in light output

Concentration: 44 mg/l Cr^{6+} (40-48)
Exposure time: 5 min

Concentration: 13 mg/l Cr^{6+} (12-14)
Exposure time: 15 min

Concentration: 6.7 mg/l Cr^{6+} (6.5-6.9)

Exposure time: 30 min

Bibliographical reference: Elnabarawy, M.T., Robideau, R.R., and Beach, S.A. (1988). *Tox. Assess.* **3**, 361-370.

1,2-PROPANEDIOL

CAS RN: 57-55-6
Synonym: Propylene glycol

Sample purity: 99%
Temperature: 15 ± 0.1°C
Test parameter: EC50
Effect: Reduction in light output

Concentration: 34800 mg/l
Exposure time: 5 min

Concentration: 28700 mg/l
Exposure time: 15 min

Concentration: 26800 mg/l
Exposure time: 30 min

Comment: The pH was not adjusted.

Bibliographical reference: Tarkpea, M., Hansson, M., and Samuelsson, B. (1986). *Ecotoxicol. Environ. Safety* **11**, 127-143.

Temperature: 15°C
Test parameter: EC50
Effect: Reduction in light output
Concentration: 45000 mg/l
Exposure time: 15 min

Bibliographical reference: Bulich, A.A., Tung, K.K., and Scheibner, G. (1990). *J. Biolumin. Chemilumin.* **5**, 71-77.

Sample purity: >99%
Temperature: 15°C
Test parameter: EC50
Effect: Reduction in light output

Concentration: 648 mg/l
Exposure times: 5 and 15 min

Concentration: 710 mg/l
Exposure time: 30 min

Comment: Mean of three assays. Methanol (<10%) was used to prepare the stock solutions. EC50 values were calculated from nominal concentrations.

Bibliographical reference: Kaiser, K.L.E., and Palabrica, V.S. (1991). *Water Poll. Res. J. Canada* **26**, 361-431.

PROPANIL

CAS RN: 709-98-8

Temperature: 15°C
Test parameter: EC50
Effect: Reduction in light output

Concentration: 28.8 mg/l
Exposure time: 5 min

Concentration: 23.9 mg/l
Exposure time: 15 min

Concentration: 20.8 mg/l
Exposure time: 30 min

Comment: Mean of four assays. Methanol (<10%) was used to prepare the stock solutions. EC50 values were calculated from nominal concentrations.

Bibliographical reference: Kaiser, K.L.E., and Palabrica, V.S. (1993). National Water Research Institute, Burlington, Ontario, Canada, unpublished results.

1-PROPANOL

CAS RN: 71-23-8

Sample purity: >98 %
Temperature: 15 ± 0.1°C
Test parameter: EC50
Effect: Reduction in light output

Concentration: 17700 mg/l
Exposure time: 5 min

Concentration: 18400 mg/l
Exposure time: 15 min

Comment: Test was performed in duplicate. Toxicity values were based on nominal test concentrations.

Bibliographical reference: de Zwart, D., and Slooff, W. (1983). *Aquat. Toxicol.* **4**, 129-138.

Sample purity: >98 %
Temperature: 15 ± 0.1°C
Test parameter: EC10
Effect: Reduction in light output

Concentration: 9500 mg/l
Exposure time: 5 min

Concentration: 9800 mg/l
Exposure time: 15 min

Comment: Test was performed in duplicate. Toxicity values were based on nominal test concentrations.

Bibliographical reference: de Zwart, D., and Slooff, W. (1983). *Aquat. Toxicol.* **4**, 129-138.

Temperature: 15°C
Test parameter: EC50
Effect: Reduction in light output
Concentration: 8687 mg/l
Exposure time: 15 min

Bibliographical reference: Hermens, J., Busser, F., Leeuwangh, P., and Musch, A. (1985). *Ecotoxicol. Environ. Safety* **9**, 17-25.

Temperature: 15°C
Test parameter: EC50
Effect: Reduction in light output

Concentration: 9862 mg/l
Exposure time: 5 min

Concentration: 9318 mg/l
Exposure time: 30 min

Bibliographical reference: Speece, R. (1987). Drexel University, Philadelphia, USA, private communication.

2-PROPANOL

CAS RN: 67-63-0
Synonym: Isopropanol

Temperature: 15°C
Test parameter: EC50
Effect: Reduction in light output
Concentration: 42000 mg/l
Exposure time: 5 min

Bibliographical reference: Bulich, A.A., Greene, M.W., and Isenberg, D.L. (1981). In: *Aquatic Toxicology and Hazard Assessment: Fourth Conference, ASTM STP 737*, D.R. Branson and K.L. Dickson (eds.), American Society for Testing and Materials, Philadelphia, p. 338-347.

Test parameter: EC50
Effect: Reduction in light output
Concentration: 35000 mg/l
Exposure time: 5 min
Comment: The EC50 value was calculated from nominal concentrations.

Bibliographical reference: Curtis, C., Lima, A., Lozano, S.J., and Veith, G.D. (1982). In: *Aquatic Toxicology and Hazard Assessment: Fifth Conference, ASTM STP 766*, J.G. Pearson, R.B. Foster, and W.E. Bishop (eds.), American Society for Testing and Materials, Philadelphia, p. 170-178.

Temperature: 15°C
Test parameter: EC50
Effect: Reduction in light output
Concentration: 31500 mg/l
Exposure time: 5 min

Bibliographical reference: Kahru, A. (1993). *ATLA* **21**, 210-215.

PROPIONITRILE

CAS RN: 107-12-0

Temperature: 15 ± 0.2°C
Test parameter: EC50
Effect: Reduction in light output
Concentration: 5200 mg/l
Exposure times: 5 and 30 min
Comment: The test was run in duplicate with a minimum of five serial concentration dilutions. Samples were pH adjusted to 7 ± 0.5 pH units.

Bibliographical reference: Indorato, A.M., Snyder, K.B., and Usinowicz, P.J. (1984). In: *Toxicity Screening Procedures Using Bacterial Systems*, D. Liu and B.J. Dutka (eds.), Marcel Dekker, New York, p. 37-53.

PROPIOPHENONE

CAS RN: 93-55-0

Sample purity: 99%
Temperature: 15°C
Test parameter: EC50
Effect: Reduction in light output

Concentration: 5.22 mg/l
Exposure time: 5 min

Concentration: 5.47 mg/l
Exposure time: 15 min

Concentration: 5.72 mg/l*
Exposure time: 30 min

Comment: Mean of three assays. Methanol (<10%) was used to prepare the stock solutions. EC50 values were calculated from nominal concentrations.

Bibliographical references: Kaiser, K.L.E., and Palabrica, V.S. (1991). *Water Poll. Res. J. Canada* **26**, 361-431.
* Kaiser, K.L.E., Palabrica, V.S., and Ribo, J.M. (1987). In: *QSAR in Environmental Toxicology - II*, K.L.E. Kaiser (ed.), D. Reidel Publishing Company, Dordrecht, p. 153-168.

PROPYL ACETATE

CAS RN: 109-60-4

Temperature: 15°C
Test parameter: EC50
Effect: Reduction in light output

Concentration: 316 mg/l
Exposure time: 5 min

Concentration: 388 mg/l*
Exposure time: 15 min

Bibliographical references: Cronin, M.T.D., Dearden, J.C., and Dobbs, A.J. (1991). *Sci. Total Environ.* **109/110**, 431-439.
* Cronin, M.T.D. (1990). Thesis, Liverpool Polytechnic, Liverpool, UK.

PROPYLAMINE

CAS RN: 107-10-8

Sample purity: 98%
Temperature: 15°C
Test parameter: EC50
Effect: Reduction in light output

Concentration: 12.6 mg/l
Exposure time: 5 min

Concentration: 9.16 mg/l*

Exposure time: 15 min

Bibliographical references: Cronin, M.T.D., Dearden, J.C., and Dobbs, A.J. (1991). *Sci. Total Environ.* **109/110**, 431-439.
* Cronin, M.T.D. (1990). Thesis, Liverpool Polytechnic, Liverpool, UK.

PROPYL(±)-1-(1-PHENYLETHYL)IMIDAZOLE-5-CARBOXYLATE HYDROCHLORIDE

CAS RN: 7036-61-5

Sample purity: 99%
Temperature: 15°C
Test parameter: EC50
Effect: Reduction in light output

Concentration: 9.32 mg/l
Exposure time: 5 min

Concentration: 6.91 mg/l
Exposure time: 15 min

Concentration: 5.36 mg/l
Exposure time: 30 min

Comment: Mean of four assays. Methanol (<10%) was used to prepare the stock solutions. EC50 values were calculated from nominal concentrations.

Bibliographical reference: Kaiser, K.L.E., and Palabrica, V.S. (1991). *Water Poll. Res. J. Canada* **26**, 361-431.

PR TOXIN

CAS RN: 56299-00-4

Temperature: 15°C
Test parameter: EC50
Effect: Reduction in light output

Concentration: 7.79 mg/l

Exposure time: 5 min

Concentration: 3.26 mg/l
Exposure time: 10 min

Concentration: 2.10 mg/l
Exposure time: 15 min

Concentration: 1.72 mg/l
Exposure time: 20 min

Comment: Freshly reconstituted bacterial suspensions. Chemical was dissolved in methanol.

Bibliographical reference: Yates, I.E., and Porter, J.K. (1982). *Appl. Environ. Microbiol.* **44**, 1072-1075.

Temperature: 15°C
Test parameter: EC20
Effect: Reduction in light output

Concentration: 3.55 mg/l
Exposure time: 5 min

Concentration: 0.92 mg/l
Exposure time: 20 min

Comment: Freshly reconstituted bacterial suspensions. Chemical was dissolved in methanol.

Bibliographical reference: Yates, I.E., and Porter, J.K. (1982). *Appl. Environ. Microbiol.* **44**, 1072-1075.

Temperature: 15°C
Test parameter: EC50
Effect: Reduction in light output

Concentration: 9.49 mg/l
Exposure time: 5 min

Concentration: 3.43 mg/l
Exposure time: 10 min

Concentration: 2.26 mg/l

Exposure time: 15 min

Concentration: 1.80 mg/l
Exposure time: 20 min

Comment: Performed on bacterial suspensions maintained at 3°C for 5 h after reconstitution. Chemical was dissolved in methanol.

Bibliographical reference: Yates, I.E., and Porter, J.K. (1982). *Appl. Environ. Microbiol.* **44**, 1072-1075.

PYRAZINE

CAS RN: 290-37-9

Sample purity: >99%
Temperature: 15°C
Test parameter: EC50
Effect: Reduction in light output

Concentration: 730 mg/l
Exposure time: 5 min

Concentration: 698 mg/l
Exposure time: 15 min

Concentration: 714 mg/l
Exposure time: 30 min

Comment: Mean of four assays. Methanol (<10%) was used to prepare the stock solutions. EC50 values were calculated from nominal concentrations.

Bibliographical reference: Kaiser, K.L.E., and Palabrica, V.S. (1991). *Water Poll. Res. J. Canada* **26**, 361-431.

PYRIDAZINE

CAS RN: 289-80-5

Sample purity: 99%
Temperature: 15°C

Test parameter: EC50
Effect: Reduction in light output

Concentration: 1490 mg/l
Exposure time: 5 min

Concentration: 1460 mg/l
Exposure times: 15 and 30 min

Comment: Mean of three assays. Methanol (<10%) was used to prepare the stock solutions. EC50 values were calculated from nominal concentrations.

Bibliographical reference: Kaiser, K.L.E., and Palabrica, V.S. (1991). *Water Poll. Res. J. Canada* **26**, 361-431.

PYRIDINE

CAS RN: 110-86-1

Sample purity: >98%
Temperature: 15 ± 0.1°C
Test parameter: EC50
Effect: Reduction in light output

Concentration: 2590 mg/l
Exposure time: 5 min

Concentration: 2120 mg/l
Exposure time: 15 min

Comment: Test was performed in duplicate. Toxicity values were based on nominal test concentrations.

Bibliographical reference: de Zwart, D., and Slooff, W. (1983). *Aquat. Toxicol.* **4**, 129-138.

Sample purity: >98%
Temperature: 15 ± 0.1°C
Test parameter: EC10
Effect: Reduction in light output

Concentration: 590 mg/l

Exposure time: 5 min

Concentration: 1000 mg/l
Exposure time: 15 min

Comment: Test was performed in duplicate. Toxicity values were based on nominal test concentrations.

Bibliographical reference: de Zwart, D., and Slooff, W. (1983). *Aquat. Toxicol.* **4**, 129-138.

Temperature: 15°C
Test parameter: EC50
Effect: Reduction in light output
Concentration: 738 mg/l
Exposure time: 30 min
Comment: Mean of three assays. Methanol (<10%) was used to prepare the stock solutions. EC50 values were calculated from nominal concentrations.

Bibliographical reference: Kaiser, K.L.E., and Ribo, J.M. (1985). In: *QSAR in Toxicology and Xenobiochemistry*, M. Tichy (ed.), Elsevier, Amsterdam, p. 27-38.

Sample purity: 99+%
Temperature: 15°C
Test parameter: EC50
Effect: Reduction in light output

Concentration: 301 mg/l
Exposure time: 5 min

Concentration: 233 mg/l
Exposure time: 15 min

Concentration: 213 mg/l
Exposure time: 30 min

Comment: Mean of three assays. Methanol (<10%) was used to prepare the stock solutions. EC50 values were calculated from nominal concentrations.

Bibliographical reference: Kaiser, K.L.E., and Palabrica, V.S. (1991). *Water Poll. Res. J. Canada* **26**, 361-431.

4-PYRIDINEALDOXIME

CAS RN: 696-54-8

Sample purity: 98%
Temperature: 15°C
Test parameter: EC50
Effect: Reduction in light output

Concentration: 21.7 mg/l
Exposure time: 5 min

Concentration: 22.2 mg/l
Exposure time: 15 min

Concentration: 24.9 mg/l
Exposure time: 30 min

Comment: Mean of three assays. Methanol (<10%) was used to prepare the stock solutions. EC50 values were calculated from nominal concentrations.

Bibliographical reference: Kaiser, K.L.E., and Palabrica, V.S. (1991). *Water Poll. Res. J. Canada* **26**, 361-431.

2-PYRIDINECARBOXALDEHYDE

CAS RN: 1121-60-4
Synonym: Picolinaldehyde

Sample purity: 99%
Temperature: 15°C
Test parameter: EC50
Effect: Reduction in light output

Concentration: 166 mg/l
Exposure time: 5 min

Concentration: 126 mg/l
Exposure time: 15 min

Concentration: 112 mg/l
Exposure time: 30 min

Comment: Mean of three assays. Methanol (<10%) was used to prepare the stock solutions. EC50 values were calculated from nominal concentrations.

Bibliographical reference: Kaiser, K.L.E., and Palabrica, V.S. (1991). *Water Poll. Res. J. Canada* **26**, 361-431.

3-PYRIDINECARBOXALDEHYDE

CAS RN: 500-22-1
Synonym: Nicotinaldehyde

Sample purity: 99%
Temperature: 15°C
Test parameter: EC50
Effect: Reduction in light output

Concentration: 331 mg/l
Exposure time: 5 min

Concentration: 269 mg/l
Exposure time: 15 min

Concentration: 245 mg/l
Exposure time: 30 min

Comment: Mean of four assays. Methanol (<10%) was used to prepare the stock solutions. EC50 values were calculated from nominal concentrations.

Bibliographical reference: Kaiser, K.L.E., and Palabrica, V.S. (1991). *Water Poll. Res. J. Canada* **26**, 361-431.

4-PYRIDINECARBOXALDEHYDE

CAS RN: 872-85-5
Synonym: Isonicotinaldehyde

Sample purity: 98%
Temperature: 15°C
Test parameter: EC50
Effect: Reduction in light output

Concentration: 141 mg/l
Exposure time: 5 min

Concentration: 110 mg/l
Exposure time: 15 min

Concentration: 91.2 mg/l
Exposure time: 30 min

Comment: Mean of four assays. Methanol (<10%) was used to prepare the stock solutions. EC50 values were calculated from nominal concentrations.

Bibliographical reference: Kaiser, K.L.E., and Palabrica, V.S. (1991). *Water Poll. Res. J. Canada* **26**, 361-431.

3,4-PYRIDINEDICARBONITRILE

CAS RN: 1633-44-9
Synonym: Cinchomeronic dinitrile

Sample purity: 97%
Temperature: 15°C
Test parameter: EC50
Effect: Reduction in light output

Concentration: 174 mg/l
Exposure time: 5 min

Concentration: 182 mg/l
Exposure time: 15 min

Concentration: 205 mg/l
Exposure time: 30 min

Comment: Mean of three assays. Methanol (<10%) was used to prepare the stock solutions. EC50 values were calculated from nominal concentrations.

Bibliographical reference: Kaiser, K.L.E., and Palabrica, V.S. (1992). National Water Research Institute, Burlington, Ontario, Canada, unpublished results.

3,4-PYRIDINEDICARBOXIMIDE

CAS RN: 4664-01-1
Synonym: Cinchomeronimide

Sample purity: 97%
Temperature: 15°C
Test parameter: EC50
Effect: Reduction in light output

Concentration: 347 mg/l
Exposure time: 5 min

Concentration: 332 mg/l
Exposure time: 15 min

Concentration: 339 mg/l
Exposure time: 30 min

Comment: Mean of three assays. Methanol (<10%) was used to prepare the stock solutions. EC50 values were calculated from nominal concentrations.

Bibliographical reference: Kaiser, K.L.E., and Palabrica, V.S. (1991). *Water Poll. Res. J. Canada* **26**, 361-431.

2,3-PYRIDINEDICARBOXYLIC ACID

CAS RN: 89-00-9
Synonym: Quinolinic acid

Sample purity: 99%
Temperature: 15°C
Test parameter: EC50
Effect: Reduction in light output

Concentration: 57.9 mg/l
Exposure time: 5 min

Concentration: 66.5 mg/l
Exposure time: 15 min

Concentration: 76.4 mg/l
Exposure time: 30 min

Comment: Mean of three assays. Methanol (<10%) was used to prepare the stock solutions. EC50 values were calculated from nominal concentrations.

Bibliographical reference: Kaiser, K.L.E., and Palabrica, V.S. (1991). *Water Poll. Res. J. Canada* **26**, 361-431.

4-PYRIDINEMETHANOL

CAS RN: 586-95-8
Synonym: 4-Pyridylcarbinol

Sample purity: 99%
Temperature: 15°C
Test parameter: EC50
Effect: Reduction in light output

Concentration: 233 mg/l
Exposure time: 5 min

Concentration: 244 mg/l
Exposure time: 15 min

Concentration: 256 mg/l
Exposure time: 30 min

Comment: Mean of four assays. Methanol (<10%) was used to prepare the stock solutions. EC50 values were calculated from nominal concentrations.

Bibliographical reference: Kaiser, K.L.E., and Palabrica, V.S. (1991). *Water Poll. Res. J. Canada* **26**, 361-431.

Temperature: 15°C
Test parameter: EC50
Effect: Reduction in light output

Concentration: 228 mg/l
Exposure time: 5 min

Concentration: 244 mg/l
Exposure time: 15 min

Concentration: 268 mg/l
Exposure time: 30 min

Comment: Mean of four assays. Methanol (<10%) was used to prepare the stock solutions. EC50 values were calculated from nominal concentrations.

Bibliographical reference: Kaiser, K.L.E., and Palabrica, V.S. (1991). *Water Poll. Res. J. Canada* **26**, 361-431.

PYRIDINE *N*-OXIDE

CAS RN: 694-59-7

Sample purity: 95%
Temperature: 15°C
Test parameter: EC50
Effect: Reduction in light output

Concentration: 2500 mg/l
Exposure time: 5 min

Concentration: 3080 mg/l
Exposure time: 15 min

Concentration: 3220 mg/l
Exposure time: 30 min

Comment: Mean of three assays. Methanol (<10%) was used to prepare the stock solutions. EC50 values were calculated from nominal concentrations.

Bibliographical reference: Kaiser, K.L.E., and Palabrica, V.S. (1991). *Water Poll. Res. J. Canada* **26**, 361-431.

3-PYRIDINEPROPANOL

CAS RN: 2859-67-8

Sample purity: 99%
Temperature: 15°C
Test parameter: EC50

Effect: Reduction in light output

Concentration: 73.7 mg/l
Exposure time: 5 min

Concentration: 80.8 mg/l
Exposure time: 15 min

Concentration: 86.6 mg/l
Exposure time: 30 min

Comment: Mean of three assays. Methanol (<10%) was used to prepare the stock solutions. EC50 values were calculated from nominal concentrations.

Bibliographical reference: Kaiser, K.L.E., and Palabrica, V.S. (1991). *Water Poll. Res. J. Canada* **26**, 361-431.

4-PYRIDYLACETONITRILE

CAS RN: 92333-25-0

Sample purity: 98%
Temperature: 15°C
Test parameter: EC50
Effect: Reduction in light output

Concentration: 34.6 mg/l
Exposure time: 5 min

Concentration: 37.9 mg/l
Exposure time: 15 min

Concentration: 43.6 mg/l
Exposure time: 30 min

Comment: Mean of three assays. Methanol (<10%) was used to prepare the stock solutions. EC50 values were calculated from nominal concentrations. Actual compound is the hydrochloride.

Bibliographical reference: Kaiser, K.L.E., and Palabrica, V.S. (1991). *Water Poll. Res. J. Canada* **26**, 361-431.

4-(2-PYRIDYLAZO)-*N*,*N*-DIMETHYLANILINE

CAS RN: 13103-75-8

Sample purity: 98%
Temperature: 15°C
Test parameter: EC50
Effect: Reduction in light output

Concentration: 0.12 mg/l
Exposure time: 5 min

Concentration: 0.11 mg/l
Exposure time: 15 min

Concentration: 0.10 mg/l
Exposure time: 30 min

Comment: Mean of three assays. Methanol (<10%) was used to prepare the stock solutions. EC50 values were calculated from nominal concentrations.

Bibliographical reference: Kaiser, K.L.E., and Palabrica, V.S. (1991). *Water Poll. Res. J. Canada* **26**, 361-431.

α-(4-PYRIDYL)BENZHYDROL

CAS RN: 1620-30-0

Sample purity: Technical grade
Temperature: 15°C
Test parameter: EC50
Effect: Reduction in light output

Concentration: 37.8 mg/l
Exposure time: 5 min

Concentration: 44.4 mg/l
Exposure time: 15 min

Concentration: 54.6 mg/l
Exposure time: 30 min

Comment: Mean of three assays. Methanol (<10%) was used to prepare the stock solutions. EC50 values were calculated from nominal concentrations.

Bibliographical reference: Kaiser, K.L.E., and Palabrica, V.S. (1991). *Water Poll. Res. J. Canada* **26**, 361-431.

PYRIMIDINE

CAS RN: 289-95-2

Sample purity: 99%
Temperature: 15°C
Test parameter: EC50
Effect: Reduction in light output

Concentration: 580 mg/l
Exposure time: 5 min

Concentration: 517 mg/l
Exposure times: 15 and 30 min

Comment: Mean of three assays. Methanol (<10%) was used to prepare the stock solutions. EC50 values were calculated from nominal concentrations.

Bibliographical reference: Kaiser, K.L.E., and Palabrica, V.S. (1991). *Water Poll. Res. J. Canada* **26**, 361-431.

PYRROLE

CAS RN: 109-97-7

Sample purity: 99%
Temperature: 15°C
Test parameter: EC50
Effect: Reduction in light output

Concentration: 464 mg/l
Exposure time: 5 min

Concentration: 433 mg/l

Exposure time: 15 min

Concentration: 414 mg/l
Exposure time: 30 min

Comment: Mean of three assays. Methanol (<10%) was used to prepare the stock solutions. EC50 values were calculated from nominal concentrations.

Bibliographical reference: Kaiser, K.L.E., and Palabrica, V.S. (1991). *Water Poll. Res. J. Canada* **26**, 361-431.

2-PYRROLIDINONE

CAS RN: 616-45-5
Synonym: 2-Pyrrolidone

Sample purity: >99%
Temperature: 15°C
Test parameter: EC50
Effect: Reduction in light output

Concentration: 646 mg/l
Exposure time: 5 min

Concentration: 676 mg/l
Exposure time: 15 min

Concentration: 759 mg/l
Exposure time: 30 min

Comment: Mean of four assays. Methanol (<10%) was used to prepare the stock solutions. EC50 values were calculated from nominal concentrations.

Bibliographical reference: Kaiser, K.L.E., and Palabrica, V.S. (1991). *Water Poll. Res. J. Canada* **26**, 361-431.

PYRUVIC ALDEHYDE

CAS RN: 78-98-8

Sample purity: 40% in water
Temperature: 15°C
Test parameter: EC50
Effect: Reduction in light output

Concentration: 43.6 ± 0.72 mg/l
Exposure time: 5 min

Concentration: 40.5 ± 1.51 mg/l
Exposure time: 15 min

Concentration: 38.9 ± 1.22 mg/l
Exposure time: 25 min

Comment: Mean of six assays. The values were converted to mg/l from the original data expressed in μM and a rounded molecular weight of 72 given by the authors. Phenol solution was used for quality control/quality assurance. The 5-min EC50 value was 18.2 mg/l. The EC50 value at 15 min was 20.7 mg/l with a relative error of <5%.

Bibliographical reference: Chou, C.C., and Que Hee, S.S. (1992). *Ecotoxicol. Environ. Safety* **23**, 355-363.

QUINOLINE

CAS RN: 91-22-5
Synonym: 1-Benzazine

Temperature: 15°C
Test parameter: EC50
Effect: Reduction in light output

Concentration: 0.32 mg/l
Exposure time: 5 min

Concentration: 0.36 mg/l
Exposure time: 15 min

Concentration: 0.44 mg/l
Exposure time: 30 min

Comment: Mean of three assays. Methanol (<10%) was used to prepare the stock solutions. EC50 values were calculated from nominal concentrations.

Bibliographical reference: Kaiser, K.L.E., and Palabrica, V.S. (1992). National Water Research Institute, Burlington, Ontario, Canada, unpublished results.

QUINOXALINE

CAS RN: 91-19-0

Sample purity: 99%
Temperature: 15°C
Test parameter: EC50
Effect: Reduction in light output

Concentration: 65.2 mg/l
Exposure time: 5 min

Concentration: 69.9 mg/l
Exposure time: 15 min

Concentration: 73.2 mg/l
Exposure time: 30 min

Comment: Mean of three assays. Methanol (<10%) was used to prepare the stock solutions. EC50 values were calculated from nominal concentrations.

Bibliographical reference: Kaiser, K.L.E., and Palabrica, V.S. (1991). *Water Poll. Res. J. Canada* **26**, 361-431.

RESORCINOL

CAS RN: 108-46-3
Synonym: 3-Hydroxyphenol

Temperature: 15°C
Test parameter: EC50
Effect: Reduction in light output
Concentration: 310 mg/l
Exposure time: 5 min

Bibliographical reference: Lebsack, M.E., Anderson, A.D., DeGraeve, G.M., and Bergman, H.L. (1981). In: *Aquatic Toxicology* *and Hazard Assessment: Fourth Conference*, *ASTM STP 737*, D.R. Branson and K.L. Dickson, (eds.), American Society for Testing and Materials, Philadelphia, p. 348-356.

Temperature: 15°C
Test parameter: EC50
Effect: Reduction in light output

Concentration: 375 mg/l
Exposure time: 5 min

Concentration: 265 mg/l
Exposure time: 30 min

Bibliographical reference: Speece, R. (1987). Drexel University, Philadelphia, USA, private communication.

RIDOMIL

CAS RN: 57837-19-1

Temperature: 15 ± 0.1°C
Test parameter: EC50
Effect: Reduction in light output
Concentration: 120 mg/l
Exposure time: 5 min

Bibliographical reference: Chang, J.C., Taylor, P.B., and Leach, F.R. (1981). *Bull. Environ. Contam. Toxicol.* **26**, 150-156.

ROTENONE

CAS RN: 83-79-4
Synonym: [2*R*-(2α,6aα,12aα)]-1,2,12,12a-Tetrahydro-8,9-dimethoxy-2-(1-methylethenyl)[1]benzopyrano[3,4-*b*]furo[2,3-*h*]benzopyran-6(6a*H*)-one

Sample purity: 97%
Temperature: 15°C
Test parameter: EC50
Effect: Reduction in light output

Concentration: 3.60 mg/l
Exposure time: 5 min

Concentration: 3.52 mg/l
Exposure time: 15 min

Concentration: 3.60 mg/l
Exposure time: 30 min

Comment: Mean of four assays. Methanol (<10%) was used to prepare the stock solutions. EC50 values were calculated from nominal concentrations.

Bibliographical reference: Kaiser, K.L.E., and Palabrica, V.S. (1992). National Water Research Institute, Burlington, Ontario, Canada, unpublished results.

ROUNDUP

CAS RN: 38641-94-0
Synonym: Glyphosate mono(isopropylammonium)

Sample purity: Agrichemical grade
Temperature: 15 ± 0.1°C
Test parameter: EC50
Effect: Reduction in light output
Concentration: 17.56 mg/l
Exposure time: 5 min
Comment: Test was performed on *Photobacterium phosphoreum* NZ11D obtained from the Scripps Institute of Oceanography (La Jolla, CA).

Bibliographical reference: McFeters, G.A., Bond, P.J., Olson, S.B., and Tchan, Y.T. (1983). *Water Res.* **17**, 1757-1762.

Sample purity: Glyphosate 350 g/l
Temperature: 15 ± 0.1°C
Test parameter: EC50
Effect: Reduction in light output

Concentration: 127 mg/l
Exposure time: 5 min

Concentration: 78 mg/l
Exposure time: 15 min

Concentration: 62.3 mg/l
Exposure time: 30 min

Comment: The pH was not adjusted.

Bibliographical reference: Tarkpea, M., Hansson, M., and Samuelsson, B. (1986). *Ecotoxicol. Environ. Safety* **11**, 127-143.

RUBRATOXIN B

CAS RN: 21794-01-4

Temperature: 15°C
Test parameter: EC50
Effect: Reduction in light output

Concentration: 31.79 mg/l
Exposure time: 5 min

Concentration: 33.36 mg/l
Exposure time: 10 min

Concentration: 35.17 mg/l
Exposure time: 15 min

Concentration: 34.73 mg/l
Exposure time: 20 min

Comment: Freshly reconstituted bacterial suspensions. Chemical was dissolved in methanol.

Bibliographical reference: Yates, I.E., and Porter, J.K. (1982). *Appl. Environ. Microbiol.* **44**, 1072-1075.

Temperature: 15°C
Test parameter: EC20
Effect: Reduction in light output

Concentration: 19.69 mg/l
Exposure time: 5 min

Concentration: 26.36 mg/l
Exposure time: 20 min

Comment: Freshly reconstituted bacterial suspensions. Chemical was dissolved in methanol.

Bibliographical reference: Yates, I.E., and Porter, J.K. (1982). *Appl. Environ. Microbiol.* **44**, 1072-1075.

Temperature: 15°C
Test parameter: EC50
Effect: Reduction in light output

Concentration: 26.68 mg/l
Exposure time: 5 min

Concentration: 31.09 mg/l
Exposure time: 10 min

Concentration: 31.82 mg/l
Exposure time: 15 min

Concentration: 32.82 mg/l
Exposure time: 20 min

Comment: Performed on bacterial suspensions maintained at 3°C for 5 h after reconstitution. Chemical was dissolved in methanol.

Bibliographical reference: Yates, I.E., and Porter, J.K. (1982). *Appl. Environ. Microbiol.* **44**, 1072-1075.

SALICYLALDEHYDE

CAS RN: 90-02-8
Synonyms: Salicylal; *o*-Hydroxybenzaldehyde

Sample purity: >98%
Temperature: 15 ± 0.1°C
Test parameter: EC50
Effect: Reduction in light output

Concentration: 16.3 mg/l
Exposure time: 5 min

Concentration: 14.3 mg/l
Exposure time: 15 min

Comment: Test was performed in duplicate. Toxicity values were based on nominal test concentrations.

Bibliographical reference: de Zwart, D., and Slooff, W. (1983). *Aquat. Toxicol.* **4**, 129-138.

Sample purity: >98%
Temperature: 15 ± 0.1°C
Test parameter: EC10
Effect: Reduction in light output

Concentration: 6.8 mg/l
Exposure time: 5 min

Concentration: 6.5 mg/l
Exposure time: 15 min

Comment: Test was performed in duplicate. Toxicity values were based on nominal test concentrations.

Bibliographical reference: de Zwart, D., and Slooff, W. (1983). *Aquat. Toxicol.* **4**, 129-138.

SALICYLIC ACID

CAS RN: 69-72-7
Synonym: 2-Hydroxybenzoic acid

Temperature: 15 ± 0.1°C
Test parameter: EC50
Effect: Reduction in light output
Concentration: 213.9 mg/l
Exposure time: 5 min

Bibliographical reference: Somasundaram, L., Coats, J.R., Racke, K.D., and Stahr, H.M. (1990). *Bull. Environ. Contam. Toxicol.* **44**, 254-259.

SILVER NITRATE

CAS RN: 7761-88-8

Temperature: 15°C
Test parameter: EC50
Effect: Reduction in light output
Result:

	15 min	30 min
Microbics strain	1.4 ± 0.6 mg/l Ag^+	0.70 ± 0.27 mg/l Ag^+
	0.86 ± 0.39 mg/l Ag^+	0.44 ± 0.17 mg/l Ag^+
Dr. Lange strain	1.53 ± 0.53 mg/l Ag^+	1.48 ± 0.75 mg/l Ag^+
	0.97 ± 0.33 mg/l Ag^+	0.94 ± 0.48 mg/l Ag^+

Comment: EC50 values were calculated from nominal concentrations.

Bibliographical reference: Vasseur, P. (1992). Centre des Sciences de l'Environnement, Metz, France, private communication.

SIMAZINE

CAS RN: 122-34-9
Synonym: 2-Chloro-4,6-bis(ethylamino)-1,3,5-triazine

Sample purity: Agrichemical grade
Temperature: 15 ± 0.1°C
Test parameter: EC50
Effect: Reduction in light output
Concentration: 238.33 mg/l
Exposure time: 5 min
Comment: Test was performed on *Photobacterium phosphoreum* NZ11D obtained from the Scripps Institute of Oceanography (La Jolla, CA).

Bibliographical reference: McFeters, G.A., Bond, P.J., Olson, S.B., and Tchan, Y.T. (1983). *Water Res.* **17**, 1757-1762.

SODIUM ARSENATE HEPTAHYDRATE

CAS RN: 10048-95-0

Temperature: 15 ± 0.3°C
Test parameter: EC50

Effect: Reduction in light output
Concentrations: 33.5 mg/l As^{5+}
36.3 mg/l As^{5+}
Exposure time: 5 min
Comment: Test solutions were analyzed.

Bibliographical reference: Qureshi, A.A., Flood, K.W., Thompson, S.R., Janhurst, S.M., Inniss, C.S., and Rokosh, D.A. (1982). In: *Aquatic Toxicology and Hazard Assessment: Fifth Conference, ASTM STP 766*, J.G. Pearson, R.B. Foster, and W.E. Bishop (eds.), American Society for Testing and Materials, Philadelphia, p. 179-195.

Sample purity: 99.5%
Temperature: 15 ± 0.1°C
Test parameter: EC50
Effect: Reduction in light output

Concentration: 2.47 mg/l As^{5+}
Exposure time: 5 min

Concentration: 1.07 mg/l As^{5+}
Exposure time: 15 min

Concentration: 1.45 mg/l As^{5+}
Exposure time: 30 min

Comment: The pH was not adjusted.

Bibliographical reference: Tarkpea, M., Hansson, M., and Samuelsson, B. (1986). *Ecotoxicol. Environ. Safety* **11**, 127-143.

Temperature: 15°C
Test parameter: EC50
Effect: Reduction in light output
Concentration: 1.5 mg/l As^{5+}
Exposure time: 5 min

Bibliographical reference: Kahru, A. (1993). *ATLA* **21**, 210-215.

SODIUM CYANIDE

CAS RN: 143-33-9

Test parameter: EC50
Effect: Reduction in light output
Concentration: 2.8 mg/l
Exposure time: 15 min
Comment: Test was performed in duplicate at pH = 6.7.

Bibliographical references: Dutka, B.J., and Kwan, K.K. (1981). *Bull. Environ. Contam. Toxicol.* **27**, 753-757.
Dutka, B.J., and Kwan, K.K. (1984). In: *Toxicity Screening Procedures Using Bacterial Systems*, D. Liu and B.J. Dutka (eds.), Marcel Dekker, New York, p. 125-138.

Effect: Percentage light inhibition
Result:

Chemical	incubation time		
ppm	5-min	10-min	15-min
2.5 sodium cyanide	18%	22%	42%
0.5 sodium lauryl sulfate (SLS)	21%	22%	22%
2.5 sodium cyanide + 0.5 SLS	17%	21%	40%

Comment: Mean of duplicated tests. Chemicals were tested at pH 6.7.

Bibliographical reference: Dutka, B.J., and Kwan, K.K. (1982). *Environ. Pollut. Ser. A* **29**, 125-134.

SODIUM DIETHYLDITHIOCARBAMATE

CAS RN: 148-18-5

Sample purity: ≥99%
Test parameter: EC50
Effect: Reduction in light output
Concentration: 1.22 mg/l (0.91-1.64)
Exposure time: 15 min

Bibliographical reference: van Leeuwen, C.J., Maas-Diepeveen, J.L., Niebeek, G., Vergouw, W.H.A., Griffioen, P.S., and Luijken, M.W. (1985). *Aquat. Toxicol.* **7**, 145-164.

SODIUM DIMETHYLDITHIOCARBAMATE

CAS RN: 128-04-1

Sample purity: ≥97%
Test parameter: EC50
Effect: Reduction in light output
Concentration: 0.51 mg/l (0.40-0.66)
Exposure time: 15 min

Bibliographical reference: van Leeuwen, C.J., Maas-Diepeveen, J.L., Niebeek, G., Vergouw, W.H.A., Griffioen, P.S., and Luijken, M.W. (1985). *Aquat. Toxicol.* **7**, 145-164.

SODIUM HYDROXIDE

CAS RN: 1310-73-2

Temperature: 15°C
Test parameter: EC50
Effect: Reduction in light output
Concentration: 22 mg/l
Exposure time: 15 min

Bibliographical reference: Bulich, A.A., Tung, K.K., and Scheibner, G. (1990). *J. Biolumin. Chemilumin.* **5**, 71-77.

SODIUM HYPOCHLORITE

CAS RN: 7681-52-9

Temperature: 15°C
Test parameter: EC50
Effect: Reduction in light output
Concentration: 0.1 mg/l
Exposure time: 15 min

Bibliographical reference: Bulich, A.A., Tung, K.K., and Scheibner, G. (1990). *J. Biolumin. Chemilumin.* **5**, 71-77.

SODIUM LAURYL SULFATE

CAS RN: 151-21-3
Synonym: Sodium dodecyl sulfate

Temperature: 15°C
Test parameter: EC50
Effect: Reduction in light output
Concentrations: 0.45 mg/l
2.2 mg/l
Exposure time: 5 min

Bibliographical reference: King, E.F., and Painter, H.A. (1981). In: *Les Tests de Toxicité Aiguë en Milieu Aquatique. Acute Aquatic Ecotoxicological Tests*, H. Leclerc and D. Dive (eds.), INSERM 106, Paris, p. 143-153.

Temperature: 15°C
Test parameter: EC50
Effect: Reduction in light output
Concentration: 1.6 mg/l
Exposure time: 5 min

Bibliographical reference: Bulich, A.A., Greene, M.W., and Isenberg, D.L. (1981). In: *Aquatic Toxicology and Hazard Assessment: Fourth Conference, ASTM STP 737*, D.R. Branson and K.L. Dickson (eds.), American Society for Testing and Materials, Philadelphia, p. 338-347.

Temperature: 15 ± 0.1°C
Test parameter: EC50
Effect: Reduction in light output
Concentration: 1.19 mg/l
Exposure time: 5 min
Comment: Test was performed on *Photobacterium phosphoreum* NZ11D obtained from the Scripps Institute of Oceanography (La Jolla, CA).

Bibliographical reference: McFeters, G.A., Bond, P.J., Olson, S.B., and Tchan, Y.T. (1983). *Water Res.* **17**, 1757-1762.

Temperature: 15°C
Test parameter: EC50

Effect: Reduction in light output

Concentration: 3.19 mg/l
Exposure time: 5 min

Concentration: 2.1 mg/l
Exposure time: 10 min

Concentration: 1.8 mg/l*
Exposure time: 15 min

Comment: Test was performed in duplicate at pH = 6.7.

Bibliographical references: Dutka, B.J., Nyholm, N., and Petersen, J. (1983). *Water Res.* **17**, 1363-1368.
* Dutka, B.J., and Kwan, K.K. (1981). *Bull. Environ. Contam. Toxicol.* **27**, 753-757.
* Dutka, B.J., and Kwan, K.K. (1984). In: *Toxicity Screening Procedures Using Bacterial Systems*, D. Liu and B.J. Dutka (eds.), Marcel Dekker, New York, p. 125-138.

Test parameter: EC50
Effect: Reduction in light output
Concentrations: 2.08 mg/l
1.82 mg/l
2.06 mg/l
1.41 mg/l
1.16 mg/l
1.26 mg/l
Exposure time: 5 min

Bibliographical reference: Herschke, B., and Lhotellier, D. (1983). *Eau Industrie Nuisances* **75**, 68-72.

Test parameter: EC50
Effect: Reduction in light output
Concentrations: 1.42 mg/l
1.45 mg/l
1.47 mg/l
0.92 mg/l
0.75 mg/l
0.93 mg/l
Exposure time: 10 min

Bibliographical reference: Herschke, B., and Lhotellier, D. (1983). *Eau Industrie Nuisances* **75**, 68-72.

Temperature: 15°C
Test parameter: EC50
Effect: Reduction in light output

Concentrations: 1.42 mg/l (time after reconstitution = 0.5 h)
2.16 mg/l (time after reconstitution = 3 h)
Exposure time: 5 min

Concentrations: 0.64 mg/l (time after reconstitution = 0.5 h)
0.93 mg/l (time after reconstitution = 3 h)
Exposure time: 15 min

Bibliographical reference: Qureshi, A.A., Coleman, R.N., and Paran, J.H. (1984). In: *Toxicity Screening Procedures Using Bacterial Systems*, D. Liu and B.J. Dutka (eds.), Marcel Dekker, New York, p. 1-22.

Temperature: 15°C
Test parameter: EC50
Effect: Reduction in light output

Concentration: 1.78 mg/l
Exposure time: 5 min

Concentration: 1.94 mg/l
Exposure time: 10 min

Concentration: 0.90 mg/l
Exposure time: 15 min

Concentration: 0.78 mg/l
Exposure time: 20 min

Concentration: 0.68 mg/l
Exposure time: 30 min

Comment: Results derived from the average of two replicates.

Bibliographical reference: Qureshi, A.A., Coleman, R.N., and Paran, J.H. (1984). In: *Toxicity Screening Procedures Using Bacterial*

Systems, D. Liu and B.J. Dutka (eds.), Marcel Dekker, New York, p. 1-22.

Sample purity: 99%
Temperature: 15°C
Test parameter: EC50
Effect: Reduction in light output
Concentration: 1.5 mg/l
Exposure time: 15 min

Bibliographical reference: King, E.F. (1984). In: *Toxicity Screening Procedures Using Bacterial Systems*, D. Liu and B.J. Dutka (eds.), Marcel Dekker, New York, p. 175-194.

Sample purity: Analytical grade
Temperature: 15°C
Test parameter: EC50
Effect: Reduction in light output
Concentration: 2.02 mg/l
Exposure time: 5 min

Bibliographical reference: Beaubien, A., Lapierre, L., Bouchard, A., and Jolicoeur, C. (1986). *Tox. Assess.* **1**, 187-200.

Sample purity: >99%
Temperature: 15 ± 0.1°C
Test parameter: EC50
Effect: Reduction in light output

Concentrations: 1.78 mg/l
1.26 mg/l
1.57 mg/l
1.38 mg/l
1.80 mg/l
Exposure time: 5 min

Concentrations: 0.93 mg/l
0.54 mg/l
0.63 mg/l
0.78 mg/l
1.08 mg/l
Exposure time: 15 min

Concentrations: 0.65 mg/l
0.28 mg/l
0.42 mg/l
0.47 mg/l
0.80 mg/l
Exposure time: 30 min

Comment: The pH was not adjusted.

Bibliographical reference: Tarkpea, M., Hansson, M., and Samuelsson, B. (1986). *Ecotoxicol. Environ. Safety* **11**, 127-143.

Sample purity: Reagent grade
Test parameter: EC50
Effect: Reduction in light output

Concentration: 3.5 mg/l (3.3-3.7)
Exposure time: 5 min

Concentration: 1.6 mg/l (1.5-1.7)
Exposure time: 15 min

Concentration: 1.2 mg/l (1.1-2.0)
Exposure time: 30 min

Bibliographical reference: Elnabarawy, M.T., Robideau, R.R., and Beach, S.A. (1988). *Tox. Assess.* **3**, 361-370.

Temperature: 15°C
Test parameter: EC50
Effect: Reduction in light output

Concentration: 3.02 mg/l
Exposure time: 5 min

Concentration: 1.80 mg/l
Exposure time: 15 min

Comment: The assays were performed in duplicate. The pH of the chemical solutions was not adjusted.

Bibliographical reference: Awong, J., Bitton, G., Koopman, B., and Morel, J.L. (1989). *Bull. Environ. Contam. Toxicol.* **43**, 118-122.

Temperature: 15°C
Test parameter: EC20
Effect: Reduction in light output
Concentration: 0.89 mg/l (0.81-0.97)
Exposure time: 15 min
Comment: Test performed in Gulf of Mexico sea water.

Bibliographical reference: Ankley, G.T., Hoke, R.A., Giesy, J.P., and Winger, P.V. (1989). *Chemosphere* **18**, 2069-2075.

Temperature: 15°C
Test parameter: EC20
Effect: Reduction in light output
Concentration: 0.45 mg/l (0.38-0.53)
Exposure time: 15 min
Comment: Test performed in distilled deionized water. Samples were osmotically adjusted with NaCl (20 g/l) before testing.

Bibliographical reference: Ankley, G.T., Hoke, R.A., Giesy, J.P., and Winger, P.V. (1989). *Chemosphere* **18**, 2069-2075.

Temperature: 15°C
Test parameter: EC50
Effect: Reduction in light output
Concentration: 2.04 ± 0.51 mg/l
Exposure time: 15 min

Bibliographical reference: Rychert, R., and Mortimer, M. (1991). *Environ. Toxicol. Water Qual.* **6**, 415-421.

SODIUM OMADINE 40%

CAS RN: 3811-73-2 (major constituent)

Sample purity: 40% Sodium-2-pyridinethiol-1-oxide
Temperature: 15°C
Test parameter: EC50
Effect: Reduction in light output
Concentration: 0.12 mg/l
Exposure time: 5 min
Comment: Chemical was tested in duplicate at 0.01, 0.05, 0.10, and 0.15 mg/l. Verification that the controls were valid was accomplished

by making sure that the difference between the duplicate blank ratios (5 minute light reading divided by the time 0 reading) was not greater than 0.02, otherwise the test was not continued. The 15-min EC50 value could not be determined due to a rapid decrease in light output at concentrations above 0.05 mg/l sodium omadine 40%.

Bibliographical reference: Mallak, F.P., and Brunker, R.L. (1984). In: *Toxicity Screening Procedures Using Bacterial Systems*, D. Liu and B.J. Dutka (eds.), Marcel Dekker, New York, p. 65-76.

SODIUM PENTACHLOROPHENATE

CAS RN: 131-52-2

Temperature: 15°C
Test parameter: EC50
Effect: Reduction in light output
Concentration: 0.5 mg/l
Exposure time: 5 min

Bibliographical reference: Bulich, A.A., Greene, M.W., and Isenberg, D.L. (1981). In: *Aquatic Toxicology and Hazard Assessment: Fourth Conference, ASTM STP 737*, D.R. Branson and K.L. Dickson, (eds.), American Society for Testing and Materials, Philadelphia, p. 338-347.

Sample purity: >98%
Temperature: 15 ± 0.1°C
Test parameter: EC50
Effect: Reduction in light output

Concentration: 0.94 mg/l
Exposure time: 5 min

Concentration: 0.76 mg/l
Exposure time: 15 min

Comment: Test was performed in duplicate. Toxicity values were based on nominal test concentrations.

Bibliographical reference: de Zwart, D., and Slooff, W. (1983). *Aquat. Toxicol.* 4, 129-138.

Sample purity: >98%
Temperature: 15 ± 0.1°C
Test parameter: EC10
Effect: Reduction in light output

Concentration: 0.2 mg/l
Exposure time: 5 min

Concentration: 0.15 mg/l
Exposure time: 15 min

Comment: Test was performed in duplicate. Toxicity values were based on nominal test concentrations.

Bibliographical reference: de Zwart, D., and Slooff, W. (1983). *Aquat. Toxicol.* **4**, 129-138.

Sample purity: 93%
Temperature: 15°C
Test parameter: EC50
Effect: Reduction in light output
Concentration: 1.2 mg/l
Exposure time: 15 min

Bibliographical reference: King, E.F. (1984). In: *Toxicity Screening Procedures Using Bacterial Systems*, D. Liu and B.J. Dutka (eds.), Marcel Dekker, New York, p. 175-194.

Temperature: 15°C
Test parameter: EC50
Effect: Reduction in light output
Concentration: 0.85 ± 0.17 mg/l
Exposure time: 15 min

Bibliographical reference: Rychert, R., and Mortimer, M. (1991). *Environ. Toxicol. Water Qual.* **6**, 415-421.

SODIUM SELENITE

CAS RN: 10102-18-8

Sample purity: Analytical grade

Temperature: 15°C
Test parameter: EC50
Effect: Reduction in light output

Concentration: 215.2 mg/l Se^{4+}
Exposure time: 5 min

Concentration: 53.9 mg/l Se^{4+}
Exposure time: 10 min

Concentration: 33.4 mg/l Se^{4+}
Exposure time: 15 min

Concentration: 29.1 mg/l Se^{4+}
Exposure time: 20 min

Concentration: 19.5 mg/l Se^{4+}
Exposure time: 30 min

Comment: Results derived from the average of two replicates.

Bibliographical reference: Qureshi, A.A., Coleman, R.N., and Paran, J.H. (1984). In: *Toxicity Screening Procedures Using Bacterial Systems*, D. Liu and B.J. Dutka (eds.), Marcel Dekker, New York, p. 1-22.

SODIUM SULFIDE

CAS RN: 1313-82-2

Sample purity: ≥99%
Test parameter: EC50
Effect: Reduction in light output
Concentration: 4.3 mg/l (2.9-6.3)
Exposure time: 15 min

Bibliographical reference: van Leeuwen, C.J., Maas-Diepeveen, J.L., Niebeek, G., Vergouw, W.H.A., Griffioen, P.S., and Luijken, M.W. (1985). *Aquat. Toxicol.* **7**, 145-164.

STYRENE

CAS RN: 100-42-5

Temperature: 15 ± 0.3°C
Test parameter: EC50
Effect: Reduction in light output
Concentrations: 5.5 mg/l
5.4 mg/l
Exposure time: 5 min
Comment: Test solutions were analyzed using gas chromatography.

Bibliographical reference: Qureshi, A.A., Flood, K.W., Thompson, S.R., Janhurst, S.M., Inniss, C.S., and Rokosh, D.A. (1982). In: *Aquatic Toxicology and Hazard Assessment: Fifth Conference, ASTM STP 766*, J.G. Pearson, R.B. Foster, and W.E. Bishop (eds.), American Society for Testing and Materials, Philadelphia, p. 179-195.

Temperature: 15°C
Test parameter: EC50
Effect: Reduction in light output
Concentration: 6 mg/l
Exposure time: 5 min

Bibliographical reference: Kahru, A. (1993). *ATLA* **21**, 210-215.

SULFANILIC ACID

CAS RN: 121-57-3

Sample purity: 99%
Temperature: 15°C
Test parameter: EC50
Effect: Reduction in light output

Concentration: 43.5 mg/l
Exposure time: 5 min

Concentration: 60.1 mg/l
Exposure time: 15 min

Concentration: 114 mg/l
Exposure time: 30 min

Comment: Mean of three assays. Methanol (<10%) was used to prepare the stock solutions. EC50 values were calculated from nominal concentrations.

Bibliographical reference: Kaiser, K.L.E., and Palabrica, V.S. (1991). *Water Poll. Res. J. Canada* **26**, 361-431.

4,4'-SULFONYLDIPHENOL

CAS RN: 80-09-1

Temperature: 15°C
Test parameter: EC50
Effect: Reduction in light output
Concentration: 19.9 mg/l
Exposure time: 30 min
Comment: Mean of three assays. Methanol (<10%) was used to prepare the stock solutions. EC50 values were calculated from nominal concentrations.

Bibliographical reference: Kaiser, K.L.E., and Palabrica, V.S. (1991). *Water Poll. Res. J. Canada* **26**, 361-431.

2,4,5-T

CAS RN: 93-76-5
Synonym: 2,4,5-Trichlorophenoxyacetic acid

Sample purity: Analytical reference standard grade
Temperature: 15 ± 0.1°C
Test parameter: EC50
Effect: Reduction in light output
Concentration: 157 mg/l
Exposure time: 5 min
Comment: Test was performed on *Photobacterium phosphoreum* NZ11D obtained from the Scripps Institute of Oceanography (La Jolla, CA).

Bibliographical reference: McFeters, G.A., Bond, P.J., Olson, S.B., and Tchan, Y.T. (1983). *Water Res.* **17**, 1757-1762.

Sample purity: Agrichemical grade
Temperature: 15 ± 0.1°C
Test parameter: EC50
Effect: Reduction in light output
Concentration: 78.75 mg/l
Exposure time: 5 min
Comment: Test was performed on *Photobacterium phosphoreum* NZ11D obtained from the Scripps Institute of Oceanography (La Jolla, CA).

Bibliographical reference: McFeters, G.A., Bond, P.J., Olson, S.B., and Tchan, Y.T. (1983). *Water Res.* **17**, 1757-1762.

Temperature: 15 ± 0.1°C
Test parameter: EC50
Effect: Reduction in light output
Concentration: 51.7 mg/l
Exposure time: 5 min

Bibliographical reference: Somasundaram, L., Coats, J.R., Racke, K.D., and Stahr, H.M. (1990). *Bull. Environ. Contam. Toxicol.* **44**, 254-259.

TEBUTHIURON

CAS RN: 34014-18-1
Synonym: 1-(5-*tert*-Butyl-1,3,4-thiadiazol-2-yl)-1,3-dimethylurea

Temperature: 15 ± 0.1°C
Test parameter: EC50
Effect: Reduction in light output
Concentration: 328 ± 47 mg/l
Exposure time: 5 min
Comment: Chemical was tested at 0, 31.3, 62.5, 125, 250, and 500 mg/l. Four assays.

Bibliographical reference: Blaise, C., and Harwood, M. (1991). *Rev. Sci. Eau* **4**, 121-134.

TEREPHTHALALDEHYDE

CAS RN: 623-27-8

Sample purity: 98%
Temperature: 15°C
Test parameter: EC50
Effect: Reduction in light output
Concentration: 7.04 mg/l
Exposure time: 30 min
Comment: Mean of three assays. Methanol (<10%) was used to prepare the stock solutions. EC50 values were calculated from nominal concentrations.

Bibliographical reference: Ribo, J.M., and Kaiser, K.L.E. National Water Research Institute, Burlington, Ontario, Canada, unpublished results (value later published by Kaiser, K.L.E., and Palabrica, V.S. (1991). *Water Poll. Res. J. Canada* **26**, 361-431).

TEREPHTHALONITRILE

CAS RN: 623-26-7
Synonym: 1,4-Dicyanobenzene

Temperature: 15°C
Test parameter: EC50
Effect: Reduction in light output

Concentration: 82.7 mg/l
Exposure time: 5 min

Concentration: 73.7 mg/l
Exposure time: 15 min

Concentration: 70.4 mg/l*
Exposure time: 30 min

Comment: Recrystallized from methanol. Mean of three assays. Methanol (<10%) was used to prepare the stock solutions. EC50 values were calculated from nominal concentrations.

Bibliographical references: Kaiser, K.L.E., and Palabrica, V.S. (1991). *Water Poll. Res. J. Canada* **26**, 361-431.

* Kaiser, K.L.E., and Gough, K.M. (1989). In: *Aquatic Toxicology and Environmental Fate: Eleventh Volume, ASTM STP 1007*, G.W. Suter and M.A. Lewis (eds.), American Society for Testing and Materials, Philadelphia, p. 424-441.

TEREPHTHALOYL CHLORIDE

CAS RN: 100-20-9

Sample purity: >99%
Temperature: 15°C
Test parameter: EC50
Effect: Reduction in light output

Concentration: 8.27 mg/l
Exposure time: 5 min

Concentration: 10.7 mg/l
Exposure time: 15 min

Concentration: 12.0 mg/l
Exposure time: 30 min

Comment: Mean of three assays. Methanol (<10%) was used to prepare the stock solutions. EC50 values were calculated from nominal concentrations.

Bibliographical reference: Kaiser, K.L.E., and Palabrica, V.S. (1992). National Water Research Institute, Burlington, Ontario, Canada, unpublished results.

TETRAALLYLTIN

CAS RN: 7393-43-3

Temperature: 15°C
Test parameter: EC50
Effect: Reduction in light output

Concentration: 5.48 ± 0.96 mg/l
Exposure time: 5 min

Concentration: 4.19 ± 1.1 mg/l
Exposure time: 15 min

Comment: Chemical was dissolved in ethanol (~0.05%).

Bibliographical reference: Dooley, C.A., and Kenis, P. (1987). In: *Oceans '87, International Organotin Symposium*, Department of Fisheries and Oceans, Canada, and the Society for Underwater Technology, William MacNab & Son, Halifax, p. 1517-1524.

TETRA-1-BUTENYLTIN

CAS RN: 125112-24-5

Temperature: 15°C
Test parameter: EC50
Effect: Reduction in light output

Concentration: 0.59 ± 0.13 mg/l
Exposure time: 5 min

Concentration: 0.33 ± 0.04 mg/l
Exposure time: 15 min

Comment: Chemical was dissolved in ethanol (~0.05%).

Bibliographical reference: Dooley, C.A., and Kenis, P. (1987). In: *Oceans '87, International Organotin Symposium*, Department of Fisheries and Oceans, Canada, and the Society for Underwater Technology, William MacNab & Son, Halifax, p. 1517-1524.

TETRA-2-BUTENYLTIN

CAS RN: 51952-14-8

Temperature: 15°C
Test parameter: EC50
Effect: Reduction in light output

Concentration: 1.60 ± 0.51 mg/l
Exposure time: 5 min

Concentration: 0.88 ± 0.23 mg/l
Exposure time: 15 min

Comment: Chemical was dissolved in ethanol (~0.05%).

Bibliographical reference: Dooley, C.A., and Kenis, P. (1987). In: *Oceans '87, International Organotin Symposium*, Department of Fisheries and Oceans, Canada, and the Society for Underwater Technology, William MacNab & Son, Halifax, p. 1517-1524.

TETRA-3-BUTENYLTIN

CAS RN: 53911-88-9

Temperature: 15°C
Test parameter: EC50
Effect: Reduction in light output

Concentration: 0.31 ± 0.03 mg/l
Exposure time: 5 min

Concentration: 0.22 ± 0.01 mg/l
Exposure time: 15 min

Comment: Chemical was dissolved in ethanol (~0.05%).

Bibliographical reference: Dooley, C.A., and Kenis, P. (1987). In: *Oceans '87, International Organotin Symposium*, Department of Fisheries and Oceans, Canada, and the Society for Underwater Technology, William MacNab & Son, Halifax, p. 1517-1524.

TETRABUTYL THIOPEROXYDICARBONIC DIAMIDE

CAS RN: 1634-02-2

Sample purity: ≥95%
Test parameter: EC50
Effect: Reduction in light output
Concentration: >60 mg/l
Exposure time: 15 min

Bibliographical reference: van Leeuwen, C.J., Maas-Diepeveen, J.L., Niebeek, G., Vergouw, W.H.A., Griffioen, P.S., and Luijken, M.W. (1985). *Aquat. Toxicol.* **7**, 145-164.

TETRABUTYLTIN

CAS RN: 1461-25-2

Temperature: 15°C
Test parameter: EC50
Effect: Reduction in light output
Concentration: 0.0011 mg/l
Exposure time: 30 min

Bibliographical reference: Steinhäuser, K.G., Amann, W., Späth, A., and Polenz, A. (1985). *Vom Wasser* **65**, 203-214.

Temperature: 15°C
Test parameter: EC50
Effect: Reduction in light output

Concentration: 1.50 ± 0.59 mg/l
Exposure time: 5 min

Concentration: 0.73 ± 0.33 mg/l
Exposure time: 15 min

Comment: Chemical was dissolved in ethanol (~0.05%).

Bibliographical reference: Dooley, C.A., and Kenis, P. (1987). In: *Oceans '87, International Organotin Symposium*, Department of Fisheries and Oceans, Canada, and the Society for Underwater Technology, William MacNab & Son, Halifax, p. 1517-1524.

2,3,4,5-TETRACHLOROANILINE

CAS RN: 634-83-3

Temperature: 15°C
Test parameter: EC50
Effect: Reduction in light output

Concentration: 1.13 mg/l
Exposure time: 5 min

Concentration: 1.18 mg/l
Exposure time: 15 min

Concentration: 0.99 mg/l
Exposure time: 30 min

Comment: Mean of three assays. Methanol (<10%) was used to prepare the stock solutions. EC50 values were calculated from nominal concentrations.

Bibliographical reference: Ribo, J.M., and Kaiser, K.L.E. (1984). In: *QSAR in Environmental Toxicology*, K.L.E. Kaiser (ed.), D. Reidel Publishing Company, Dordrecht, p. 319-336.

2,3,5,6-TETRACHLOROANILINE

CAS RN: 3481-20-7

Temperature: 15°C
Test parameter: EC50
Effect: Reduction in light output

Concentration: 1.46 mg/l
Exposure time: 5 min

Concentration: 1.63 mg/l
Exposure time: 15 min

Concentration: 1.60 mg/l
Exposure time: 30 min

Comment: Mean of three assays. Methanol (<10%) was used to prepare the stock solutions. EC50 values were calculated from nominal concentrations.

Bibliographical reference: Ribo, J.M., and Kaiser, K.L.E. (1984). In: *QSAR in Environmental Toxicology*, K.L.E. Kaiser (ed.), D. Reidel Publishing Company, Dordrecht, p. 319-336.

1,2,3,4-TETRACHLOROBENZENE

CAS RN: 634-66-2

Temperature: 15°C
Test parameter: EC50
Effect: Reduction in light output

Concentration: 2.26 mg/l
Exposure time: 5 min

Concentration: 3.34 mg/l
Exposure time: 15 min

Concentration: 4.02 mg/l
Exposure time: 30 min

Comment: Mean of three assays. Methanol (<10%) was used to prepare the stock solutions. EC50 values were calculated from nominal concentrations.

Bibliographical reference: Ribo, J.M., and Kaiser, K.L.E. (1983). *Chemosphere* **12**, 1421-1442.

Temperature: 15°C
Test parameter: EC50
Effect: Reduction in light output
Concentration: 1.88 mg/l
Exposure time: 15 min
Comment: Value derived by extrapolation.

Bibliographical reference: Hermens, J., Busser, F., Leeuwangh, P., and Musch, A. (1985). *Ecotoxicol. Environ. Safety* **9**, 17-25.

1,2,3,5-TETRACHLOROBENZENE

CAS RN: 634-90-2

Temperature: 15°C
Test parameter: EC50
Effect: Reduction in light output

Concentration: 3.27 mg/l

Exposure time: 5 min

Concentration: 3.50 mg/l
Exposure time: 15 min

Concentration: 2.48 mg/l
Exposure time: 30 min

Comment: Mean of three assays. Methanol (<10%) was used to prepare the stock solutions. EC50 values were calculated from nominal concentrations.

Bibliographical reference: Ribo, J.M., and Kaiser, K.L.E. (1983). *Chemosphere* **12**, 1421-1442.

1,2,4,5-TETRACHLOROBENZENE

CAS RN: 95-94-3

Temperature: 15°C
Test parameter: EC50
Effect: Reduction in light output

Concentration: 10.1 mg/l
Exposure time: 5 min

Concentration: 6.52 mg/l
Exposure time: 15 min

Concentration: 4.51 mg/l
Exposure time: 30 min

Comment: Mean of three assays. Methanol (<10%) was used to prepare the stock solutions. EC50 values were calculated from nominal concentrations.

Bibliographical reference: Ribo, J.M., and Kaiser, K.L.E. (1983). *Chemosphere* **12**, 1421-1442.

1,2,3,4-TETRACHLOROBUTANE

CAS RN: 52134-24-4

Temperature: 15°C
Test parameter: EC50
Effect: Reduction in light output

Concentration: 5.17 mg/l*
Exposure time: 5 min

Concentration: 5.40 mg/l*
Exposure time: 30 min

* EC50 values from linear regression fit to data.

Bibliographical reference: Speece, R. (1987). Drexel University, Philadelphia, USA, private communication.

TETRACHLOROETHANE

CAS RN: 25322-20-7

Test parameter: EC50
Effect: Reduction in light output
Concentration: 8.6 mg/l
Exposure time: 5 min
Comment: Isomer not specified. Concentrations in the test were measured.

Bibliographical reference: Curtis, C., Lima, A., Lozano, S.J., and Veith, G.D. (1982). In: *Aquatic Toxicology and Hazard Assessment: Fifth Conference, ASTM STP 766*, J.G. Pearson, R.B. Foster, and W.E. Bishop (eds.), American Society for Testing and Materials, Philadelphia, p. 170-178.

1,1,1,2-TETRACHLOROETHANE

CAS RN: 630-20-6

Temperature: 15°C
Test parameter: EC50
Effect: Reduction in light output

Concentration: 2.03 mg/l
Exposure time: 5 min

Concentration: 2.81 mg/l
Exposure time: 30 min

Bibliographical reference: Speece, R. (1987). Drexel University, Philadelphia, USA, private communication.

1,1,2,2-TETRACHLOROETHANE

CAS RN: 79-34-5

Temperature: 15°C
Test parameter: EC50
Effect: Reduction in light output

Concentration: 5.37 mg/l
Exposure time: 5 min

Concentration: 7.94 mg/l
Exposure time: 30 min

Bibliographical reference: Speece, R. (1987). Drexel University, Philadelphia, USA, private communication.

TETRACHLOROETHYLENE

CAS RN: 127-18-4

Temperature: 20°C

Test parameter: EC20
Effect: Reduction in light output
Concentration: 4.18 mg/l (3.36-5.19)
Exposure time: 5 min

Test parameter: EC50
Effect: Reduction in light output
Concentration: 17.19 mg/l (15.67-18.86)
Exposure time: 5 min

Bibliographical reference: Casseri, N.A., Ying, W, and Sojka, S.A. (1983). Proceedings 38th Industrial Waste Conference, USA, May 1983, p. 867-878.

Temperature: 20°C

Test parameter: EC20
Effect: Reduction in light output
Concentration: 4.47 mg/l
Exposure time: 15 min

Test parameter: EC50
Effect: Reduction in light output
Concentration: 19.7 mg/l
Exposure time: 15 min

Bibliographical reference: Casseri, N.A., Ying, W, and Sojka, S.A. (1983). Proceedings 38th Industrial Waste Conference, USA, May 1983, p. 867-878.

Temperature: 15°C
Test parameter: EC50
Effect: Reduction in light output
Concentration: 68 mg/l
Exposure time: 10 min
Comment: Mean of six assays. Test solutions were analyzed using gas chromatography.

Bibliographical reference: Bazin, C., Chambon, P., Bonnefille, M., and Larbaigt, G. (1987). *Sci. Eau* **6**, 403-413.

Temperature: 15°C
Test parameter: EC50
Effect: Reduction in light output

Concentration: 89.6 mg/l
Exposure time: 5 min

Concentration: 118 mg/l
Exposure time: 30 min

Bibliographical reference: Speece, R. (1987). Drexel University, Philadelphia, USA, private communication.

Sample purity: >99%
Temperature: 15°C
Test parameter: EC50

Effect: Reduction in light output

Concentration: 67.6 mg/l
Exposure time: 5 min

Concentration: 63.0 mg/l
Exposure time: 15 min

Concentration: 64.5 mg/l
Exposure time: 30 min

Comment: Mean of two assays. Methanol (<10%) was used to prepare the stock solutions. EC50 values were calculated from nominal concentrations.

Bibliographical reference: Kaiser, K.L.E., and Palabrica, V.S. (1992). National Water Research Institute, Burlington, Ontario, Canada, unpublished results.

TETRACHLOROGUAIACOL

CAS RN: 2539-17-5
Synonym: 2,3,4,5-Tetrachloro-6-methoxyphenol

Temperature: 15°C
Test parameter: EC50
Effect: Reduction in light output

Concentration: 4.77 mg/l
Exposure time: 5 min

Concentration: 3.87 mg/l
Exposure times: 15 and 30 min

Comment: Mean of three assays. Methanol (<10%) was used to prepare the stock solutions. EC50 values were calculated from nominal concentrations.

Bibliographical reference: Kaiser, K.L.E., and Palabrica, V.S. (1992). National Water Research Institute, Burlington, Ontario, Canada, unpublished results.

1,2,5,6-TETRACHLOROHEXANE

CAS RN: 89281-87-8

Test parameter: EC50
Effect: Reduction in light output
Concentration: 3.1 mg/l (2.9-3.4)
Exposure time: 5 min

Bibliographical reference: Dorn, P.B., van Compernolle, R., Meyer, C.L., and Crossland, N.O. (1991). *Environ. Toxicol. Chem.* **10**, 691-703.

2,3,4,5-TETRACHLORONITROBENZENE

CAS RN: 879-39-0

Temperature: 15°C
Test parameter: EC50
Effect: Reduction in light output

Concentration: 3.14 mg/l
Exposure time: 5 min

Concentration: 2.17 mg/l
Exposure time: 15 min

Concentration: 1.40 mg/l*
Exposure time: 30 min

Comment: Mean of three assays. Methanol (<10%) was used to prepare the stock solutions. EC50 values were calculated from nominal concentrations.

Bibliographical references: Kaiser, K.L.E., and Palabrica, V.S. (1991). *Water Poll. Res. J. Canada* **26**, 361-431.
* Kaiser, K.L.E., and Ribo, J.M. (1985). In: *QSAR in Toxicology and Xenobiochemistry*, M. Tichy (ed.), Elsevier, Amsterdam, p. 27-38.

2,3,5,6-TETRACHLORONITROBENZENE

CAS RN: 117-18-0

Temperature: 15°C
Test parameter: EC50
Effect: Reduction in light output

Concentration: 7.19 mg/l
Exposure time: 5 min

Concentration: 7.02 mg/l
Exposure time: 15 min

Concentration: 8.25 mg/l*
Exposure time: 30 min

Comment: Mean of three assays. Methanol (<10%) was used to prepare the stock solutions. EC50 values were calculated from nominal concentrations.

Bibliographical references: Kaiser, K.L.E., and Palabrica, V.S. (1991). *Water Poll. Res. J. Canada* **26**, 361-431.
* Kaiser, K.L.E., and Ribo, J.M. (1985). In: *QSAR in Toxicology and Xenobiochemistry*, M. Tichy (ed.), Elsevier, Amsterdam, p. 27-38.

2,3,4,5-TETRACHLOROPHENOL

CAS RN: 4901-51-3

Temperature: 15°C
Test parameter: EC50
Effect: Reduction in light output

Concentration: 0.34 mg/l
Exposure time: 5 min

Concentration: 0.21 mg/l
Exposure time: 15 min

Concentration: 0.18 mg/l
Exposure time: 30 min

Comment: Mean of three assays. Methanol (<10%) was used to prepare the stock solutions. EC50 values were calculated from nominal concentrations.

Bibliographical reference: Ribo, J.M., and Kaiser, K.L.E. (1983). *Chemosphere* **12**, 1421-1442.

2,3,4,6-TETRACHLOROPHENOL

CAS RN: 58-90-2

Temperature: 15°C
Test parameter: EC50
Effect: Reduction in light output

Concentration: 1.88 mg/l
Exposure time: 5 min

Concentration: 1.46 mg/l
Exposure time: 15 min

Concentration: 1.27 mg/l
Exposure time: 30 min

Comment: Mean of three assays. Methanol (<10%) was used to prepare the stock solutions. EC50 values were calculated from nominal concentrations.

Bibliographical reference: Ribo, J.M., and Kaiser, K.L.E. (1983). *Chemosphere* **12**, 1421-1442.

2,3,5,6-TETRACHLOROPHENOL

CAS RN: 935-95-5

Temperature: 15°C
Test parameter: EC50
Effect: Reduction in light output

Concentration: 2.79 mg/l
Exposure time: 5 min

Concentration: 2.54 mg/l
Exposure time: 15 min

Concentration: 2.21 mg/l

Exposure time: 30 min

Comment: Mean of three assays. Methanol (<10%) was used to prepare the stock solutions. EC50 values were calculated from nominal concentrations.

Bibliographical reference: Ribo, J.M., and Kaiser, K.L.E. (1983). *Chemosphere* **12**, 1421-1442.

Temperature: 15°C
Test parameter: EC50
Effect: Reduction in light output

Concentration: 2.92 mg/l
Exposure time: 5 min

Concentration: 2.08 mg/l
Exposure time: 30 min

Bibliographical reference: Speece, R. (1987). Drexel University, Philadelphia, USA, private communication.

TETRADECYL SULFATE SODIUM SALT

CAS RN: 1191-50-0

Sample purity: 99%
Temperature: 15°C
Test parameter: EC50
Effect: Reduction in light output

Concentration: 56.3 mg/l
Exposure time: 5 min

Concentration: 51.3 mg/l
Exposure time: 15 min

Concentration: 67.7 mg/l
Exposure time: 30 min

Comment: Mean of three assays. Methanol (<10%) was used to prepare the stock solutions. EC50 values were calculated from nominal concentrations.

Bibliographical reference: Kaiser, K.L.E., and Palabrica, V.S. (1991). *Water Poll. Res. J. Canada* **26**, 361-431.

2,3,4,5-TETRAFLUOROACETOPHENONE

CAS RN: 66286-21-3

Sample purity: 99%
Temperature: 15°C
Test parameter: EC50
Effect: Reduction in light output

Concentration: 116 mg/l
Exposure time: 5 min

Concentration: 113 mg/l
Exposure time: 15 min

Concentration: 106 mg/l
Exposure time: 30 min

Comment: Mean of three assays. Methanol (<10%) was used to prepare the stock solutions. EC50 values were calculated from nominal concentrations.

Bibliographical reference: Kaiser, K.L.E., and Palabrica, V.S. (1991). *Water Poll. Res. J. Canada* **26**, 361-431.

2,3,4,6-TETRAFLUOROANILINE

CAS RN: 363-73-5
Synonym: 2,3,4,6-Tetrafluorobenzenamine

Sample purity: 98%
Temperature: 15°C
Test parameter: EC50
Effect: Reduction in light output

Concentration: 137 mg/l
Exposure time: 5 min

Concentration: 151 mg/l

Exposure time: 15 min

Concentration: 169 mg/l
Exposure time: 30 min

Comment: Mean of three assays. Methanol (<10%) was used to prepare the stock solutions. EC50 values were calculated from nominal concentrations.

Bibliographical reference: Kaiser, K.L.E., and Palabrica, V.S. (1991). *Water Poll. Res. J. Canada* **26**, 361-431.

2,3,5,6-TETRAFLUOROPYRIDINE

CAS RN: 2875-18-5

Sample purity: 95%
Temperature: 15°C
Test parameter: EC50
Effect: Reduction in light output

Concentration: 379 mg/l
Exposure time: 5 min

Concentration: 397 mg/l
Exposure time: 15 min

Concentration: 426 mg/l
Exposure time: 30 min

Comment: Mean of three assays. Methanol (<10%) was used to prepare the stock solutions. EC50 values were calculated from nominal concentrations.

Bibliographical reference: Kaiser, K.L.E., and Palabrica, V.S. (1991). *Water Poll. Res. J. Canada* **26**, 361-431.

TETRAHYDROFURAN

CAS RN: 109-99-9

Sample purity: >99%

Temperature: 15°C
Test parameter: EC50
Effect: Reduction in light output

Concentration: 455 mg/l
Exposure times: 5 and 15 min

Concentration: 466 mg/l
Exposure time: 30 min

Comment: Mean of four assays. Methanol (<10%) was used to prepare the stock solutions. EC50 values were calculated from nominal concentrations.

Bibliographical reference: Kaiser, K.L.E., and Palabrica, V.S. (1992). National Water Research Institute, Burlington, Ontario, Canada, unpublished results.

TETRAHYDROFURFURYL ALCOHOL

CAS RN: 97-99-4

Temperature: 15°C
Test parameter: EC50
Effect: Reduction in light output
Concentration: 1600 mg/l
Exposure time: 15 min

Bibliographical reference: Bulich, A.A., Tung, K.K., and Scheibner, G. (1990). *J. Biolumin. Chemilumin.* **5**, 71-77.

TETRAHYDROXYQUINONE

CAS RN: 319-89-1
Synonym: 2,3,5,6-Tetrahydroxy-2,5-cyclohexadiene-1,4-dione

Temperature: 15°C
Test parameter: EC50
Effect: Reduction in light output
Concentration: 5.57 mg/l
Exposure time: 30 min

Comment: Mean of four assays. Methanol (<10%) was used to prepare the stock solutions. EC50 values were calculated from nominal concentrations. Values calculated from that for bishydrate.

Bibliographical reference: Kaiser, K.L.E., and Palabrica, V.S. (1991). *Water Poll. Res. J. Canada* **26**, 361-431.

TETRAHYDROXYQUINONE BISHYDRATE

CAS RN: 5676-48-2

Temperature: 15°C
Test parameter: EC50
Effect: Reduction in light output

Concentration: 5.87 mg/l
Exposure time: 5 min

Concentration: 6.14 mg/l
Exposure time: 15 min

Concentration: 6.74 mg/l
Exposure time: 30 min

Comment: Mean of three assays. Methanol (<10%) was used to prepare the stock solutions. EC50 values were calculated from nominal concentrations.

Bibliographical reference: Kaiser, K.L.E., and Palabrica, V.S. (1991). *Water Poll. Res. J. Canada* **26**, 361-431.

TETRAMETHYLAMMONIUM BROMIDE

CAS RN: 64-20-0

Temperature: 15°C
Test parameter: EC50
Effect: Reduction in light output

Concentration: 847 mg/l
Exposure times: 5 and 15 min

Concentration: 737 mg/l
Exposure time: 30 min

Comment: Mean of three assays. Methanol (<10%) was used to prepare the stock solutions. EC50 values were calculated from nominal concentrations.

Bibliographical reference: Kaiser, K.L.E., and Palabrica, V.S. (1991). *Water Poll. Res. J. Canada* **26**, 361-431.

N,N,N',N'-TETRAMETHYL-1,4-PHENYLENEDIAMINE.$(HCl)_2$

CAS RN: 637-01-4
Synonym: Wurster's reagent

Temperature: 15°C
Test parameter: EC50
Effect: Reduction in light output
Concentration: 25.4 mg/l
Exposure time: 30 min
Comment: Mean of three assays. Methanol (<10%) was used to prepare the stock solutions. EC50 values were calculated from nominal concentrations.

Bibliographical reference: Kaiser, K.L.E. (1987). In: *QSAR in Environmental Toxicology - II*, K.L.E. Kaiser (ed.), D. Reidel Publishing Company, Dordrecht, p. 169-188.

Sample purity: >99%
Temperature: 15°C
Test parameter: EC50
Effect: Reduction in light output

Concentration: 17.6 mg/l
Exposure time: 5 min

Concentration: 19.7 mg/l
Exposure time: 15 min

Concentration: 20.7 mg/l
Exposure time: 30 min

Comment: Mean of three assays. Methanol (<10%) was used to prepare the stock solutions. EC50 values were calculated from nominal concentrations.

Bibliographical reference: Kaiser, K.L.E., and Palabrica, V.S. (1991). *Water Poll. Res. J. Canada* **26**, 361-431.

TETRAMETHYLTHIOUREA

CAS RN: 2782-91-4

Sample purity: ≥97%
Test parameter: EC50
Effect: Reduction in light output
Concentration: 72.1 mg/l (63.5-81.8)
Exposure time: 15 min

Bibliographical reference: van Leeuwen, C.J., Maas-Diepeveen, J.L., Niebeek, G., Vergouw, W.H.A., Griffioen, P.S., and Luijken, M.W. (1985). *Aquat. Toxicol.* **7**, 145-164.

Sample purity: ≥97%
Test parameter: EC50
Effect: Reduction in light output
Concentration: 72.5 mg/l
Exposure time: 15 min
Comment: EC50 value converted to mg/l from the original data expressed in log(1/μmol l^{-1}) and a rounded molecular weight of 132 given by the authors.

Bibliographical reference: Govers, H., Ruepert, C., Stevens, T., and van Leeuwen, C.J. (1986). *Chemosphere* **15**, 383-393.

TETRAMETHYLTHIURAM MONOSULFIDE

CAS RN: 97-74-5

Sample purity: ≥95%
Test parameter: EC50
Effect: Reduction in light output
Concentration: 1.9 mg/l (1.7-2.1)

Exposure time: 15 min

Bibliographical reference: van Leeuwen, C.J., Maas-Diepeveen, J.L., Niebeek, G., Vergouw, W.H.A., Griffioen, P.S., and Luijken, M.W. (1985). *Aquat. Toxicol.* **7**, 145-164.

TETRAMETHYLUREA

CAS RN: 632-22-4

Sample purity: ≥98%
Test parameter: EC50
Effect: Reduction in light output
Concentration: 1100 mg/l (1000-1300)
Exposure time: 15 min

Bibliographical reference: van Leeuwen, C.J., Maas-Diepeveen, J.L., Niebeek, G., Vergouw, W.H.A., Griffioen, P.S., and Luijken, M.W. (1985). *Aquat. Toxicol.* **7**, 145-164.

TETRAOCTYLTIN

CAS RN: 3590-84-9

Temperature: 15°C
Test parameter: EC50
Effect: Reduction in light output
Concentration: 0.00063 mg/l
Exposure time: 30 min

Bibliographical reference: Steinhäuser, K.G., Amann, W., Späth, A., and Polenz, A. (1985). *Vom Wasser* **65**, 203-214.

TETRAPHENYLTIN

CAS RN: 595-90-4

Temperature: 15°C
Test parameter: EC50
Effect: Reduction in light output

Concentration: >0.041 mg/l
Exposure time: 30 min

Bibliographical reference: Steinhäuser, K.G., Amann, W., Späth, A., and Polenz, A. (1985). *Vom Wasser* **65**, 203-214.

TETRA-*n*-PROPYLTHIURAM DISULFIDE

CAS RN: 2556-42-5

Sample purity: ≥95%
Test parameter: EC50
Effect: Reduction in light output
Concentration: >100 mg/l
Exposure time: 15 min

Bibliographical reference: van Leeuwen, C.J., Maas-Diepeveen, J.L., Niebeek, G., Vergouw, W.H.A., Griffioen, P.S., and Luijken, M.W. (1985). *Aquat. Toxicol.* **7**, 145-164.

TETRAPROPYLTIN

CAS RN: 2176-98-9

Temperature: 15°C
Test parameter: EC50
Effect: Reduction in light output

Concentration: 11.96 ± 1.50 mg/l
Exposure time: 5 min

Concentration: 8.77 ± 1.39 mg/l
Exposure time: 15 min

Comment: Chemical was dissolved in ethanol (~0.05%).

Bibliographical reference: Dooley, C.A., and Kenis, P. (1987). In: *Oceans '87, International Organotin Symposium*, Department of Fisheries and Oceans, Canada, and the Society for Underwater Technology, William MacNab & Son, Halifax, p. 1517-1524.

1-*H*-TETRAZOLE

CAS RN: 288-94-8

Temperature: 15°C
Test parameter: EC50
Effect: Reduction in light output

Concentration: 47.4 mg/l
Exposure time: 5 min

Concentration: 48.5 mg/l
Exposure times: 15 and 30 min

Comment: Mean of three assays. Methanol (<10%) was used to prepare the stock solutions. EC50 values were calculated from nominal concentrations.

Bibliographical reference: Kaiser, K.L.E., and Palabrica, V.S. (1991). *Water Poll. Res. J. Canada* **26**, 361-431.

THENOYLTRIFLUOROACETONE

CAS RN: 326-91-0

Temperature: 15 ± 0.1°C
Test parameter: EC50
Effect: Reduction in light output
Concentration: 3.5 mg/l
Exposure time: 5 min

Bibliographical reference: Chang, J.C., Taylor, P.B., and Leach, F.R. (1981). *Bull. Environ. Contam. Toxicol.* **26**, 150-156.

THIABENDAZOLE

CAS RN: 148-79-8

Temperature: 15 ± 0.1°C
Test parameter: EC50
Effect: Reduction in light output
Concentration: 3400 mg/l

Exposure time: 5 min

Bibliographical reference: Chang, J.C., Taylor, P.B., and Leach, F.R. (1981). *Bull. Environ. Contam. Toxicol.* **26**, 150-156.

THIAZOLE

CAS RN: 288-47-1

Sample purity: 99%
Temperature: 15°C
Test parameter: EC50
Effect: Reduction in light output

Concentration: 11.5 mg/l
Exposure time: 5 min

Concentration: 12.0 mg/l
Exposure time: 15 min

Concentration: 13.5 mg/l
Exposure time: 30 min

Comment: Mean of three assays. Methanol (<10%) was used to prepare the stock solutions. EC50 values were calculated from nominal concentrations.

Bibliographical reference: Kaiser, K.L.E., and Palabrica, V.S. (1991). *Water Poll. Res. J. Canada* **26**, 361-431.

THIOBENZAMIDE

CAS RN: 2227-79-4

Temperature: 15°C
Test parameter: EC50
Effect: Reduction in light output
Concentration: 30.7 mg/l
Exposure time: 30 min
Comment: Mean of three assays. Methanol (<10%) was used to prepare the stock solutions. EC50 values were calculated from nominal concentrations.

Bibliographical reference: Kaiser, K.L.E., Palabrica, V.S., and Ribo, J.M. (1987). In: *QSAR in Environmental Toxicology - II*, K.L.E. Kaiser (ed.), D. Reidel Publishing Company, Dordrecht, p. 153-168.

THIOISONICOTINAMIDE

CAS RN: 2196-13-6

Sample purity: 98%
Temperature: 15°C
Test parameter: EC50
Effect: Reduction in light output

Concentration: 4.58 mg/l
Exposure time: 5 min

Concentration: 4.90 mg/l
Exposure time: 15 min

Concentration: 8.92 mg/l
Exposure time: 30 min

Comment: Mean of three assays. Methanol (<10%) was used to prepare the stock solutions. EC50 values were calculated from nominal concentrations.

Bibliographical reference: Kaiser, K.L.E., and Palabrica, V.S. (1991). *Water Poll. Res. J. Canada* **26**, 361-431.

THIONIN ACETATE

CAS RN: 78338-22-4

Temperature: 15°C
Test parameter: EC50
Effect: Reduction in light output

Concentration: 2.08 mg/l
Exposure time: 5 min

Concentration: 0.95 mg/l
Exposure time: 15 min

Concentration: 0.69 mg/l
Exposure time: 30 min

Comment: Mean of three assays. Methanol (<10%) was used to prepare the stock solutions. EC50 values were calculated from nominal concentrations.

Bibliographical reference: Kaiser, K.L.E., and Palabrica, V.S. (1991). *Water Poll. Res. J. Canada* **26**, 361-431.

THIOPHENE

CAS RN: 110-02-1

Sample purity: >99%
Temperature: 15°C
Test parameter: EC50
Effect: Reduction in light output

Concentration: 150 mg/l
Exposure times: 5 and 15 min

Concentration: 180 mg/l
Exposure time: 30 min

Comment: Mean of four assays. Methanol (<10%) was used to prepare the stock solutions. EC50 values were calculated from nominal concentrations.

Bibliographical reference: Kaiser, K.L.E., and Palabrica, V.S. (1991). *Water Poll. Res. J. Canada* **26**, 361-431.

2-THIOPHENECARBOXYLIC ACID SODIUM SALT

CAS RN: 25112-68-9

Sample purity: 98%
Temperature: 15°C
Test parameter: EC50
Effect: Reduction in light output

Concentration: 598 mg/l

Exposure time: 5 min

Concentration: 612 mg/l
Exposure times: 15 and 30 min

Comment: Mean of three assays. Methanol (<10%) was used to prepare the stock solutions. EC50 values were calculated from nominal concentrations.

Bibliographical reference: Kaiser, K.L.E., and Palabrica, V.S. (1991). *Water Poll. Res. J. Canada* **26**, 361-431.

THIOPHENOL

CAS RN: 108-98-5

Temperature: 15 ± 0.1°C
Test parameter: EC50
Effect: Reduction in light output
Concentration: 4.8 mg/l
Exposure time: 5 min

Bibliographical reference: Somasundaram, L., Coats, J.R., Racke, K.D., and Stahr, H.M. (1990). *Bull. Environ. Contam. Toxicol.* **44**, 254-259.

Sample purity: 99%
Temperature: 15°C
Test parameter: EC50
Effect: Reduction in light output

Concentration: 1.27 mg/l
Exposure time: 5 min

Concentration: 0.86 mg/l
Exposure time: 15 min

Concentration: 0.88 mg/l*
Exposure time: 30 min

Comment: Mean of two assays. Methanol (<10%) was used to prepare the stock solutions. EC50 values were calculated from nominal concentrations.

Bibliographical references: Kaiser, K.L.E., and Palabrica, V.S. (1991). *Water Poll. Res. J. Canada* **26**, 361-431.
* Kaiser, K.L.E., Palabrica, V.S., and Ribo, J.M. (1987). In: *QSAR in Environmental Toxicology - II*, K.L.E. Kaiser (ed.), D. Reidel Publishing Company, Dordrecht, p. 153-168.

THIOUREA

CAS RN: 62-56-6

Sample purity: ≥97%
Test parameter: EC50
Effect: Reduction in light output
Concentration: 3395 mg/l
Exposure time: 15 min
Comment: EC50 value converted to mg/l from the original data expressed in log(1/μmol l^{-1}) and a rounded molecular weight of 76 given by the authors.

Bibliographical reference: Govers, H., Ruepert, C., Stevens, T., and van Leeuwen, C.J. (1986). *Chemosphere* **15**, 383-393.

THIRAM

CAS RN: 137-26-8
Synonym: Tetramethylthioperoxydicarbonic diamide $[(Me_2N)C(S)]_2S_2$

Temperature: 15°C
Test parameter: EC50
Effect: Reduction in light output

Concentration: 0.45 ± 0.11 mg/l
Exposure time: 5 min
Comment: Mean of four assays. Chemical dissolved in DMSO (0.1%). Toxicity values were calculated from nominal concentrations.

Concentration: 0.32 ± 0.06 mg/l
Exposure time: 10 min
Comment: Mean of three assays. Chemical dissolved in DMSO (0.1%). Toxicity values were calculated from nominal concentrations.

Bibliographical reference: Ferard, J.F., Vasseur, P., Danoux, L., and Larbaigt, G. (1983). *Rev. Fr. Sci. Eau* **2**, 221-237.

Sample purity: ≥98%
Test parameter: EC50
Effect: Reduction in light output
Concentration: 0.10 mg/l (0.08-0.11)
Exposure time: 15 min

Bibliographical reference: van Leeuwen, C.J., Maas-Diepeveen, J.L., Niebeek, G., Vergouw, W.H.A., Griffioen, P.S., and Luijken, M.W. (1985). *Aquat. Toxicol.* **7**, 145-164.

TIRON

CAS RN: 149-45-1
Synonym: 4,5-Dihydroxy-1,3-benzenedisulfonic acid disodium salt monohydrate

Temperature: 15°C
Test parameter: EC50
Effect: Reduction in light output

Concentration: 480 mg/l
Exposure time: 5 min

Concentration: 469 mg/l
Exposure time: 15 min

Concentration: 816 mg/l
Exposure time: 30 min

Comment: Mean of three assays. Methanol (<10%) was used to prepare the stock solutions. EC50 values were calculated from nominal concentrations.

Bibliographical reference: Kaiser, K.L.E., and Palabrica, V.S. (1991). *Water Poll. Res. J. Canada* **26**, 361-431.

4-TOLUAMIDE

CAS RN: 619-55-6
Synonym: 4-Methylbenzamide

Sample purity: 99%
Temperature: 15°C
Test parameter: EC50
Effect: Reduction in light output

Concentration: 34.7 mg/l
Exposure time: 5 min

Concentration: 35.6 mg/l
Exposure times: 15 and 30 min

Comment: Mean of four assays. Methanol (<10%) was used to prepare the stock solutions. EC50 values were calculated from nominal concentrations.

Bibliographical reference: Kaiser, K.L.E., and Palabrica, V.S. (1991). *Water Poll. Res. J. Canada* **26**, 361-431.

TOLUENE

CAS RN: 108-88-3
Synonym: Methylbenzene

Temperature: 15 ± 0.1°C
Test parameter: EC50
Effect: Reduction in light output
Concentration: 50 mg/l
Exposure time: 5 min

Bibliographical reference: Chang, J.C., Taylor, P.B., and Leach, F.R. (1981). *Bull. Environ. Contam. Toxicol.* **26**, 150-156.

Temperature: 15°C
Test parameter: EC50
Effect: Reduction in light output
Concentration: 48 mg/l
Exposure time: 5 min

Bibliographical reference: Samak, Q.M., and Noiseux, R. (1981). *Can. Tech. Rep. Fish. Aquat. Sci.* **990**, 288-308.

Temperature: 15°C
Test parameter: EC50
Effect: Reduction in light output

Concentration: 218 mg/l
Exposure time: 5 min

Concentration: 216 mg/l
Exposure time: 10 min

Comment: Mean of two assays. Toxicity values were calculated from nominal concentrations.

Bibliographical reference: Ferard, J.F., Vasseur, P., Danoux, L., and Larbaigt, G. (1983). *Rev. Fr. Sci. Eau* **2**, 221-237.

Temperature: 15°C
Test parameter: EC50
Effect: Reduction in light output
Concentration: 18.0 mg/l
Exposure time: 15 min

Bibliographical reference: Hermens, J., Busser, F., Leeuwangh, P., and Musch, A. (1985). *Ecotoxicol. Environ. Safety* **9**, 17-25.

Temperature: 15°C
Test parameter: EC50
Effect: Reduction in light output
Concentration: 23.1 mg/l
Exposure time: 30 min
Comment: Mean of three assays. Methanol (<10%) was used to prepare the stock solutions. EC50 values were calculated from nominal concentrations.

Bibliographical reference: Kaiser, K.L.E., Palabrica, V.S., and Ribo, J.M. (1987). In: *QSAR in Environmental Toxicology - II*, K.L.E. Kaiser (ed.), D. Reidel Publishing Company, Dordrecht, p. 153-168.

Sample purity: Reagent grade

Test parameter: EC50
Effect: Reduction in light output

Concentration: 17 mg/l (16.5-17.5)
Exposure time: 5 min

Concentration: 19.7 mg/l (19.0-20.4)
Exposure times: 15 and 30 min

Bibliographical reference: Elnabarawy, M.T., Robideau, R.R., and Beach, S.A. (1988). *Tox. Assess.* **3**, 361-370.

Sample purity: 99.5%
Temperature: 15°C
Test parameter: EC50
Effect: Reduction in light output

Concentration: 23 mg/l
Exposure time: 5 min

Concentration: 28 mg/l
Exposure time: 15 min

Bibliographical reference: Cronin, M.T.D. (1993). Liverpool John Moores University, UK, private communication.

Temperature: 15°C
Test parameter: EC50
Effect: Reduction in light output
Concentration: 33 mg/l
Exposure time: 5 min

Bibliographical reference: Kahru, A. (1993). *ATLA* **21**, 210-215.

TOLUENE-4-SULFONAMIDE

CAS RN: 70-55-3

Sample purity: 99+%
Temperature: 15°C
Test parameter: EC50
Effect: Reduction in light output

Concentration: 43.0 mg/l
Exposure time: 5 min

Concentration: 49.4 mg/l
Exposure time: 15 min

Concentration: 56.7 mg/l
Exposure time: 30 min

Comment: Mean of three assays. Methanol (<10%) was used to prepare the stock solutions. EC50 values were calculated from nominal concentrations.

Bibliographical reference: Kaiser, K.L.E., and Palabrica, V.S. (1991). *Water Poll. Res. J. Canada* **26**, 361-431.

TOLUENE-4-SULFONYL CHLORIDE

CAS RN: 98-59-9

Sample purity: 99+%
Temperature: 15°C
Test parameter: EC50
Effect: Reduction in light output

Concentration: 2.00 mg/l
Exposure time: 5 min

Concentration: 2.29 mg/l
Exposure time: 15 min

Concentration: 2.40 mg/l
Exposure time: 30 min

Comment: Mean of three assays. Methanol (<10%) was used to prepare the stock solutions. EC50 values were calculated from nominal concentrations.

Bibliographical reference: Kaiser, K.L.E., and Palabrica, V.S. (1991). *Water Poll. Res. J. Canada* **26**, 361-431.

o-TOLUIDINE

CAS RN: 95-53-4

Temperature: 15°C
Test parameter: EC50
Effect: Reduction in light output
Concentration: 13.2 mg/l
Exposure time: 30 min
Comment: Mean of three assays. Methanol (<10%) was used to prepare the stock solutions. EC50 values were calculated from nominal concentrations.

Bibliographical reference: Kaiser, K.L.E., and Palabrica, V.S. (1991). *Water Poll. Res. J. Canada* **26**, 361-431.

m-TOLUIDINE

CAS RN: 108-44-1

Temperature: 15°C
Test parameter: EC50
Effect: Reduction in light output
Concentration: 11.7 mg/l
Exposure time: 30 min
Comment: Mean of three assays. Methanol (<10%) was used to prepare the stock solutions. EC50 values were calculated from nominal concentrations.

Bibliographical reference: Kaiser, K.L.E., and Palabrica, V.S. (1991). *Water Poll. Res. J. Canada* **26**, 361-431.

p-TOLUIDINE

CAS RN: 106-49-0

Temperature: 15°C
Test parameter: EC50
Effect: Reduction in light output
Concentration: 8.13 mg/l
Exposure time: 30 min

Comment: Mean of three assays. Methanol (<10%) was used to prepare the stock solutions. EC50 values were calculated from nominal concentrations.

Bibliographical reference: Kaiser, K.L.E. (1987). In: *QSAR in Environmental Toxicology - II*, K.L.E. Kaiser (ed.), D. Reidel Publishing Company, Dordrecht, p. 169-188.

4-TOLUNITRILE

CAS RN: 104-85-8

Sample purity: 98%
Temperature: 15°C
Test parameter: EC50
Effect: Reduction in light output

Concentration: 3.38 mg/l
Exposure time: 5 min

Concentration: 3.88 mg/l
Exposure time: 15 min

Concentration: 4.56 mg/l*
Exposure time: 30 min

Comment: Mean of four assays. Methanol (<10%) was used to prepare the stock solutions. EC50 values were calculated from nominal concentrations.

Bibliographical references: Kaiser, K.L.E., and Palabrica, V.S. (1991). *Water Poll. Res. J. Canada* **26**, 361-431.
* Kaiser, K.L.E., and Gough, K.M. (1989). In: *Aquatic Toxicology and Environmental Fate: Eleventh Volume, ASTM STP 1007*, G.W. Suter and M.A. Lewis (eds.), American Society for Testing and Materials, Philadelphia, p. 424-441.

4-TOLUOYL CHLORIDE

CAS RN: 874-60-2

Sample purity: 98%

Temperature: 15°C
Test parameter: EC50
Effect: Reduction in light output

Concentration: 2.88 mg/l
Exposure time: 5 min

Concentration: 3.23 mg/l
Exposure times: 15 and 30 min

Comment: Mean of three assays. Methanol (<10%) was used to prepare the stock solutions. EC50 values were calculated from nominal concentrations.

Bibliographical reference: Kaiser, K.L.E., and Palabrica, V.S. (1991). *Water Poll. Res. J. Canada* **26**, 361-431.

TOLYLENE 2,6-DIISOCYANATE

CAS RN: 91-08-7
Synonym: 2-Methyl-1,3-phenylene diisocyanate

Sample purity: 97%
Temperature: 15°C
Test parameter: EC50
Effect: Reduction in light output

Concentration: 48.0 mg/l
Exposure time: 5 min

Concentration: 44.8 mg/l
Exposure time: 15 min

Concentration: 41.8 mg/l
Exposure time: 30 min

Comment: Mean of four assays. Methanol (<10%) was used to prepare the stock solutions. EC50 values were calculated from nominal concentrations.

Bibliographical reference: Kaiser, K.L.E., and Palabrica, V.S. (1991). *Water Poll. Res. J. Canada* **26**, 361-431.

2-(4-TOLYL)ETHYLAMINE

CAS RN: 3261-62-9
Synonym: 4-Methylphenethylamine

Sample purity: 97%
Temperature: 15°C
Test parameter: EC50
Effect: Reduction in light output

Concentration: 7.10 mg/l
Exposure time: 5 min

Concentration: 3.81 mg/l
Exposure time: 15 min

Concentration: 3.03 mg/l
Exposure time: 30 min

Comment: Mean of three assays. Methanol (<10%) was used to prepare the stock solutions. EC50 values were calculated from nominal concentrations.

Bibliographical reference: Kaiser, K.L.E., and Palabrica, V.S. (1991). *Water Poll. Res. J. Canada* **26**, 361-431.

4-TOLYL ISOCYANATE

CAS RN: 622-58-2

Sample purity: 99%
Temperature: 15°C
Test parameter: EC50
Effect: Reduction in light output

Concentration: 3.50 mg/l
Exposure time: 5 min

Concentration: 3.34 mg/l
Exposure time: 15 min

Concentration: 3.05 mg/l
Exposure time: 30 min

Comment: Mean of three assays. Methanol (<10%) was used to prepare the stock solutions. EC50 values were calculated from nominal concentrations.

Bibliographical reference: Kaiser, K.L.E., and Palabrica, V.S. (1991). *Water Poll. Res. J. Canada* **26**, 361-431.

TRIADINE 10

CAS RN: 72103-18-5

Sample purity: Hexahydro-1,3,5-tris(2-hydroxyethyl)-*s*-triazine (63.6%), sodium-2-pyridinethiol-1-oxide (6.4%)
Temperature: 15°C
Test parameter: EC50
Effect: Reduction in light output
Concentration: 0.6 mg/l
Exposure time: 5 min
Comment: Chemical was tested in duplicate at 0.1, 0.2, 0.5, and 1 mg/l. Verification that the controls were valid was accomplished by making sure that the difference between the duplicate blank ratios (5 minute light reading divided by the time 0 reading) was not greater than 0.02, otherwise the test was not continued. The 15-min EC50 could not be determined due to a rapid decrease in light output at concentrations above 0.1 mg/l triadine 10.

Bibliographical reference: Mallak, F.P., and Brunker, R.L. (1984). In: *Toxicity Screening Procedures Using Bacterial Systems*, D. Liu and B.J. Dutka (eds.), Marcel Dekker, New York, p. 65-76.

1,2,4-TRIAZOLE

CAS RN: 288-88-0

Sample purity: 98%
Temperature: 15°C
Test parameter: EC50
Effect: Reduction in light output

Concentration: 5000 mg/l
Exposure time: 5 min

Concentration: 5120 mg/l
Exposure time: 15 min

Concentration: 5240 mg/l
Exposure time: 30 min

Comment: Mean of three assays. Methanol (<10%) was used to prepare the stock solutions. EC50 values were calculated from nominal concentrations.

Bibliographical reference: Kaiser, K.L.E., and Palabrica, V.S. (1991). *Water Poll. Res. J. Canada* **26**, 361-431.

2,4,6-TRIBROMOPHENOL

CAS RN: 118-79-6

Test parameter: EC50
Effect: Reduction in light output
Concentration: 2.7 mg/l
Exposure time: 5 min
Comment: The EC50 value was calculated from nominal concentrations.

Bibliographical reference: Curtis, C., Lima, A., Lozano, S.J., and Veith, G.D. (1982). In: *Aquatic Toxicology and Hazard Assessment: Fifth Conference, ASTM STP 766*, J.G. Pearson, R.B. Foster, and W.E. Bishop (eds.), American Society for Testing and Materials, Philadelphia, p. 170-178.

Temperature: 15°C
Test parameter: EC50
Effect: Reduction in light output

Concentration: 6.89 mg/l
Exposure time: 5 min

Concentration: 6.42 mg/l
Exposure time: 30 min

Bibliographical reference: Speece, R. (1987). Drexel University, Philadelphia, USA, private communication.

Temperature: 15°C
Test parameter: EC50
Effect: Reduction in light output
Concentration: 13 mg/l
Exposure time: 15 min
Comment: pH = 7.2.

Bibliographical reference: Neilson, A.H., Allard, A.S., Fischer, S., Malmberg, M., and Viktor, T. (1990). *Ecotoxicol. Environ. Safety* **20**, 82-97.

TRIBUT-1-ENYLTIN BROMIDE

CAS RN: 125112-25-6

Temperature: 15°C
Test parameter: EC50
Effect: Reduction in light output

Concentration: 0.30 ± 0.06 mg/l
Exposure time: 5 min

Concentration: 0.16 ± 0.01 mg/l
Exposure time: 15 min

Comment: Chemical was dissolved in ethanol (~0.05%).

Bibliographical reference: Dooley, C.A., and Kenis, P. (1987). In: *Oceans '87, International Organotin Symposium*, Department of Fisheries and Oceans, Canada, and the Society for Underwater Technology, William MacNab & Son, Halifax, p. 1517-1524.

TRIBUT-3-ENYLTIN BROMIDE

CAS RN: 125112-27-8

Temperature: 15°C
Test parameter: EC50
Effect: Reduction in light output

Concentration: 0.16 ± 0.02 mg/l
Exposure time: 5 min

Concentration: 0.10 ± 0.01 mg/l
Exposure time: 15 min

Comment: Chemical was dissolved in ethanol (~0.05%).

Bibliographical reference: Dooley, C.A., and Kenis, P. (1987). In: *Oceans '87, International Organotin Symposium*, Department of Fisheries and Oceans, Canada, and the Society for Underwater Technology, William MacNab & Son, Halifax, p. 1517-1524.

N,N,N-TRIBUTYL-1-BUTANAMINIUM CHLORIDE

CAS RN: 1112-67-0

Temperature: 15°C
Test parameter: EC50
Effect: Reduction in light output

Concentration: 327 mg/l
Exposure time: 5 min

Concentration: 136 mg/l
Exposure time: 15 min

Concentration: 78.3 mg/l
Exposure time: 30 min

Comment: Mean of three assays. Methanol (<10%) was used to prepare the stock solutions. EC50 values were calculated from nominal concentrations.

Bibliographical reference: Kaiser, K.L.E., and Palabrica, V.S. (1991). *Water Poll. Res. J. Canada* **26**, 361-431.

2,4,6-TRI-*tert*-BUTYLPHENOL

CAS RN: 732-26-3

Sample purity: 96%
Temperature: 15°C
Test parameter: EC50
Effect: Reduction in light output

Concentration: 10.9 mg/l
Exposure time: 5 min

Concentration: 10.4 mg/l
Exposure time: 15 min

Concentration: 12.3 mg/l
Exposure time: 30 min

Comment: Mean of four assays. Methanol (<10%) was used to prepare the stock solutions. EC50 values were calculated from nominal concentrations.

Bibliographical reference: Kaiser, K.L.E., and Palabrica, V.S. (1992). National Water Research Institute, Burlington, Ontario, Canada, unpublished results.

TRIBUTYLTIN BROMIDE

CAS RN: 1461-23-0

Temperature: 15°C
Test parameter: EC50
Effect: Reduction in light output

Concentration: 0.048 ± 0.004 mg/l
Exposure time: 5 min

Concentration: 0.022 ± 0.007 mg/l
Exposure time: 15 min

Comment: Chemical was dissolved in ethanol (~0.05%).

Bibliographical reference: Dooley, C.A., and Kenis, P. (1987). In: *Oceans '87, International Organotin Symposium*, Department of Fisheries and Oceans, Canada, and the Society for Underwater Technology, William MacNab & Son, Halifax, p. 1517-1524.

TRIBUTYLTIN CHLORIDE

CAS RN: 1461-22-9

Temperature: 15°C
Test parameter: EC50
Effect: Reduction in light output

Concentration: 0.02 ± 0.01 mg/l
Exposure time: 5 min

Concentration: 0.007 ± 0.003 mg/l
Exposure time: 15 min

Comment: Chemical was dissolved in ethanol (~0.05%).

Bibliographical reference: Dooley, C.A., and Kenis, P. (1987). In: *Oceans '87, International Organotin Symposium*, Department of Fisheries and Oceans, Canada, and the Society for Underwater Technology, William MacNab & Son, Halifax, p. 1517-1524.

TRIBUTYLTIN LINOLEATE

CAS RN: 24124-25-2

Temperature: 15°C
Test parameter: EC50
Effect: Reduction in light output
Concentration: 0.0159 mg/l
Exposure time: 30 min

Bibliographical reference: Steinhäuser, K.G., Amann, W., Späth, A., and Polenz, A. (1985). *Vom Wasser* **65**, 203-214.

TRICHLOROACETIC ACID

CAS RN: 76-03-9

Temperature: 15°C
Test parameter: EC50
Effect: Reduction in light output
Concentration: 35 mg/l
Exposure time: 15 min

Bibliographical reference: Bulich, A.A., Tung, K.K., and Scheibner, G. (1990). *J. Biolumin. Chemilumin.* **5**, 71-77.

2,3,4-TRICHLOROANILINE

CAS RN: 634-67-3

Temperature: 15°C
Test parameter: EC50
Effect: Reduction in light output

Concentration: 1.88 mg/l
Exposure time: 5 min

Concentration: 2.15 mg/l
Exposure time: 15 min

Concentration: 2.36 mg/l
Exposure time: 30 min

Comment: Mean of three assays. Methanol (<10%) was used to prepare the stock solutions. EC50 values were calculated from nominal concentrations.

Bibliographical reference: Ribo, J.M., and Kaiser, K.L.E. (1984). In: *QSAR in Environmental Toxicology*, K.L.E. Kaiser (ed.), D. Reidel Publishing Company, Dordrecht, p. 319-336.

2,4,5-TRICHLOROANILINE

CAS RN: 636-30-6

Temperature: 15°C
Test parameter: EC50
Effect: Reduction in light output

Concentration: 1.75 mg/l
Exposure time: 5 min

Concentration: 1.63 mg/l
Exposure time: 15 min

Concentration: 1.49 mg/l
Exposure time: 30 min

Comment: Mean of three assays. Methanol (<10%) was used to prepare the stock solutions. EC50 values were calculated from nominal concentrations.

Bibliographical reference: Ribo, J.M., and Kaiser, K.L.E. (1984). In: *QSAR in Environmental Toxicology*, K.L.E. Kaiser (ed.), D. Reidel Publishing Company, Dordrecht, p. 319-336.

2,4,6-TRICHLOROANILINE

CAS RN: 634-93-5

Temperature: 15°C
Test parameter: EC50
Effect: Reduction in light output

Concentration: 4.30 mg/l
Exposure time: 5 min

Concentration: 4.40 mg/l
Exposure time: 15 min

Concentration: 4.61 mg/l
Exposure time: 30 min

Comment: Mean of three assays. Methanol (<10%) was used to prepare the stock solutions. EC50 values were calculated from nominal concentrations.

Bibliographical reference: Ribo, J.M., and Kaiser, K.L.E. (1984). In: *QSAR in Environmental Toxicology*, K.L.E. Kaiser (ed.), D. Reidel Publishing Company, Dordrecht, p. 319-336.

3,4,5-TRICHLOROANILINE

CAS RN: 634-91-3

Temperature: 15°C
Test parameter: EC50
Effect: Reduction in light output

Concentration: 3.34 mg/l

Exposure time: 5 min

Concentration: 4.01 mg/l
Exposure time: 15 min

Concentration: 3.34 mg/l
Exposure time: 30 min

Comment: Mean of three assays. Methanol (<10%) was used to prepare the stock solutions. EC50 values were calculated from nominal concentrations.

Bibliographical reference: Ribo, J.M., and Kaiser, K.L.E. (1984). In: *QSAR in Environmental Toxicology*, K.L.E. Kaiser (ed.), D. Reidel Publishing Company, Dordrecht, p. 319-336.

1,2,3-TRICHLOROBENZENE

CAS RN: 87-61-6

Temperature: 15°C
Test parameter: EC50
Effect: Reduction in light output

Concentration: 1.86 mg/l
Exposure time: 5 min

Concentration: 2.62 mg/l
Exposure time: 15 min

Concentration: 3.15 mg/l
Exposure time: 30 min

Comment: Mean of three assays. Methanol (<10%) was used to prepare the stock solutions. EC50 values were calculated from nominal concentrations.

Bibliographical reference: Ribo, J.M., and Kaiser, K.L.E. (1983). *Chemosphere* **12**, 1421-1442.

Temperature: 15°C
Test parameter: EC50
Effect: Reduction in light output

Concentration: 2.50 mg/l
Exposure time: 15 min

Bibliographical reference: Hermens, J., Busser, F., Leeuwangh, P., and Musch, A. (1985). *Ecotoxicol. Environ. Safety* **9**, 17-25.

1,2,4-TRICHLOROBENZENE

CAS RN: 120-82-1

Temperature: 15°C
Test parameter: EC50
Effect: Reduction in light output

Concentration: 2.34 mg/l
Exposure time: 5 min

Concentration: 3.70 mg/l
Exposure time: 15 min

Concentration: 3.97 mg/l
Exposure time: 30 min

Comment: Mean of three assays. Methanol (<10%) was used to prepare the stock solutions. EC50 values were calculated from nominal concentrations.

Bibliographical reference: Ribo, J.M., and Kaiser, K.L.E. (1983). *Chemosphere* **12**, 1421-1442.

Temperature: 15°C
Test parameter: EC50
Effect: Reduction in light output
Concentration: 3 mg/l
Exposure time: 10 min
Comment: Mean of three assays. Chemical was dissolved in 0.1 ml/l acetone. Test solutions were analyzed using gas chromatography.

Bibliographical reference: Bazin, C., Chambon, P., Bonnefille, M., and Larbaigt, G. (1987). *Sci. Eau* **6**, 403-413.

Sample purity: 99+%

Temperature: 15°C
Test parameter: EC50
Effect: Reduction in light output

Concentration: 1.6 mg/l
Exposure time: 5 min

Concentration: 1.9 mg/l
Exposure time: 15 min

Comment: Chemical was prepared in an initial solution of 2% methanol.

Bibliographical reference: Cronin, M.T.D. (1993). Liverpool John Moores University, UK, private communication.

1,3,5-TRICHLOROBENZENE

CAS RN: 108-70-3

Temperature: 15°C
Test parameter: EC50
Effect: Reduction in light output

Concentration: 12.8 mg/l
Exposure time: 5 min

Concentration: 14.1 mg/l
Exposure times: 15 and 30 min

Comment: Mean of three assays. Methanol (<10%) was used to prepare the stock solutions. EC50 values were calculated from nominal concentrations.

Bibliographical reference: Ribo, J.M., and Kaiser, K.L.E. (1983). *Chemosphere* **12**, 1421-1442.

TRICHLOROCYANURIC ACID

CAS RN: 87-90-1

Temperature: 15°C

Test parameter: EC50
Effect: Reduction in light output

Concentration: 0.45 mg/l
Exposure time: 5 min

Concentration: 0.53 mg/l
Exposure time: 15 min

Concentration: 0.63 mg/l
Exposure time: 30 min

Comment: Mean of three assays. Methanol (<10%) was used to prepare the stock solutions. EC50 values were calculated from nominal concentrations.

Bibliographical reference: Kaiser, K.L.E., and Palabrica, V.S. (1991). *Water Poll. Res. J. Canada* **26**, 361-431.

TRICHLOROETHANE

CAS RN: 25323-89-1

Test parameter: EC50
Effect: Reduction in light output
Concentration: 105 mg/l
Exposure time: 5 min
Comment: Isomer not specified. Concentrations in the test were measured.

Bibliographical reference: Curtis, C., Lima, A., Lozano, S.J., and Veith, G.D. (1982). In: *Aquatic Toxicology and Hazard Assessment: Fifth Conference, ASTM STP 766*, J.G. Pearson, R.B. Foster, and W.E. Bishop (eds.), American Society for Testing and Materials, Philadelphia, p. 170-178.

1,1,1-TRICHLOROETHANE

CAS RN: 71-55-6

Temperature: 15°C
Test parameter: EC50

Effect: Reduction in light output
Concentration: 8.04 mg/l
Exposure time: 15 min

Bibliographical reference: Hermens, J., Busser, F., Leeuwangh, P., and Musch, A. (1985). *Ecotoxicol. Environ. Safety* **9**, 17-25.

1,1,2-TRICHLOROETHANE

CAS RN: 79-00-5

Temperature: 15°C
Test parameter: EC50
Effect: Reduction in light output

Concentration: 106 mg/l
Exposure time: 5 min

Concentration: 175 mg/l
Exposure time: 30 min

Bibliographical reference: Speece, R. (1987). Drexel University, Philadelphia, USA, private communication.

2,2,2-TRICHLOROETHANOL

CAS RN: 115-20-8

Test parameter: EC50
Effect: Reduction in light output
Concentration: 1800 mg/l
Exposure time: 5 min
Comment: The EC50 value was calculated from nominal concentrations.

Bibliographical reference: Curtis, C., Lima, A., Lozano, S.J., and Veith, G.D. (1982). In: *Aquatic Toxicology and Hazard Assessment: Fifth Conference, ASTM STP 766*, J.G. Pearson, R.B. Foster, and W.E. Bishop (eds.), American Society for Testing and Materials, Philadelphia, p. 170-178.

Temperature: 15°C
Test parameter: EC50
Effect: Reduction in light output
Concentration: 43.9 mg/l (36.9-52.2)
Exposure time: 15 min
Comment: Single batches of Microtox® reagent were used for less than 2 h before being discarded. Four concentrations were tested. Concentrations were unmeasured.

Bibliographical reference: Nacci, D., Jackim, E., and Walsh, R. (1986). *Environ. Toxicol. Chem.* **5**, 521-525.

Temperature: 15°C
Test parameter: EC50
Effect: Reduction in light output

Concentration: 52.7 mg/l
Exposure time: 5 min

Concentration: 66.7 mg/l
Exposure time: 30 min

Bibliographical reference: Speece, R. (1987). Drexel University, Philadelphia, USA, private communication.

TRICHLOROETHYLENE

CAS RN: 79-01-6

Sample purity: >98%
Temperature: 15 ± 0.1°C
Test parameter: EC50
Effect: Reduction in light output

Concentrations: 156 mg/l
164 mg/l
Exposure time: 5 min

Concentrations: 115 mg/l
118 mg/l
Exposure time: 15 min

Comment: Test was performed in duplicate. Toxicity values were based on nominal test concentrations.

Bibliographical reference: de Zwart, D., and Slooff, W. (1983). *Aquat. Toxicol.* **4**, 129-138.

Sample purity: >98%
Temperature: 15 ± 0.1°C
Test parameter: EC10
Effect: Reduction in light output

Concentration: 87 mg/l
Exposure time: 5 min

Concentration: 75 mg/l
Exposure time: 15 min

Comment: Test was performed in duplicate. Toxicity values were based on nominal test concentrations.

Bibliographical reference: de Zwart, D., and Slooff, W. (1983). *Aquat. Toxicol.* **4**, 129-138.

Temperature: 20°C

Test parameter: EC20
Effect: Reduction in light output
Concentration: 42.75 mg/l (38.07-48.02)
Exposure time: 5 min

Test parameter: EC50
Effect: Reduction in light output
Concentration: 97.82 mg/l (88.97-107.54)
Exposure time: 5 min

Bibliographical reference: Casseri, N.A., Ying, W, and Sojka, S.A. (1983). Proceedings 38th Industrial Waste Conference, USA, May 1983, p. 867-878.

Temperature: 20°C

Test parameter: EC20
Effect: Reduction in light output

Concentration: 50.53 mg/l
Exposure time: 15 min

Test parameter: EC50
Effect: Reduction in light output
Concentration: 117.14 mg/l
Exposure time: 15 min

Bibliographical reference: Casseri, N.A., Ying, W, and Sojka, S.A. (1983). Proceedings 38th Industrial Waste Conference, USA, May 1983, p. 867-878.

Temperature: 15°C
Test parameter: EC50
Effect: Reduction in light output
Concentration: 190 mg/l
Exposure time: 15 min

Bibliographical reference: Hermens, J., Busser, F., Leeuwangh, P., and Musch, A. (1985). *Ecotoxicol. Environ. Safety* **9**, 17-25.

Temperature: 20°C
Test parameter: EC50
Effect: Reduction in light output

Concentration: 96.84 mg/l (88.08-106.46)
Exposure time: 5 min

Concentration: 115.97 mg/l (102.19-131.62)
Exposure time: 15 min

Bibliographical reference: Casseri, N.A. (1985). Occidental Chemical Corporation, Grand Island, New York, USA, private communication.

Temperature: 15°C
Test parameter: EC50
Effect: Reduction in light output
Concentration: 602 mg/l
Exposure time: 10 min
Comment: Mean of five assays. Test solutions were analyzed using gas chromatography.

Bibliographical reference: Bazin, C., Chambon, P., Bonnefille, M., and Larbaigt, G. (1987). *Sci. Eau* **6**, 403-413.

Temperature: 15°C
Test parameter: EC50
Effect: Reduction in light output

Concentration: 964 mg/l
Exposure time: 5 min

Concentration: 1070 mg/l
Exposure time: 30 min

Bibliographical reference: Speece, R. (1987). Drexel University, Philadelphia, USA, private communication.

3,4,5-TRICHLOROGUAIACOL

CAS RN: 57057-83-7
Synonym: 3,4,5-Trichloro-2-methoxyphenol

Temperature: 15°C
Test parameter: EC50
Effect: Reduction in light output

Concentration: 1.50 mg/l
Exposure time: 5 min

Concentration: 1.25 mg/l
Exposure times: 15 and 30 min

Comment: Mean of three assays. Methanol (<10%) was used to prepare the stock solutions. EC50 values were calculated from nominal concentrations.

Bibliographical reference: Kaiser, K.L.E., and Palabrica, V.S. (1992). National Water Research Institute, Burlington, Ontario, Canada, unpublished results.

4,5,6-TRICHLOROGUAIACOL

CAS RN: 2668-24-8
Synonym: 2,3,4-Trichloro-6-methoxyphenol

Sample purity: >98%
Temperature: 15°C
Test parameter: EC50
Effect: Reduction in light output
Concentration: 11 mg/l
Exposure time: 15 min
Comment: pH = 7.2.

Bibliographical reference: Neilson, A.H., Allard, A.S., Fischer, S., Malmberg, M., and Viktor, T. (1990). *Ecotoxicol. Environ. Safety* **20**, 82-97.

Temperature: 15°C
Test parameter: EC50
Effect: Reduction in light output

Concentration: 5.58 mg/l
Exposure time: 5 min

Concentration: 5.71 mg/l
Exposure time: 15 min

Concentration: 6.56 mg/l
Exposure time: 30 min

Comment: Mean of three assays. Methanol (<10%) was used to prepare the stock solutions. EC50 values were calculated from nominal concentrations.

Bibliographical reference: Kaiser, K.L.E., and Palabrica, V.S. (1992). National Water Research Institute, Burlington, Ontario, Canada, unpublished results.

2,5,6-TRICHLOROHEXYL-1',3'-DICHLOROISOPROPYL ETHER

CAS RN: 134984-68-2

Test parameter: EC50
Effect: Reduction in light output
Concentration: 5.1 mg/l (4.6-5.7)
Exposure time: 5 min

Bibliographical reference: Dorn, P.B., van Compernolle, R., Meyer, C.L., and Crossland, N.O. (1991). *Environ. Toxicol. Chem.* **10**, 691-703.

2,5,6-TRICHLOROHEXYL-2',3'-DICHLOROPROPYL ETHER

CAS RN: 134984-69-3

Test parameter: EC50
Effect: Reduction in light output
Concentration: 1.5 mg/l (1.3-1.8)
Exposure time: 5 min

Bibliographical reference: Dorn, P.B., van Compernolle, R., Meyer, C.L., and Crossland, N.O. (1991). *Environ. Toxicol. Chem.* **10**, 691-703.

2,3,4-TRICHLORONITROBENZENE

CAS RN: 17700-09-3
Synonym: 1,2,3-Trichloro-4-nitrobenzene

Temperature: 15°C
Test parameter: EC50
Effect: Reduction in light output

Concentration: 2.60 mg/l
Exposure time: 5 min

Concentration: 2.72 mg/l
Exposure time: 15 min

Concentration: 2.85 mg/l*
Exposure time: 30 min

Comment: Mean of three assays. Methanol (<10%) was used to prepare the stock solutions. EC50 values were calculated from nominal concentrations.

Bibliographical references: Kaiser, K.L.E., and Palabrica, V.S. (1991). *Water Poll. Res. J. Canada* **26**, 361-431.
* Kaiser, K.L.E., and Ribo, J.M. (1985). In: *QSAR in Toxicology and Xenobiochemistry*, M. Tichy (ed.), Elsevier, Amsterdam, p. 27-38.

2,4,5-TRICHLORONITROBENZENE

CAS RN: 89-69-0
Synonym: 1,2,4-Trichloro-5-nitrobenzene

Temperature: 15°C
Test parameter: EC50
Effect: Reduction in light output

Concentration: 8.03 mg/l
Exposure time: 5 min

Concentration: 7.00 mg/l
Exposure time: 15 min

Concentration: 5.31 mg/l*
Exposure time: 30 min

Comment: Mean of three assays. Methanol (<10%) was used to prepare the stock solutions. EC50 values were calculated from nominal concentrations.

Bibliographical references: Kaiser, K.L.E., and Palabrica, V.S. (1991). *Water Poll. Res. J. Canada* **26**, 361-431.
* Kaiser, K.L.E., and Ribo, J.M. (1985). In: *QSAR in Toxicology and Xenobiochemistry*, M. Tichy (ed.), Elsevier, Amsterdam, p. 27-38.

2,4,6-TRICHLORONITROBENZENE

CAS RN: 18708-70-8
Synonym: 1,3,5-Trichloro-2-nitrobenzene

Temperature: 15°C

Test parameter: EC50
Effect: Reduction in light output

Concentration: 0.72 mg/l
Exposure time: 5 min

Concentration: 0.79 mg/l
Exposure time: 15 min

Concentration: 0.88 mg/l*
Exposure time: 30 min

Comment: Mean of three assays. Methanol (<10%) was used to prepare the stock solutions. EC50 values were calculated from nominal concentrations.

Bibliographical references: Kaiser, K.L.E., and Palabrica, V.S. (1991). *Water Poll. Res. J. Canada* **26**, 361-431.
* Kaiser, K.L.E., and Ribo, J.M. (1985). In: *QSAR in Toxicology and Xenobiochemistry*, M. Tichy (ed.), Elsevier, Amsterdam, p. 27-38.

2,3,4-TRICHLOROPHENOL

CAS RN: 15950-66-0

Temperature: 15°C
Test parameter: EC50
Effect: Reduction in light output

Concentration: 1.76 mg/l
Exposure time: 5 min

Concentration: 1.60 mg/l
Exposure time: 15 min

Concentration: 1.25 mg/l
Exposure time: 30 min

Comment: Mean of three assays. Methanol (<10%) was used to prepare the stock solutions. EC50 values were calculated from nominal concentrations.

Bibliographical reference: Ribo, J.M., and Kaiser, K.L.E. (1983). *Chemosphere* **12**, 1421-1442.

2,3,5-TRICHLOROPHENOL

CAS RN: 933-78-8

Temperature: 15°C
Test parameter: EC50
Effect: Reduction in light output

Concentration: 1.76 mg/l
Exposure time: 5 min

Concentration: 1.37 mg/l
Exposure time: 15 min

Concentration: 1.11 mg/l
Exposure time: 30 min

Comment: Mean of three assays. Methanol (<10%) was used to prepare the stock solutions. EC50 values were calculated from nominal concentrations.

Bibliographical reference: Ribo, J.M., and Kaiser, K.L.E. (1983). *Chemosphere* **12**, 1421-1442.

Temperature: 15°C
Test parameter: EC50
Effect: Reduction in light output

Concentration: 1.90 mg/l
Exposure time: 5 min

Concentration: 1.45 mg/l
Exposure time: 30 min

Bibliographical reference: Speece, R. (1987). Drexel University, Philadelphia, USA, private communication.

2,3,6-TRICHLOROPHENOL

CAS RN: 933-75-5

Temperature: 15°C
Test parameter: EC50

Effect: Reduction in light output

Concentration: 14.0 mg/l
Exposure time: 5 min

Concentration: 13.3 mg/l
Exposure time: 15 min

Concentration: 12.7 mg/l
Exposure time: 30 min

Comment: Mean of three assays. Methanol (<10%) was used to prepare the stock solutions. EC50 values were calculated from nominal concentrations.

Bibliographical reference: Ribo, J.M., and Kaiser, K.L.E. (1983). *Chemosphere* **12**, 1421-1442.

Temperature: 15°C
Test parameter: EC50
Effect: Reduction in light output

Concentration: 21.1 mg/l
Exposure time: 5 min

Concentration: 21.0 mg/l
Exposure time: 30 min

Bibliographical reference: Speece, R. (1987). Drexel University, Philadelphia, USA, private communication.

2,4,5-TRICHLOROPHENOL

CAS RN: 95-95-4

Temperature: 15°C
Test parameter: EC50
Effect: Reduction in light output

Concentration: 1.19 mg/l
Exposure time: 5 min

Concentration: 1.22 mg/l

Exposure time: 15 min

Concentration: 1.27 mg/l
Exposure time: 30 min

Comment: Mean of three assays. Methanol (<10%) was used to prepare the stock solutions. EC50 values were calculated from nominal concentrations.

Bibliographical reference: Ribo, J.M., and Kaiser, K.L.E. (1983). *Chemosphere* **12**, 1421-1442.

Temperature: 15 ± 0.1°C
Test parameter: EC50
Effect: Reduction in light output
Concentration: 1.8 mg/l
Exposure time: 5 min

Bibliographical reference: Somasundaram, L., Coats, J.R., Racke, K.D., and Stahr, H.M. (1990). *Bull. Environ. Contam. Toxicol.* **44**, 254-259.

Temperature: 15°C
Test parameter: EC50
Effect: Reduction in light output
Concentration: 1.2 mg/l
Exposure time: 15 min
Comment: pH = 7.2.

Bibliographical reference: Neilson, A.H., Allard, A.S., Fischer, S., Malmberg, M., and Viktor, T. (1990). *Ecotoxicol. Environ. Safety* **20**, 82-97.

Test parameter: EC50
Effect: Reduction in light output
Concentration: 1.18 mg/l
Exposure times: 5 and 15 min

Bibliographical reference: Ribo, J.M., and Rogers, F. (1990). *Tox. Assess.* **5**, 135-152.

2,4,6-TRICHLOROPHENOL

CAS RN: 88-06-2

Test parameter: EC50
Effect: Reduction in light output
Concentration: 7.2 mg/l
Exposure time: 5 min
Comment: The EC50 value was calculated from nominal concentrations.

Bibliographical reference: Curtis, C., Lima, A., Lozano, S.J., and Veith, G.D. (1982). In: *Aquatic Toxicology and Hazard Assessment: Fifth Conference, ASTM STP 766*, J.G. Pearson, R.B. Foster, and W.E. Bishop (eds.), American Society for Testing and Materials, Philadelphia, p. 170-178.

Temperature: 15°C
Test parameter: EC50
Effect: Reduction in light output

Concentration: 5.96 mg/l
Exposure time: 5 min

Concentration: 8.23 mg/l
Exposure time: 15 min

Concentration: 7.68 mg/l
Exposure time: 30 min

Comment: Mean of three assays. Methanol (<10%) was used to prepare the stock solutions. EC50 values were calculated from nominal concentrations.

Bibliographical reference: Ribo, J.M., and Kaiser, K.L.E. (1983). *Chemosphere* **12**, 1421-1442.

Temperature: 15°C
Test parameter: EC50
Effect: Reduction in light output

Concentration: 13.1 mg/l
Exposure time: 5 min

Concentration: 12.3 mg/l
Exposure time: 30 min

Bibliographical reference: Speece, R. (1987). Drexel University, Philadelphia, USA, private communication.

Temperature: 15°C
Test parameter: EC50
Effect: Reduction in light output
Concentration: 23 mg/l
Exposure time: 15 min
Comment: pH = 7.2.

Bibliographical reference: Neilson, A.H., Allard, A.S., Fischer, S., Malmberg, M., and Viktor, T. (1990). *Ecotoxicol. Environ. Safety* **20**, 82-97.

3,4,5-TRICHLOROPHENOL

CAS RN: 609-19-8

Temperature: 15°C
Test parameter: EC50
Effect: Reduction in light output

Concentration: 0.44 mg/l
Exposure time: 5 min

Concentration: 0.38 mg/l
Exposure time: 15 min

Concentration: 0.36 mg/l
Exposure time: 30 min

Comment: Mean of three assays. Methanol (<10%) was used to prepare the stock solutions. EC50 values were calculated from nominal concentrations.

Bibliographical reference: Ribo, J.M., and Kaiser, K.L.E. (1983). *Chemosphere* **12**, 1421-1442.

Sample purity: >98%

Temperature: 15°C
Test parameter: EC50
Effect: Reduction in light output
Concentration: 0.47 mg/l
Exposure time: 15 min
Comment: pH = 7.2.

Bibliographical reference: Neilson, A.H., Allard, A.S., Fischer, S., Malmberg, M., and Viktor, T. (1990). *Ecotoxicol. Environ. Safety* **20**, 82-97.

1,2,3-TRICHLOROPROPANE

CAS RN: 96-18-4

Temperature: 15°C
Test parameter: EC50
Effect: Reduction in light output

Concentration: 18.7 mg/l
Exposure time: 5 min

Concentration: 24.6 mg/l
Exposure time: 30 min

Bibliographical reference: Speece, R. (1987). Drexel University, Philadelphia, USA, private communication.

3,5,6-TRICHLORO-2(1*H*)-PYRIDINONE

CAS RN: 6515-38-4

Temperature: 15 ± 0.1°C
Test parameter: EC50
Effect: Reduction in light output
Concentration: 18.6 mg/l
Exposure time: 5 min

Bibliographical reference: Somasundaram, L., Coats, J.R., Racke, K.D., and Stahr, H.M. (1990). *Bull. Environ. Contam. Toxicol.* **44**, 254-259.

α,α,α-TRICHLOROTOLUENE

CAS RN: 98-07-7

Sample purity: 99+%
Temperature: 15°C
Test parameter: EC50
Effect: Reduction in light output

Concentration: 11.8 mg/l
Exposure time: 5 min

Concentration: 15.9 mg/l
Exposure time: 15 min

Concentration: 17.8 mg/l*
Exposure time: 30 min

Comment: Mean of three assays. Methanol (<10%) was used to prepare the stock solutions. EC50 values were calculated from nominal concentrations.

Bibliographical references: Kaiser, K.L.E., and Palabrica, V.S. (1991). *Water Poll. Res. J. Canada* **26**, 361-431.
* Kaiser, K.L.E., Palabrica, V.S., and Ribo, J.M. (1987). In: *QSAR in Environmental Toxicology - II*, K.L.E. Kaiser (ed.), D. Reidel Publishing Company, Dordrecht, p. 153-168.

1',1',1'-TRICHLORO-*p*-TOLUNITRILE

CAS RN: 2179-45-5

Temperature: 15°C
Test parameter: EC50
Effect: Reduction in light output
Concentration: 0.012 mg/l
Exposure times: 5, 15, and 30 min
Comment: Mean of three assays. Methanol (<10%) was used to prepare the stock solutions. EC50 values were calculated from nominal concentrations.

Bibliographical reference: Kaiser, K.L.E., and Palabrica, V.S. (1991). *Water Poll. Res. J. Canada* **26**, 361-431.

TRIDECANOIC ACID

CAS RN: 638-53-9

Sample purity: 98%
Temperature: 15°C
Test parameter: EC50
Effect: Reduction in light output

Concentration: 4.60 ± 1.69 mg/l
Exposure time: 5 min

Concentration: 4.45 ± 0.90 mg/l
Exposure time: 15 min

Concentration: 4.62 ± 1.33 mg/l
Exposure time: 25 min

Comment: Mean of five assays. The values were converted to mg/l from the original data expressed in µM and a rounded molecular weight of 214 given by the authors. The EC50 values were determined in 0.45% (v/v) methanol in 2% (w/w) sodium chloride solution. The 5-min EC50 value for methanol was 43000 mg/l. Phenol solution was used for quality control/quality assurance. The 5-min EC50 value was 18.2 mg/l. The EC50 value at 15 min was 20.7 mg/l with a relative error of <5%.

Bibliographical reference: Chou, C.C., and Que Hee, S.S. (1992). *Ecotoxicol. Environ. Safety* **23**, 355-363.

TRIDECYL ALDEHYDE

CAS RN: 10486-19-8
Synonym: Tridecanal

Sample purity: 90%
Temperature: 15°C
Test parameter: EC50
Effect: Reduction in light output

Concentration: 1.08 ± 0.11 mg/l
Exposure time: 15 min

Concentration: 0.96 ± 0.10 mg/l

Exposure time: 25 min

Comment: Mean of six assays. The values were converted to mg/l from the original data expressed in μM and a rounded molecular weight of 198 given by the authors. The EC50 values were determined in 0.45% (v/v) methanol in 2% (w/w) sodium chloride solution. The 5-min EC50 value for methanol was 43000 mg/l. Phenol solution was used for quality control/quality assurance. The 5-min EC50 value was 18.2 mg/l. The EC50 value at 15 min was 20.7 mg/l with a relative error of <5%. Not enough light inhibition to calculate a 5-min EC50 value.

Bibliographical reference: Chou, C.C., and Que Hee, S.S. (1992). *Ecotoxicol. Environ. Safety* **23**, 355-363.

TRIETHANOLAMINE

CAS RN: 102-71-6

Temperature: 15°C
Test parameter: EC50
Effect: Reduction in light output
Concentration: 110 mg/l
Exposure time: 15 min

Bibliographical reference: Bulich, A.A., Tung, K.K., and Scheibner, G. (1990). *J. Biolumin. Chemilumin.* **5**, 71-77.

Sample purity: 98%
Temperature: 15°C
Test parameter: EC50
Effect: Reduction in light output

Concentration: 184 mg/l
Exposure time: 5 min

Concentration: 175 mg/l
Exposure time: 15 min

Concentration: 197 mg/l
Exposure time: 30 min

Comment: Mean of three assays. Methanol (<10%) was used to prepare the stock solutions. EC50 values were calculated from nominal concentrations.

Bibliographical reference: Kaiser, K.L.E., and Palabrica, V.S. (1992). National Water Research Institute, Burlington, Ontario, Canada, unpublished results.

TRIETHYLENE GLYCOL

CAS RN: 112-27-6

Test parameter: EC50
Effect: Reduction in light output
Concentration: 33000 mg/l
Exposure time: 5 min
Comment: The EC50 value was calculated from nominal concentrations.

Bibliographical reference: Curtis, C., Lima, A., Lozano, S.J., and Veith, G.D. (1982). In: *Aquatic Toxicology and Hazard Assessment: Fifth Conference, ASTM STP 766*, J.G. Pearson, R.B. Foster, and W.E. Bishop (eds.), American Society for Testing and Materials, Philadelphia, p. 170-178.

TRIFLUOROACETAMIDE

CAS RN: 354-38-1

Temperature: 15°C
Test parameter: EC50
Effect: Reduction in light output

Concentration: 15200 mg/l
Exposure time: 5 min

Concentration: 11600 mg/l
Exposure time: 15 min

Concentration: 11000 mg/l
Exposure time: 30 min

Comment: Mean of three assays. Methanol (<10%) was used to prepare the stock solutions. EC50 values were calculated from nominal concentrations.

Bibliographical reference: Kaiser, K.L.E., and Palabrica, V.S. (1991). *Water Poll. Res. J. Canada* **26**, 361-431.

2,2,2-TRIFLUOROACETOPHENONE

CAS RN: 434-45-7

Sample purity: 99%
Temperature: 15°C
Test parameter: EC50
Effect: Reduction in light output

Concentration: 138 mg/l
Exposure time: 5 min

Concentration: 120 mg/l
Exposure time: 15 min

Concentration: 85.3 mg/l*
Exposure time: 30 min

Comment: Mean of two assays. Methanol (<10%) was used to prepare the stock solutions. EC50 values were calculated from nominal concentrations.

Bibliographical references: Kaiser, K.L.E., and Palabrica, V.S. (1991). *Water Poll. Res. J. Canada* **26**, 361-431.
* Kaiser, K.L.E., Palabrica, V.S., and Ribo, J.M. (1987). In: *QSAR in Environmental Toxicology - II*, K.L.E. Kaiser (ed.), D. Reidel Publishing Company, Dordrecht, p. 153-168.

2,2,2-TRIFLUOROETHANOL

CAS RN: 75-89-8

Sample purity: >99.5%
Temperature: 15°C
Test parameter: EC50

Effect: Reduction in light output

Concentration: 1290 mg/l
Exposure time: 5 min

Concentration: 1100 mg/l
Exposure time: 15 min

Concentration: 1000 mg/l
Exposure time: 30 min

Comment: Mean of three assays. Methanol (<10%) was used to prepare the stock solutions. EC50 values were calculated from nominal concentrations.

Bibliographical reference: Kaiser, K.L.E., and Palabrica, V.S. (1992). National Water Research Institute, Burlington, Ontario, Canada, unpublished results.

4'-(TRIFLUOROMETHYL)ACETOPHENONE

CAS RN: 709-63-7

Sample purity: 95%
Temperature: 15°C
Test parameter: EC50
Effect: Reduction in light output

Concentration: 6.52 mg/l
Exposure time: 5 min

Concentration: 7.15 mg/l
Exposure time: 15 min

Concentration: 7.49 mg/l*
Exposure time: 30 min

Comment: Mean of three assays. Methanol (<10%) was used to prepare the stock solutions. EC50 values were calculated from nominal concentrations.

Bibliographical references: Kaiser, K.L.E., and Palabrica, V.S. (1991). *Water Poll. Res. J. Canada* **26**, 361-431.

* Kaiser, K.L.E., and Gough, K.M. (1989). In: *Aquatic Toxicology and Environmental Fate: Eleventh Volume, ASTM STP 1007*, G.W. Suter and M.A. Lewis (eds.), American Society for Testing and Materials, Philadelphia, p. 424-441.

3-(TRIFLUOROMETHYL)BENZONITRILE

CAS RN: 368-77-4

Sample purity: 99%
Temperature: 15°C
Test parameter: EC50
Effect: Reduction in light output

Concentration: 6.81 mg/l
Exposure time: 5 min

Concentration: 7.64 mg/l
Exposure time: 15 min

Concentration: 8.00 mg/l
Exposure time: 30 min

Comment: Mean of three assays. Methanol (<10%) was used to prepare the stock solutions. EC50 values were calculated from nominal concentrations.

Bibliographical reference: Kaiser, K.L.E., and Palabrica, V.S. (1991). *Water Poll. Res. J. Canada* **26**, 361-431.

4-(TRIFLUOROMETHYL)BENZOPHENONE

CAS RN: 728-86-9

Sample purity: 97%
Temperature: 15°C
Test parameter: EC50
Effect: Reduction in light output

Concentration: 0.16 mg/l
Exposure time: 5 min

Concentration: 0.19 mg/l
Exposure time: 15 min

Concentration: 0.24 mg/l
Exposure time: 30 min

Comment: Mean of three assays. Methanol (<10%) was used to prepare the stock solutions. EC50 values were calculated from nominal concentrations.

Bibliographical reference: Kaiser, K.L.E., and Palabrica, V.S. (1991). *Water Poll. Res. J. Canada* **26**, 361-431.

4-(TRIFLUOROMETHYL)BENZYL ALCOHOL

CAS RN: 349-95-1

Sample purity: 98%
Temperature: 15°C
Test parameter: EC50
Effect: Reduction in light output

Concentration: 2.61 mg/l
Exposure time: 5 min

Concentration: 2.86 mg/l
Exposure times: 15 and 30 min

Comment: Mean of three assays. Methanol (<10%) was used to prepare the stock solutions. EC50 values were calculated from nominal concentrations.

Bibliographical reference: Kaiser, K.L.E., and Palabrica, V.S. (1991). *Water Poll. Res. J. Canada* **26**, 361-431.

4-(TRIFLUOROMETHYL)BENZYLAMINE

CAS RN: 3300-51-4

Sample purity: 98%
Temperature: 15°C
Test parameter: EC50

Effect: Reduction in light output

Concentration: 41.1 mg/l
Exposure time: 5 min

Concentration: 33.4 mg/l
Exposure time: 15 min

Concentration: 32.6 mg/l
Exposure time: 30 min

Comment: Mean of three assays. Methanol (<10%) was used to prepare the stock solutions. EC50 values were calculated from nominal concentrations.

Bibliographical reference: Kaiser, K.L.E., and Palabrica, V.S. (1991). *Water Poll. Res. J. Canada* **26**, 361-431.

trans-4-(TRIFLUOROMETHYL)CINNAMIC ACID

CAS RN: 16642-92-5

Sample purity: 99%
Temperature: 15°C
Test parameter: EC50
Effect: Reduction in light output

Concentration: 32.0 mg/l
Exposure times: 5 and 10 min

Concentration: 31.2 mg/l
Exposure time: 30 min

Comment: Mean of four assays. Methanol (<10%) was used to prepare the stock solutions. EC50 values were calculated from nominal concentrations.

Bibliographical reference: Kaiser, K.L.E., and Palabrica, V.S. (1991). *Water Poll. Res. J. Canada* **26**, 361-431.

3-TRIFLUOROMETHYL-4-NITROPHENOL

CAS RN: 88-30-2

Temperature: 15°C
Test parameter: EC50
Effect: Reduction in light output

Concentration: 2.02 mg/l
Exposure time: 5 min

Concentration: 1.72 mg/l
Exposure time: 15 min

Concentration: 1.65 mg/l
Exposure time: 30 min

Comment: Mean of four assays. Methanol (<10%) was used to prepare the stock solutions. EC50 values were calculated from nominal concentrations.

Bibliographical reference: Kaiser, K.L.E., and Palabrica, V.S. (1991). *Water Poll. Res. J. Canada* **26**, 361-431.

4-(TRIFLUOROMETHYL)PHENYLACETONITRILE

CAS RN: 2338-75-2

Sample purity: 98%
Temperature: 15°C
Test parameter: EC50
Effect: Reduction in light output

Concentration: 0.10 mg/l
Exposure time: 5 min

Concentration: 0.11 mg/l
Exposure times: 15 and 30 min

Comment: Mean of three assays. Methanol (<10%) was used to prepare the stock solutions. EC50 values were calculated from nominal concentrations.

Bibliographical reference: Kaiser, K.L.E., and Palabrica, V.S. (1991). *Water Poll. Res. J. Canada* **26**, 361-431.

1,3,5-TRIFLUORO-2-NITROBENZENE

CAS RN: 315-14-0

Sample purity: 98%
Temperature: 15°C
Test parameter: EC50
Effect: Reduction in light output

Concentration: 128 mg/l
Exposure times: 5 and 15 min

Concentration: 134 mg/l
Exposure time: 30 min

Comment: Mean of four assays. Methanol (<10%) was used to prepare the stock solutions. EC50 values were calculated from nominal concentrations.

Bibliographical reference: Kaiser, K.L.E., and Palabrica, V.S. (1991). *Water Poll. Res. J. Canada* **26**, 361-431.

α,α,α-TRIFLUOROTOLUENE

CAS RN: 98-08-8

Sample purity: 99%
Temperature: 15°C
Test parameter: EC50
Effect: Reduction in light output

Concentration: 19.7 mg/l
Exposure time: 5 min

Concentration: 23.2 mg/l
Exposure time: 15 min

Concentration: 32.0 mg/l*
Exposure time: 30 min

Comment: Mean of three assays. Methanol (<10%) was used to prepare the stock solutions. EC50 values were calculated from nominal concentrations.

Bibliographical references: Kaiser, K.L.E., and Palabrica, V.S. (1991). *Water Poll. Res. J. Canada* **26**, 361-431.
* Kaiser, K.L.E., Palabrica, V.S., and Ribo, J.M. (1987). In: *QSAR in Environmental Toxicology - II*, K.L.E. Kaiser (ed.), D. Reidel Publishing Company, Dordrecht, p. 153-168.

α,α,α-TRIFLUORO-4-TOLUIC ACID

CAS RN: 455-24-3

Sample purity: 98%
Temperature: 15°C
Test parameter: EC50
Effect: Reduction in light output

Concentration: 61.5 mg/l
Exposure time: 5 min

Concentration: 69.0 mg/l
Exposure time: 15 min

Concentration: 83.0 mg/l
Exposure time: 30 min

Comment: Mean of three assays. Methanol (<10%) was used to prepare the stock solutions. EC50 values were calculated from nominal concentrations.

Bibliographical reference: Kaiser, K.L.E., and Palabrica, V.S. (1991). *Water Poll. Res. J. Canada* **26**, 361-431.

α,α,α-TRIFLUORO-4-TOLUNITRILE

CAS RN: 455-18-5

Sample purity: 99%
Temperature: 15°C
Test parameter: EC50

Effect: Reduction in light output

Concentration: 7.30 mg/l
Exposure time: 5 min

Concentration: 7.47 mg/l
Exposure time: 15 min

Concentration: 8.00 mg/l*
Exposure time: 30 min

Comment: Mean of three assays. Methanol (<10%) was used to prepare the stock solutions. EC50 values were calculated from nominal concentrations.

Bibliographical references: Kaiser, K.L.E., and Palabrica, V.S. (1991). *Water Poll. Res. J. Canada* **26**, 361-431.
* Kaiser, K.L.E., and Gough, K.M. (1989). In: *Aquatic Toxicology and Environmental Fate: Eleventh Volume, ASTM STP 1007*, G.W. Suter and M.A. Lewis (eds.), American Society for Testing and Materials, Philadelphia, p. 424-441.

3-(3,4,5-TRIMETHOXYPHENYL)PROPIONITRILE

CAS RN: 49621-50-3

Temperature: 15°C
Test parameter: EC50
Effect: Reduction in light output

Concentration: 157 mg/l
Exposure time: 5 min

Concentration: 150 mg/l
Exposure time: 15 min

Concentration: 160 mg/l
Exposure time: 30 min

Comment: Mean of three assays. Methanol (<10%) was used to prepare the stock solutions. EC50 values were calculated from nominal concentrations.

Bibliographical reference: Kaiser, K.L.E., and Palabrica, V.S. (1991). *Water Poll. Res. J. Canada* **26**, 361-431.

2,3,5-TRIMETHYLPHENOL

CAS RN: 697-82-5

Sample purity: 98+%
Temperature: 15°C
Test parameter: EC50
Effect: Reduction in light output

Concentration: 8.79 mg/l
Exposure time: 5 min

Concentration: 9.21 mg/l
Exposure time: 15 min

Concentration: 9.64 mg/l
Exposure time: 30 min

Comment: Mean of three assays. Methanol (<10%) was used to prepare the stock solutions. EC50 values were calculated from nominal concentrations.

Bibliographical reference: Kaiser, K.L.E., and Palabrica, V.S. (1991). *Water Poll. Res. J. Canada* **26**, 361-431.

2,3,6-TRIMETHYLPHENOL

CAS RN: 2416-94-6

Sample purity: 97+%
Temperature: 15°C
Test parameter: EC50
Effect: Reduction in light output

Concentration: 6.52 mg/l
Exposure time: 5 min

Concentration: 6.98 mg/l
Exposure time: 15 min

Concentration: 6.83 mg/l
Exposure time: 30 min

Comment: Mean of three assays. Methanol (<10%) was used to prepare the stock solutions. EC50 values were calculated from nominal concentrations.

Bibliographical reference: Kaiser, K.L.E., and Palabrica, V.S. (1991). *Water Poll. Res. J. Canada* **26**, 361-431.

2,4,6-TRIMETHYLPHENOL

CAS RN: 527-60-6

Sample purity: 99%
Temperature: 15°C
Test parameter: EC50
Effect: Reduction in light output

Concentration: 11.3 mg/l
Exposure time: 5 min

Concentration: 11.9 mg/l
Exposure times: 15 and 30 min

Comment: Mean of three assays. Methanol (<10%) was used to prepare the stock solutions. EC50 values were calculated from nominal concentrations.

Bibliographical reference: Kaiser, K.L.E., and Palabrica, V.S. (1991). *Water Poll. Res. J. Canada* **26**, 361-431.

TRIMETHYL PHOSPHATE

CAS RN: 512-56-1

Sample purity: 99+%
Temperature: 15°C
Test parameter: EC50
Effect: Reduction in light output

Concentration: 255 mg/l

Exposure time: 5 min

Concentration: 293 mg/l
Exposure time: 15 min

Concentration: 352 mg/l
Exposure time: 30 min

Comment: Mean of three assays. Methanol (<10%) was used to prepare the stock solutions. EC50 values were calculated from nominal concentrations.

Bibliographical reference: Kaiser, K.L.E., and Palabrica, V.S. (1992). National Water Research Institute, Burlington, Ontario, Canada, unpublished results.

TRINITROTOLUENE

CAS RN: 118-96-7

Temperature: 15°C
Test parameter: EC50
Effect: Reduction in light output
Concentration: 20 mg/l
Exposure time: 5 min

Bibliographical reference: Bulich, A.A., Greene, M.W., and Isenberg, D.L. (1981). In: *Aquatic Toxicology and Hazard Assessment: Fourth Conference*, *ASTM STP 737*, D.R. Branson and K.L. Dickson, (eds.), American Society for Testing and Materials, Philadelphia, p. 338-347.

TRIPHENYLAMINE

CAS RN: 603-34-9

Temperature: 15°C
Test parameter: EC50
Effect: Reduction in light output
Concentration: 2.19 mg/l
Exposure time: 30 min

Comment: Mean of three assays. Methanol (<10%) was used to prepare the stock solutions. EC50 values were calculated from nominal concentrations.

Bibliographical reference: Kaiser, K.L.E., Palabrica, V.S., and Ribo, J.M. (1987). In: *QSAR in Environmental Toxicology - II*, K.L.E. Kaiser (ed.), D. Reidel Publishing Company, Dordrecht, p. 153-168.

Sample purity: 98+%
Temperature: 15°C
Test parameter: EC50
Effect: Reduction in light output

Concentration: 2.19 mg/l
Exposure time: 5 min

Concentration: 1.70 mg/l
Exposure time: 15 min

Concentration: 1.66 mg/l
Exposure time: 30 min

Comment: Mean of three assays. Methanol (<10%) was used to prepare the stock solutions. EC50 values were calculated from nominal concentrations.

Bibliographical reference: Kaiser, K.L.E., and Palabrica, V.S. (1991). *Water Poll. Res. J. Canada* **26**, 361-431.

TRIPHENYLARSINE OXIDE

CAS RN: 1153-05-5

Temperature: 15°C
Test parameter: EC50
Effect: Reduction in light output

Concentration: 144 mg/l
Exposure time: 5 min

Concentration: 137 mg/l
Exposure time: 15 min

Concentration: 137 mg/l*
Exposure time: 30 min

Comment: Mean of three assays. Methanol (<10%) was used to prepare the stock solutions. EC50 values were calculated from nominal concentrations.

Bibliographical references: Kaiser, K.L.E., and Palabrica, V.S. (1991). *Water Poll. Res. J. Canada* **26**, 361-431.
* Kaiser, K.L.E., Palabrica, V.S., and Ribo, J.M. (1987). In: *QSAR in Environmental Toxicology - II*, K.L.E. Kaiser (ed.), D. Reidel Publishing Company, Dordrecht, p. 153-168.

TRIPHENYLMETHANOL

CAS RN: 76-84-6

Sample purity: 98+%
Temperature: 15°C
Test parameter: EC50
Effect: Reduction in light output

Concentration: 4.52 mg/l
Exposure time: 5 min

Concentration: 3.59 mg/l
Exposure time: 15 min

Concentration: 3.35 mg/l
Exposure time: 30 min

Comment: Mean of three assays. Methanol (<10%) was used to prepare the stock solutions. EC50 values were calculated from nominal concentrations.

Bibliographical reference: Kaiser, K.L.E., and Palabrica, V.S. (1991). *Water Poll. Res. J. Canada* **26**, 361-431.

TRIPHENYLPHOSPHINE

CAS RN: 603-35-0

Sample purity: 99%
Temperature: 15°C
Test parameter: EC50
Effect: Reduction in light output

Concentration: 2.04 mg/l
Exposure time: 5 min

Concentration: 1.54 mg/l
Exposure times: 15 and 30 min

Comment: Mean of three assays. Methanol (<10%) was used to prepare the stock solutions. EC50 values were calculated from nominal concentrations.

Bibliographical reference: Kaiser, K.L.E., and Palabrica, V.S. (1991). *Water Poll. Res. J. Canada* **26**, 361-431.

TRIPHENYLTIN CHLORIDE

CAS RN: 639-58-7

Temperature: 15°C
Test parameter: EC50
Effect: Reduction in light output
Concentration: 0.0157 mg/l
Exposure time: 30 min

Bibliographical reference: Steinhäuser, K.G., Amann, W., Späth, A., and Polenz, A. (1985). *Vom Wasser* **65**, 203-214.

Temperature: 15°C
Test parameter: EC50
Effect: Reduction in light output

Concentration: 0.14 ± 0.03 mg/l
Exposure time: 5 min

Concentration: 0.05 ± 0.01 mg/l
Exposure time: 15 min

Comment: Chemical was dissolved in ethanol (~0.05%).

Bibliographical reference: Dooley, C.A., and Kenis, P. (1987). In: *Oceans '87, International Organotin Symposium*, Department of Fisheries and Oceans, Canada, and the Society for Underwater Technology, William MacNab & Son, Halifax, p. 1517-1524.

TRIS(HYDROXYMETHYL)NITROMETHANE

CAS RN: 126-11-4

Sample purity: 98%
Temperature: 15°C
Test parameter: EC50
Effect: Reduction in light output

Concentration: 5.49 mg/l
Exposure time: 5 min

Concentration: 2.95 mg/l
Exposure time: 15 min

Concentration: 2.34 mg/l
Exposure time: 30 min

Comment: Mean of three assays. Methanol (<10%) was used to prepare the stock solutions. EC50 values were calculated from nominal concentrations.

Bibliographical reference: Kaiser, K.L.E., and Palabrica, V.S. (1991). *Water Poll. Res. J. Canada* **26**, 361-431.

TRITON X 100

CAS RN: 9002-93-1

Temperature: 15°C
Test parameter: EC50
Effect: Reduction in light output

Concentrations: 0.12 mg/l (time after reconstitution = 0.5 h)
0.31 mg/l (time after reconstitution = 1.75 h)
Exposure time: 5 min

Concentrations: 0.08 mg/l (time after reconstitution = 0.5 h)
0.18 mg/l (time after reconstitution = 1.75 h)
Exposure time: 15 min

Bibliographical reference: Qureshi, A.A., Coleman, R.N., and Paran, J.H. (1984). In: *Toxicity Screening Procedures Using Bacterial Systems*, D. Liu and B.J. Dutka (eds.), Marcel Dekker, New York, p. 1-22.

TROPAEOLIN O

CAS RN: 547-57-9

Temperature: 15°C
Test parameter: EC50
Effect: Reduction in light output

Concentration: 26.3 mg/l
Exposure time: 5 min

Concentration: 25.7 mg/l
Exposure time: 15 min

Concentration: 30.9 mg/l
Exposure time: 30 min

Comment: Mean of five assays. Methanol (<10%) was used to prepare the stock solutions. EC50 values were calculated from nominal concentrations.

Bibliographical reference: Kaiser, K.L.E., and Palabrica, V.S. (1992). National Water Research Institute, Burlington, Ontario, Canada, unpublished results.

TWEEN 80

CAS RN: 9005-65-6
Synonym: Polyoxyethylene (20) sorbitan monooleate

Temperature: 15°C
Test parameter: EC50
Effect: Reduction in light output

Concentration: 1100 mg/l
Exposure time: 15 min

Bibliographical reference: Bulich, A.A., Tung, K.K., and Scheibner, G. (1990). *J. Biolumin. Chemilumin.* **5**, 71-77.

TYRAMINE

CAS RN: 51-67-2

Sample purity: 97%
Temperature: 15°C
Test parameter: EC50
Effect: Reduction in light output

Concentration: 29.3 mg/l
Exposure time: 5 min

Concentration: 27.4 mg/l
Exposure time: 15 min

Concentration: 28.7 mg/l*
Exposure time: 30 min

Comment: Mean of four assays. Methanol (<10%) was used to prepare the stock solutions. EC50 values were calculated from nominal concentrations.

Bibliographical references: Kaiser, K.L.E., and Palabrica, V.S. (1991). *Water Poll. Res. J. Canada* **26**, 361-431.
* Kaiser, K.L.E. (1987). In: *QSAR in Environmental Toxicology - II*, K.L.E. Kaiser (ed.), D. Reidel Publishing Company, Dordrecht, p. 169-188.

UNDECYLENIC ACID

CAS RN: 112-38-9

Sample purity: 99%
Temperature: 15°C
Test parameter: EC50
Effect: Reduction in light output

Concentration: 8.43 ± 3.31 mg/l
Exposure time: 5 min

Concentration: 10.32 ± 2.21 mg/l
Exposure time: 15 min

Concentration: 7.38 ± 0.85 mg/l
Exposure time: 25 min

Comment: Mean of five assays. The values were converted to mg/l from the original data expressed in µM and a rounded molecular weight of 184 given by the authors. The EC50 values were determined in 0.45% (v/v) methanol in 2% (w/w) sodium chloride solution. The 5-min EC50 value for methanol was 43000 mg/l. Phenol solution was used for quality control/quality assurance. The 5-min EC50 value was 18.2 mg/l. The EC50 value at 15 min was 20.7 mg/l with a relative error of <5%.

Bibliographical reference: Chou, C.C., and Que Hee, S.S. (1992). *Ecotoxicol. Environ. Safety* **23**, 355-363.

UNDECYLIC ALDEHYDE

CAS RN: 112-44-7
Synonym: Undecanal

Temperature: 15°C
Test parameter: EC50
Effect: Reduction in light output

Concentration: 4.32 ± 0.27 mg/l
Exposure time: 5 min

Concentration: 4.78 ± 0.71 mg/l
Exposure time: 15 min

Concentration: 5.05 ± 0.63 mg/l
Exposure time: 25 min

Comment: Mean of seven assays. The values were converted to mg/l from the original data expressed in µM and a rounded molecular weight of 170 given by the authors. The EC50 values were determined in 0.45% (v/v) methanol in 2% (w/w) sodium chloride solution. The 5-

min EC50 value for methanol was 43000 mg/l. Phenol solution was used for quality control/quality assurance. The 5-min EC50 value was 18.2 mg/l. The EC50 value at 15 min was 20.7 mg/l with a relative error of <5%.

Bibliographical reference: Chou, C.C., and Que Hee, S.S. (1992). *Ecotoxicol. Environ. Safety* **23**, 355-363.

UREA

CAS RN: 57-13-6

Temperature: 15°C
Test parameter: EC50
Effect: Reduction in light output
Concentration: 24000 mg/l
Exposure time: 5 min

Bibliographical reference: Bulich, A.A., Greene, M.W., and Isenberg, D.L. (1981). In: *Aquatic Toxicology and Hazard Assessment: Fourth Conference, ASTM STP 737*, D.R. Branson and K.L. Dickson, (eds.), American Society for Testing and Materials, Philadelphia, p. 338-347.

URETHANE

CAS RN: 51-79-6
Synonym: Ethyl carbamate

Sample purity: 99%
Temperature: 15°C
Test parameter: EC50
Effect: Reduction in light output

Concentration: 1700 mg/l
Exposure time: 5 min

Concentration: 1620 mg/l
Exposure times: 15 and 30 min

Comment: Mean of four assays. Methanol (<10%) was used to prepare the stock solutions. EC50 values were calculated from nominal concentrations.

Bibliographical reference: Kaiser, K.L.E., and Palabrica, V.S. (1992). National Water Research Institute, Burlington, Ontario, Canada, unpublished results.

VANILLIN

CAS RN: 121-33-5

Sample purity: 99%
Temperature: 15°C
Test parameter: EC50
Effect: Reduction in light output

Concentration: 60.0 mg/l
Exposure time: 5 min

Concentration: 68.0 mg/l
Exposure time: 15 min

Comment: Chemical was prepared in an initial solution of 5% methanol.

Bibliographical reference: Cronin, M.T.D. (1993). Liverpool John Moores University, UK, private communication.

VANILLIN AZINE

CAS RN: 1696-60-2

Sample purity: 99%
Temperature: 15°C
Test parameter: EC50
Effect: Reduction in light output

Concentration: 7.37 mg/l
Exposure time: 5 min

Concentration: 7.90 mg/l

Exposure time: 15 min

Concentration: 8.46 mg/l
Exposure time: 30 min

Comment: Mean of three assays. Methanol (<10%) was used to prepare the stock solutions. EC50 values were calculated from nominal concentrations.

Bibliographical reference: Kaiser, K.L.E., and Palabrica, V.S. (1991). *Water Poll. Res. J. Canada* **26**, 361-431.

9-VINYLCARBAZOLE

CAS RN: 1484-13-5

Sample purity: 98%
Temperature: 15°C
Test parameter: EC50
Effect: Reduction in light output

Concentration: 1.31 mg/l
Exposure time: 5 min

Concentration: 2.07 mg/l
Exposure time: 15 min

Concentration: 2.43 mg/l
Exposure time: 30 min

Comment: Mean of four assays. Methanol (<10%) was used to prepare the stock solutions. EC50 values were calculated from nominal concentrations.

Bibliographical reference: Kaiser, K.L.E., and Palabrica, V.S. (1991). *Water Poll. Res. J. Canada* **26**, 361-431.

1-VINYLIMIDAZOLE

CAS RN: 1072-63-5

Sample purity: >99%

Temperature: 15°C
Test parameter: EC50
Effect: Reduction in light output

Concentration: 211 mg/l
Exposure time: 5 min

Concentration: 206 mg/l
Exposure time: 15 min

Concentration: 216 mg/l
Exposure time: 30 min

Comment: Mean of four assays. Methanol (<10%) was used to prepare the stock solutions. EC50 values were calculated from nominal concentrations.

Bibliographical reference: Kaiser, K.L.E., and Palabrica, V.S. (1991). *Water Poll. Res. J. Canada* **26**, 361-431.

4-VINYLPYRIDINE

CAS RN: 100-43-6

Sample purity: 95%
Temperature: 15°C
Test parameter: EC50
Effect: Reduction in light output

Concentration: 8.55 mg/l
Exposure time: 5 min

Concentration: 9.59 mg/l
Exposure time: 15 min

Concentration: 9.81 mg/l
Exposure time: 30 min

Comment: Mean of three assays. Methanol (<10%) was used to prepare the stock solutions. EC50 values were calculated from nominal concentrations.

Bibliographical reference: Kaiser, K.L.E., and Palabrica, V.S. (1991). *Water Poll. Res. J. Canada* **26**, 361-431.

WARFARIN

CAS RN: 81-81-2

Sample purity: 98%
Temperature: 15°C
Test parameter: EC50
Effect: Reduction in light output

Concentration: 48.9 mg/l
Exposure time: 5 min

Concentration: 47.8 mg/l
Exposure times: 15 and 30 min

Comment: Mean of three assays. Methanol (<10%) was used to prepare the stock solutions. EC50 values were calculated from nominal concentrations.

Bibliographical reference: Kaiser, K.L.E., and Palabrica, V.S. (1991). *Water Poll. Res. J. Canada* **26**, 361-431.

WITACLOR 149

Sample purity: 49% Cl, carbon no C_{10-13}
Temperature: 15 ± 0.1°C
Test parameter: EC50
Effect: Reduction in light output

Concentration: 3.02 mg/l
Exposure time: 5 min

Concentration: 1.42 mg/l
Exposure time: 15 min

Concentration: 2.94 mg/l
Exposure time: 30 min

Comment: The pH was not adjusted.

Bibliographical reference: Tarkpea, M., Hansson, M., and Samuelsson, B. (1986). *Ecotoxicol. Environ. Safety* **11**, 127-143.

XANTHONE

CAS RN: 90-47-1

Sample purity: 99%
Temperature: 15°C
Test parameter: EC50
Effect: Reduction in light output

Concentration: 7.81 mg/l
Exposure time: 5 min

Concentration: 6.96 mg/l
Exposure time: 15 min

Concentration: 7.46 mg/l
Exposure time: 30 min

Comment: Mean of four assays. Methanol (<10%) was used to prepare the stock solutions. EC50 values were calculated from nominal concentrations.

Bibliographical reference: Kaiser, K.L.E., and Palabrica, V.S. (1991). *Water Poll. Res. J. Canada* **26**, 361-431.

o-XYLENE

CAS RN: 95-47-6
Synonym: 1,2-Dimethylbenzene

Temperature: 15°C
Test parameter: EC50
Effect: Reduction in light output
Concentration: 9.25 mg/l
Exposure time: 15 min

Bibliographical reference: Hermens, J., Busser, F., Leeuwangh, P., and Musch, A. (1985). *Ecotoxicol. Environ. Safety* **9**, 17-25.

m-XYLENE

CAS RN: 108-38-3

Synonym: 1,3-Dimethylbenzene

Temperature: 15°C
Test parameter: EC50
Effect: Reduction in light output

Concentration: 75 mg/l
Exposure time: 5 min

Concentration: 73 mg/l
Exposure time: 10 min

Comment: Mean of two assays. Toxicity values were calculated from nominal concentrations.

Bibliographical reference: Ferard, J.F., Vasseur, P., Danoux, L., and Larbaigt, G. (1983). *Rev. Fr. Sci. Eau* **2**, 221-237.

Sample purity: 99%
Temperature: 15°C
Test parameter: EC50
Effect: Reduction in light output

Concentration: 2.61 mg/l
Exposure time: 5 min

Concentration: 3.36 mg/l
Exposure time: 15 min

Concentration: 7.18 mg/l
Exposure time: 30 min

Comment: Mean of three assays. Methanol (<10%) was used to prepare the stock solutions. EC50 values were calculated from nominal concentrations.

Bibliographical reference: Kaiser, K.L.E., and Palabrica, V.S. (1992). National Water Research Institute, Burlington, Ontario, Canada, unpublished results.

p-XYLENE

CAS RN: 106-42-3

Synonym: 1,4-Dimethylbenzene

Sample purity: 99.5%
Temperature: 15°C
Test parameter: EC50
Effect: Reduction in light output
Concentration: 5.70 mg/l
Exposure time: 30 min
Comment: Mean of three assays. Methanol (<10%) was used to prepare the stock solutions. EC50 values were calculated from nominal concentrations.

Bibliographical reference: Ribo, J.M., and Kaiser, K.L.E. National Water Research Institute, Burlington, Ontario, Canada, unpublished results (value later published by Kaiser, K.L.E., and Palabrica, V.S. (1991). *Water Poll. Res. J. Canada* **26**, 361-431).

XYLENES

CAS RN: 1330-20-7

Temperature: 15°C
Test parameter: EC50
Effect: Reduction in light output
Concentration: 16 mg/l
Exposure time: 5 min

Bibliographical reference: Samak, Q.M., and Noiseux, R. (1981). *Can. Tech. Rep. Fish. Aquat. Sci.* **990**, 288-308.

ZEARALENONE

CAS RN: 17924-92-4

Temperature: 15°C
Test parameter: EC50
Effect: Reduction in light output

Concentration: 14.37 mg/l
Exposure time: 5 min

Concentration: 13.70 mg/l

Exposure time: 10 min

Concentration: 13.21 mg/l
Exposure time: 15 min

Concentration: 12.29 mg/l
Exposure time: 20 min

Comment: Freshly reconstituted bacterial suspensions. Chemical was dissolved in DMSO and methanol.

Bibliographical reference: Yates, I.E., and Porter, J.K. (1982). *Appl. Environ. Microbiol.* **44**, 1072-1075.

Temperature: 15°C
Test parameter: EC20
Effect: Reduction in light output

Concentration: 9.35 mg/l
Exposure time: 5 min

Concentration: 9.69 mg/l
Exposure time: 20 min

Comment: Freshly reconstituted bacterial suspensions. Chemical was dissolved in DMSO and methanol.

Bibliographical reference: Yates, I.E., and Porter, J.K. (1982). *Appl. Environ. Microbiol.* **44**, 1072-1075.

Temperature: 15°C
Test parameter: EC50
Effect: Reduction in light output

Concentration: 11.59 mg/l
Exposure times: 5 and 10 min

Concentration: 11.37 mg/l
Exposure time: 15 min

Concentration: 11.30 mg/l
Exposure time: 20 min

Comment: Performed on bacterial suspensions maintained at 3°C for 5 h after reconstitution. Chemical was dissolved in DMSO and methanol.

Bibliographical reference: Yates, I.E., and Porter, J.K. (1982). *Appl. Environ. Microbiol.* **44**, 1072-1075.

Temperature: 10°C
Test parameter: EC50
Effect: Reduction in light output

Concentration: 8.44 mg/l
Exposure time: 5 min

Concentration: 7.25 mg/l
Exposure time: 10 min

Concentration: 6.66 mg/l
Exposure time: 15 min

Concentration: 6.33 mg/l
Exposure time: 20 min

Comment: Chemical was dissolved in methanol. Test was performed at pH = 6.0 units.

Bibliographical reference: Yates, I.E., and Porter, J.K. (1984). In: *Toxicity Screening Procedures Using Bacterial Systems*, D. Liu and B.J. Dutka (eds.), Marcel Dekker, New York, p. 77-88.

ZINC ACETATE DIHYDRATE

CAS RN: 5970-45-6

Temperature: 20°C
Test parameter: EC50
Effect: Reduction in light output
Concentration: 0.67 ± 0.14 mg/l Zn^{++}
Exposure time: 30 min
Comment: Toxicity values were calculated from nominal concentrations.

Bibliographical reference: Vasseur, P., Bois, F., Ferard, J.F., Rast, C., and Larbaigt, G. (1986). *Tox. Assess.* **1**, 283-300.

ZINC CHLORIDE

CAS RN: 7646-85-7

Temperature: 20°C
Test parameter: EC50
Effect: Reduction in light output
Concentration: 0.62 ± 0.16 mg/l Zn^{++}
Exposure time: 30 min
Comment: Toxicity values were calculated from nominal concentrations.

Bibliographical reference: Vasseur, P., Bois, F., Ferard, J.F., Rast, C., and Larbaigt, G. (1986). *Tox. Assess.* **1**, 283-300.

Temperature: 15°C

Test parameter: EC20
Effect: Reduction in light output
Concentration: 0.2 mg/l Zn^{++}
Exposure time: 30 min

Test parameter: EC50
Effect: Reduction in light output
Concentration: 0.8 mg/l Zn^{++}
Exposure time: 30 min

Comment: Toxicity values were calculated from nominal concentrations.

Bibliographical reference: Reteuna, C. (1988). Thesis, University of Metz, Metz, France.

ZINC DIETHYLDITHIOCARBAMATE

CAS RN: 14324-55-1

Sample purity: ≥90%
Test parameter: EC50
Effect: Reduction in light output
Concentration: 1.70 mg/l (1.33-2.17)
Exposure time: 15 min

Bibliographical reference: van Leeuwen, C.J., Maas-Diepeveen, J.L., Niebeek, G., Vergouw, W.H.A., Griffioen, P.S., and Luijken, M.W. (1985). *Aquat. Toxicol.* **7**, 145-164.

ZINC SULFATE HEPTAHYDRATE

CAS RN: 7446-20-0

Temperature: 15°C
Test parameter: EC50
Effect: Reduction in light output

Concentration: 13.8 mg/l Zn^{++}
Exposure time: 5 min

Concentration: 6.1 mg/l Zn^{++}
Exposure time: 10 min

Concentration: 3.45 mg/l Zn^{++}
Exposure time: 15 min

Comment: Chemical was tested at pH 6.7.

Bibliographical references: Dutka, B.J., and Kwan, K.K. (1982). *Environ. Pollut. Ser. A* **29**, 125-134.
Dutka, B.J., Nyholm, N., and Petersen, J. (1983). *Water Res.* **17**, 1363-1368.

Temperature: 15°C
Test parameter: EC50
Effect: Reduction in light output
Concentration: 3.7 ± 0.9 mg/l Zn^{++}
Exposure time: 10 min
Comment: Mean of five assays. Toxicity values were calculated from nominal concentrations.

Bibliographical reference: Ferard, J.F., Vasseur, P., Danoux, L., and Larbaigt, G. (1983). *Rev. Fr. Sci. Eau* **2**, 221-237.

Temperature: 20°C
Test parameter: EC50
Effect: Reduction in light output

Concentration: 1.22 ± 0.29 mg/l Zn^{++}
Exposure time: 10 min
Comment: Mean of seven assays. Toxicity values were calculated from nominal concentrations.

Bibliographical reference: Ferard, J.F., Vasseur, P., Danoux, L., and Larbaigt, G. (1983). *Rev. Fr. Sci. Eau* **2**, 221-237.

Sample purity: Analytical grade
Temperature: 15°C
Test parameter: EC50
Effect: Reduction in light output

Concentrations: 36.1 mg/l Zn^{++} (time after reconstitution = 0.75 h)
18.5 mg/l Zn^{++} (time after reconstitution = 4 h)
Exposure time: 5 min

Concentrations: 5.77 mg/l Zn^{++} (time after reconstitution = 0.75 h)
3.34 mg/l Zn^{++} (time after reconstitution = 4 h)
Exposure time: 15 min

Bibliographical reference: Qureshi, A.A., Coleman, R.N., and Paran, J.H. (1984). In: *Toxicity Screening Procedures Using Bacterial Systems*, D. Liu and B.J. Dutka (eds.), Marcel Dekker, New York, p. 1-22.

Sample purity: Analytical grade
Temperature: 15°C
Test parameter: EC50
Effect: Reduction in light output

Concentration: 55.5 mg/l Zn^{++}
Exposure time: 5 min

Concentration: 64.7 mg/l Zn^{++}
Exposure time: 10 min

Concentration: 6.08 mg/l Zn^{++}
Exposure time: 15 min

Concentration: 3.60 mg/l Zn^{++}
Exposure time: 20 min

Concentration: 1.44 mg/l Zn^{++}

Exposure time: 30 min

Comment: Results derived from the average of two replicates.

Bibliographical reference: Qureshi, A.A., Coleman, R.N., and Paran, J.H. (1984). In: *Toxicity Screening Procedures Using Bacterial Systems*, D. Liu and B.J. Dutka (eds.), Marcel Dekker, New York, p. 1-22.

Sample purity: >99.5%
Temperature: 15 ± 0.1°C
Test parameter: EC50
Effect: Reduction in light output

Concentration: 18.7 mg/l Zn^{++}
Exposure time: 5 min

Concentration: 3.91 mg/l Zn^{++}
Exposure time: 15 min

Concentration: 1.18 mg/l Zn^{++}
Exposure time: 30 min

Comment: The pH was not adjusted.

Bibliographical reference: Tarkpea, M., Hansson, M., and Samuelsson, B. (1986). *Ecotoxicol. Environ. Safety* **11**, 127-143.

Temperature: 20°C
Test parameter: EC50
Effect: Reduction in light output
Concentration: 0.65 ± 0.18 mg/l Zn^{++}
Exposure time: 30 min
Comment: Toxicity values were calculated from nominal concentrations.

Bibliographical reference: Vasseur, P., Bois, F., Ferard, J.F., Rast, C., and Larbaigt, G. (1986). *Tox. Assess.* **1**, 283-300.

Sample purity: Reagent grade
Test parameter: EC50
Effect: Reduction in light output

Concentration: 23 mg/l Zn^{++} (21-26)
Exposure time: 5 min

Concentration: 1.69 mg/l Zn^{++} (1.65-1.74)
Exposure time: 15 min

Concentration: 0.68 mg/l Zn^{++} (0.53-0.87)
Exposure time: 30 min

Bibliographical reference: Elnabarawy, M.T., Robideau, R.R., and Beach, S.A. (1988). *Tox. Assess.* **3**, 361-370.

Temperature: 15°C
Test parameter: EC50
Effect: Reduction in light output
Result:

	15 min	30 min
Microbics strain	3.17 ± 1.15 mg/l Zn^{++}	2.31 ± 1.1 mg/l Zn^{++}
	0.72 ± 0.26 mg/l Zn^{++}	0.53 ± 0.25 mg/l Zn^{++}
Dr. Lange strain	8.21 ± 3.96 mg/l Zn^{++}	5.9 ± 1.9 mg/l Zn^{++}
	1.9 ± 0.9 mg/l Zn^{++}	1.35 ± 0.42 mg/l Zn^{++}

Comment: EC50 values were calculated from nominal concentrations.

Bibliographical reference: Vasseur, P. (1992). Centre des Sciences de l'Environnement, Metz, France, private communication.

ZINEB

CAS RN: 12122-67-7
Synonym: Zinc ethylenebis(dithiocarbamate) (polymeric)

Sample purity: ≥95%
Test parameter: EC50
Effect: Reduction in light output
Concentration: 6.2 mg/l (4.8-8.0)
Exposure time: 15 min

Bibliographical reference: van Leeuwen, C.J., Maas-Diepeveen, J.L., Niebeek, G., Vergouw, W.H.A., Griffioen, P.S., and Luijken, M.W. (1985). *Aquat. Toxicol.* **7**, 145-164.

Sample purity: Pestanal
Temperature: 15°C
Test parameter: EC50
Effect: Reduction in light output
Concentration: 2.1 ± 0.04 mg/l
Exposure time: 30 min
Comment: The assay was run in triplicate (pH = 6.2-6.6) using different bacterial reagents. The toxicity values were calculated from nominal concentrations. Synergistic effects were observed with $CuSO_4.5H_2O$ (20 µg/l Cu^{++} and 9 µg/l zineb). 90 % inhibition of the bacterial luminescence was obtained with solutions containing 40 µg/l Cu^{++} and 18 µg/l zineb.

Bibliographical reference: Vasseur, P., Dive, D., Sokar, Z., and Bonnemain, H. (1988). *Chemosphere* **17**, 767-782.

Temperature: 15°C
Test parameter: EC50
Effect: Reduction in light output
Concentration: 4 ± 0.3 mg/l
Exposure time: 15 min
Comment: Toxicity values were calculated from nominal concentrations.

Bibliographical reference: Vasseur, P. (1992). Centre des Sciences de l'Environnement, Metz, France, private communication.

ZIRAM

CAS RN: 137-30-4
Synonym: Zinc bis(dimethyldithiocarbamate)

Sample purity: ≥95%
Test parameter: EC50
Effect: Reduction in light output
Concentration: 0.15 mg/l (0.12-0.19)
Exposure time: 15 min

Bibliographical reference: van Leeuwen, C.J., Maas-Diepeveen, J.L., Niebeek, G., Vergouw, W.H.A., Griffioen, P.S., and Luijken, M.W. (1985). *Aquat. Toxicol.* **7**, 145-164.

INDEX OF CAS RN

INDEX OF NAMES

For Product Safety Concerns and Information please contact our EU representative GPSR@taylorandfrancis.com Taylor & Francis Verlag GmbH, Kaufingerstraße 24, 80331 München, Germany

Batch number: 10731063

Printed by Printforce, the Netherlands